CRYSTALLINE SILICON – PROPERTIES AND USES

Edited by **Sukumar Basu**

Crystalline Silicon – Properties and Uses
Edited by Sukumar Basu

Published by InTech
Janeza Trdine 9, 51000 Rijeka, Croatia

Publishing Process Manager Iva Lipovic

Technical Editor Teodora Smiljanic

Cover Designer InTech Design Team

Additional hard copies can be obtained from orders@intechopen.com

Crystalline Silicon – Properties and Uses, Edited by Sukumar Basu
p. cm.
ISBN 978-953-307-587-7

For EU product safety concerns:
IN TECH d.o.o., Prolaz Marije Krucifikse Kozulić 3, 51000 Rijeka, Croatia,
info@intechopen.com or visit our website at intechopen.com.

Meet the editor

Sukumar Basu obtained his Ph.D. in solid-state chemistry from Indian Institute of Technology (IIT), Kharagpur, India, in 1973. He spent some period at the University of Vienna, Austria and Max Planck Institut fuer Kohlenforschung, Germany as post doctoral research fellow. He joined IIT Kharagpur, India as the faculty member in 1979. He attended a large number of national and international conferences and gave lectures on his research findings. Prof. Basu's research specializations are electronic materials and chemical sensors. He has published more than 200 scientific papers in the peer reviewed journals and several book chapters. Prof. Basu joined Jadavpur University, Kolkata, India in 2005 as research advisor to continue his research activity on sensors. Recently he has started working on the synthesis and characterizations of the novel material, graphene for gas sensor applications. Prof. Basu is the editorial board member of a number of national and international journals on sensors and materials science.

Contents

Preface

The importance of crystalline silicon and the emergence of nanocrystalline material are heading towards miniaturization of silicon based devices. The entire device technology is getting a radical transformation through bottom up approach and corresponding increase in density of integration that is a challenge in processes and materials via top down approach. The availability of macro-micro-nano phases of silicon is a boom to the silicon based technology for the third generation electronic and optoelectronic devices and their integration for ICs, solar cells, sensors and biomedical devices. So, it can be said that silicon is the heart of both modern & future technology. The crystalline silicon is a store house of developing innumerable human friendly technology. For example, the evolution of green energy to avoid the global contamination from petroleum and its related products is possible only by silicon and silicon related devices. The rich abundance of silicon in nature and its minimum toxic property is a distinct commercial advantage over other synthetic materials. An extensive research & development on silicon materials and devices is a continuing process to study & clearly understand the fundamental changes in the crystalline structure and the defect states with the decrease of the crystallite dimensions from macro to nano sizes. The quantization effect in silicon that has already revealed some interesting properties needs further investigations for more vital information. Along with the theory more advanced experimental techniques are to be employed for this purpose.

The book 'Crystalline Silicon: Properties and Uses' presents fifteen chapters in all with the examples of different forms of silicon material, their properties and uses. Formation of silicon thin films through solution route via organic precursors has been described in Chapter 1. The modern techniques to study the oxide –silicon interface in different crystalline forms have been highlighted in chapter 2 and the behaviour of the native oxide on silicon has been demonstrated in chapter 3 of this book. Chapter 4 deals with the characterizations of dislocations in silicon in an elaborate fashion. Doping of silicon by high mass molecular ion implantation is treated in detail and an ellipsometric investigation of doping by ion implantation is discussed in chapters 5 and 6 respectively. Silicon nanocrystals, in general, are presented in chapter 7.The cathodoluminescent characterization of silicon nanoclusters in silicon dioxide has been discussed in depth in two chapters e.g. chapters 8 and 9. Nanocrystalline porous silicon, a novel material for nano-electronic, optoelectronic and sensor applications are presented in three chapters (10, 11 & 12) that cover different novel methods of preparations, structural & optical properties and porous silicon integrated photonic devices for bio-applications. Chapter 13 has been devoted to silicon based photovoltaic solar cells and their life cycle assessment. The use of silicon based MEMS devices in the microphone technology is an interesting addition to this book and the details are dealt in chapter 14. The commercial aspects of the availability & consumption of silicon on global perspective have been taken into consideration in chapter 15. In fact, this book presents different basic and applied aspects of crystalline silicon. It is a unique combination of conventional and novel approaches to understand the behaviour of silicon in different crystalline states for potential applications in the present scenario and in near future.

The valuable contributions of the renowned researchers from different parts of the globe working on various aspects of crystalline silicon are magnificent and deserve great appreciations. It is once again proved that knowledge knows no bounds. The credit goes to the entire InTech publishing group members for their tireless efforts to work on this project to publish the book in time. The editorial assistance of the process manager, Ms. Iva Lipovic needs special mention for the success of this book project. The help of Dr. (Ms.) Jayita Kanungo, the research associate of Jadavpur University, India is sincerely appreciated.

Prof. Sukumar Basu
IC Design & Fabrication Centre,
Dept. of Electronics and Telecommunication Engineering,
Jadavpur University,
India

1

Amorphous and Crystalline Silicon Films from Soluble Si-Si Network Polymers

Michiya Fujiki[1] and Giseop Kwak[2]
[1]*Graduate School of Materials Science, Nara Institute of Science and Technology*
[2]*Department of Polymer Science, Kyungpook National University*
[1]*Japan*
[2]*Korea*

1. Introduction

Silicon (Si) is the second most abundant element (Clarke number ~26%) on Earth and exists mainly in the oxidized silicate (SiO_2) form. Si sources are neither localized in very specific regions nor are they noble. However, crystalline (*c*-Si) and amorphous (*a*-Si) silicons remain the most fundamental, purely inorganic materials used for microelectronics, optoelectronics, and photonics because the lithographic and *p*-*n* doping processes are already well-established in industry. To produce these materials, vacuum and vapor-phase deposition processes and mechanical slicing/polishing techniques of Si-wafers are invariably utilized. However, these techniques require the use of an expensive XeCl excimer laser for annealing of *a*-Si; this step is followed by a crystallization process to prepare a poly-Si thin film transistor (TFT) from a *a*-Si thin film, which is deposited using a highly dangerous SiH_4–Si_2H_6 chemical vapor deposition (CVD) process.

1.1 Physical and chemical approaches for controlling the band gap of crystalline silicon

There are many types of Si-based materials ranging from zero-dimensional (0D) nanocrystalline silicon (*nc*-Si) and nanoparticles, one-dimensional (1D) polysilane and nanowire, and two-dimensional (2D) Si-skeletons, including Si-Si bonded network polysilyne (SNP), Wöhler siloxene, and Si/SiO_2 superlattice, to three-dimensional (3D) Si-skeletons, including *c*-Si and *a*-Si (Table 1).

The fundamental materials for microelectronics, *c*-Si and *a*-Si, are poor UV-visible-near IR emitters with low quantum yields (Φ_F) of ~10^{-2}% at 300 K because of their narrow band gap (1127 nm, 1.1 eV) with indirect electronic transitions (Lockwood, 1998; Yu & Cardona, 2005). Since the first reports of fairly efficient photoluminescence (PL) in the visible–near IR region from *nc*-Si (Furukawa & Miyasato, 1988; Takagi et al., 1990; Kanemitsu et al., 1993; Kanemitsu et al., 1995; Kanemitsu et al., 1996; Wilson et al., 1993) and porous Si (Cullis & Canham, 1991; Cullis et al., 1997; Lehmann & Gösele, 1991; Heitmann et al., 2005), extensive research efforts have been expended to produce Si with efficient, tunable UV-visible emission. To effectively confine a photoexcited electron-hole pair (exciton) within Si's Bohr radius (r_B) of ~5 nm (Lockwood, 1998; Yu & Cardona, 2005), various low-dimensional Si-based materials have been theoretically (Takeda & Shiraishi, 1997; Takeda & Shiraishi, 1998;

Brus, 1994; Alivisatos, 1996) and experimentally explored as follows: (a) 0D and 1D materials as visible-near IR emitters, including *nc*-Si and nanoparticles (Holmes et al., 2001; Grom et al., 2000; Kovalev et al., 1998; Gelloz et al., 2005; Walters et al., 2006; Jurbergs et al., 2006; English et al., 2002; Fojtik & Henglein, 1994; Li et al., 2004; Liu et al., 2005; Choi et al., 2007; Watanabe, 2003; Pi et al., 2008; Liu, 2008; Bley & Kauzlarich, 1996; Mayeri et al., 2001; Zou et al., 2004; Zhang et al., 2007; Wilcoxon et al., 1999; Hua et al., 2006; Nayfeh & Mitas, 2008) and Si nanowires (Qi et al., 2003; Ma et al., 2005); (b) 1D materials as exitonic UV emitters, including chain-like polysilane (Fujiki, 2001; Hasegawa et al., 1996); (c) 2D Si skeletons as visible emitters, including Si-Si bonded network polymers (SNP) (Takeda & Shiraishi, 1997; Bianconi et al., 1989; Bianconi & Weidman, 1988; Furukawa et al., 1990; Wilson & Weidman, 1991), Wöhler siloxenes (Brandt et al., 2003), and a Si/SiO_2 superlattice (Lu et al., 1995). Although SNPs have been regarded as soluble precursor polymers of *a*-Si (Wilson & Weidman, 1991) and 2D-Si nanosheets (saturated, bonded "sila-graphene") (Brandt et al., 2003; Nesper, 2003), further studies on the pyrolytic products of SNP derivatives and their inherent photophysical properties in a vacuum at low temperature have not yet been reported.

4A-elements	Dimensionality and E_g (eV) (Brus, 1994; Takeda & Shiraishi, 1989)		
	3D 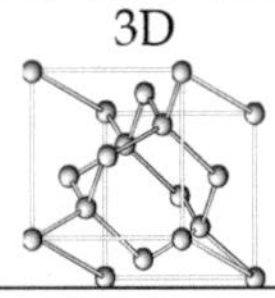	2D 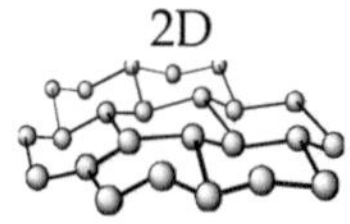	1D 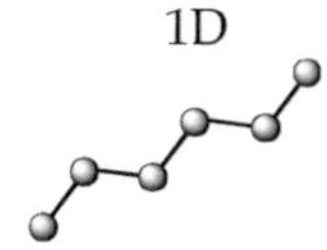
C	5.5 (*IG*)	-	~8
Si	1.1 (*IG*)	2.3 (*IG* and *DG*)	~4 (*DG*)
Ge	0.7 (*IG*)	1.8 (*IG* and *DG*)	~4 (*DG*)
Sn	0.1	-	~ 3 (*DG*)
Pb	~0	-	-

Table 1. Schematic concept of skeleton dimensionality and elements. *DG*: Direct gap, *IG*: Indirect gap.

Solution processing of metal chalcogenide semiconductors to fabricate stable and high-performance transistors has recently been developed (Alivisatos, 1996). To produce *c*-Si, *a*-Si, and new Si-based materials with controlled optical band gaps, low-cost solution and thermal production methods that are environmentally friendly and safe and can deposit Si on a plastic film at lower temperatures (below 250 °C) using soluble Si-source materials are greatly preferable. Among the Si source materials, organosilicon compounds may be some of the most promising candidates to satisfy the above criteria in actual Si-device fabrication processes because organosilicon compounds are usually air-stable, non-toxic, non-flammable, non-explosive, and soluble in common organic solvents.

Through re-evaluation of previously reported research, we endeavor to advance our knowledge and understanding of Si-related materials science. Among the Si-related materials mentioned above, SNPs are especially interesting as soluble precursor polymers to pyrolytically transform into 3D Si-skeleton materials. In this chapter, we establish strong evidence that SNP is one of the most promising, air-stable, soluble Si-source materials for the straightforward production of *c*-Si, *a*-Si, and controlled bang-gap Si-based materials via simple control of the organic side groups of SNP as well as the vacuum pyrolysis conditions, including the pyrolysis temperature, pyrolysis time, and the presence of a trace amount of air.

1.2 Soluble silicon network polymers bearing organic groups

Various SNPs can be prepared via a one-step condensation reaction of the corresponding, non-flammable, non-toxic alkyltrichlorosilanes with sodium in 50–60% yield at 120 °C in inert organic solvent. A liquid NaK alloy and ultrasonic wave (USW) irradiation were applied in the preparation of the first SNP (Bianconi et al., 1989; Bianconi and Weidman, 1988). Subsequently, the use of Na metal with catalytic amounts of crown ethers readily afforded these SNPs in milder and safer conditions without USW irradiation, as shown in Scheme 1 (Furukawa et al., 1990).

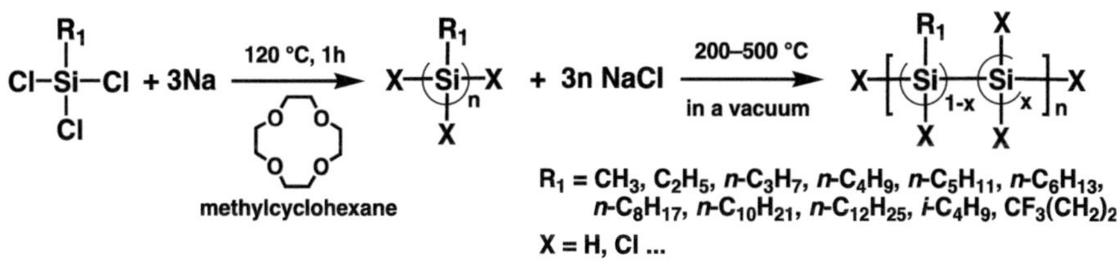

Scheme 1. General synthesis and vacuum-pyrolysis procedures for the preparation of SNPs.

In the present study, SNPs were prepared by the modified Na-mediated reduction (Wurtz-Kipping reaction) of the corresponding alkyltrichlorosilanes in the presence of 12-crown-4-ether as co-catalyst under a N_2 atmosphere (Fujiki et al., 2009). The SNPs were protected from contact with air and moisture during all of the synthetic procedures, including preparation, isolation, and sample enclosure in a glass tube. The SNPs were typically synthesized as shown in Scheme 1. For example, methylcyclohexane (4 mL, dried over 4 Å molecular sieves) containing Na (0.43 g, 19 mmol) and 12-crown-4-ether (0.02 g, 0.11 mmol) was placed in a four-necked 100 mL flask and refluxed at the relatively high temperature of 120 °C with vigorous mechanical stirring. To this mixture, *n*-butyltrichlorosilane (0.98 g, 5.1 mmol) dissolved in methylcyclohexane (4 mL) was added drop-wise. After the addition was complete, the solution was stirred for 1 h and then allowed to cool to room temperature. The reaction vessel was transferred to a glove box filled with 99.9% N_2 gas. To remove excess Na and NaCl, the reaction mixture was filtered using a fluorinated membrane filter (0.50 μm pore size) under pressure to yield a clear yellow solution containing *n*-butyl-substituted SNP (*n*-BSNP). The polymer was isolated by precipitating the solution in dry acetone and then dried in the glove box via connection to an external vacuum pump. Ethyl, *n*-propyl, *i*-butyl, *n*-pentyl, *n*-hexyl, *n*-octyl, *n*-decyl, *n*-dodecyl, and 3,3,3-trifluoropropyl-substituted SNP derivatives were similarly prepared as soluble polymers. Only methyl-substituted SNP was insoluble in all solvents, due to its very short alkyl group. The yields of SNPs ranged from 40–50%. Weight-averaged and number-averaged molecular weights (M_w and M_n) of the soluble SNPs ranged from 3–43 x 10^3 g mol^{-1}. Freshly-prepared SNPs did not exhibit any IR absorption due to the Si-O-Si stretching vibration of the oxidized Si-Si bond around 1000–1100 cm^{-1}.

A methylcyclohexane solution of SNPs was transferred into glass tubes (ID 5 mm, OD 7 mm); the inner wall of the tube was manually coated with the solution, and the solution was dried by blowing with N_2 gas. The SNP films deposited in the glass tubes were connected to a two-way vacuum bulb. The glass tubes coated with the SNP films were removed from the glove box and sealed using a hand-burner using vacuum techniques (0.3 Torr by rotary pump or 5 x 10^{-5} Torr by a Pfeiffer turbo molecular pump). For pyrolysis experiments and photoluminescence (PL)/PL excitation (PLE) measurements, the glass tubes were placed into a housing made of an aluminum block and then onto a digitally-controlled hotplate (Thermolyne), and the temperature of the housing was monitored with a chromel-alumel thermocouple.

1.3 Pyrolysis of Si-containing polymers

Pyrolysis of several Si-containing polymers, including poly(dimethylsilane) (Yajima et al., 1978), polycarbosilane (PCS) (Liu et al., 1999; Schmidt et al., 1991), and polysilane-containing chlorine (Martin et al., 1997), is well-known to produce β-silicon carbide exclusively (β-SiC). This fact engendered the idea that SNPs might also produce β-SiC as a result of pyrolysis. However, thermogravimetric (TG) and isothermal thermogravimetric (ITG) analyses of ten SNPs in a stream of pure nitrogen gas indicated that elemental Si was produced.

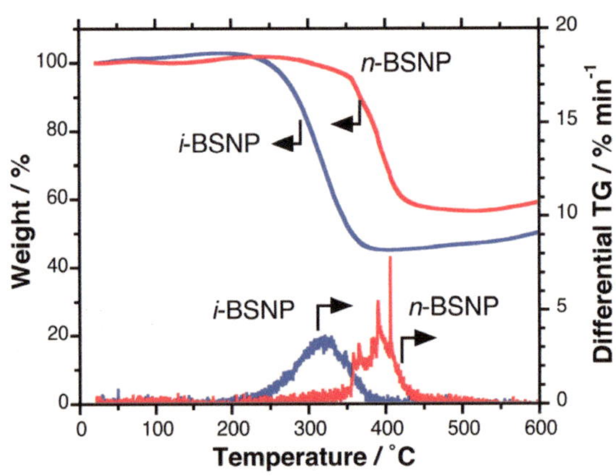

Fig. 1. Thermogravimetric (TG) and differential thermogravimetric (DTG) analysis curves of *n*-BSNP and *i*-BSNP in a N_2 atmosphere (heating rate = 5 °C min^{-1}).

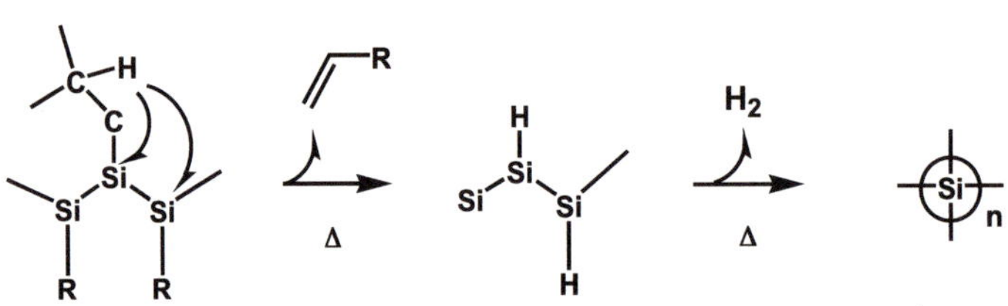

Scheme 2. Proposed mechanism of the β-H shift from the alkyl group to the Si-Si bonded skeleton.

The TG analysis data showed that SNPs undergo degradation in two steps (Fig. 1). With the exception of sterically-overcrowded isobutyl- and trifluoropropyl-substituted SNP derivatives (*i*-BSNP and FPSNP), most SNPs began to degrade at ~300 °C (not shown here), suggesting that an alkylene moiety is eliminated due to a β-H shift from the alkyl group to the Si skeleton, followed by the release of H_2 gas from the Si-H bond at ~450 °C. In contrast, the *i*-BSNP and FPSNP derivatives began to degrade at temperatures as low as

~250 °C. The proposed SNP pyrolysis mechanism is shown in Scheme 2. ITG data for SNPs at 500 °C as a function of pyrolysis time (Fig. 2) suggests that the observed weight loss of the SNP after prolonged (90 min) pyrolysis corresponds to a residue of pure Si (*not* SiC), regardless of alkyl side group (Fig. 3). This result was further confirmed by scanning electron microscopy (SEM)/X-ray photoelectron spectroscopy (XPS)/energy dispersive X-ray spectroscopy (EDS) analyses of the product yielded by pyrolysis of *n*-propyl-SNP and *n*-butyl-SNP (*n*-BSNP) at 900 °C, which showed that the surface was oxidized, as evidenced by 1:1 signals of Si and O but no detectable C signal due to SiC (Fig. 4 and Fig. 5).

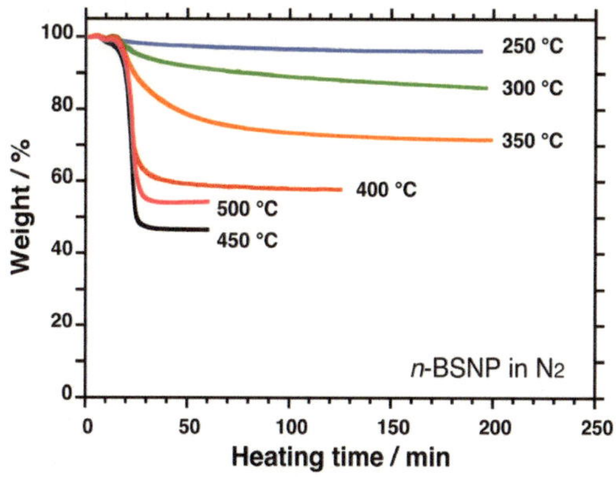

Fig. 2. Isothermal thermogravimetric (ITG) analysis curves of *n*-BSNP in a stream of N_2 gas at selected temperatures. ITG curves between 450 °C and 500 °C appeared to be reversed. This result was reproducible, probably due to rapid evolution of some volatile products from *n*-BSNP during prolonged heating at 450 °C.

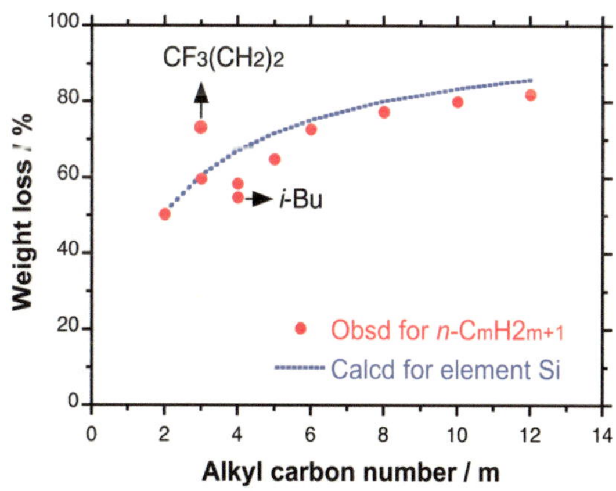

Fig. 3. Weight loss values of ten SNPs bearing different alkyl side chains upon pyrolysis at 500 °C for 90 min. Filled circles and dotted lines are experimental and calculated weight values for elemental Si, respectively.

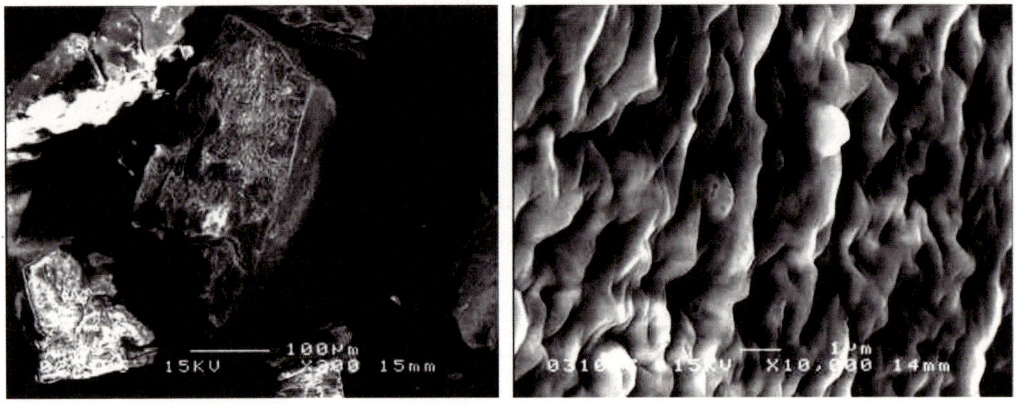

Fig. 4. Scanning electron microscope (SEM) images of *n*-propyl SNP pyrolyzed at 900 °C in a N_2 atmosphere (scale bars: left, 100 μm, and right, 1 μm).

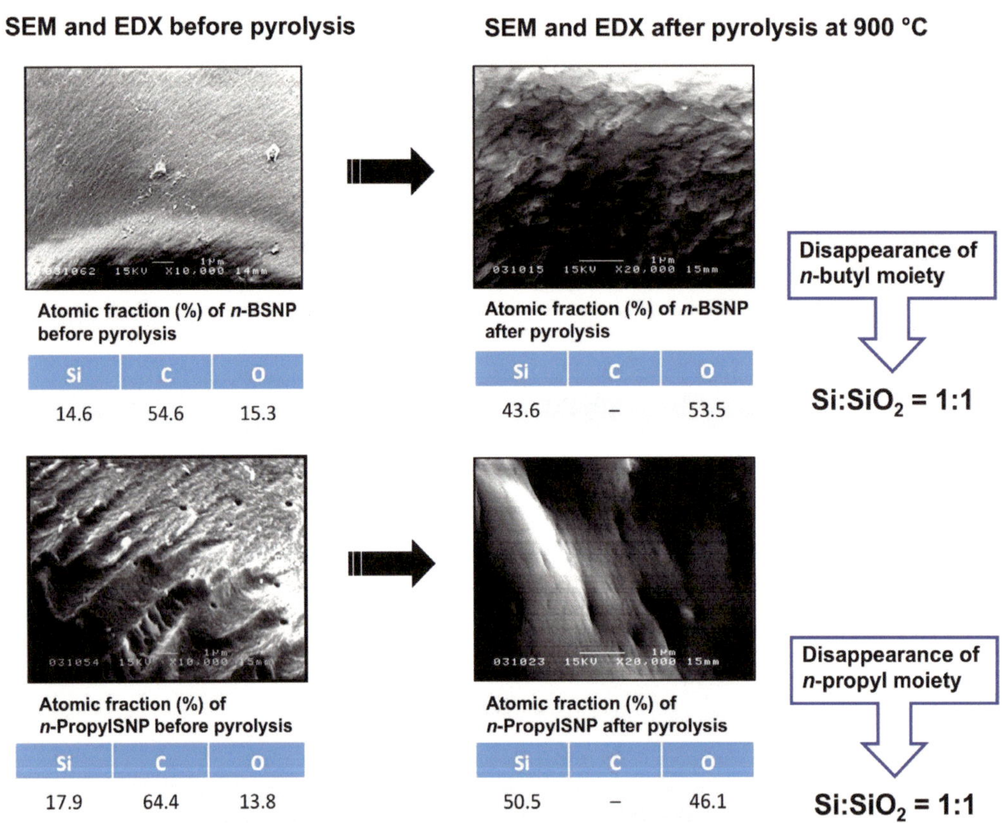

Fig. 5. Surface analysis of *n*-BSNP and *n*-propyl SNP by SEM and energy dispersive X-ray spectroscopy (EDS) before and after pyrolysis at 900 °C.

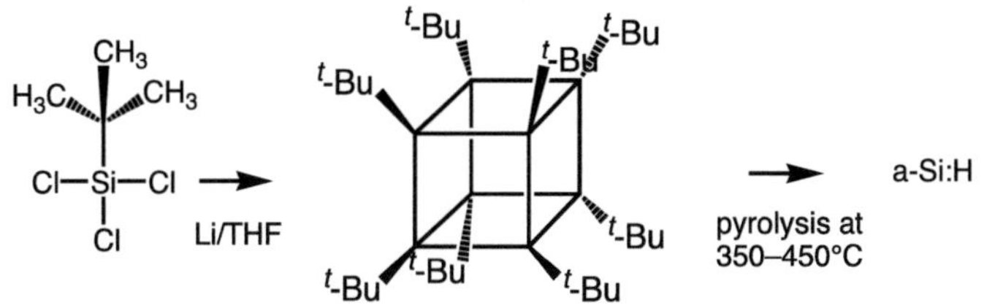

Scheme 3. The pyrolysis of octa(*tert*-butyl)octasilacubane (Furukawa, 2000).

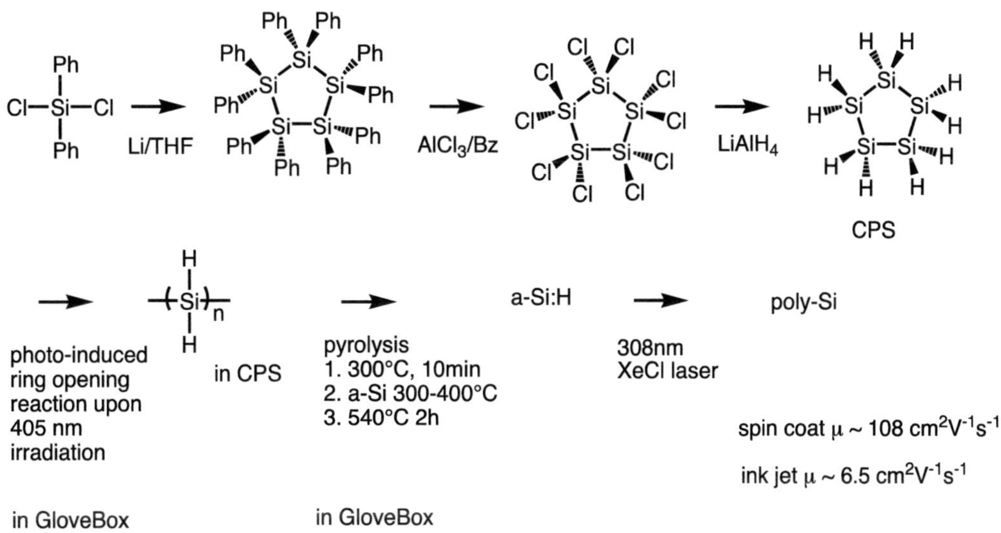

Scheme 4. The pyrolysis of hydrogenated polysilane prepared via a four-step synthesis from diphenyldichlorosilane as a starting material (Shimoda et al., 2006).

Based on pyrolysis experiments with PCS (Liu et al., 1999; Schmidt et al., 1991), β-H elimination from alkyl C-H (Scheme 2) is postulated to be the key for producing elemental Si without significant interference by the Kumada rearrangement reaction responsible for β-SiC production (Shini and Kumada, 1958). Peripheral SNP structures may be terminated with Si-Cl and Si-H, as exemplified in chlorine-containing polysilane (Martin et al., 1997). For most SNPs, the existence of Si-Cl and Si-H bonds were evidenced by ^{29}Si-NMR signals at ~30 ppm and from the FT-IR signal around 2080 cm^{-1}, respectively (Fujiki et al., 2009). The free Cl atom from the Si-Cl bond may catalyze efficient β-H elimination. This unexpected result may be a common feature of reactions involving Si-Si bonded molecules and polymers bearing appropriate side groups in an oxygen-free environment. Indeed, an Si-Si bonded cubic molecule with bulky organic groups, *tert*-butyl-containing octasilacubane was transformed into an *a*-Si film via pyrolysis in a vacuum at 350–450 °C (Scheme 3) (Furukawa, 2000). A polycrystalline Si thin film with a high-carrier mobility was prepared via pyrolysis of a Si-Si bonded linear polymer, $(SiH_2)_n$, at 300–550 °C in an oxygen-free glove box (Shimoda et al., 2006). This hydrogenated polysilane was produced by a four-step synthesis, including a photo-induced ring-opening process, using diphenyldichlorosilane as a starting material (Scheme 4).

2. A new family of silicon network polymers as a precursor for *c*-Si, *a*-Si, and other Si-based materials

2.1 Controlled vacuum pyrolysis with a trace amount of air

TGA and ITGA data for the ten SNPs in pure N_2 gas (99.99%) indicated that elemental Si was produced. To confirm this result, laser Raman scattering spectra and microscope imaging of the *n*-BSNP pyrolyzed at 500 °C for 10 and 90 min were compared with the freshly-prepared *n*-BSNP film (Fig. 6). Several particles with a size of ~30 μm were observed in the *n*-BSNP pyrolyzed at 500 °C for 90 min. The regions with a metallic luster showed a sharp Raman resonance at 508 cm^{-1} due to *nc*-like-Si particles, whereas regions with a non-metallic luster (amorphous regions) showed a very broad Raman shift at ~480 cm^{-1}, similar to the freshly-prepared *n*-BSNP film.

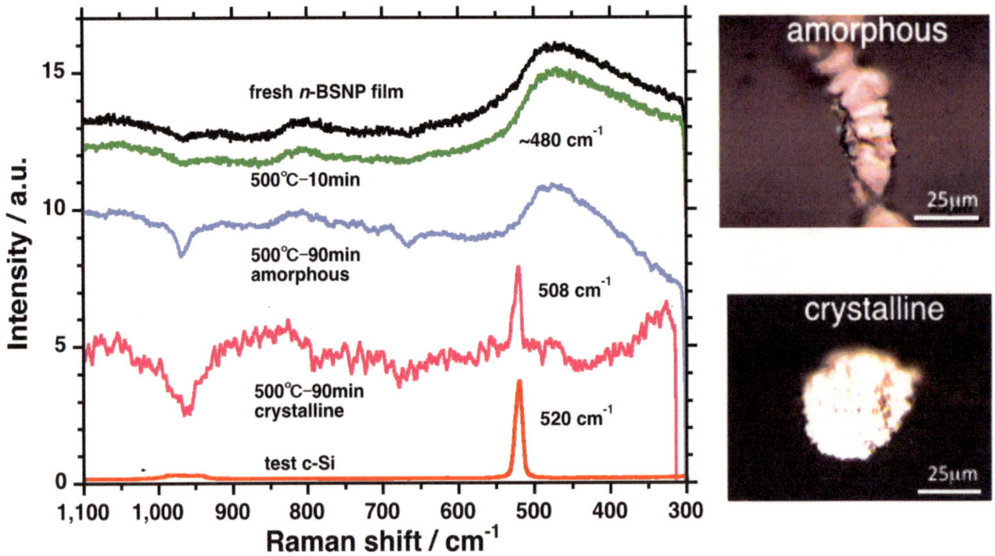

Fig. 6. Laser Raman scattering spectra and optical microscopy images of *n*-BSNP pyrolyzed at 500 °C.

2.2 Photophysical properties: photoluminescence and absorption spectra

The photoluminescence (PL) emission spectra of the ten soluble SNPs (Scheme 1) were measured between 460 nm and 740 nm (2.70-1.68 eV) at 77 K and at room temperature with controlled pyrolysis temperature (200~500 °C) and time (10~90 min) in a vacuum. The changes in the PL spectra of *n*-BSNP and *i*-BSNP films excited at 360 nm and 77 K, treated at several different pyrolysis temperatures for 10 and 90 min, are summarized in Fig. 7 (bottom) along with several color photographs of these *n*-BSNP films (Fig. 7, top).

Before pyrolysis, the freshly-prepared *n*-BSNP film emitted a yellowish PL, corresponding to a band maximum at 560 nm, with two fast lifetime components of 1.3 and 5.8 nsec. When the film was treated at 250 °C for 90 min, the PL band was markedly blue-shifted, exhibiting a blue PL band at 460 nm.

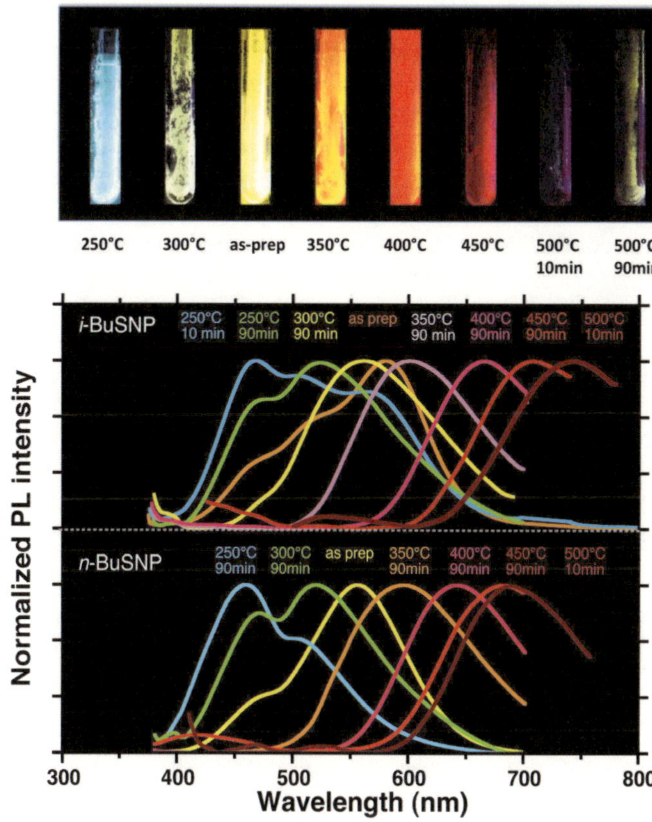

Fig. 7. Emission color photographs (top) and PL spectra (bottom) of *n*-BSNP and *i*-BSNP films pyrolyzed at different pyrolysis temperatures for different time periods (excited at 360 nm at 77 K).

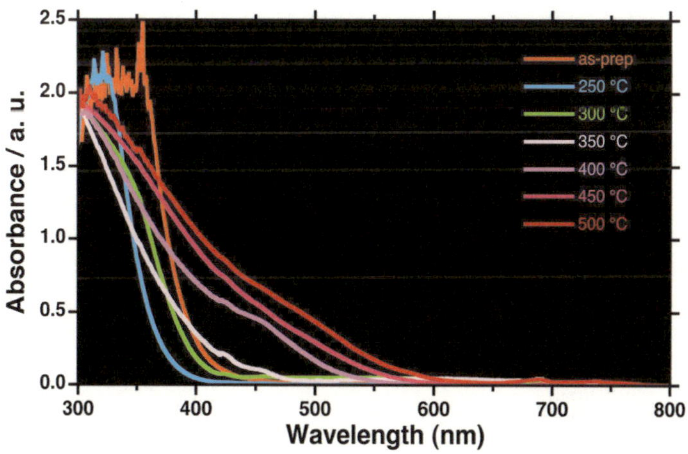

Fig. 8. UV-visible absorption spectra (normalized at 300 nm) of *n*-BSNP films pyrolyzed at different temperatures.

When the virgin films were treated at 300 °C for 90 min, the PL band was shifted slightly to green, corresponding to a peak maximum at 520 nm. Conversely, when the freshly-prepared films were treated at 350 °C, 400 °C, and 450 °C for 90 min, the PL wavelength was progressively red-shifted to orange at 580 nm, to red at 640 nm, and to deep-red at 680 nm. Similarly, the *i*-BSNP film progressively shifted first towards the blue region and then towards the red region. When the freshly-prepared *i*-BSNP film was treated at 500 °C for 10 min, the deepest red PL, corresponding to a peak maximum at 740 nm, was observed. Both *n*-BSNP or *i*-BSNP treated at 250–300 °C for 90 min exhibited a greenish-white emission to the naked eye.

When the freshly-prepared *n*-BSNP film was extensively heated at 500 °C for 90 min, it became a lustrous, metallic film that emitted a very weak, deep-red PL at approximately 680 nm with a marked decrease in intensity that was one-sixth that of the sample treated at 500 °C for 10 min. Figure 8 shows these marked, progressive blue- and red-shifts in the UV-visible absorption edge of pyrolyzed *n*-BSNP films maintained in sealed tubes at increasing pyrolysis temperatures. This change in the UV-visible absorption edge corresponds well to the blue- and red-shifts of the PL bands. It is noteworthy that the SNP film treated at 500 °C clearly showed broad absorption bands in the range of 350–600 nm, indicating that a significant change occurred in the Si-Si bonded skeleton.

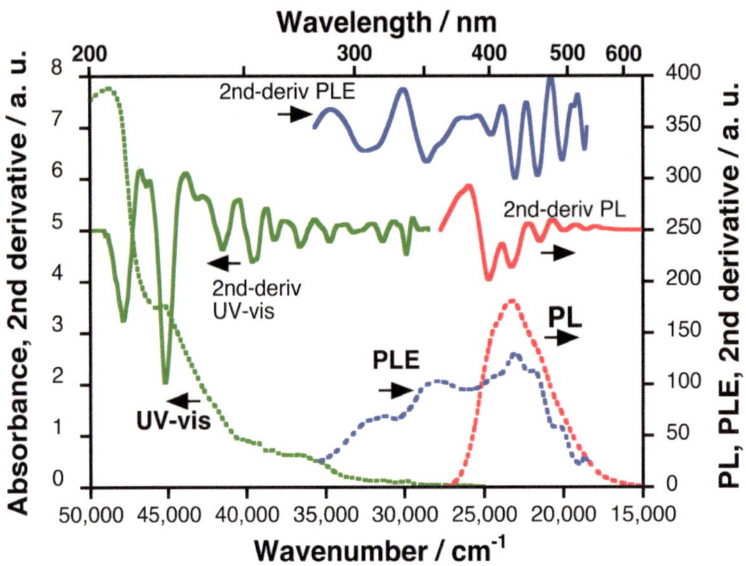

Fig. 9. PL excited at 360 nm, PL excitation (PLE) monitored at 540 nm, and UV-visible absorption spectra with their corresponding second derivative spectra for the air-exposed *nc*-like-Si particles dispersed in *n*-hexane.

Although exact Φ_F values for these films were not determined, they were assumed to be several% (not exceeding 10%) at 77 K, and the PL intensities of the films at room temperature decreased by one-sixth compared to those at 77 K. This estimation was based on the fact that the Φ_F value of virgin *n*-BSNP in THF solution at room temperature is ~1% using 9,10-diphenylanthracene as the reference (Φ_F ~97% in methylcyclohexane). As for the *n*-BSNP film pyrolyzed at 300 °C, the PL band at 560 nm with excitation at 370 nm had fast and slow lifetime components of ~5 nsec and >10 nsec, respectively, at 77 K. This short lifetime can be compared to the long lifetime of *c*-Si of 4.6 hrs (Lockwood, 1998; Yu and Cardona, 2005). An oxygen molecule may be inserted into the SNP skeleton because the SNP film was sealed in the presence of a small amount of air (~3x10^{-1} Torr). In fact, the film sealed in a trace amount of air (~5x10^{-5} Torr) exhibited a major PL band around 550 nm that was almost unchanged, even after thermal treatment at 200 and 300 °C. Controlling the time and temperature of the air-oxidation and pyrolysis of the virgin SNP film may thus facilitate fine tuning of the PL wavelength between 460 nm (2.70 eV) and 740 nm (1.66 eV).

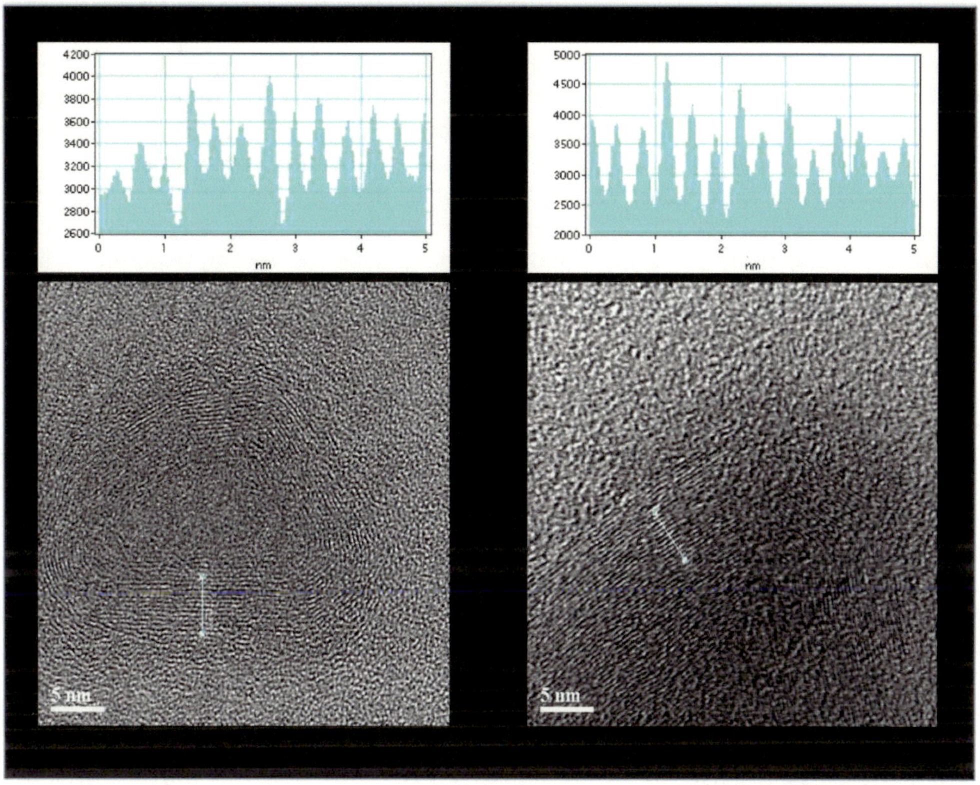

Fig. 10. HRTEM images of the air-exposed *nc*-Si particles. Left-side image: Baum-küchen-like, multi-layered structures with circular shapes and ~0.34 nm spacing. Right-side image: Baum-küchen-like, lamellar shapes with ~0.37 nm spacing. Surface profile at indicated location. Scale bar is 5 nm.

When the very weakly emitting Si particles in deep-red, which were produced by the pyrolysis of *n*-BSNP at 500 °C for 90 min, were exposed to air, the PL switched abruptly to an intense sky-blue color (λ = 430 nm). The air-exposed Si particles, dispersed in common organic solvents at room temperature, exhibited an extremely high Φ_F of 20–25% and a short lifetime of ~5 nsec, probably due to the presence of siloxene-like, multi-layered Si-sheet structures.

$$E_n = \hbar^2/2m \bullet (\pi/na)^2, \text{ purely electronic transitions with } n \quad (1)$$

$$E_n = (n + 1/2)\,\hbar\,\omega, \square \text{ purely vibronic transitions with } n \quad (2)$$

The PL (excited at 360 nm), PL excitation (PLE, monitored at 540 nm), and UV-visible absorption spectra of the air-exposed Si particles dispersed in *n*-hexane are shown with the corresponding second derivative spectra in Fig. 9. Based on the second derivative spectra, the apparent broad PL, PLE, and absorption spectra consist of at least five well-resolved bands with almost equal energy spacing (1650 ± 100 cm^{-1} for the UV-visible absorption spectra, 1580 ± 200 cm^{-1} for the PL spectra, 1470 ± 70 cm^{-1} for the PLE spectra).

This periodic behavior may be related to the combination of bands arising from the Si-Si stretching mode (~460 cm^{-1}) of the 2D-like Si skeleton and the Si-O-Si stretching mode (~1100 cm^{-1}). Specifically, a coupling between an electron (from the Si-Si skeleton) and a phonon (from the Si-O-Si stretching vibration) is responsible for the strongly blue emission, due to the loss of translational symmetry. If the multiple electronic transitions in the PL, PLE, and UV-visible spectra came from purely electronic origins within the 2D structure, the energy separation (E_n) should obey the inverse square of the quantum number, n (Eq. 1). If the transitions were connected to vibronic transitions, the separation would be related to n in a parabolic potential well (Eq. 2). The unexposed *nc*-like-Si samples may be of the former type due to very weak electron-phonon coupling, whereas the air-exposed *nc*-like-Si particles are presumably an example of the latter case [Yu and Cardona, 2005; Konagai, 1987; Davies, 1998, Colvin et al., 1994].

2.3 Structural analysis of the pyrolytic products by high-resolution transmission electron microscopy and laser Raman spectroscopy

The blue- and red-emissive structures of the air-exposed Si particles in the pyrolysis products were investigated by high-resolution transmission electron microscopy (HRTEM) with EDS. The HRTEM images of the air-exposed Si particles are shown in Fig. 10. The majority of the image regions show finely featured *nc*-Si particles with a diameter of ~1 nm and a lattice spacing of ~2.5 Å. EDS analysis revealed the existence of oxygen in the image regions with a Si/O ratio of ~1/3. It is interesting that the two HRTEM images clearly show 'Baum-küchen-like' multi-layered structures; the left-side image in Fig. 10 shows circular shapes with ~3.4 Å spacing while the right-side image in Fig. 10 shows lamellar shapes with ~3.7 Å spacing. The *n*-BSNP exhibited a *d*-spacing of 5.5 nm (2θ~16°, Cu$K\alpha$) in a WAXD pattern, indicating that an interlayer spacing was present between *n*-BSNP multi-sheets, whereas the pyrolyzed *n*-BSNP does not show any ordered structures. These layered structures imply that the air-exposed Si particles have a 2D-Wöhler-siloxene structure (Brandt et al., 2003, Nesper, 2003) separated by a highly stretched Si-O-Si bond with an opened Si-O-Si bond angle, which is regarded to be a chemically well-controlled Zintl phase. The origin of the blue-shifted PL band at 250 °C is assumed to be due to the partial oxidation of the SNP single-sheet when oxygen gas in the sealed tube is consumed during pyrolysis.

In nearly oxygen-free sealed conditions, a blue-shifted PL was not observed when the SNPs were treated at 200–300 °C. The origin of the progressively red-shifted PL band at more elevated temperatures can be conjectured to be multi-layered with the spontaneous, stacked structure of 2D-SNP single-sheets, formed through the elimination of the organic moieties and hydrogen during pyrolysis. The PL wavelength is variable according to the thickness of the Si film (Lu et al., 1995). When an Si-Si bond length projected in the stacking direction is assumed to be 1.85 Å, the Si layer number extrapolated from the PL peak wavelength can be calculated using E_{PL} (eV) = $1.6 + 0.7/d_{Si}^2$, which is given for the Si/SiO_2 superlattice (Lu et al., 1995).

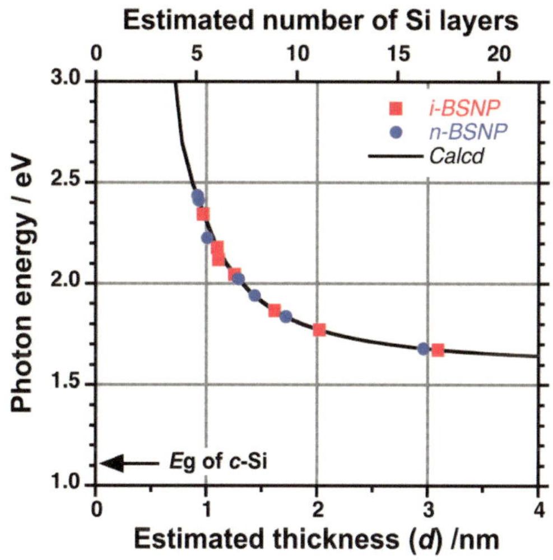

Fig. 11. Estimated numbers of Si multi-layers extrapolated from the PL peak energy of *n*-BSNP and *i*-BSNP, based on the equation (E_{PL}(eV) = $1.6 + 0.7/d_{Si}^2$) for the Si/SiO_2 superlattice (Lu et al., 1995).

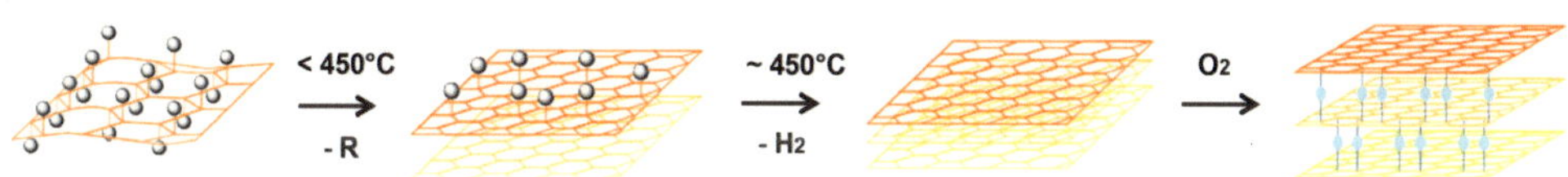

Raman shift / cm^{-1}	~ 460	~ 460	508	~ 480
Estimated Si layers	1	3−7	> 8−15	1 (Si-O-Si)
Band gap Eg / eV	2.1	1.8 − 2.7	~ 1.1 ?	2.9

Fig. 12. Proposed scheme for changes in the structural hierarchy of SNP from 2D single-sheet to quasi-3D multi-sheets based on a Wöhler-siloxene-like structure with and without Si-O-Si interlayer spacers.

The PL energy in eV is plotted as a function of estimated thickness and number of Si layers in Fig. 11. When the pyrolyzed SNP without organic or H moieties was exposed to air, the spontaneous insertion of oxygen atoms into the multi-layer Si ultrathin films occurred, resulting in the formation of a periodic $(Si)_1/(SiO_2)_1$ superlattice (Lu et al., 1995) identical to the Wöhler-siloxene multi-layers (Brandt et al., 2003) with Si-O-Si interspacing. A previous study predicted that a Wöhler-siloxene structure bearing oxygen moieties would be highly emissive due to changes in electronic transitions from indirect- to direct-type band structures (Kanemitsu et al., 1993). This change results from characteristic σ–n orbital mixing of the 2D-Si σ electrons with the lone pair electrons of oxygen at the band-edge states for an ideal 2D-Si polymer bearing OH and H side groups (Takeda and Shiraishi, 1993).

Based on the results shown above, a change in hierarchical structure based on a model of Wöhler-siloxene multi-sheet layers separated by an Si-O-Si linkage at elevated pyrolysis temperatures, followed by exposure to air, is proposed in Fig. 12.

2.4 Circularly polarized light from chiral SNPs

The generation, amplification, and switching of circularly polarized luminescence (CPL) and circular dichroism (CD) by polymers (Chen et al., 1999; Oda et al., 2000; Kawagoe et al., 2010), small molecules (Lunkley et al., 2008; Harada et al., 2009), and solid surface crystals (Furumi and Sakka, 2006; Krause & Brett, 2008; Iba et al., 2011) have received considerable theoretical and experimental attention.

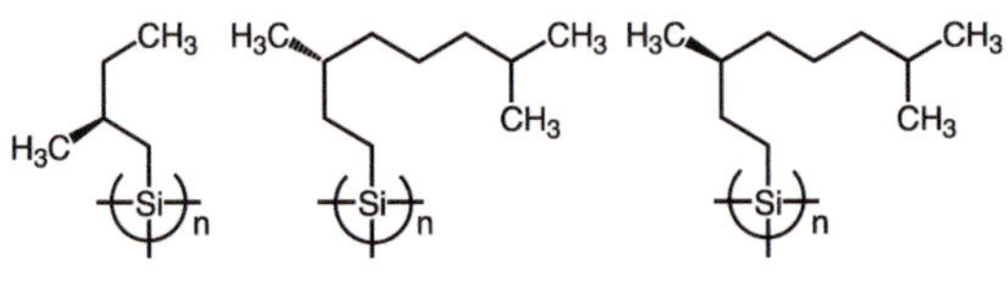

1*S* **2*R*** **2*S***

Scheme 5. Soluble, optically-active SNPs bearing chiral organic groups.

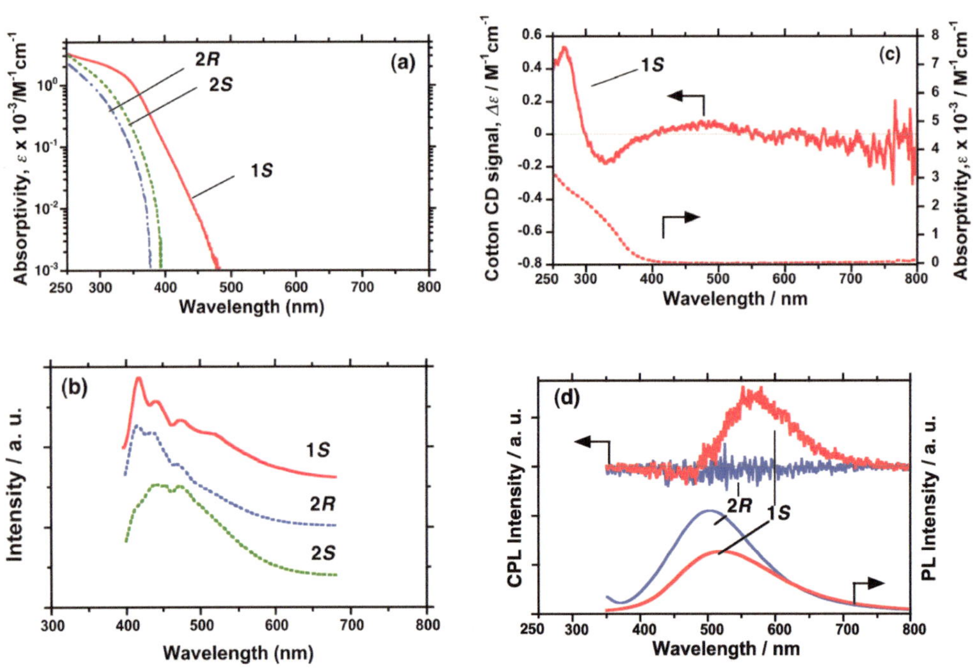

Fig. 13. UV-visible, PL, CD, and CPL spectra of **1*S***, **2*S***, and **2*R*** in THF at 25 °C.

CPL is inherent to asymmetric luminophores in the excited state, whereas CD is due to asymmetric chromophores in the ground state. The first chiroptical (CPL and CD) properties of three new SNPs bearing chiral alkyl side groups (Fukao & Fujiki, 2009) were recently demonstrated for poly[(*S*)-2-methylbutylsilyne] (**1*S***), poly[(*R*)-3,7-dimethyloctylsilyne] (**2*R***), and poly[(S)-3,7-dimethyloctylsilyne] (**2*S***) (Scheme 5).

This study revealed that only **1S**, bearing β-branched chiral groups, clearly showed an intense CPL signal at ~570 nm with Φ_F of ~1% along with corresponding Cotton CD signals in THF solution at room temperature (Fig. 13). In contrast, **2*R*** and **2*S***, which possess γ-branched chiral groups, did not exhibit any CPL signals although they did exhibit CD bands. By analogy to the optically inactive SNPs described above, optically active SNPs might be candidates for use as Si-source materials in the production of *a*-Si and *c*-Si films that exhibit circular polarization via controlled vacuum pyrolysis.

2.5 A Ge–Ge bonded network polymer (GNP) as an SNP analogue

Our understanding of the Si-Si bonded network polymeric materials led us to investigate a 2D Ge–Ge bonded network polymer (GNP) as a soluble model of insoluble polygermyne. A common approach for studying Si- and Ge-based materials is to effectively confine a photoexcited electron-hole pair within the Bohr radius (r_B) for Si (r_B ~5 nm) and for Ge (r_B ~24 nm) (Gu et al., 2001). However, research on low-dimensional Ge-based materials has been delayed due to the limited synthetic approaches available for preparing soluble Ge–Ge bonded materials using organogermanium sources, which are 1000 times more expensive than the corresponding organosilane sources. Several Ge-based materials were recently fabricated using the molecular beam epitaxy (MBE) technique in an ultrahigh vacuum using inexpensive Ge-based inorganic sources, rapidly increasing their potential use in the fields of physics and applied physics.

In the area of solid-state physics, Kanemitsu, Masumoto, and coworkers observed a broad PL band at 570 nm (2.18 eV) for microcrystalline Ge (μc-Ge) embedded into SiO_2 glass at room temperature (Maeda et al., 1991). Stutzmann, Brandt, and coworkers reported a near infrared PL band at 920 nm (1.35 eV) for multi-layered Ge sheets produced on a solid surface, which is a pseudo-2D multi-layered Ge crystal known as polygermyne synthesized from Zintl-phase $CaGe_2$ (Vogg et al., 2000). However, μc-Ge, polygermyne, and polysiloxene are purely inorganic and are thus insoluble in any organic solvent.

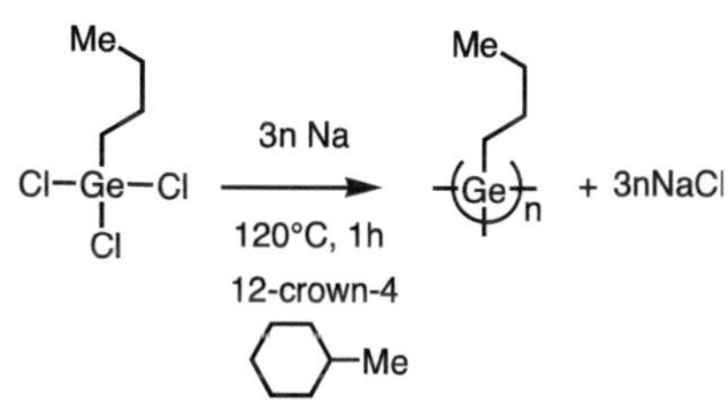

Scheme 6. Synthesis of soluble *n*-butyl GNP.

In 1993, Bianconi et al. reported the first synthesis of GNP via reduction of *n*-hexyltrichlorogermane with a NaK alloy under ultrasonic irradiation (Hymanclki et al., 1993). However, the photophysical properties of GNP have not yet been reported in detail. In 1994, Kishida et al. reported that poly(*n*-hexylgermyne) at 77 K possesses a green PL band with a maximum at 560 nm (2.21 eV) whereas poly(*n*-hexylsilyne) exhibits a blue PL band around 480 nm (2.58 eV) (Kishida et al., 1994).

By applying our modified technique to a soluble GNP bearing *n*-butyl groups (*n*-BGNP) and through careful polymer synthesis (Scheme 6) and measurement of the PL, we briefly demonstrated that *n*-BGNP exhibits a very brilliant red PL band at 690 nm (1.80 eV). This result was obtained using a vacuum at 77 K without the pyrolysis process; under these conditions, *n*-BSNP reveals a very brilliant green-colored PL band at 540 nm (2.30 eV) (Fig. 14) (Fujiki et al., 2009). This result differs from that of a previous report of green PL from poly(*n*-hexylgermyne) (Kishida et al., 1994).

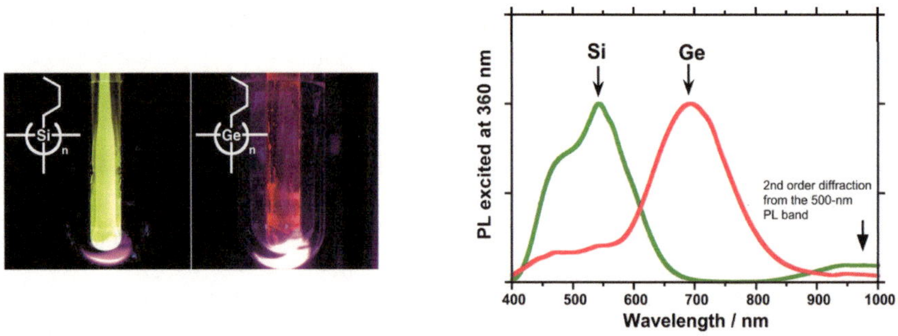

Fig. 14. Photographs (left) and PL spectra (right) of *n*-BSNP and *n*-BGNP films excited at 365 nm at 77 K.

By analogy with the SNPs described above, GNP may have potential uses as NIR emitters and narrow band gap materials with a loss of organic moieties by the pyrolysis process. In recent years, several studies have demonstrated the preparation and characterization of Ge nanoclusters capped with organic groups. Watanabe et al. elucidated that pyrolysis products of soluble Ge-Ge bonded nanoclusters capped with organic groups offer high-carrier mobility and optical waveguide with a high-refractive index value in semiconducting materials (Watanabe et al., 2005). Klimov et al. recently reported the presence of a near IR PL band at 1050 nm (1.18 eV) with a fairly high Φ_F of 8% for *nc*-Ge capped with 1-octadecene, enabling a great reduction in Ge surface oxidation due to formation of strong Si–C bonds (Lee et al., 2009). The study of GNP pyrolysis is in progress and will be reported in the future.

2.6 Scope and perspectives

In recent years, solution processes for the fabrication of electronic and optoelectronic devices, as alternative methods to the conventional vacuum and vapor phase deposition processes, have received significant attention in a wide range of applications due to their many advantages, including processing simplicity, reduction in total production costs, and safety of chemical treatments. Particularly, the utilization of liquefied source material of an air-stable, non-toxic, non-flammable, non-explosive solid may be essential in some potential applications in printed semiconductor devices for large-area flexible displays, solar cells, and thin-film transistors (TFTs). Recent progress in this area has largely been focused on organic semiconductors with π-conjugated polymers due to their ease of processing, some of which have a relatively high carrier mobility that is comparable to that of *a*-Si.

Because of their ease of coating and dispersion in the form of 'Si-ink' in comparison to II-VI group nanocrystals [Colvin et al., 1994], soluble SNP, GNP, and their pyrolysis products can serve as Si-/Ge-source materials for the production of variable range Si-based and/or Si-Ge alloyed semiconductors at room temperature. The ionization potential of the pyrolyzed Si materials range between 5.2 and 5.4 eV while the electron affinity ranges between 4.0 and 3.2 eV (Lu et al., 1995). These values are well-matched with the work-functions of ITO and

Al/Ag/Mg electrodes. Recently, air stable red-green-blue emitting *nc*-Si was achieved using a SiH_4 plasma following CF_4 plasma etching (Pi et al., 2008). As an alternative method, laser ablation of bulk *c*-Si in supercritical CO_2 after excitation with a 532-nm nanosecond pulsed laser yielded *nc*-Si that could produce blue, green, and red emitters. (Saitow & Yamamura, 2009). As we have demonstrated, controlled vacuum pyrolysis using a single SNP source material, possibly including GNP source material, should offer a new, environmentally friendly, safer process to efficiently produce red-green-blue-near infrared emitters, thin films for TFTs, and solar cells because the required technology is largely compatible with XeCl excimer laser annealing and the crystallization process for making poly-Si TFTs from *a*-Si thin films deposited using the SiH_4–Si_2H_6 CVD process.

The dimensionality of inorganic materials makes it possible to tailor the band gap value, as shown in Table 1. Soluble SNP and GNP, because of their ease of coating and dispersion in the form of "Si-ink" and "Ge-ink", may serve as controlled soluble Si/Ge source materials without the need for the SiH_4/GeH_4 CVD process. Our results provide a better understanding of the intrinsic nature of pseudo-2D Si electronic structure by varying Si layer numbers. The chemistry of SNP vacuum pyrolysis opens a new methodology to safely produce *a*-Si, *c*-Si, Si-based semiconductors, and alloys with Ge.

3. Summary

Although *c*-Si is the most archetypal semiconducting material for microelectronics, it is a poor visible emitter with a quantum yield of 0.01% at 300 K and a long PL lifetime of several hours. Pyrolysis of chain-like Si-containing polysilane and polycarbosilane has previously been shown to efficiently produce β-SiC; however, our TGA and ITGA pyrolysis experiments with various soluble SNPs indicated that elemental Si is produced. The SNP was transformed into a visible emitter that is tunable from 460 nm (2.7 eV) to 740 nm (1.68 eV) through control of the pyrolysis temperature and time (200–500 °C, 10-90 min). Moreover, air-exposed *nc*-like-Si, produced by pyrolyzing SNP at 500 °C, showed an intense blue PL with a maximum at 430 nm, a quantum yield of 20–25%, and a short lifetime of ~5 nsec; furthermore, these particles disperse in common organic solvents at room temperature. HRTEM, laser-Raman, and second-derivative UV-visible, PL, and PLE spectra indicated that the siloxene-like, multi-layered Si-sheet structures are responsible for the wide range of visible PL colors with high quantum yields. Circular polarization for SNPs bearing chiral side groups was also demonstrated for the first time. Through an analogous synthesis to that of green photoluminescent SNPs, the Ge-Ge bonded network polymer, GNP, was determined to be a red photoluminescent material.

4. Acknowledgements

This work was fully supported by the Nippon Sheet Glass Foundation for Materials Science and Engineering and partially supported by a Grant-in-Aid for Scientific Research (B) from MEXT (22350052, FY2010–FY2013). The authors thank Prof. Kyozaburo Takeda, Prof. Kenji Shiraishi, Prof. Nobuo Matsumoto, Prof. Masaie Fujino, Prof. Akira Watanabe, Prof. Masanobu Naito, Prof. Kotohiro Nomura, Prof. Akiharu Satake, Dr. Kazuaki Furukawa, Dr. Anubhav Saxena, and our students, Dr. Masaaki Ishikawa, Satoshi Fukao, Dr. Takuma Kawabe, Yoshiki Kawamoto, Masahiko Kato, Yuji Fujimoto, Tomoki Saito, and Shin-ichi Hososhima for their helpful discussions and contributions.

5. References

Alivisatos, A. P. (1996). Perspectives on the physical chemistry of semiconductor nanocrystals. *The Journal of Physical Chemistry*, Vol. 100, No. 31, 13226-13239.

Bianconi, P. A.; Schilling, F. C. & Weidman, T. W. (1989). Ultrasound-mediated reductive condensation synthesis of silicon-silicon-bonded network polymers. *Macromolecules*, Vol. 22, No. 4, 1697–1704.

Bianconi, P. A. & Weidman, T. W. (1988). Poly(*n*-hexylsilyne): Synthesis and properties of the first alkyl silicon $[RSi]_n$ network polymer. *Journal of the American Chemical Society*, 1988, Vol. 110, No. 7, 2342–2344.

Brus, L. (1994). Luminescence of silicon materials: chains, sheets, nanocrystals, nanowires, microcrystals, and porous silicon. *The Journal of Physical Chemistry*, Vol. 98, No. 14, 3575–3581.

Bley, R. A. & Kauzlarich, S. M. (1996). A low-temperature solution phase route for the synthesis of silicon nanoclusters. *Journal of the American Chemical Society*, Vol. 118, No. 49, 12461–12462.

Brandt, M. S.; Vogg, G. & Stutzmann, M. (2003). *Silicon- and Germanium-Based Sheet Polymers and Zintl Phases, in* Jutzi, P. & Schubert, U. (eds.), *Silicon Chemistry*, ISBN-13: 978-3527306473, Wiley-VCH, Weinheim, Chapter 15, pp. 194–213.

Chen, S. H., Katsis, D., Schmid, A. W., Mastrangelo, J. C., Tsutsui, T. & Blanton, T. N. (1999). Circularly polarized light generated by photoexcitation of luminophores in glassy liquid-crystal films. *Nature*, Vol. 397, 11 February 1999, 506–508.

Choi, J.; Wang, N. S. & Reipa, V. (2007). Photoassisted tuning of silicon nanocrystal photoluminescence. *Langmuir*, Vol. 23, No. 6, 3388–3394.

Colvin, V. L.; Schlamp, M. C. & Alivisatos, A. P. (1994). Light-emitting diodes made from cadmium selenide nanocrystals and a semiconducting polymer, *Nature*, Vol. 370, 4 August 1994, 354–357.

Cullis, A. G. & Canham, L. T. (1991). Visible light emission due to quantum size effects in highly porous crystalline silicon. *Nature*, Vol. 353, 26 September 1991, 335–338.

Cullis, A. G.; Canham, L. T. & Calcott, P. E. J. (1997). The structural and luminescence properties of porous silicon. *Journal of Applied Physics*, Vol. 82, No. 3, 909–965.

Davies, J. H. (1998). *The Physics of Low-dimensional Semiconductors: An Introduction*, ISBN-13: 978-0521484916, Cambridge University Press, Cambridge.

Diener, J.; Kovalev, D.; Koch, F. & Tsybeskov, L. Ordering and self-organization in nanocrystalline silicon. *Nature* 2000, Vol. *407*, 21 September 2000, 358–361.

English, D. S.; Pell, L. E.; Yu, Z.; Barbara, P. F. & Korgel, B. A. (2002). Size tunable visible luminescence from individual organic monolayer stabilized silicon nanocrystal quantum dots. *Nano Letters* Vol. 2, No. 7, 681-685.

Fujiki, M. (2001). Optically active polysilylenes: State-of-the-art chiroptical polymers. *Macromolecular Rapid Communications*, Vol. 22, No. 8, 539–563.

Fujiki, M.; Kawamoto, Y.; Kato, M.; Fujimoto, Y.; Saito, T.; Hososhima, S.-i. & Kwak, G. (2009). Full-visible-spectrum emitters from pyrolysis of soluble Si-Si bonded network polymers. *Chemistry of Materials*, Vol. 21, No. 12, 2459–2466.

Fujiki, M.; Kato, M.; Kawamoto, Y. & Kwak, G. (2011). Green-and-red photoluminescence from Si–Si and Ge–Ge bonded network homopolymers and copolymers. *Polymer Chem*istry, Vol. 2, No. 4, 914–922.

Fojtik, A. & Henglein, A. (1994). Luminescent colloidal silicon particles. *Chemical Physics Letters*, Vol. 221, No. 5-6, 363-367.
Fukao, S. & Fujiki, M. (2009). Circularly polarized luminescence and circular dichroism from Si-Si-bonded network polymers. *Macromolecules*, Vol. 42, No. 21, 8062-8067.
Furukawa, K.; Fujino, M. & Matsumoto, N. (1990). Optical properties of silicon network polymers. *Macromolecules*, Vol. 23, No. 14, 3423-3426.
Furukawa, K. (2000). *Synthesis and Optical Properties of Silicon-backbone Materials: (RSi)n (R = Organic group)*, Doctor thesis, Waseda University, Tokyo, Japan, Chapter 7 (in Japanese).
Furukawa, S. & Miyasato, T. (1988). Quantum size effects on the optical band-gap of microcrystalline Si-H. *Physical Review B*, Vol. 38, No. 8, 5726-5729.
Furumi, S. & Sakka, Y. (2006). Chiroptical properties induced in chiral photonic-bandgap liquid crystals leading to a highly efficient laser-feedback effect. *Advanced Materials*, Vol. 18, No. 6, 775-780.
Gelloz, B.; Kojima, A. & Koshida, N. (2005). Highly efficient and stable luminescence of nanocrystalline porous silicon treated by high-pressure water vapor annealing. *Applied Physics Letters*, Vo. 87, No. 3, 031107.
Gu, G.; Burghard, M.; Kim, G. T.; Düsberg, G. S.; Chiu, P. W.; Krstic, V.; Roth S. & Han, W. Q. (2001). Growth and electrical transport of germanium nanowires. *Journal of Applied Physics*, 2001, Vol. 90, No. 11, 5747-5751.
Harada, T., Nakano, Y., Fujiki, M., Naito, M., Kawai, T. & Hasegawa, Y. (2009). Circularly polarized luminescence of Eu(III) complexes with point- and axis-chiral ligands dependent on coordination structures. *Inorganic Chemistry*, Vol. 48, No. 23, 11242-11250.
Hasegawa, T.; Iwasa, Y.; Koda, T.; Kishida, H.; Tokura, Y.; Wada, S.; Tashiro, H.; Tachibana, H. & Matsumoto, M. (1996). Nature of one-dimensional excitons in polysilanes. *Physical Review B*, Vol. 54, No. 16, 11365-11374.
Heitmann, J.; Möller, F.; Zacharias, M. & Gösele, U. (2005). Silicon nanocrystals: size matters. *Advanced Materials*, Vol. 17, No. 7, 795-803.
Holmes, J. D.; Ziegler, K. J.; Doty, R. C.; Pell, L. E.; Johnston, K. P. & Korgel, B. A. (2001). Highly luminescent silicon nanocrystals with discrete optical transitions. *Journal of the American Chemical Society*, Vol. 123, No. 16, 3743-3748.
Hua, F.; Erogbogbo, F.; Swihart, M. T. & Ruckenstein, E. (2006). Organically capped silicon nanoparticles with blue photoluminescence prepared by hydrosilylation followed by oxidation. *Langmuir*, Vol. 22, No. 9, 4363-4370.
Hymanclki, W. J.; Viclscher, G. T. & Bianconi, P. A. (1993). Polygermynes: Synthesis and properties of germanium-germanium bonded network polymers. *Macromolecules*, 1993, Vol. 26, No. 4, 869-871.
Iba, S., Koh, S., Ikeda, K. & Kawaguchi, H. (2011). Room temperature circularly polarized lasing in an optically spin injected vertical-cavity surface-emitting laser with (110) GaAs quantum wells. *Applied Physics Letters*, Vol. 98, No. 8, 081113 (2011).
Jurbergs, D.; Rogojina, E.; Mangolini, L. & Kortshagen, U. (2006). Silicon nanocrystals with ensemble quantum yields exceeding 60%. *Applied Physics Letters*, Vol. 88, No. 23, 233116.

Kanemitsu, Y.; Ogawa, T.; Shiraishi, K. & Takeda, K. (1993). Visible photoluminescence from oxide Si nanometer-sized spheres - Exciton confinement on a spherical-shell. *Physical Review B*, Vol. 48, No. 7, 4883–4886.

Kanemitsu, Y. (1996). Photoluminescence spectrum and dynamics in oxidized silicon nanocrystals: A nanoscopic disorder system. *Physical Review B*, Vol. 53, No. 20, 13515–13520.

Kawagoe, Y., Fujiki, M. & Nakano, Y. (2000). Limonene magic: noncovalent molecular chirality transfer leading to ambidextrous circularly polarised luminescent π-conjugated polymers. *New Journal of Chemistry*, Vol. 34, No. 4, 637–647.

Kishida, H.; Tachibana, H.; Matsumoto, M. & Tokura, Y. (1994). Optical spectra of Si/Ge-network copolymers: $[Si(C_6H_{13})]_{1-x}[Ge(C_6H_{13})]_x$. *Applied Physics Letters*, 1994, Vol. 65, No. 11, 1358–1360.

Kovalev, D.; Heckler, H.; Ben-Chorin, M.; Polisski, G.; Schwartzkopff, M. & Koch, F. (1998). Breakdown of the *k*-conservation rule in Si nanocrystals. *Physical Review Letters*, Vol. 81, No. 13, 2803-2806.

Kovalev, D. & Fujii, M. (2005). Silicon nanocrystals: photosensitizers for oxygen molecules. *Advanced Materials*, Vol. 17, No. 21, 2531–2544.

Konagai, M. (1987). *Handotai Chokoshi Nyumon (Introduction to Semiconductor Superlattice)*, ISBN-13: 978-4563034351, Baifukan, Tokyo (in Japanese).

Krause, K. M. & Brett, M. J. (2008). Spatially graded nanostructured chiral films as tunable circular polarizers. *Advanced Functional Materials*, Vol. 18, No. 20, 3111–3118.

Lee, D. C.; Pietryga, J. M.; Robel, I.; Werder, D. J.; Schaller, R. D. & Klimov, V. I. (2009). Colloidal synthesis of infrared-emitting germanium nanocrystals. *Journal of the American Chemical Society*, Vol. 131, No. 10, 3436–3437.

Lehmann, V. & Gösele, U. (1991). The structural and luminescence properties of porous silicon. *Applied Physics Letters*, Vol. 58, No. 8, 856–858.

Li, X.; He, Y. & Swihart, M. T. (2004). Surface functionalization of silicon nanoparticles produced by laser-driven pyrolysis of silane followed by HF-HNO_3 etching. *Langmuir*, Vol. 20, No. 11, 4720–4727.

Liu, Q.; Wu, H.-J.; Lewis, R.; Maciel, G. E. & Interrante, L. V. (1999). Investigation of the pyrolytic conversion of poly(silylenemethylene) to silicon carbide. *Chemistry of Materials*, Vol. 11, No. 8, 2038–2048.

Liu, S.-M. (2008). Luminescent silicon nanoparticles formed in solution. *Journal of Nanoscience and Nanotechnology*, Vol. 8, No. 3, 1110–1125.

Liu, S.-M.; Sato, S. & Kimura, K. (2005). Synthesis of luminescent silicon nanopowders redispersible to various solvents. *Langmuir*, Vol. 21, No. 14, 6324–6329.

Liu, S.-M.; Yang, Y.; Sato, S. & Kimura, K. (2006). Enhanced photoluminescence from Si nano-organosols by functionalization with alkenes and their size evolution. *Chemistry of Materials*, Vol. 18, No. 3, 637–642.

Lockwood, D. J. (1998). *Light Emission in Silicon, in* Lockwood, D. J. (ed.), *Light Emission in Silicon From Physics to Devices*, Academic Press, ISBN-13: 978-0127521572, New York, 1998. Chapter 1, pp 1–34.

Lu, Z. H.; Lockwood, D. J. & Baribeau, J.-M. (1995). Quantum confinement and light emission in SiO_2/Si superlattices. *Nature*, Vol. 378, 16 November 1995, 258–260.

Lunkley, J. L., Shirotani, D., Yamanari, K., Kaizaki, S. & Muller, G. (2008). Extraordinary circularly polarized luminescence activity exhibited by cesium tetrakis(3-

heptafluoro-butylryl-(+)-camphorato) Eu(III) complexes in EtOH and $CHCl_3$ Solutions. *Journal of the American Chemical Society*, Vol. 130, No. 42, 13814–13815.

Maeda, Y.; Tsukamoto, N.; Yazawa, Y.;Kanemitsu, Y. & Masumoto, Y. (1991). Visible photoluminescence of Ge microcrystals embedded in SiO_2 glassy matrices. *Applied Physics Letters*, 1991, Vol 59, No. 24, 3168–3170.

Ma, D. D. D.; Lea, S. T. & Shinar, J. (2005). Strong polarization-dependent photoluminescence from silicon nanowire fibers. *Applied Physics Letters*, Vol. 87, No. 3, 033107.

Martin, H.-P.; Müller, E.; Richter, R.; Roewer, G. & Brendler, E. (1997). Methylchlorooligosilanes as products of the basecatalysed disproportionation of various methylchlorodisilanes. *Journal of Organometallic Chemistry*, Vol. 32, No. 1381–1387.

Mayeri, D.; Phillips, B. L.; Augustine, M. P. & Kauzlarich, S. M. (2001). NMR study of the synthesis of alkyl-terminated silicon nanoparticles from the reaction of $SiCl_4$ with the Zintl salt, NaSi. *Chemistry of Materials*. Vol. 13, No. 3, 765–770.

Nayfeh, M. & Mitas, L. (2008). *Silicon Nanoparticles: New Photonic and Electronic Material at the Transition Between Solid and Molecules, in* Kumar, V. (ed.), *Nanosilicon*, Elsevier, Oxford, ISBN-13: 978-0080445281, Chapter 1, pp. 1–78.

Nesper, R. (2003). *Structural and Electrnic Systematics in Zintl Phases of the Tetrels, in* Jutzi, P. & Schubert, U. (eds.), *Silicon Chemistry*, ISBN-13: 978-3527306473, Wiley-VCH, Weinheim, Chapter 13, pp. 171–180.

Oda, M.; Nothofer, H.G.; Lieser, G.; Scherf, U.; Meskers, S. C. & Neher, D. (2000). Circularly polarized electroluminescence from liquid-crystalline chiral polyfluorenes. *Advanced Materials*, Vol. 12, No. 5, 362-365.

Pi, X. D.; Liptak, R. W.; Nowak, J. D.; Wells, N. P.; Carter, C. B.; Campbell, S. A. & Kortshage, U. (2008). Air-stable full-visible-spectrum emission from silicon nanocrystals synthesized by an all-gas-phase plasma approach. *Nanotechnology*, Vol. 19, No. 24, 245603.

Qi, J.; Belcher, A. M. & White, J. M. (2003). Spectroscopy of individual silicon nanowires. *Applied Physics Letters*, Vol. 82, No. 16, 2616–2618.

Saitow, K. & Yamamura, T. (2009). Effective cooling generates efficient rmission: Blue, green, and red light-emitting Si nanocrystals. *The Journal of Physical Chemistry C*, Vol. 113, No. 19, 8465–8470.

Schmidt, W. R.; Interrante, L. V.; Doremus, R. H.; Trout, T. K.; Marchetti, P. S. & Maciels, G. E. (1991). Pyrolysis chemistry of an organometallic precursor to silicon carbide. *Chemistry of Materials*, Vol. 3, No. 2, 257–267.

Shimoda, T.; Matsuki, Y.; Furusawa, M.; Aoki, T.; Yudasaka, I.; Tanaka, H.; Iwasawa, H.; Wang, D.; Miyasaka, M. & Takeuchi, Y. Solution-processed silicon films and transistors. *Nature* 2006, Vol. 440, 6 April 2006, 783–786.

Shini, K. & Kumada, M. (1958). Thermal rearrangement of hexamethyldisilane to trimethyl-(dimethylsilylmethyl)silane. *The Journal of Organic Chem*istry, Vol. 23, No. 1, 139-139.

Smith, D. A.; Joray, S. J. & Bianconi, P. A. (2005). Synthetic method development and molecular weight control for homo- and co-polysilynes, silicon-based network-backbone polymers. *Journal of Polymer Research*, 2005, Vol. 12, No. 5, 393–401.

Takagi, H.; Ogawa, H.; Yamazaki, A.; Ishizaki, A. & Nakagiri, T. (1990). Quantum size effects on photoluminescence in ultrafine Si particles. *Applied Physics Letters*, Vol. 56, No. 24, 2379–2381.

Takeda, K.; Teramae, H. & Matsumoto, N. (1986). Electronic structure of chainlike polysilane. *Journal of the American Chemical Society*, Vol. 108, No. 26, 8186–8190.

Takeda, K. & Shiraishi, K. (1989). Electronic structure of Si-skeleton materials. *Physical Review B*, Vol. 39, No. 15, 11028–11037.

Takeda, K. & Shiraishi, K. (1993). Electronic structure of silicon-oxygen high polymers. *Solid State Communications*, Vol. 85, No. 4, 301–305.

Teramae, H. & Takeda, K. (1989). Ab initio studies on silicon compounds. Part II. The gauche structure of the parent polysilane. *Journal of the American Chemical Society*, Vol. 111, No. 4, 1281-1285.

Vogg, G.; Brandt, M. S. & Stutzmann, M. (2000). Polygermyne—A prototype system for layered germanium polymers. *Advanced Materials*, 2000, Vol. 12, No. 17, 1278–1281.

Walters, R. J.; Kalkman, J.; Polman, A.; Atwater, H. A. & de Dood, M. J. A. (2006). Photoluminescence quantum efficiency of dense silicon nanocrystal ensembles in SiO_2. *Physical Review B*, Vol. 73, No. 13, 132302.

Watanabe, A. (2003). Optical properties of polysilanes with various silicon skeletons. *Journal of Organometallic Chemistry*, Vol. 685, No. 1-2, 122–133.

Watanabe, A.; Hojo, F. & Miwa, T. (2005). Field-effect transistor based on organosoluble germanium nanoclusters. *Applied Organometallic Chemistry*, Vol. 19, No. 4, 530–537.

Wilcoxon, J. P.; Samara, G. A. & Provencio, P. N. (1999). Optical and electronic properties of Si nanoclusters synthesized in inverse micelles. *Physical Review B*, Vol. 60, No. 4, 2704–2714.

Wilson, W. L. & Weidman, T. W. (1991). Excited-state dynamics of one- and two-dimensional σ-conjugated silicon frame polymers: dramatic effects of branching in a series of hexylsilyne-branched poly(hexylmethylsilylene) copolymers. *The Journal of Physical Chemistry*, Vol. 95, No. 11, 4568–4572.

Wilson, W. L.; Szajowski, P. F. & Brus, L. E. (1993). Quantum confinement in size-selected, surface-oxidized silicon nanocrystals. *Science*, Vol. 262, No. 5137, 1242–1244.

Yu, P. Y. & Cardona, M. (2005). *Fundamentals of Semiconductors: Physics and Materials Properties; 3rd Ed.*, Springer-Verlag, ISBN-13: 978-3540254706, Chapter 7, 345–426.

Yajima, S.; Hasegawa, Y.; Hayashi, J. & Iimura, M. (1978). Synthesis of continuous silicon carbide fibre with high tensile strength and high Young's modulus: Part 1. Synthesis of polycarbosilane as precursor. *Journal of Material Science*, Vol. 13, No. 12, 2569–2576.

Zou,J.; Baldwin, R. K.; Pettigrew, K. A. & Kauzlarich, S. M. (2004). Solution synthesis of ultrastable luminescent siloxane-coated silicon nanoparticles. *Nano Letters*, Vol. 4, No. 7, 1181–1186.

Zhang, X.; Brynda, M.; Britt, R. D.; Carroll, E. C.; Larsen, D. S.; Louie, A. Y. & Kauzlarich, S. M. (2007). Synthesis and characterization of manganese-doped silicon nanoparticles: Bifunctional paramagnetic-optical nanomaterial. *Journal of the American Chemical Society*, Vol. 129, No. 35, 10668–10669.

2

Study of SiO_2/Si Interface by Surface Techniques

Constantin Logofatu, Catalin Constantin Negrila,
Rodica V. Ghita, Florica Ungureanu, Constantin Cotirlan,
Cornelui Ghica Adrian Stefan Manea and Mihai Florin Lazarescu
National Institute of Materials Physics, Bucharest
Romania

1. Introduction

Due to its dominant role in silicon devices technologies [1, 2] the SiO_2/Si interface has been intensively studied in the last five decades. The ability to form a chemically stable protective layer of silicon dioxide (SiO_2) at the surface of silicon is one of the main reasons that make silicon the most widely used semiconductor material. This silicon oxide layer is a high quality electrically insulating layer on the silicon surface, serving as a dielectric in numerous devices that can also be a preferential masking layer in many steps during device fabrication. Native oxidation of silicon is known to have detrimental effects on ultra-large-scale integrated circuit (ULSIC) processes and properties including metal/silicon ohmic contact, the low-temperature epitaxy of silicide and dielectric breakdown of thin SiO_2 [3]. The use of thermal oxidation of Si(100) to grow very thin SiO_2 layers (~ 100Å) with extremely high electrical quality of both film and interface is a key element on which has been built the success of modern MOS (metal-oxide-semiconductor) device technology [4]. At the same time the understanding of the underlying chemical and physical mechanisms responsible for such perfect structures represents a profound fundamental challenge, one which has a particular scientific significance in that the materials (Si, O) and chemical reaction processes (e.g. thermal oxidation and annealing) are so simple conceptually.
As a result of extreme decrease in the dimensions of Si metal-oxide-semiconductor field effect transistor device (MOSFET), the electronic states in Si/SiO interfacial transition region playa vital role in device operation [5]. The existence of abrupt interfaces, atomic displacements of interface silicon and intermediate oxidation states of silicon are part of different experiments [6, 7]. The chemical bonding configurations deduced from the observed oxidation states of silicon at the interface are the important basis for the understanding of the electronic states. The distribution of the intermediate oxidation states in the oxide film and the chemical bonding configuration at the interface for Si(100) and Si(111) were investigated [5] using measurements of Si 2p photoelectron spectra. One of the X-ray photoelectron spectroscopy (XPS) results is that the difference for <100> and <111> orientations is observed in the intermediate oxidation state spectra. Ultra thin SiO_2 films are critical for novel nanoelectronic devices as well as for conventional deep submicron ULSIC where the gate oxide is reduced to less than 30Å. Precise thickness measurement of these

ultra thin films is very critical in the development of Si- based devices. Oxide thickness is commonly measured by ellipsometry [8] but as film thicknesses is scaled down to several atomic layers, surface analytical techniques such as XPS become applicable tools to quantify these films [9]. An XPS measurement offers the additional advantage of providing information such as surface contamination and chemical composition of the film.

The purpose of the present section is to study the chemical structure modifications at the surface on semiconductors (e.g. Si, GaAs) by XPS, (angle resolved XPS) ARXPS and (scanning tunneling microscopy) STM techniques. It will be studied the variation of the interface for native oxides and for thermally grown oxides. This analysis will be the base for *in situ* procedures in the development of different devices as Schottky diodes or in the technique of local anodic oxidation (LAO) [10] for fabricating electronic devices on a nanometer scale.

A silicon dioxide layer is often thermally formed in the presence of oxygen compounds at a temperature in the range 900 to 1300^{0}C. There exist two basic means of supplying the necessary oxygen into the reaction chamber. The first is in gaseous pure oxygen form (dry oxidation) through the reaction: $Si+O_2 \rightarrow SiO_2$. The second is in the form of water vapor (wet oxidation) through the reaction: $Si+2H_2O \rightarrow SiO_2+2H_2$. For both means of oxidation, the high temperature allows the oxygen to diffuse easily through the silicon dioxide and the silicon is consumed as the oxide grows. A typical oxidation growth cycle consists of dry-wet-dry oxidations, where most of the oxide is grown in the wet oxidation phase. Dry oxidation is slower and results in more dense, higher quality oxides. This type of oxidation method is used mostly for MOS gate oxides. Wet oxidation results in much more rapid growth and is used mostly for thicker masking layers. Before thermal oxidation, the silicon is usually preceded by a cleaning sequence designed to remove all contaminants. Sodium contamination is the most harmful and can be reduced by incorporating a small percentage of chlorine into the oxidizing gas. The cleaned wafers are dried and loaded into a quartz wafer holder and introduced in a furnace. The furnace is suitable for either dry or wet oxidation film growth by turning a control valve. In the dry oxidation method, oxygen gas is introduced into the quartz tube. High-purity gas is used to ensure that no impurities are incorporated in the oxide layer as it forms. The oxygen gas can also be mixed with pure nitrogen in order to decrease the total cost of oxidation process. In the wet oxidation method, the water vapor introduced into the furnace system is usually creating by passage a carrier gas into a container with ultra pure water and maintained at a constant temperature below its boiling point (100^{0}C). The carrier gas can be either nitrogen or oxygen and both result in equivalent oxide thickness growth rates.

The structure of SiO_2/Si interface has been elusive despite many efforts to come up with models. Previous studies [11-13] generally agree in identifying two distinct regions. The near interface consists of a few atomic layers containing Si atoms in intermediate oxidation states i.e. Si^{1+} (Si_2O), Si^{2+} (SiO) and Si^{3+} (Si_2O_3). A second region extends about 30Å into SiO_2 overlayer. The SiO_2 in this layer is compressed because the density of Si atoms is higher for Si than for SiO_2. Different structural models [14-17] have been proposed for SiO_2 on Si (100), each predicting a characteristic distribution of oxidation states, and most of the models assume an atomically abrupt interface. From experiments was observed [1] at interface the existence of a large portion of Si^{3+}, and the model in accord this observation is that of an extended-interface for SiO_2/Si (100) by minimizing the strain energy [17]. Relatively new models ('90 years) are based for SiO_2/Si (100) and SiO_2/Si (111) on the distribution and intensity of intermediate oxidation states. These models are characterized by an extended interface with protrusions of Si^{3+} reaching about 3 Å into the SiO_2 overlayer.

Experimental techniques as the one presented in this work were used to determine the structure of the interface, its extend and to appreciate its roughness.

2. Investigation techniques

X-ray Photoelectron Spectroscopy (XPS) technique offers several key features which makes it ideal for structural and morphological characterization of ultra-thin oxide films. The relatively low kinetic energy of photoelectrons (< 1.5 keV) makes XPS inherently surface sensitive in the range (1-10 nm). Secondly, the energy of the photoelectron is not only characteristic of the atom from which it was ejected, but also in many cases is characteristic of the oxidation state of the atom (as an example the electrons emitted from $2p_{3/2}$ shell in SiO_2 are present approximately 4 eV higher in binding energy than electrons from the same shell originating from Si^0 (bulk Si). In the third place the XPS has the advantage that is straightforward to quantify through the use of relative sensitivity factors that are largely independent of the matrix.
The XPS recorded spectra were obtained using SPECS XPS spectrometer based on Phoibus analyzer with monochromatic X-rays emitted by an anti-cathode of Al (1486.7 eV). The complex system of SPECS spectrometer presented in Fig.1 allows the ARXPS analysis, UPS and STM as surface investigation techniques.

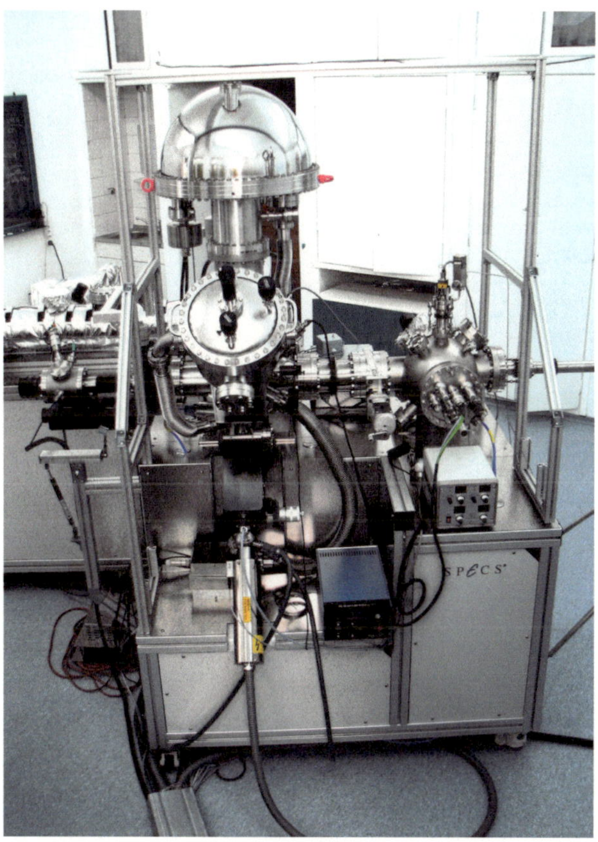

Fig. 1. SPECS complex system for surface analysis

A hemispherical analyzer was operated in constant energy mode with a pass energy of 5 eV giving an energy resolution of 0.4 eV, which was established as FWHM (full width half maximum) of the Ag 3d5/2 peak. The analysis chamber was maintained in ultra high vacuum conditions (~ 10^{-9} torr). As a standard practice in some XPS studies the C (1s) line (285 eV) corresponding to the C-C line bond had been used as reference Binding Energy (BE) [18]. The recorded XPS spectra were processed using Spectral Data Procesor v 2.3 (SDP) software. In its structure the SDP soft uses the deconvolution of a XPS line as a specific ratio between Lorentzian and Gaussian line shape and these characteristics ensures a good fit of experimental data.

Angle resolved X-ray Photoelectron Spectroscopy (ARXPS) is related to a XPS analysis of recorded spectra on the same surface at different detection angles θ of photoelectrons measured to the normal at the surface. The analysis chamber is maintained at ultra-high vacuum (~ 10^{-9} torr) and the take-off-angle (TOA) was defined in accord to ASTM document E 673-03 related to standard terminology related to surface analysis that describes TOA as the angle at which particles leave a specimen relative to the plane of specimen surface; it is worth to mention that our experimental measured angle is congruent with TOA as angles with correspondingly perpendicular sides. For a detection angle θ, the depth λ from where it proceeds the XPS signal is given by the projection of photoelectrons pass λ_m (the maximum escape depth) on the detection direction:

$$\lambda = \lambda_m \cos\theta$$

In Fig.2 is presented the TOA angle considered in the equation for oxide thickness evaluation as presented in [3, 19, 20, and 21].

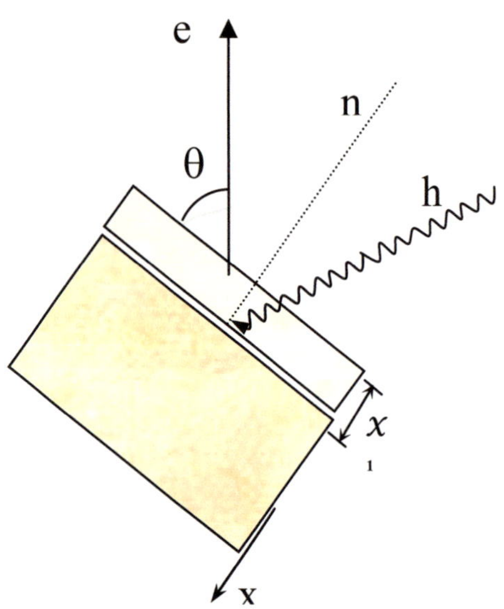

Fig. 2. Sample characteristics in ARXPS measurement

The oxide film thickness d_{oxy} is determined by the Si 2p core level intensity ratio of the oxidized silicon film I_{oxy} and substrate silicon I_{si} by:

$$d_{oxy}=\lambda_{oxy}\sin\theta[I_{oxy}/(\alpha I_{Si})+1] \quad (1) \text{ reference [9]}$$

where

$$\alpha=I_{oxide\infty}/I_{si\infty}=0.76 \quad (2)$$

The value for this ratio was experimentally obtained taking into account the intensity for the line SiO_2 (Si^{4+}) in a thick layer of oxide (where the signal for the bulk silicon is not present) reported to the intensity of Si^0 line in bulk silicon (where the oxide do not exists e.g. after Ar^+ ion sputtering).

It is well known however that large discrepancies exist for the photoelectron effective attenuation length in SiO_2 where values from 2 to 4 nm have been reported and compared to theoretical prediction for the inelastic mean free path. The ARXPS measurements are dependent on the value of sinθ, and the ratio I_{oxy}/I_{Si} will be computed only for SiO_2 oxide (Si^{4+}). The electron inelastic mean free path (IMFP) λ is analyzed and computed in terms of the Bethe equation for inelastic scattering which can be written [22]:

$$\lambda=E/\ [E_p^2\beta \ln(\gamma E)]\ \text{Å} \quad (3)$$

For electrons in the range (50-200) eV [7, 8] the computed IMFP is presented in the form of TPP-2M formula:

$$\lambda=E/\ \{E_p^2\ [\beta\ln(\gamma E)-C/E+D/E^2]\} \quad (4)$$

E-electron energy (in eV)

$\beta= -0.10+0.944/\ (E_p^2 + E_g^2) + 0.069\ \rho^{0.1}$

$$\gamma= 0.191\ \rho^{-0.50} \quad (5)$$

C= 1.97- 0.91 U

D= 53.4 - 20.8 U

$U = N_V\rho/\ A$ (N_V- total number of valence electrons per atom or molecule, ρ- density (gcm^{-3}), A- atomic or molecular weight, E_g-band gap, E_p- plasmon energy)

For E_p which is the free- electron plasmon energy (in eV) it was used the formula

$$E_p= 28.8\ (\rho N_V/A)^{1/2}\ \text{eV} \quad (6)$$

as is mentioned in reference [21].

For compounds N_V is calculated from the sum of contributions from each constituent element (i.e. N_V for each element multiplied by the chemical or estimated stoichiometric coefficient for that element) [19].

The explored depth of surface layers by XPS technique can be adjusted by the variation of θ angle.

Scanning Tunneling Microscopy (*STM*) is based on the quantum mechanical effect of tunneling. If two metals are brought in close contact and a small voltage is applied between them, a tunneling current can be measured which is:

$$I_t \sim \exp(-2kd) \quad (7)$$

where d is the distance between the conductors, and as an important message the reason why STM works, is the exponential dependence of the tunneling current on the distance between conductors. Typical values for tunneling voltage are from few mV to several V, and for the current from 0.5 to 5 nA. The tip-sample distance is a few Angstrom; the tunneling current depends very strongly on this distance. A change of 1 Å causes a change in the tunneling current by a factor of ten. In practice, the tunneling voltage is not always very small. Especially for semiconductors materials a small tunneling voltage can be impossible because there are no carriers in the gap which can be involved in the tunneling. This means that STM can look at both, occupied and unoccupied states of the sample depending on the bias voltage.

Transmission Electron Microscopy (TEM)-is a microscopy technique whereby a beam of electrons is transmitted through an ultra thin specimen, interacting with the specimen as it passes through. TEMs are capable of imaging at a significantly higher resolution owing to the small de Broglie wavelength of electrons. This enables the instrument's user to examine fine details-even as small as a single column of atoms. At smaller magnifications TEM image contrast is due to absorption of electrons in the material, due to the thickness and composition of the material.

UV-Photoelectron Spectroscopy (UPS)-is the most powerful technique available for probing surface electronic structure. UPS in the laboratory requires a He gas discharge line source which can be operated to maximize the output of either He I (21.2 eV) or He II (40.8 eV) radiation. The use of these photon energies makes accessible only valence levels and very shallow core levels. UPS refers to the measurement of kinetic energy spectra of photoelectrons emitted by ultraviolet photons, to determine molecular energy levels in the valence region [23]. The kinetic energy E_K of an emitted photoelectron is given by (Einstein law applied to a free molecule):

$$E_K=hv-I \tag{8}$$

Where h is Planck's constant, v is the frequency of the ionizing light, and I is an ionization energy corresponding to the energy of an occupied molecular orbital. In the study of solid surfaces in particular is sensitive to the surface region (to 10 nm depth) due to the short range of the emitted photoelectrons (compared to X-rays). A useful result from characterization of solids by UPS is the determination of the work function of the material. The work function Φ can be defined in terms of the minimum energy $e\Phi$ required to remove an electron from the highest occupied level of a solid to a specified final state. The value of Φ may depend on distance from the surface on account of the varying electrostatic potential associated with different crystal surfaces.

3. Native oxides

Silicon samples Si (100) were exposed to a naturally oxidation process for a long time decade (years) in atmosphere. A thin layer of native oxide was grown, that was firstly put into evidence by a color surface change. In Fig.3 the Si 2p spectra present two lines where the lower binding energy is associated with Si^0 (bulk) and the higher binding energy is associated with Silicon dioxide. The Si 2p oxide line intensity increases with the increase in oxide thickness while the Si 2p substrate line intensity decreases [24].

In Fig.4 -are presented the Si 2p lines in crystalline Silicon and in Silicon oxides for the spectra taken at TOA:25^0, 55^0 and 75^0 for native oxides. As a general remark: in superimposed spectra as the TOA increases the signal from Si^0) bulk) is more prominent. For the sample Si (oxidized) the deconvolution for Si core levels 2p lines are related to specific Binding Energies (BE) for Si^0(A), Si^{1+} (Si_2O-B), Si^{2+}(SiO-C), Si^{3+}(Si_2O_3-D) and Si^{4+}(SiO_2-E) as presented in Fig.5 at TOA= 25^0.

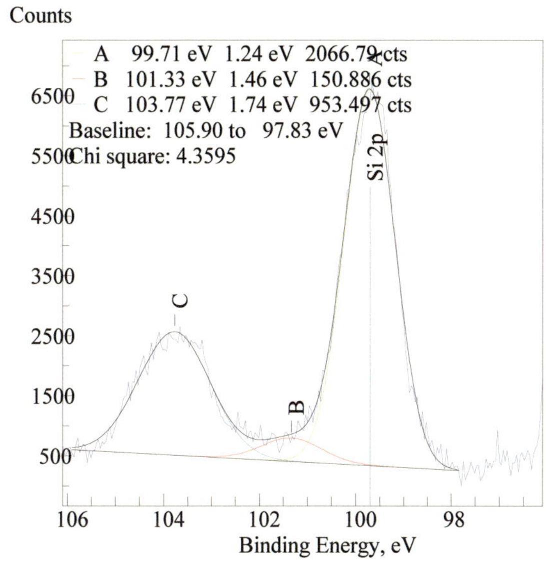

Fig. 3. Si 2p spectra for Si^{4+}(SiO_2-A) and Si^0 (bulk-C)

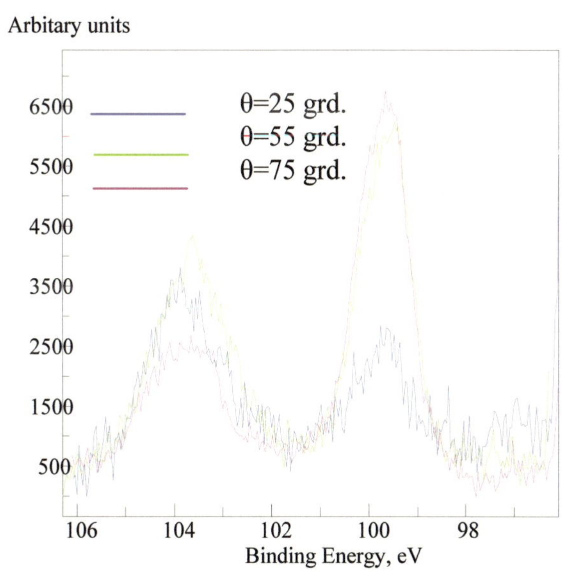

Fig. 4. SiO_2/Si native oxides XPS proportional spectra

As it was mentioned [1] the structure of the interface Silicon Oxide/Silicon consists of two regions. The near interface contains few atomic layers of Si atoms in intermediate oxidation states i.e Si^{1+}(Si_2O), Si^{2+}(SiO) and Si^{3+}(Si_2O_3). A second region extends about 30Å into SiO_2 overlayer [1]. As a general remark we assert that although the deconvolution has a slight arbitrary degree in S0 sample, it prevails the sub oxides with high oxidation states (e.g. Si^{3+}). Regarding the measurement of oxide thickness the ratio I_{oxy}/I_{Si} in the oxidation state Si^{4+} (SiO_2) can be evaluated for the sample S0 and the computed value is in accord with the presented theory of Electron Inelastic Mean Free Path (IMFP) as presented in literature since 1994, and conduced to the value d_{oxy}~23.6Å for native oxide.

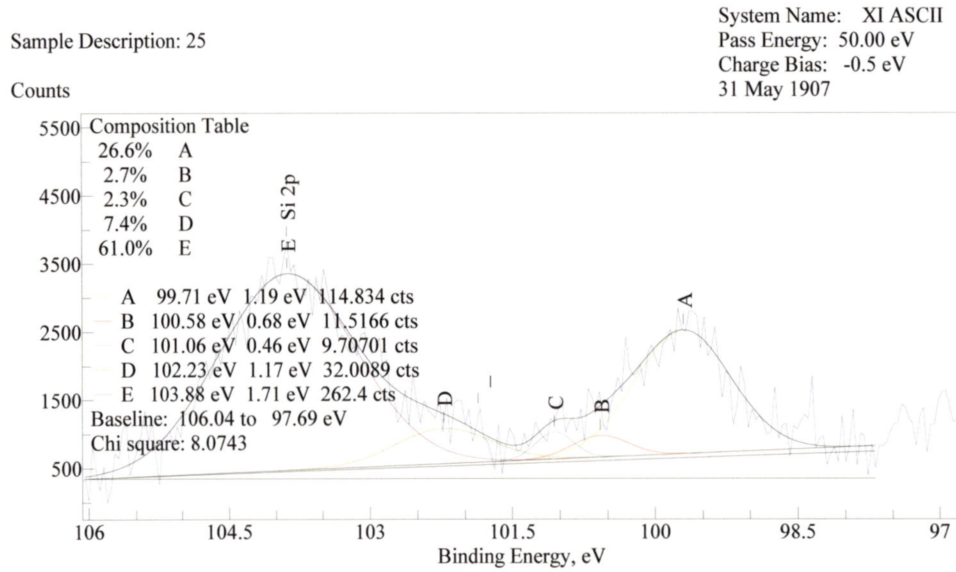

Fig. 5. XPS spectra for Si (oxidized) S0 sample at TOA=25⁰

Oxidation states for Si	Results in our experiment BE(eV)	BE(eV) Reference[1]	Composition table (%)
Si^0	99.7	99.5	26.6
Si^{1+}	100.58	100.45	2.7
Si^{2+}	101.06	101.25	2.3
Si^{3+}	102.23	101.98	7.4
Si^{4+}	103.88	103.40	61

Table 1. Comparative contribution for Silicon oxides

As an example of a native oxide analysis on semiconductors we present a less extended technological case, the one of GaAs. The GaAs (100) surfaces have a high surface energy [25], and as a consequence they are very reactive and chemically unstable. Due to the reactivity of the native oxides is difficult to reach the passivation of GaAs surfaces. Bare arsenic atoms are thought to be one of the species present within the native oxides responsible for pinning the Fermi level [26]. The arsenic atoms result from chemistry that occurs at the oxide/GaAs interface. Both As_2O_3 and Ga_2O_3 will form when a clean GaAs surface is exposed to oxygen and light. The formation of Ga_2O_3 is thermodynamically favored and results in the reaction: $As_2O_3+2GaAs \rightarrow Ga_2O_3+4As$ leaving bare arsenic atoms embedded within the oxide near the oxide/GaAs interface. The As_2O_3 is also mobile at grain boundaries, resulting in a nonuniform oxide in which an As_2O_3-rich layer is found near the oxide/air interface, and the bare arsenic atoms are found embedded within the Ga_2O_3-rich layer near the oxide/ GaAs interface. In native oxide layer, both Ga_2O_3 and As_2O_3 are somewhat soluble in water and their solubility is dependent on pH. The complicated chemistry of the GaAs native oxides has prevented the development of a simple and robust surface passivation scheme for this surface. IN the ARXPS measurement the Ga and As 3d spectra from the as received (naturally oxidized) GaAs surface taken at TOA: 15^0,20^0,30^0,50^0,90^0 as presented in Fig. 6 in a proportional ratios. For As, the signal from 41 eV corresponds to As in the volume (GaAs matrix) and the signal near the broad peak of 44 eV is related to As oxides: As_2O_3 together with As_2O_5. The As signal from TOA: 15^0 is the most sensitive to the surface structure, due to a similar peak intensities arising from oxide (interface) and volume (GaAs).As presented in Fig.7 at small analysis angles the As and Ga

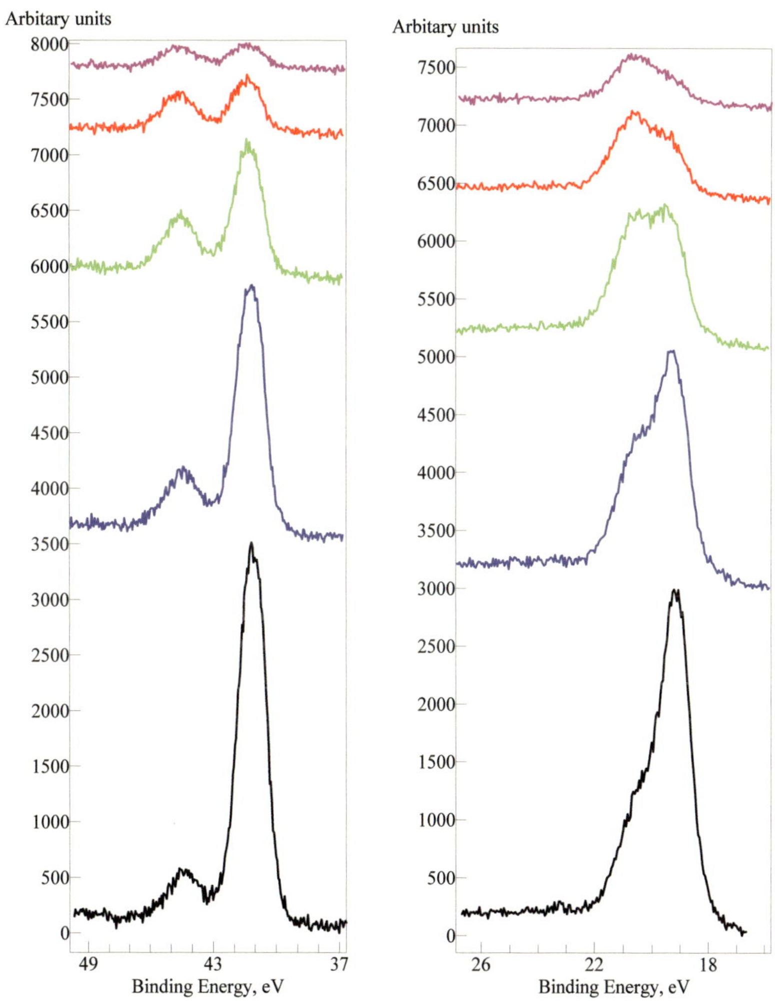

Fig. 6. ARXPS spectra of As 3d (left) and Ga 3d (right) for TOA angles:90^0-black, 50^0 blue, 30^0 -green, 20^0-red, 15^0-pink

concentrations grows and at the most surface sensitive angle the concentration of C and O is higher than the concentrations for As and Ga. For the native oxidized sample the atomic surface composition Ga/As ratio is related to the entire signal arisen from surface and volume. The Ga signal arises from 19.1 eV (GaAs) and 20.3 eV (Ga_2O_3) [27]. The As to Ga ratio in the bulk is close to a stoichiometric value of 1.05. For Ga to As ratio in naturally oxidized sample (storage for years) this ratio revealed the contribution of As and Ga from native oxides in the surface layer. The concentration of Ga oxide is greater than of As oxide and from this fact resulted a Ga enrichment in the air exposed GaAs, an observation in accord to reference [28].

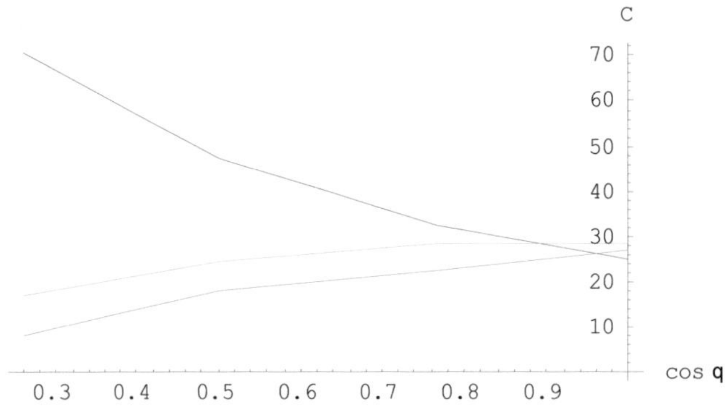

Fig. 7. Relative variation of concentration for C-black, O-blue, As-green and Ga-red as a function of analysis angle (q) (TOA=90^0-q)

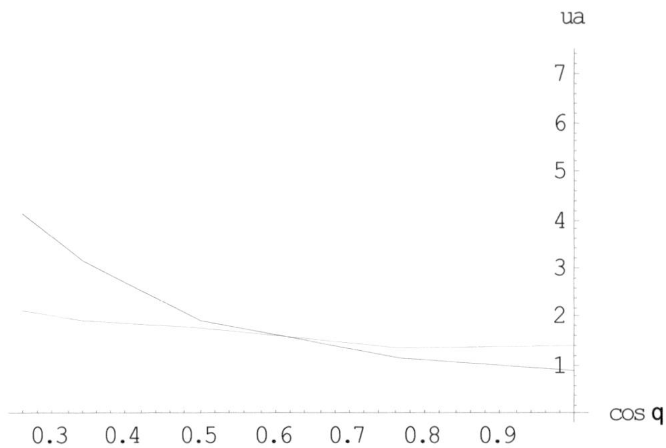

Fig. 8. Variation of different concentration ratios for C_C/C_O-black, C_{Ga}/C_{As}-red, C_c+C_O/C_{Ga}+C_{As}-green as a function of analysis angle (q) (TOA=90-q)

As can be observed at high TOA angles the concentration ratio C_C/C_O is low and it grows to low TOA angles corresponding to the surface layer. For concentration ratio C_{Ga}/C_{As} on an extended range of angles is constant in a prime approximation, with a slight increase at small TOA angles. The evolution of ratio for contaminants to GaAs (C_C+C_O/C_{Ga}+C_{As}) is related to an increase at low TOA angles, as a main result of oxidation to the GaAs surface. We conclude that the surface native oxide comprise a mixture of Ga_2O_3, As_2O_3 and As_2O_5 phases

4. Silicon/oxide interface

The Silicon oxide samples were prepared for different analysis by cleaning in organic solvents, and chemical etching in aqueous solution of hydrofluoric acid. There were examined samples exposed to air oxidation for a long period of time together with samples maintained for 2-3 hours in atmosphere after a chemical etching as well as chemical etched fresh samples. In this experiment there were used as substrates p-Si (100) and p-Si (111) of medium resistivity. The ARXPS spectrum of O 1s and C 1s are presented in Fig.9

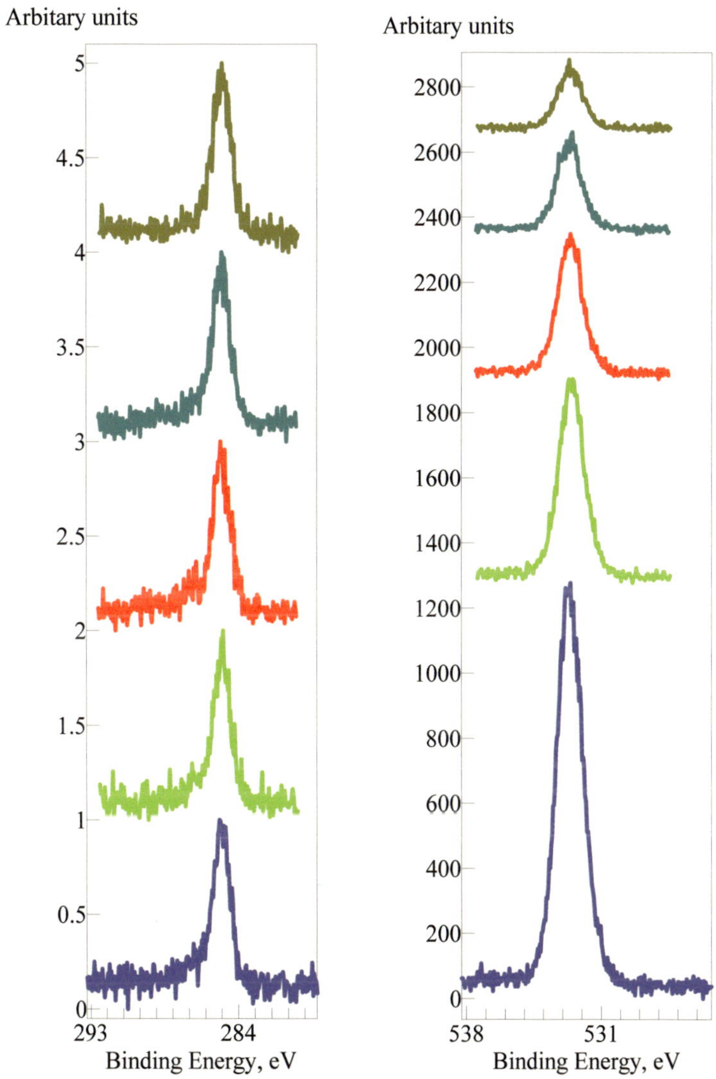

Fig. 9. ARXPS spectra for C1s(left) and O1s (right) at TOA: 90^0(blue), 50^0(green),30^0(red), 20^0(turquoise), 15^0(olive)

The ARXPS spectra of Si 2s and Si 2p for the same sample exposed to natural oxidation is presented in Fig.10

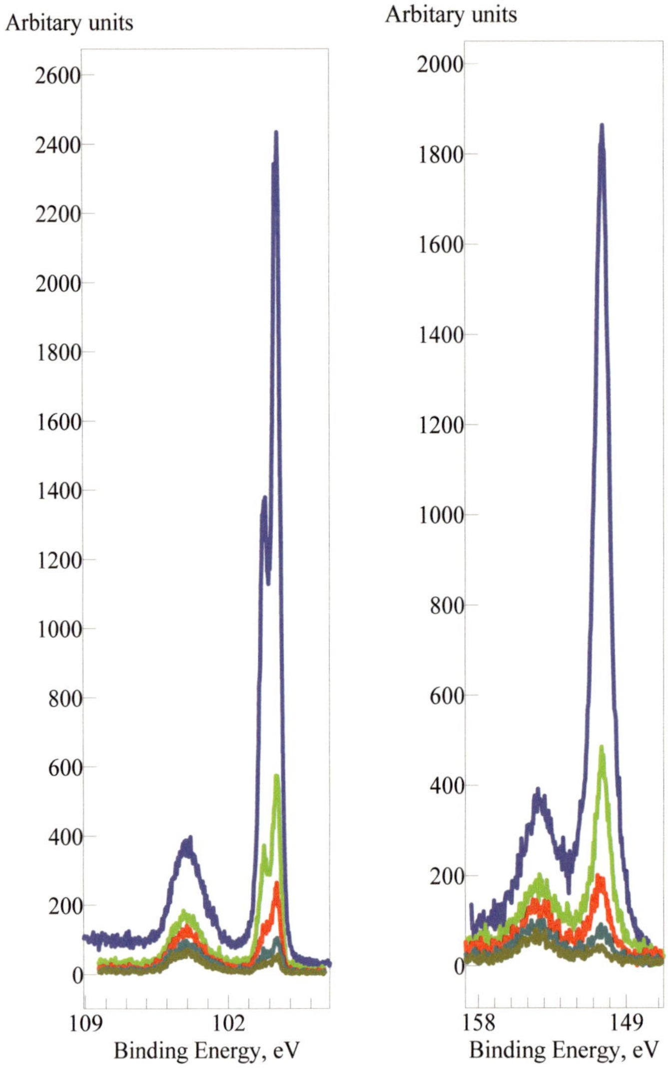

Fig. 10. ARXPS spectra of Si 2s(right) and Si 2p (left) at TOA: 90⁰(blue), 50⁰(green), 30⁰(red), 20⁰(turquoise), 15⁰(olive)

The proposed deconvolution of Si 2p XPS spectra for TOA: 90⁰ and TOA: 15⁰ (the most sensitive angle for the surface composition are presented in Fig.11 (a and b).

The deconvolution for the XPS spectra of Si 2s are presented in Fig.12 (a and b) firstly at TOA:90⁰, secondly at TOA:15⁰ as surface sensitive angle.

In Fig.11 (a) the A and B peaks are related to the signal of Si^0 2p ½ and Si^0 2p 3/2, and the peaks of C, D, E and F are related to the signals of sub-oxides as it follows: Si^{1+}, Si^{2+}, Si^{3+} and Si^{4+}. In Fig.11(b) is presented the surface composition for the XPS signal for Si^0 2p1/2 and 2p 3/2 (A and B peaks) and for Si^{1+}(C), Si^{2+}(D), Si^{3+}(E) and Si^{4+}(F). The most interesting part of the presented deconvolution is related to the signal of Si 2s that has important similarities with Si 2p spectra, that means in Fig. 12(a) and Fig.12 (b) the presence of Si^0 for A peak, and related sub-oxides as follows: Si^{1+} (B), Si^{2+}(C), Si^{3+}(D), Si^{4+}(E). As can be observed in Fig.12

(b) the surface composition is similar as order of magnitude for the signal of Si 2p and Si 2s, and we also notice that the information of XPS spectra related to Silicon/Oxide interface for Si 2s is rare in literature experimental data. For the Binding Energy BE Si 2p3/2= 99.67 eV the shift BE Si 2p3/2 (Si)-BE Si2p3/2 (SiO_x) we have the following results: Si^{1+}-0.9 eV, Si^{2+}-2.1eV, Si^{3+}- 3.5eV and Si^{4+}-4.5 eV For the Binding Energy BE Si 2s (Si)=150.51 eV the shift BE Si 2s-BE Si 2s (SiO_x) we have the following results: Si^{1+}-0.97 eV, Si^{2+}- 1.79 eV, Si^{3+}-2.96 eV, and Si^{4+}-4.11 eV.

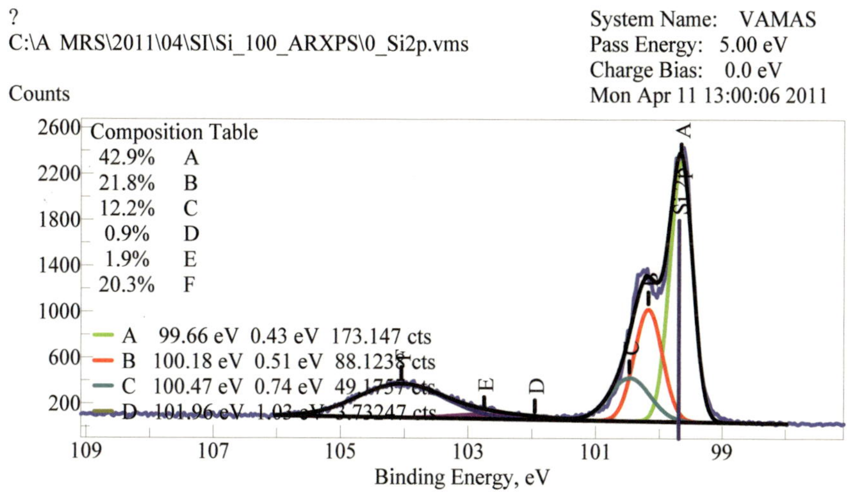

Fig. 11. (a) XPS spectrum of Si 2p at TOA: 90^0

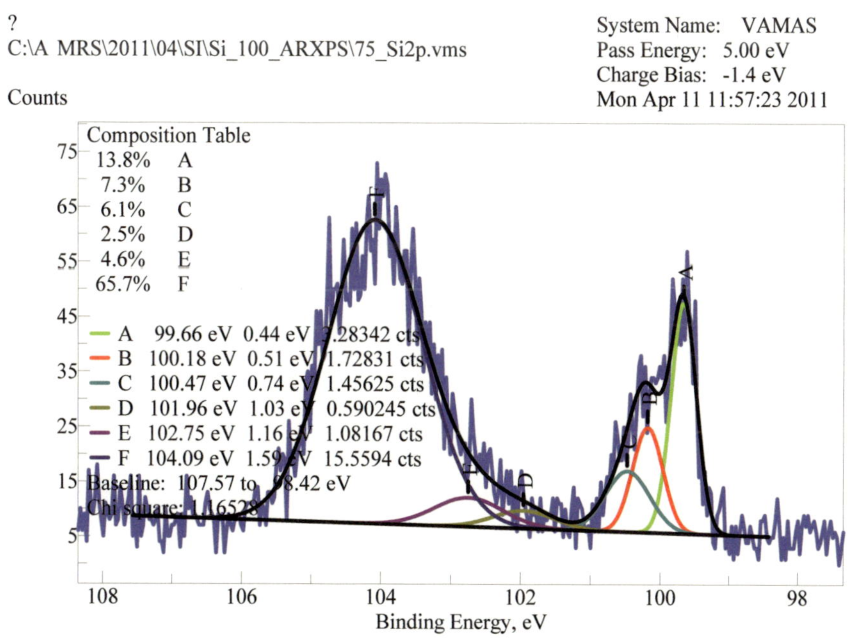

Fig. 11. (b) XPS spectrum of Si 2p at TOA: 15^0

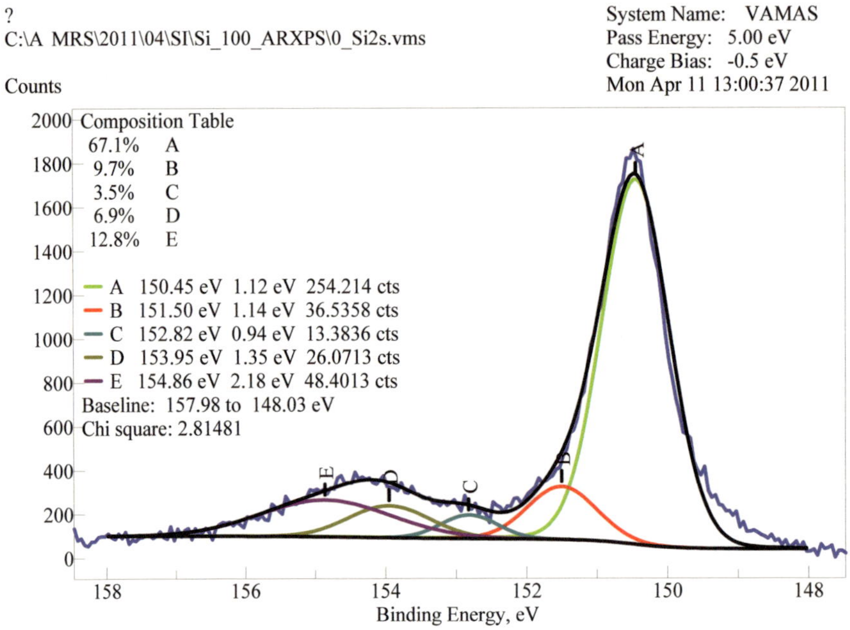

Fig. 12. (a) XPS spectrum 2s at TOA:90^0

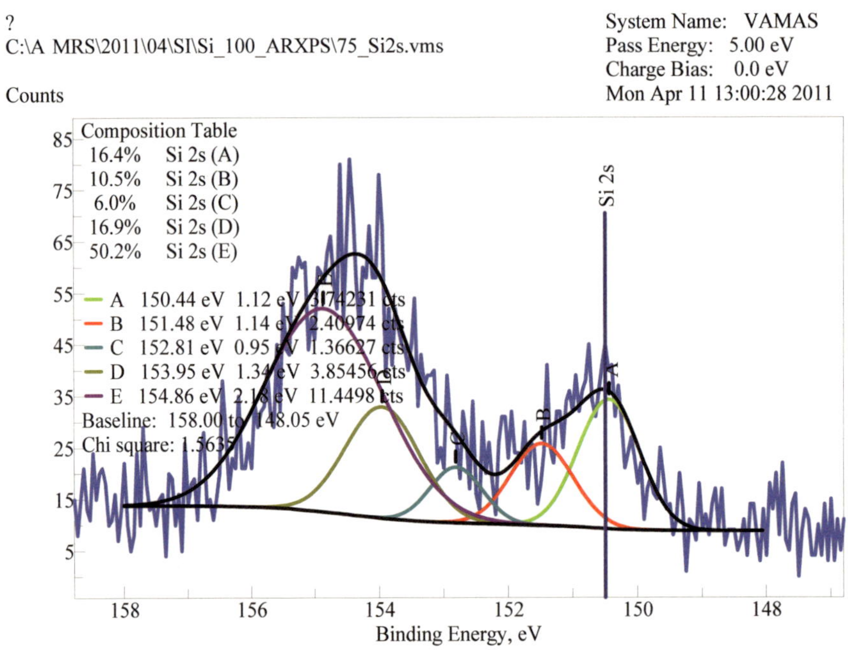

Fig. 12. (b) XPS spectrum 2s at TOA: 15^0

In Fig.13 (a) we present the variation of composition with TOA angle that means from bulk (right) to surface (left) for Si, SiO_x, and SiO_2 for the signal of Si 2p. As can be observed at TOA:90^0 the signal of Si^0 from the bulk is high and is decreasing near the surface at a

TOA: 15^0. At this angle of surface sensitivity the ARXPS signal is related to the presence of oxides, firstly for SiO_2 (green) and secondly for SiO_x (blue). In Fig.13 (b) is presented the concentration variation from the bulk to the surface of ARXPS signal for Si 2s. As can be observed the concentration curves for Si^0, SiO_2 and SiO_x are similar for Si 2p and Si 2s signal. For the ARXPS deconvolutions for Si 2p and Si 2s the positions for Si^{1+}- Si^{4+} are matching in the limit of experimental errors.

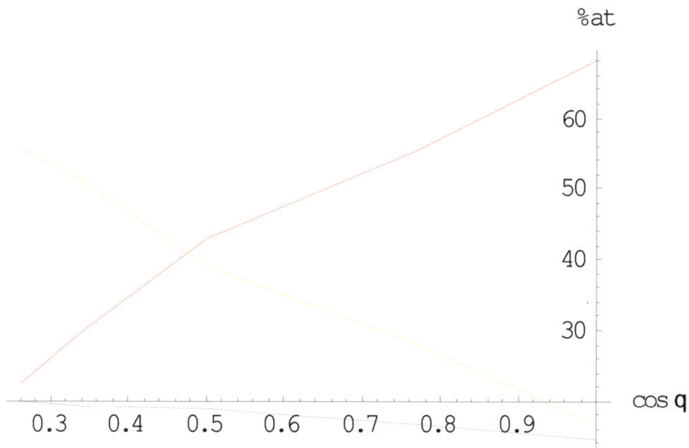

Fig. 13. (a) Concentration variation of Si^0(red), SiO_2 (green) and SiO_x (blue) at TOA(90^0-q) for the ARXPS signal of Si 2p

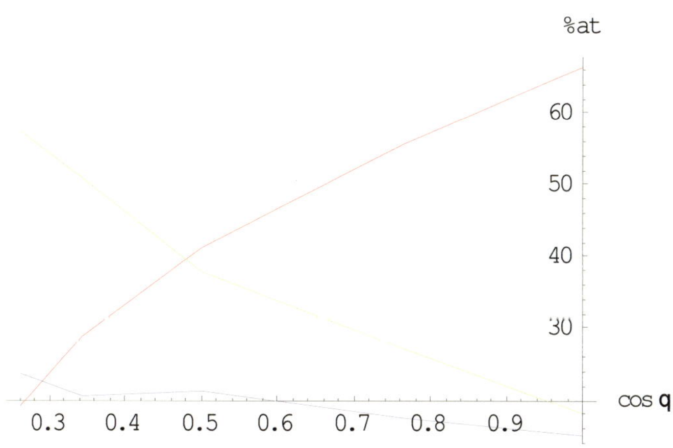

Fig. 13. (b) Concentration variation of Si^0(red), SiO_2 (green) and SiO_x(blue) at TOA (90^0-q) for the ARXPS signal of Si 2s

In Fig.14 is presented the variations for different concentration ratios in the surface area for the XPS signal Si (2p) and Si (2s). The proposed experimental concentration ratios are C_{Si^0}/C_{ox}, and $C_{Si^{4+}}/C_{SiOx}$ at different angle orientations in ARXPS signal. As can be observed there is an experimental accord between the ARXPS data between the Si (2p) and Si (2s) data starting from the bulk to the surface, the concentration of Si^{4+} is higher in the surface region, a surface with natural oxidation.

In Fig.15 is present an experimental case of comparison between the XPS spectra of native oxide SiO_2 in Si and native SiO_2 (quartz) and quartz exposed to different ion etching. In the case of second ion etching it can be observed a shoulder in the XPS signal both on Si (2p) line and Si (2s) line, this shoulder is related to the appearance of a sub-oxide, probably Si^{3+}.

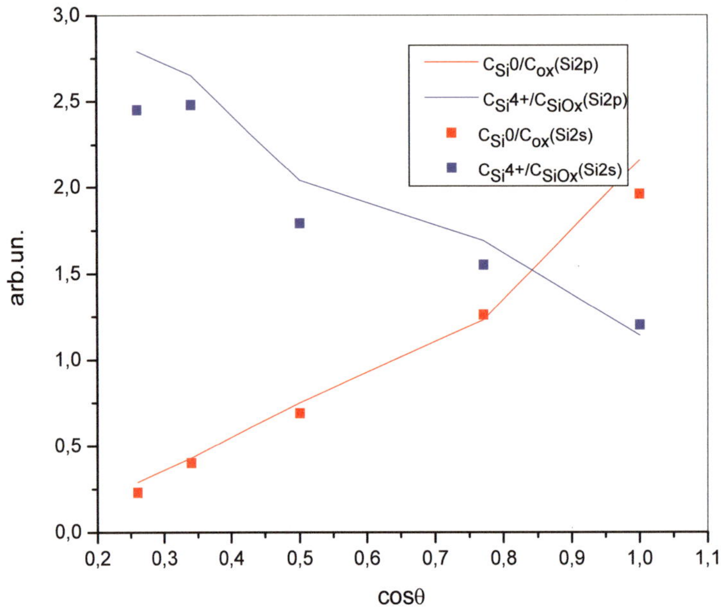

Fig. 14. Variation of concentration ratios for Si (2p) and Si (2s) as a function of TOA (90^0-θ)

In principle, the thickness of the native amorphous oxide layer on top of silicon wafers (usually in the range of 2-3 nm) may be directly measured on cross-section images of transmission electron microscopy (TEM). Although the task seems easily achievable, we would like to explain the technical difficulty of the procedure. Cross-section specimens for TEM observations are prepared by gluing against each other fine stripes of Si diced from the original wafer. The obtained sandwich is afterwards mechanically grinded followed by ion milling until a hole is produced in the interface region. The thin border of the created hole represents the useful area for TEM investigations. It is expected that TEM images at high magnification of the wafer surface to reveal the native amorphous layer. The impediment consists in the fact that the assembling resin holding together the two stripes of Si is also amorphous, showing the same contrast as the native amorphous silicon oxide layer, which makes it rather difficult to distinguish the limit between the two amorphous materials in contact.

In our case, a cross section specimen has been prepared from a Si(100) wafer by mechanical grinding and lapping on the two sides of the assembled Si-Si sandwich followed by ion milling at low incidence angle (7 degrees) and 4 kV ion accelerating voltage in a Gatan PIPS installation. The ion milling procedure has been ended with a fine milling step at low voltage (2 kV) in order to remove the amorphous layer created by ion milling and enveloping the surfaces exposed to ion beam. The TEM observation of the prepared specimen has been performed on a JEOL 200CX electron microscope.

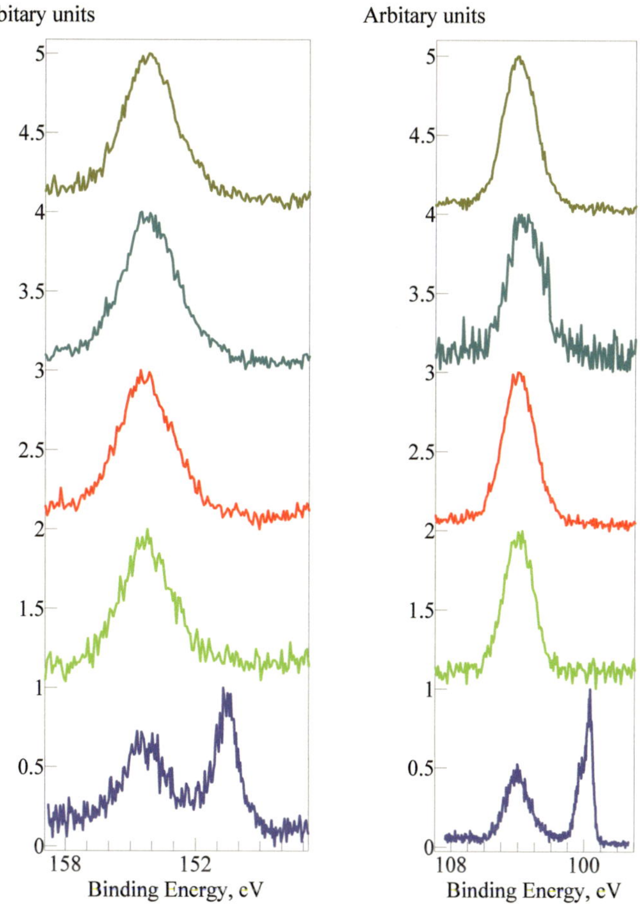

Fig. 15. XPS signal for Si-2p (right) and Si-2s (left) for SiO_2/Si (blue), SiO_2 (quartz- green), SiO_2 ion etching 1(red), SiO_2 ion etching 2 (turquoise), SiO_2 ion etching 3 (olive)

For a better observation of the amorphous surface layer, the cross-section specimen has been oriented in the microscope along the [110] zone axis as shown in the Selected Area Electron Diffraction pattern inserted in Fig. 16 (a). This way, the strongly diffracting crystalline object, the Si wafer, shows a strong dark contrast, allowing to clearly seeing the interface between the crystalline Si and the amorphous layer on the surface. In the thicker areas of the TEM specimen, the assembling resin has not been removed during the ion milling preparation stage (Fig. 16(a)). Here, the limit between the amorphous SiO_2 layer and the amorphous assembling resin is rather difficult to notice. However, the contrast difference between the two amorphous materials allows one to measure the thickness of the SiO_2 layer. One can notice the roughness of the crystalline Si wafer and the amorphous band with a rather constant thickness (about 2.5±0.5 nm) running along the surface.

In the thinner areas of the specimen (Fig. 16 (b)), the assembling resin has been removed by ion milling while a band of amorphous material with the same thickness (2.5±0.5 nm) running parallel to the crystalline surface is still observable.

We conclude, therefore, that the thickness of the amorphous Si layer on top of the Si(001) wafer measured by TEM is 2.5±0.5 nm.

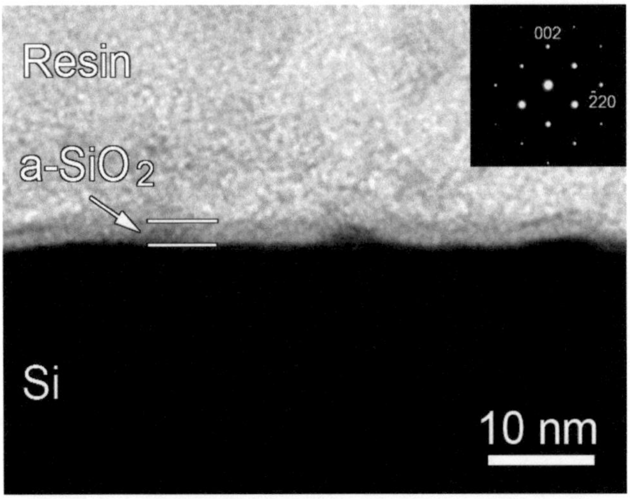

(a)

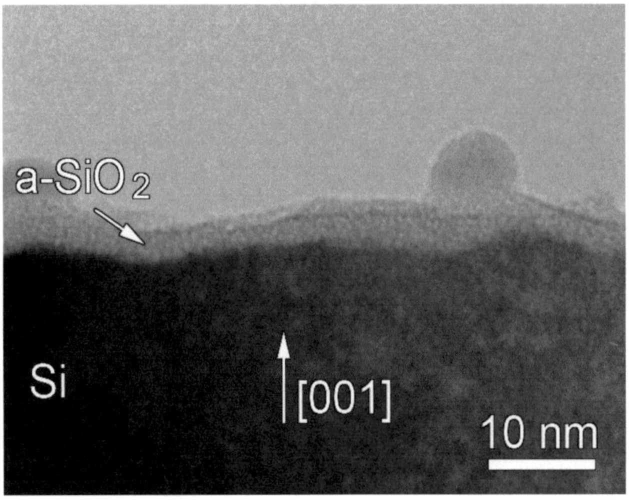

(b)

Fig. 16. (a) Cross-section TEM image of the Si surface in a thicker area of the specimen where the assembling resin is still visible after the ion milling. Inset shows the (b) Cross-section TEM image of the Si surface in a thinner of the specimen, where the assembling resin has been removed by ion milling.

As it was stated in previous works [29, 30, 31] the interface between crystalline Si and its amorphous native oxide SiO_2 is the basis for most current computer technology, although its structure is poorly understood. In this line, the study of the structural properties of water near a silica interface by classical and *ab-initio* molecular dynamics simulations is a part of this effort. The orientation of water molecules at the interface determined in classical force fields and quantum simulations [30] show that near the interface the water molecules are oriented such that at least one of the hydrogen atoms are nearer the silica than the oxygen of the water molecule. The importance of characterizing the atomic structure of the silicon/silicon dioxide interface as an essential component in highly integrated circuits has steadily increased as a result of continuing miniaturization of silicon chips.

5. Conclusions

The surface investigations techniques put into evidence the characteristics of Silicon/Oxide interface as it follows:

- the most important result is the XPS analysis of Si (2p) and Si (2s) signals that are similar in the interface region
- the XPS signals of Silicon oxides are related to the oxidation states:Si^{1+}, Si^{2+}, Si^{3+} and Si^{4+}
- the concentration of Si^{4+} is higher in the surface region of natural oxidation
- the result of Ion etching of natural SiO_2 (quartz) present the oxidation state Si^{3+}
- TEM result put into evidence a region of oxide at the surface that has the properties of the interface including its irregularities, at a thickness of the amorphous Si layer of the Si (001) wafer measured by TEM is 2.5±0.5 nm.

6. References

[1] F. J. Himpsel, F.R.Mc Feely, A.Taleb-Ibrahimi and J.A.Yarmoff, *Physical Review B,* Vol.38, No.9, pp.6084-6095 (1988)
[2] M. Razeghi *Technology of Quantum Devices* pp.42 LLC (2010), Springer, ISBN 978-1-4419-1055-4
[3] F. Yano, A.Hiroaka, T.Itoga, H.Kojima and K.Kanehori ,*J.Vac. Sci.Technol A,* Vol.13, No.6 pp.2671 (1995)
[4] G. W. Rubloff, *J.Vac.Sci.Technol. A,* Vol.8, No.3, pp.1857 (1990)
[5] T. Hattori and T.Suzuki, *Appl.Phys.Lett,* Vol.43, No.5 pp.470 (1983)
[6] R. Haight and L.C.Feldman, *J.Appl.Phys,* Vol.53, pp.4884 (1982)
[7] F.J.Grunthaner, P.J. Grunthaner, R.P.Vasquez , B.F.Lewis and J.Maserjian, *J.Vac.Sci.Technol* 16 pp.1443 (1979)
[8] A. Kalnitshi, S.P.Tay, J.P.Ellul, S.Chongsawangvirod, J.W.Andrews and E.A Irene *J.Electrochem. Soc.* 137, pp.235 (1990)
[9] Z. H. Lu, J.P.Mc Caffrey, B.Brar, G.D.Wilk, R.M. Wallace, L.C.Feldman and S.P. Tay, *Appl.Phys Lett.* Vol.71 No.19, pp.2764 (1997)
[10] R. Held, T.Vancura, T.Heinzel, K.Ensslin, M.Holland, W.Wegscheider, *Appl.Phys.Lett,*Vol.73, No.2 pp.262 (1998)
[11] *The physics of SiO_2 and its Interfaces* edited by Sokrates T.Pantelides (Pergamon, New York, 1978)
[12] F. J. Grunthaner and P.J.Grunthaner, *Mater, Sci Rep.* 1, pp.65 (1986)
[13] *Proceedings of the 173-rd meeting of the Electrochemical Society,* Atlanta, Georgia, 1988, edited by C.R.Helms
[14] F. Rochet, S.Rigo, M.frament, C.D'Anterroches, C.Maillot, H.Roulet and G.Dufour, *Adv.Phys.* 35, pp.237 (1986)
[15] F. Herman, R.V.Kasowski *J.Vac.Sci.Technol,* 19,pp.395 (1981)
[16] A. Ourmazd, D.W.Taylor, J.A.Rentschles and J.Bevk, *Phys.Rev.Lett,* 59, pp.213 (1987)
[17] L. Ohdomari, H.Akatsu, Y.Yamakoshi and K.Kishimoto *J.Appl.Phys* 62, 3751 (1987)
[18] R. V. Ghita, C.Negrila, A.S.Manea, C.Logofatu, M.Cernea, M.F.Lazarescu, *J.Optoelectron.Adv.Mater,* 5, pp.859 (2003)

[19] S. Tanuma, C.J.Powell and D.R.Penn, *Surf.Interface Anal* 21,pp.165 (1994)
[20] S. Tanuma, C.J.Powell and D.R.Penn,*Journal of Electron Spectroscopy and Related Phenomena* 52, pp.285 (1990)
[21] S. Tanuma, C.J.Powell and D.R.Penn, *Surface Science* 192, L 849 (1987)
[22] H. Bethe, *Ann.der Physik*, 5 pp.325 (1930)
[23] http://en.wikipedia.org
[24] C. C. Negrila, C.Cotirlan, F.Ungureanu, C.Logofatu, R.V.Ghita, M.F.Lazarescu, *J.Optoelectron. Adv.Mater*, 10 (6), pp.1379 (2008)
[25] Freiberger General Specifications, issue 200
[26] (www.fem-semcond.com/pdf/gen.spec.pdf)
[27] T. Hou, C.M.Greenlief, S.W.Keller, L.Nelen and J.F.Kauffman, *Chem. Mater.* 9,pp.3181 (1997)
[28] C. C. Negrila, C.Logofatu, R.V.Ghita, C.Cotirlan, F.Ungureanu, A.S.Manea, M.F.Lazarescu, *J.Crystal Growth*, Vol.310, No.7t-9, pp.1576 (2008).
[29] J. L. Sullivan, W.Yu and S.O.Saied, *Surface and Interface Analysis*, Vol.22, pp.515 (1994)
[30] Y. Tu and J. Tersoff, *Thin Solid Films*, Vol.400, No.1-2, pp.95(2001)
[31] Ch. D. Lorenz, M.Tsige, Susan B.Rempe, M.Chandross, M.J.Stevens, G.S.Grest *Journal of Computational and Theoretica* Nanoscience, Vol.7, No.12, pp.2586 (2010)
[32] S. Bergfeld, B.Braunschweig, W.Daum, *Physical Review Letters*, Vol.93, No.9 (2004)

3

Effect of Native Oxide on the Electric Field-induced Characteristics of Device-Quality Silicon at Room Temperature

Khlyap Halyna, Laptev Viktor, Pankiv Lyudmila and Tsmots Volodymyr
[1]*State Pedagogical University, Drohobych*
[2]*Russian New University, Moscow*
[1]*Ukraine*
[2]*Russian Federation*

1. Introduction

There is no needing emphasize about the importance of silicon (*Si*) as a material of choice for almost all fields of the new nano- and microelectronics. Due to its unique structural and physical properties, polycrystalline *Si* seems to be of special interest as a base for creating so-called 3D-integrated circuits.

Various studies have established the main processes of carrier transport in the structures based on this material. In particular, it was shown that tunneling and diffusion recombination processes dominate under room temperature and applied low electric fields. Nevertheless, the analysis and numerical simulation of the experimental data do not always take into account the finite dimensions of the investigated structure and the appearance of carrier depletion as an important component of the tunneling current observed experimentally. Besides that, the fabrication of any device based on polycrystalline *Si* requires high-temperature treatment. Therefore, the effect of such a treatment on the electric properties of polycrystalline, amorphous and monocrystalline *Si* is also seemed to be important. Regardless of the huge number of publications describing numerous characteristics of the material and structures based on polycrystalline *Si* of various types of conductivity, the question about room temperature carrier depletion (exclusion from the contact regions) in polycrystalline material is still open.

As is known, native oxides of about 5-10 nm thickness are formed on surfaces after finishing growth of semiconductor bulk materials or deposition (by molecular beam epitaxy, modified liquid phase epitaxy, laser ablation, high-temperature treatment, etc.) of thin films immediately after excluding the samples from the technological chamber. These ultrathin layers form additional potential barriers which can sufficiently affect the performance of active elements.

This chapter reports experimental data resulted from the investigations of room-temperature current-voltage (IVC) and capacitance-voltage (CVC) characteristics performed on amorphous silicon thin films fabricated by the magnetron sputtering technique and bulk crystalline silicon of device quality grown by Czochralsky method. The low-resistive contact pads were placed on front and faceplate surfaces of the samples. Studies of room-temperature electric field-induced characteristics for these structures are seemed to be important for analyzing operation of multi-element devices (for example, integrated circuits). It was found out that experimental IVC's and CBC's are similar to those of metal-insulator-semiconductor structures. These results are analyzed in framework of semiclassical theory of semiconductor devices.

2. Photosensitivity of amorphous silicon thin films prepared by magnetron sputtering

Amorphous silicon is a unique material for design of a large number of novel optoelectronic and photovoltaic devices. Structures *Me/α-Si* and α-*Si* thin films are the elements of choice not only for fundamental studies but also for practical applications and numerical simulations of their properties.
Examination of photosensitive and external electric field-induced characteristics of these structures is of particular interest. Metal-semiconductor junctions *Al/α-Si* were chosen as an object of the room temperature investigations. Amorphous silicon thin films (thickness up to 300 nm) were manufactured by magnetron sputtering technology in the range of the current density (10^{-9}-10^{-7}) A/cm^2 at T = 300 K.
Current-voltage characteristics nd photosensitivity of the samples was carried out under normal atmospheric conditions before and after the treatment of the structures in molecular hydrogen. The hydrogenation of the samples was provided by the special chamber filled in with molecular H2 during 24 hours at T = 400^{0}C and the gas pressure $P_H = 2500\ Pa$ (Khlyap, 2003).

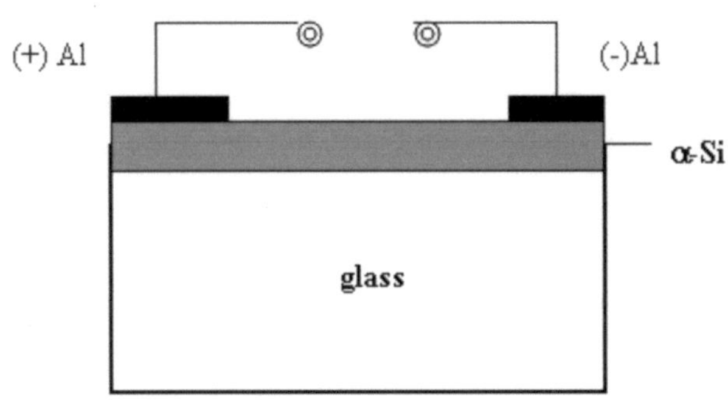

Fig. 1. Sketch of the experimental sample.

The experimental setup is plotted in Fig. 1. α-*Si* layers of 1 μm thickness were deposited on the glass substrate by magnetron sputtering under activation of SiH_4 (silane) plasma dissociation at alternate pulse bias with 55 Hz frequency. Pressure and temperature in the growth chamber were *P* = 70 Pa and 225^{0}C, respectively. Aluminum (*Al*) contacts doped with silicon (1% *Si*) were manufactured through the mask of 1 mm diameter. The investigated structure had been connected to the experimental measurement equipment. Current-voltage characteristics were measured at room temperature under illumination by UV-, near-IR and visual spectral ranges.
The experimental current-voltage characteristics (IVC) of the investigated samples are illustrated by Fig. 2. The experiment was carried out under various illuminations. The IVCs obtained under the background illumination (daylight, curve 1) and under irradiation by the light source with 100 W power (curve 2) are approximated by the following expression:

$$I_{exper} \sim (V_a)^m, \tag{1}$$

where $m < 1$.

IVCs of the structures obtained after irradiation by the light source with wavelengths in the near-IR spectral region (curve 3) and by the UV source (curve 4) can be expressed as

$$I_{exper} \sim I_s \exp(eV_a/k_B T)^m, \tag{2}$$

where I_s is a saturation current defined by the parameters of the film (charge carrier mobility and the dangling bonds density as well as by the tunneling transparency coefficient of the *Al* - α-*Si* barrier (Terukov, 2000&2001).

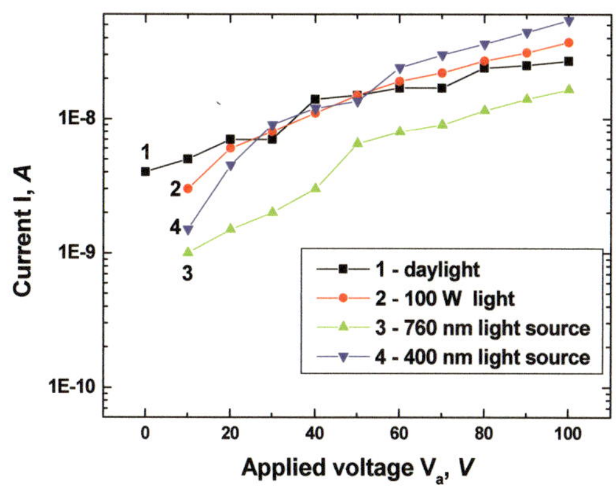

Fig. 2. Current-voltage characteristic of the investigated sample (T = 300 K) (Khlyap, 2003).

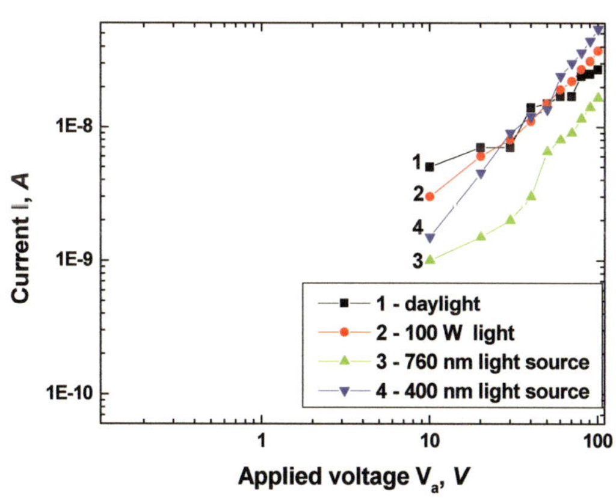

Fig. 3. Current-voltage characteristics of the investigated structure in double-log scale (Khlyap, 2003).

Re-building the experimental IVC in double-log scale (Fig. 3) allows obtaining more detail information about current mechanisms in the structures investigated.

It is obvious that all the experimental current-voltage dependencies are approximated by straight lines. According to the model (Terukov, 2001; Sze, 2007) one can suggest the following explanation: the investigated samples are high-resistive films with one group of the trap centers localized up the bottom of the conduction band (Fig. 4, Terukov, 2002). Appearance of these centers causes the space charge limited current (SCLC).

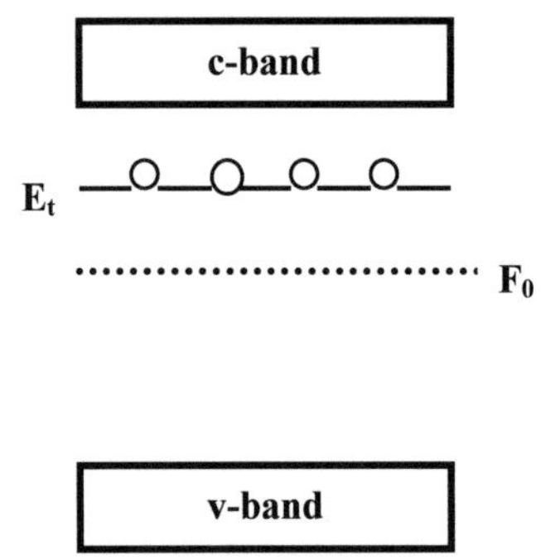

Fig. 4. Schematic drawing of the energy levels in the forbidden gap of amorphous silicon under thermodynamic equilibrium. E_t is the trap level, F_0 is the Fermi level position (Terukov, 2000&2001).

In absence of the external electric field the initial electron concentration in the investigated films is low and determined by the localization of the Fermi level of the material. In turn, the Fermi level localization depends on the concentration and the ionization energy of the trap centers E_t. Under small applied bias the electrons injected from the *Al* contacts are confined by the traps E_t. As the applied voltage increases, the centers E_t receive more and more electrons; at the same time, the concentration of the injected charge carriers is also increasing. This process is experimentally observed in the linear sections of the IVCs with different slopes *m*. UV-radiation accelerates the interaction between the injected charge carriers and the ones accumulated by the trap centers [Terukov, 2000; Khlyap, 2003).

The IR-photosensitivity of the films is of particular importance. The challenge is that the as-grown films are quite not photosensitive. One of the simplest ways to make the layers photosensitive is hydrogenation treatment of the films under certain temperatures. The as-grown layers were placed in the special chamber filled with the molecular hydrogen for 24 hours at 400^{0}C (the gas pressure in the chamber was 2500 Pa). Fig. 5 shows the experimental current-voltage dependencies.

The experiment showed a sufficient reduction of the films resistance compared with original values. The slope m has also been changed down to: m ~ 0.6 - 0.7. The photosensitivity in the near-IR spectral region (~1600 nm) is also sufficiently improved at the applied bias 0-50 V (Khlyap, 2003).

3. Charge carriers exclusion in electronic polycrystalline silicon

The simple and reliable technique of current-voltage characteristics measurements was applied for studying processes of carrier transport in the electronic polycrystalline silicon (Reich; Akopian; Khlyap, 2004). The best samples of polycrystalline *Si* grown by the

Czochralsky method were chosen for the investigations. Specimens of columnar and granular crystal structure with dimensions 8mm×2mm×2mm of *n*-type conductivity were polished in the solution $HNO_3:HF:CH_3COOH$ = 3:1:1 and rinsed in unionized water in order to maximally avoid the possible influence of surface effects on the results of electrical measurements. The studies were carried out at room temperature under applied electric fields 0 - 104 Vm^{-1}, corresponding to applied biases in the range of 0 - 190 V.

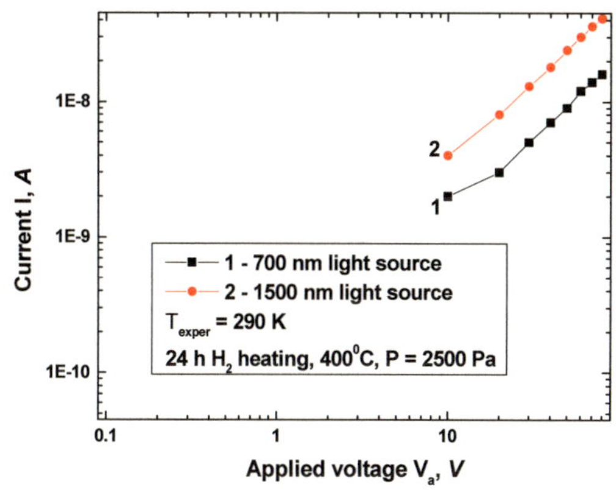

Fig. 5. Experimental current-voltage characteristics of the investigated samples after hydrogenation (Khlyap, 2003).

High-temperature (up to 1200^{0}C) heat treatment of the samples was performed under normal atmospheric conditions during 6 h in the furnace of the special construction providing a stationary temperature gradient along the sample. The measurements of current-voltage characteristics (IVC) were performed by means of the traditional bridge method (Sze). Indium contacts were thermally deposited on the lateral facets of the sample. The left and right contacts will be referred further as the first and the second ones, respectively. All experimental dependencies are represented in the coordinates of $ln\ j \sim (V_a)^{1/2}$, where j is the current density and V_a stands for the applied voltage. Fig. 6 shows the IVC of the sample of the columnar polycrystalline-like structure. As one can see, both curves ("forward" and "reverse") have no considerable difference, indicating a good quality of metallic contacts. This IVC demonstrates the domination of at least two-step tunneling with the threshold voltage $V_{TR} \sim 9$ V (Khlyap, 2004).

On the contrary, the IVC of the sample with the granular structure exhibited no asymmetry between the forward and reverse currents (Fig. 7) (Khlyap, 2004).

High-temperature treatment (1100^{0}C) of both samples does not change the IVCs qualitatively (Fig. 8). However, the resistance of the samples becomes lower and the threshold voltage of the sample with the columnar structure reduces down to 4 V. Increase of the treatment temperature up to 1100^{0}C does not lead to significant changes of the IVCs in neither sample.

As we have noted, the dominant process in carrier transport is the tunneling. Nevertheless, the attempts of numerical simulations of the experimental data according to the theoretical models developed specifically for tunneling currents (Sze) failed to describe the observed results, so that we have been forced to take into consideration the phenomena of carrier

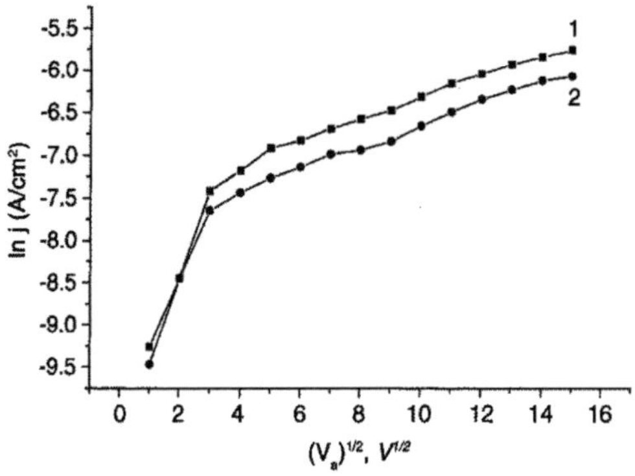

Fig. 6. Forward (curve 1) and reverse (curve 2) currents of the sample with the columnar structure before high temperature treatment (Khlyap, 2004).

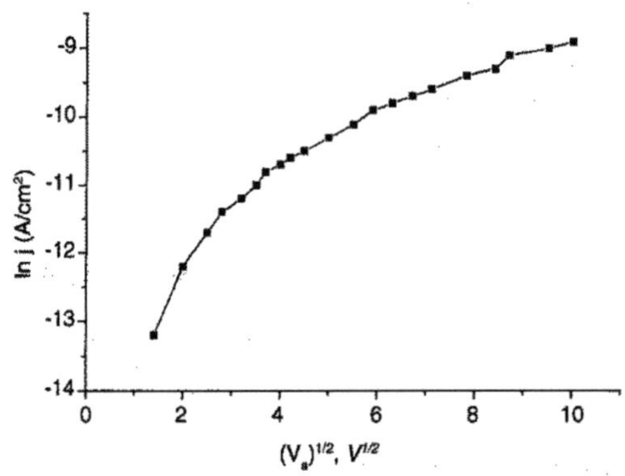

Fig. 7. Current-voltage characteristics of the sample with granular structure before high-temperature treatment (Khlyap, 2004).

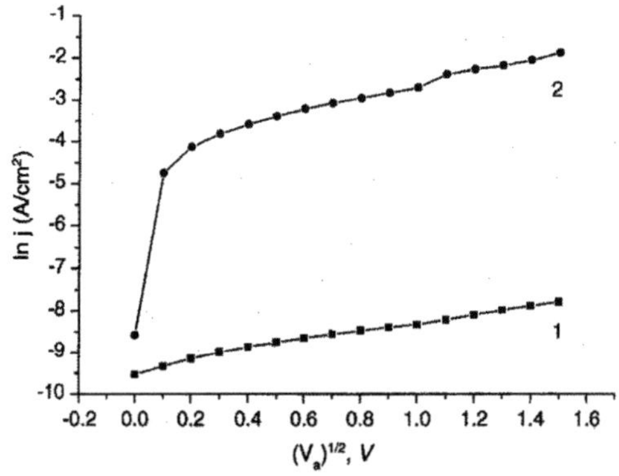

Fig. 8. Current-voltage characteristics of both samples (curve1 corresponds to the sample with granular structure and curve 2 corresponds to the sample with columnar structure) after high-temperature (900°C) treatment (Khlyap, 2004).

depletion and fluctuation of the carrier concentration on the inter-grain boundaries in the bulk of the sample (Reich, Akopian).

The depletion of charge carriers was observed experimentally and analyzed in the Ge-based monocrystalline diodes of finite length (Akopian). The effect strongly depends on the surface recombination velocity, sample length and temperature. The problem is somewhat more complicated for the structures based on polycrystalline materials, because it is necessary to take into account the processes of charge transfer along the sub-grain boundaries. The most important is to estimate the potential distribution in the bulk of the sample in order o determine the regions of carrier depletion. The potential distribution caused by the movement of carriers from the first contact toward the second one is described by the following expression (Akopian):

$$U = (k_BT/e)[lj(D_{ni})\text{-}1 - 6l^2(n_0)3/L^2n_i(2n_0 + n_i)], \quad (3)$$

where l is the length of the sample, j is the charge carriers flow, $L = [(2D_nD_p/(D_n + D_p)\tau]^{1/2}$, $D_{n,p}$ are the diffusion coefficients for electrons and holes and $\tau = 10^{-8}$ s is the lifetime of the carriers (this value is accepted to be the same for both electrons and holes), $n_0 = 10^{10}$ cm^{-3} stands for the intrinsic electron concentration, and $n_i = 10^{18}$ cm^{-3} takes care of the carrier concentration immediately involved in the charge transfer. Numerical estimations were carried out for both the samples. The depletion as an almost completely sweep out of the carriers was observed only for the sample of the columnar structure after the heat treatment at 900^{0}C. The results obtained for this sample are plotted in Fig. 9 for the range of applied biases 0.2-1.8 V, which seems to be of particular interest for device operation. The linear character of the calculated potential distribution shows (Khlyap, 2004) a considerable accumulation of carriers near he second contact region increasing with the increase of the applied bias.

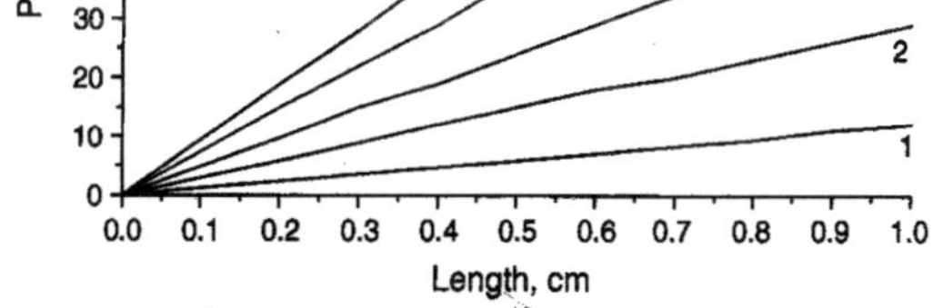

Fig. 9. Potential distribution for the sample with columnar structure after high-temperature (900^{0}C) treatment at the applied voltage V_a, V: (1) 0.2, (2) 0.6, (3) 1.0, (4) 1.4, (5) 1.8 (Khlyap, 2004).

According to the theory developed in (Reich et al.), the tunneling current j reads as follows:

$$\ln(j/j_0) = (-1/5)(2/\pi)^{1/2}(U_0/E_B)^{5/4}[n_i(a_B)^3]^{-1/2}, \quad (4)$$

where U_0 is the height of the barrier, $E_B = me^4/(h^2/2\pi^2)\varepsilon^2$ and $a_B = (h^2/4\pi^2)^2\varepsilon/me^2$ are the Bohr radius and energy for the electron, m_e is the effective mass, ε is the dielectric constant of *Si*, and j_0 stands for the saturation current. The calculation based on experimental data has demonstrated that the barrier height U_0 before the heat treatment of the samples is 0.48 eV and 0.36 eV for the granular and columnar samples, respectively. After heat treatment under 900^{0}C these values are 0.12 and 0.9 eV, respectively (Khlyap, 2004).

In summary, current-voltage characteristics and the effect of the high-temperature heat treatment (900 - 1100^{0}C) on carrier transfer in bulk polycrystalline *Si* of granular and columnar structures have been investigated. The temperature 900^{0}C has been shown to be optimal for i) reduction of the barrier height in samples of granular structure and ii) a considerable accumulation of carriers in the region of the second contact. The first experimental results reported in (Khlyap, 2004) demonstrated the possibility of additional accumulation of charge carriers in bulk polycrystalline *Si* of *n*-type conductivity after high temperature treatment without sufficient increase of the applied electric field (Khlyap, 2004).

4. Electric characteristics of the structure bulk silicon – native oxide

As we have mentioned above, the native oxide formed immediately after the sample preparation (a bulk specimen or a thin film) is an unavoidable factor of any technological process and the following design of the active element. We have investigated room-temperature electrical (current-voltage, IVC, and capacitance-voltage, CVC) characteristics of the structure bulk silicon-native oxide. The scheme of the contacts (idium pads) deposited on the bulk silicon sample is illustrated in Fig. 10.

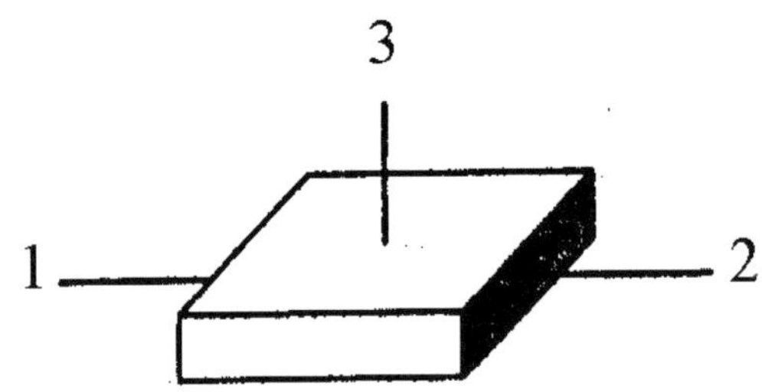

Fig. 10. Schematic image of In-contact pads deposited on the bulk crystalline silicon sample for the room-temperature electric investigations.

We have focused on examining the current-voltage functions registered under the application of external electric field in directions '1-2" and "2-1" as well as in directions "1-3, 2-3" and "3-1, 3-2". The sets of the device-quality crystalline silicon of n-type conductivity were chosen for this experiment. The samples were cut off from the as-grown ingots.

The experimental electric field-induced characteristics are plotted in Fig.11, a-c. Obvious that all the experimental current-voltage functions are described by the power law $I \sim (F_a)^m$, where F_a is an applied electric field, and m is an exponential factor determining the mode of the charge carriers transfer through the sample as a finite volume (directions 1-2 and 2-1) and through the sample volume - sample surface space (direction 1-3). The numerical analysis performed in the frame of the semiclassical model (Sze) demonstrated that the carriers flow through the volume of the sample according to the ballistic - diffusion mode (the forward current, Fig.11, a), and the dominant tunneling current is observed for the reverse current (Fig.11, b). The tunneling current is also observed for the direction 1-3 (Fig.11, c), but as the applied electric field increases, the tunneling process begins to be suppressed by the diffusion of the curries due to lowering the potential barrier formed by the native oxide.

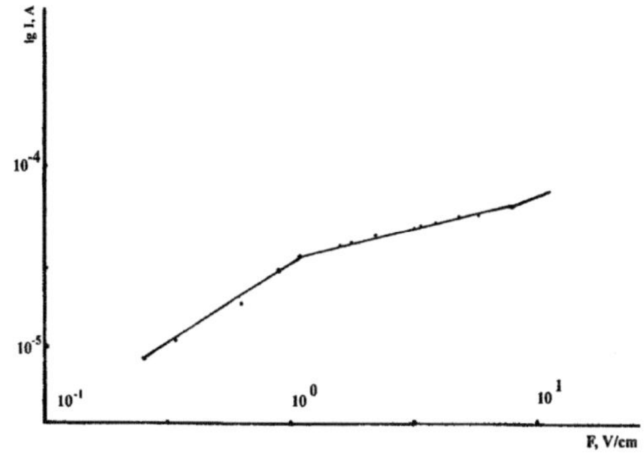

a)

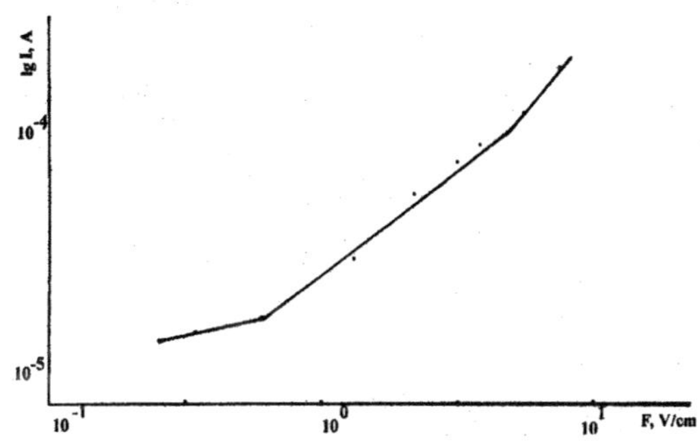

b)

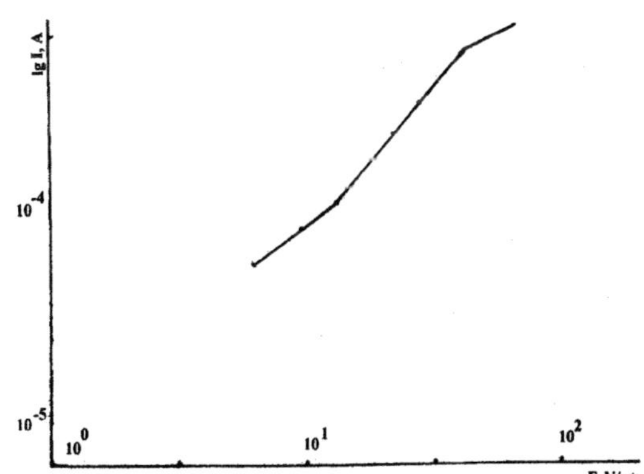

c)

Fig. 11. Room-temperature electric field-induced characteristics of the structure bulk crystalline silicon-native oxide; a) contacts 1-2, b) contacts 2-1, c) contacts 1-3 (see notes in Fig.10).

Thus, these experimental data are of the barrier type which is more typical for as-prepared metal-semiconductor structures. To confirm this conclusion we have made the capacitance-voltage measurements at T = 290 k and the test signal frequency f = 1 kHz. The experimental results are plotted in Fig.12, a-c.

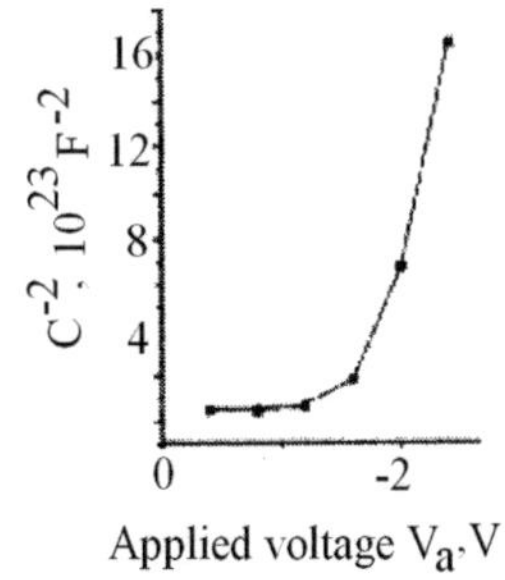

a)

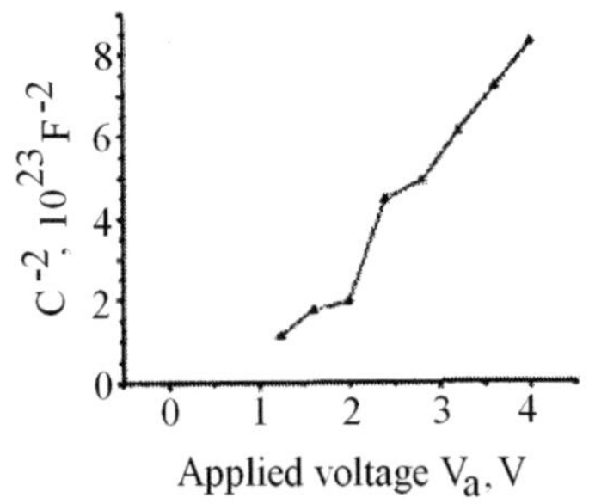

b)

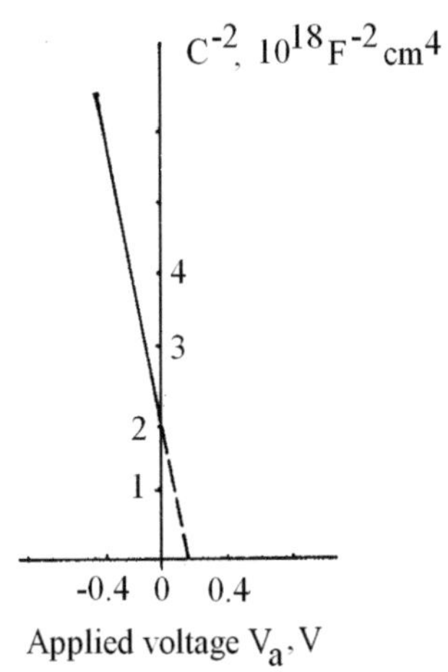

c)

Fig. 12. Room-temperature capacitance-voltage characteristics of the investigated structure under the test signal frequency f = 1 kHz: a) contacts 1-2; b) contacts 1-3 (see notes in Fig.10); c) a control metal-semiconductor structure (In-mono-n-Si).

The experimental data have allowed calculating some main parameters of the structure bulk crystalline silicon-native oxide according to the theory (Sze). The results are listed in Table 1.

Charge centers concentration	Contact	Space charge region width	Diffusion potential
N_{01}=1.26×10^{12} cm^{-2} N_{02}=7.05×10^{12} cm^{-2}	1-2	W_{01}=14.8 μm W_{02}=14.9 μm	V_{d1} = 0.53 V V_{d2} =1.40 V
N_{01}=1.62×10^{13} cm^{-2} N_{02}=1.94×10^{13} cm^{-2}	2-3	W_{01}=15.5 μm W_{02}=15.5 μm	V_{d1} = 3.50 V, V_{d2} = 4.20 V

Table 1. Electric parameters of the investigated structure.

5. Electric parameters of the structure recrystallized nanocrystalline silicon-Cu/Ag-nanocluster contacts

The unique room-temperature electrical characteristics of the porous metallic nanocluster-based structures deposited by the wet chemical technology on conventional silicon-based solar cells were described in (Laptev & Khlyap, 2008). We have analyzed the current-voltage characteristics of Cu-Ag-metallic nanocluster contact stripes and we have registered for the first time dark currents in metallic structures. Morphological investigations (Laptev & Khlyap, Kozar et al., 2010) demonstrated that copper particles are smaller than 0.1 μm and smaller than the pore diameter in silver. The contacts were deposited on nanocrystalline silicon structures obtained by the pulse laser recrystallization of the silicon thin films grown on insulator substrates. The experimental results are illustrated in Fig.13.
The numerical analysis showed the following results: the first section of forward current

$$I = T_{tun}A_{el}(4\varepsilon/9L^2)(2e/m^*)^{1/2}(V_a)^{3/2}$$

(ballistic mode) and the second one as

$$I = T_{tun}A_{el}(2\varepsilon v_s/L^2)V_a,$$

and the reverse current is

$$I = T_{tun}A_{el}(2\varepsilon v_s/L^2)V_a$$

(velocity saturation mode). Here T_{tun} is a tunneling transparency coefficient of the potential barrier formed by the ultrathin native oxide films, A_{el} and L are the electrical area and the length of the investigated structure, respectively, ε is the electrical permittivity of the structure, m* is the effective mass of the charge carriers in the metallic Cu-Ag-nanoclucter structure, and v_s is the carrier velocity (Kozar et al., 2010). These experimental data lead to the conclusion that the charge carriers can be ejected from the pores of the Cu-Ag-nanocluster wire in the potential barrier and drift under applied electric field (Sze & Ng, 2007; Peleshchak & Yatsyshyn, 1996; Datta, 2006; Ferry & Goodnick, 2005; Rhoderick, 1978).

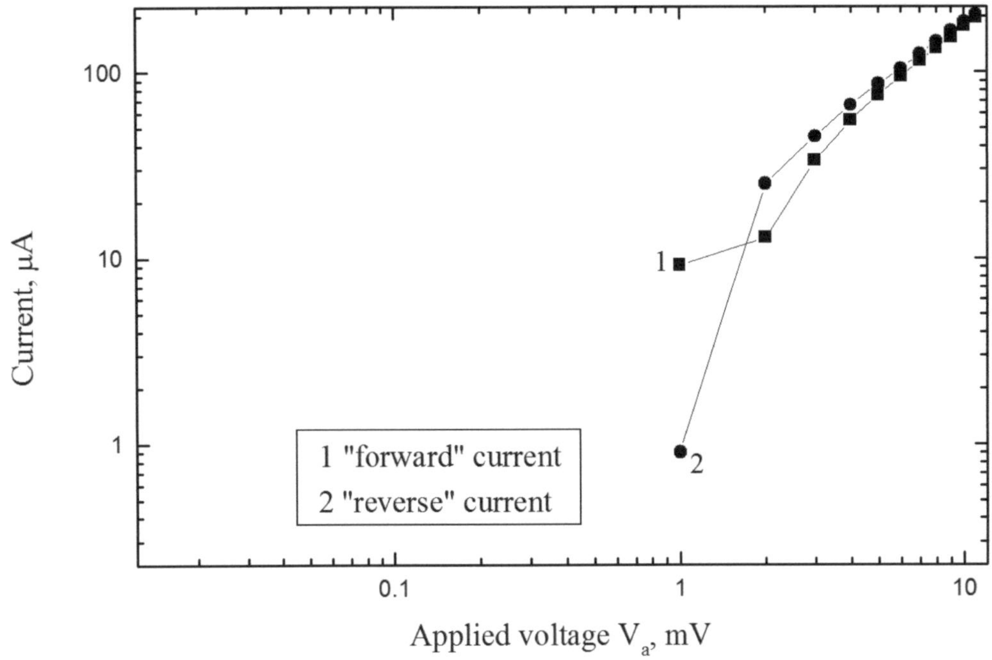

Fig. 13. Room-temperature current-voltage characteristics of the structure recrystallized nanocrystalline silicon-Ag/Cu-nanocluster contacts.

6. Conclusions

The chapter presented here reviews the principal experimental results obtained under simple and reliable room-temperature electric measurements of the structures based on various type of device-quality silicon with taking into account the effect of ultrathin native oxide films unavoidably formed after the preparation of the active element. Amorphous silicon thin films prepared by the magnetron sputtering technology have showed good photosensitive properties. Polycrystalline bulk silicon samples were investigated after the high-temperature treatment in order to clarify this effect on electrical performance of the active elements (in particular, future ICs). The necessity of different placement of contact pads on bulk crystalline silicon samples have forced us to concentrate on the examination of the IVC and CVc of the structures formed by the silicon samples and the native oxide. These investigations demonstrated that the influence of the native oxide ultrathin films forming additional (sharp or graded) potential barriers is to be taken into account under many technological processes using for the device design. And, the attempting to solve the problem of the performance of low-cost high-effective solar cells based on conventional silicon resulted in unique chemical wet deposition technology which has allowed producing Ag/Cu-nanoclustered structures on recrystallized nanocrystalline silicon. Their current-voltage characteristics are similar to those of semiclassical metal-semiconductor structures

7. References

Akopian A.A. et al. (1987), Charge Carrier Exclusion in *Ge*-Diodes, *Semiconductors* (Russia), Vol. 21, p. 1783.

Datta S. (2006). *Quantum transport: Atom to Transistor,* Cambridge Univ. Press, ISBN 0-521-63145-9, Cambridge, Great Britain.

Ferry D. & Goodnick S. (2005). *Transport in Nanostructures,* Cambridge Univ. Press, ISBN 0-521-66365-2, Cambridge, Great Britain

Khlyap H. et al. (2003), Photosensitive Amorphous Si Thin Films Prepared by Magnetron Technology, *Proceedings of the Materials Research Society, Fall 2002, Boston, USA,* Vol. 744, paper No. M5.20.1.

Khlyap H. et al. (2004), Depletion of charge carriers in electronic polycrystalline silicon, *Mater Science in Semicond Processing,* Vol. 7, p. 443-446.

Kozar T. V., Karapuzova N. A. & Laptev G. V., Laptev V. I., Khlyap G. M., Demicheva O. V., Tomishko A. G., Alekseev A. M. (2010). Silicon Solar Cells: Electrical Properties of Copper Nanoclusters Positioned in Micropores of Silver Stripe-Geometry Elements, *Nanotechnologies in Russia,* Vol. 5, № 7-8,
p.549-553, DOI: 10.1134/S1995078010070165, ISSN: 1995-0780 (print), ISSN: 1995-0799 (online).

Laptev V.I. & Khlyap H. (2008). High-Effective Solar Energy Conversion: Thermodynamics, Crystallography and Clusters, In: *Solar Cell Research Progress,* Carson J.A. (Ed.), pp. 181–204, Nova Sci. Publ., ISBN 978-1-60456-030-5, New York, USA.

Martin I. et al. (2004), Improvement of Crystalline Silicon Surface Pasivation by Hydrogen Plasma Treatment, *Appl. Phys. Lett.,* Vol. 85, p. 1474-1476.

Peleshchak R.M. & Yatsyshyn V.P. (1996). About effect of inhomogeneous deformation on electron work function of metals, *Physics of Metals and Metallography,* MAIK Nauka Publishers - Springer, vol. 82, No. 3, pp.18-26, ISSN Print: 0031-918X, ISSN Online: 1555-6190.

Reich M. et al. (1988), Effect of Barrier Localizd States on Fluctuations of tunneling Current Through Metal-semicondutor Contact, *Semiconductors* (Russia), Vol. 22, p. 1979.

Rhoderick E. H., (1978). *Metal-semiconductor contacts.* Clarendon Press, ISBN 0198593236, Oxford, Great Britain.

Ristova M. et al. (2003), Study of Hydrogeneted Amorphous Silicon Thin Films As a Potential Sensor for He-Ne Laser Light Detection, *Appl.. Surf Sci.,* Vol. 218, p. 44.

Sze S.M. & Ng, K.K. (2007). *Physics of semiconductor devices,* J. Wiley & Sons, Inc., ISBN 0-471-14323-5, Hoboken, New Jersey, USA.

Terukov E.I. et al. (2000), Current-Voltage Characteristics of Electroluminescent *Me/a-Si:H<Er>/c-Si* Structures Prepared by Magnetron Sputtering, *Semiconductors* (Russia), Vol. 34, p. 617-620.

Terukov E.I. et al. (2001), Interrelationship between the Recombination on Interfaces and Ubnormally Weak Dependence of photoconductivity on Illumination Intensity, *Semiconductors* (Russia), Vol.35, No.6, p.643-647.

Vasilev I. (2003), Optical Excitations in Small Hydrogenated Silicon Clusters: Comparison of Theory and Experiment, *Phys Stat Sol B* , Vol.239, p.19.

4

Structure and Properties of Dislocations in Silicon

Manfred Reiche[1] and Martin Kittler[2]
[1]*Max Planck Institute of Microstructure Physics, Halle*
[2]*IHP microelectronics, Frankfurt (Oder)*
Germany

1. Introduction

Defects in crystalline materials modify locally the periodic order in a crystal structure. They characterize the real structure and modify numerous physical and mechanical properties of a crystal. Crystal defects are generally divided by their dimension: point defects are also known as zero-dimensional (0-D) defects, while dislocations are 1-D, twins and grain boundaries are 2-D, and precipitates are denoted as 3-D defects. Dislocations were implemented for the first time in the early 1900th to explain the elastic behavior of homogeneous, isotropic media. Based on Volterra´s "distorsioni" (Volterra, 1907), Love has introduced the term "dislocation" to describe a discontinuity of displacement in an elastic body (Love, 1927). The application of this term to denote a particular elementary type of deviation from the ideal crystal lattice structure was due to Orowan (1934), Polanyi (1934), and Taylor (1934a, 1934b).

A dislocation is characterized by a vector parallel to the dislocation line and a displacement or Burgers vector which is a certain finite increment $\mathfrak{b}$ induced by the elastic displacement vector $\mathfrak{u}$. The Burgers vector is equal to one of the lattice vectors in magnitude and direction and may be written as (Hirth & Lothe, 1982)

$$\oint du_i = \int \frac{\partial u_i}{\partial s} ds = -b_i \, . \tag{1}$$

The direction along the contour s is that of a right-hand screw relative to the chosen direction along the dislocation line ℓ, that is, relative to the unit vector $\mathbf{t}$ tangent to the dislocation line (Frank, 1951). The edge dislocation, introduced by Orowan (1934), Polanyi (1934), and Taylor (1934a, 1934b), is represented by the line ℓ along which the vectors $\mathbf{t}$ and $\mathfrak{b}$ are perpendicular. If the vectors $\mathbf{t}$ and $\mathfrak{b}$ are parallel, then the corresponding dislocation is called a screw dislocation (Burgers, 1939, 1940). In many materials, dislocations are found where the line direction and $\mathfrak{b}$ are neither perpendicular nor parallel and these dislocations are called mixed dislocations, consisting of both edge and screw character.

In the elastic theory of isotropic media a dislocation is a line representing the boundary of the slipped region. Its strength is characterized by the displacement. The strain field around the dislocation is depicted as a cylinder. Among other things, the model explains the strain distribution around the dislocation, but cannot describe the strain in the center, i.e. in the core of the dislocation. Furthermore, the model does also not regard the influence of the lattice periodicity of real crystals. Burgers & Burgers (1935) as well as Taylor (1934a, b), Polanyi (1934), and Kochendörfer (1938) already pointed out that a dislocation moves by skipping individual atoms via potential walls. A first phenomenological model considering a potential energy of displacement that reflects the lattice periodicity was proposed by Frenkel and Kontorova (see Dehlinger and Kochendörfer, 1940). The model was modified by Peierls (1940) and extended by Nabarro (1947). Here, the displacement of the crystal lattice and the associated stress are considered to be caused by a number of infinitesimal dislocations originally suggested by Eshelby (1949). For edge dislocations, the width, or core region, in the Peierls-Nabarro model is given by

$$2\xi = d/(1-\nu) \qquad (2)$$

where d is the lattice plane distance and ν is the Poisson ratio. The introduction of the parameter ξ has the effect of removing the singularity at the origin of the dislocation that is present in model of Volterra (Hirth & Lothe, 1982). For screw dislocations the Peierls-Nabarro model assumes a stress component near the core which spreads out of the plane. This phenomenon anticipates the dissociation of a dislocation. The model also explains the motion of dislocations and results in the introduction of the Peierls energy, which represents the periodic displacement potential energy, as well as the Peierls stress required to overcome this potential barrier. The concept of kinks and jogs in dislocation lines is also a consequence of the model (Friedel, 1979). The Peierls-Nabarro model has been influential in the development of dislocation theory of more than 60 years. It was, for instance, modified to explain the dislocation motion (Hirth & Lothe, 1982), or to understand the structure of the dislocation core (Duesbery & Richardson, 1991; Bulatov & Cai, 2006).

Early investigations on semiconductor materials indicated the presence of electrically charged dislocations. It was already proved by Gallagher (1952) that plastic deformation of silicon and germanium increases their resistivity. Hall effect measurements suggested the introduction of acceptor-type levels in n-type Ge by deformation which was explained by negatively charged dislocation lines screened by a positive space charge region (Pearson et al., 1954). Based on these results and a remark of Shockley that dangling bonds in the core of an edge dislocation exist, Read (1954a,b) formulated a phenomenological theory of charged dislocations. He introduced the concept of dislocation electron levels, the occupation ratio of dislocation levels, and the radius of a Read cylinder surrounding each charged dislocation and screening the linear charge localized on it. Read (1954a,b) assumed that the dislocation states are represented by a single level or a one-dimensional band which is empty when the dislocation is in the neutral state. This assumption is applicable only at low temperatures (Labusch & Schröter, 1980). On the other hand, Schröter and Labusch (1969) argue that even at higher temperatures the dislocation band is half filled in the neutral state. Furthermore, dangling bonds does not exist in real dislocations. Numerous theoretical and experimental investigation particularly on dislocations in silicon refer to reconstructed dislocation cores. Therefore the electrical activity is related to defects on the dislocation core, such as kinks, jogs, and also by point defects bound to the core or in the elastic or electric field of the dislocation (Schröter & Cerva, 2002). While different types of dislocations are distinguished by different core defects their electrical activity is different (Alexander & Teichler, 1991). In addition, the concentration of point defects interacting with dislocations is doubtful even in the case of elemental semiconductors.

The present chapter reviews the current understanding about the structure and properties of dislocations in silicon and is based on earlier reviews given for instance by Bulatov et al. (1995), Alexander & Teichler (2000), Schröter & Cerva (2002), Spence (2007), and Kveder & Kittler (2008). All these papers demonstrate a substantial progress over the years but show also that a number of problems such as dislocation mobility, structure of the dislocation core, or electronic properties are not completely solved (George & Yip, 2001; Spence 2007). For instance, *ab-initio* computer simulations using different approaches result in a large number of models of the core structure of different dislocations which are not verified experimentally. The experimental data of dislocation motion can only be partially simulated by limiting the number of atoms, etc. (Bulatov & Cai, 2006). Another problem is the fundamental difference between theoretical calculations and experiment. While only individual dislocations are regarded in most of the calculations, a large number of dislocations is involved in experimental measurements such as plastic deformation. These measurements integrate not only over a number of dislocations but may also include data of different dislocation types and the interaction with a more or less unknown concentration of point defects. A further paragraph of this chapter is therefore dedicated to the preparation and characterization of only a small number of defined dislocations.

2. Structure of dislocations in silicon

Silicon crystallizes in the cubic diamond structure (space group Fd3m). The lattice constant is a = 0.543 nm. The glide plane is {111} and perfect dislocations have Burgers vectors of the type $\mathfrak{b} = a/2\langle 110\rangle$. Hornstra (1958) has introduced two types of perfect dislocations in the diamond lattice: a pure screw dislocation and the so-called 60° dislocation, where the Burgers vector is inclined at an angle of 60° to the dislocation line. The diamond structure corresponds to two face-centered cubic (fcc) lattices displaced by $(1/4, 1/4, 1/4)$. Hence, atoms in both lattices do not have identical surroundings. Due to this fact, there are two distinct sets of {111} lattice planes; the closely spaced glide subset and the widely spaced shuffle subset (Hirth & Lothe, 1982). There is a long controversial discussion about the dominant dislocation type in the diamond structure. Early publications suggest the presence of dislocations in the shuffle set because movement through one repeat distance on a shuffle plane breaks one covalent bond per atomic length of dislocation (e.g. Seitz, 1952). The equivalent step on a glide plane involves the breaking of three bonds (Amelinckx, 1982). The idea of splitting or dissociation of perfect dislocations in the diamond structure has been commented for the first time by Shockley (1953) and was experimentally proved later on by electron microscopy. The introduction of the weak-beam method by Cockayne et al. (1969) has particularly shown that dislocations in silicon are in general dissociated and glide in this extended configuration. Both the screw and 60° dislocation belonging to the glide set can dissociate into pairs of partial dislocations bounding an intrinsic stacking fault ribbon (Ray & Cockayne, 1971; Gomez et al., 1975; Gomez & Hirsch, 1977). On the other hand, screw and 60° dislocations of the shuffle set can only dissociate into partials bounding an intrinsic stacking fault if there is a row of either vacancies or interstitials associated with one of the partials (Amelinckx, 1982). Most of the evidence indicates that the dislocations found in plastically deformed silicon belong to the glide set (Hirsch, 1985; Alexander, 1986; Duesbery & Joós, 1996).

For the 60° dislocation a 30° partial and a 90° partial dislocation are formed through dissociation, while the screw dislocation dissociates into two 30° partials (Gomez et al., 1974; Heggie and Jones, 1982). These is described by the dissociation reaction (Marklund, 1979)

$$\mathfrak{b} \rightarrow \mathfrak{b}_1 + \mathfrak{b}_2 \, , \tag{3}$$

where in the case of a 60° dislocation

$$\mathfrak{b} = \frac{a}{2}[011] \qquad \mathfrak{b}_1 = \frac{a}{6}[121] \qquad \mathfrak{b}_2 = \frac{a}{6}[\bar{1}11] \tag{4a}$$

and for a screw dislocation

$$\mathfrak{b} = \frac{a}{2}[011] \qquad \mathfrak{b}_1 = \frac{a}{6}[121] \qquad \mathfrak{b}_2 = \frac{a}{6}[21\bar{1}] \tag{4b}$$

holds. The 30° as well as the 90° dislocations are of the Shockley type. The dissociation result as well in the formation of a stacking fault between both partial dislocations. The size of the stacking fault, i.e. the width of the splitting of the perfect dislocations d_0, depends in a stress free crystal on the stacking fault energy γ_{SF} and the repulsion force F of the partial dislocations

$$d_0 = \frac{F}{\gamma_{SF}} \tag{5}$$

The repulsion force is calculated using elastic constants given by the linear theory of elasticity resulting in (Amelinckx, 1982)

$$d_0 = \frac{Gb^2}{8\pi\gamma_{SF}} \cdot \frac{2-\nu}{1-\nu} \cdot \left(1 - \frac{2\nu}{2-\nu}\right) \cdot cos2\Theta \tag{6}$$

where G is the shear modulus, and Θ the angle between the dislocation line and the Burgers vector of the perfect dislocation. In a stressed crystal the two partials are exposed to additional forces which are in general different. Depending on the crystallographic orientation the external stress causes an increase or a decrease of d_0. Therefore the width of the splitting of a dislocation d_D by applying a resolved shear stress τ_s is given by (Wessel & Alexander, 1977)

$$d_D = \frac{d_0}{1 + \left[\mathfrak{f} - \frac{1-\alpha}{1+\alpha}\right] \cdot \frac{b\tau_s}{2\gamma_{SF}}} \tag{7}$$

with $\mathfrak{f}$ being a geometric factor and $\alpha = \mu_1/\mu_2$ as the ratio of mobilities μ_j of both partial dislocations.

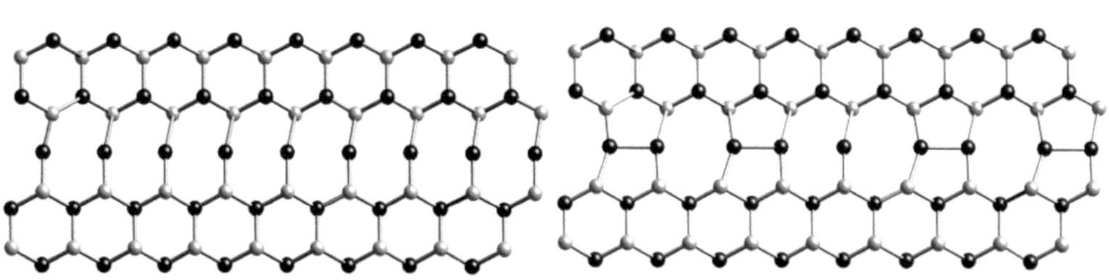

(a) (b)

Fig. 1. Models of the core structure of an unreconstructed (a) and a reconstructed 30° partial dislocation (b) according to Northrup et al. (1981) and Marklund (1983).

The initial models of perfect dislocations assumed dangling bonds in their core (Shockley, 1953; Hornstra, 1958). Experimental data, however, obtained mainly by electron paramagnetic resonance (EPR) spectroscopy refer to a low density of such dangling bonds (Alexander & Teichler, 2000). Dislocations of the glide set reconstruct by dissociation (Heggie & Jones, 1983; Marklund, 1983; Alexander, 1991), while dislocations of the shuffle set, which may exist at high applied shear stress, can stabilized by interaction with vacancies (Li et al. 2008). Different models of the core structure of partial dislocations have been suggested (figure 1, 2). The model of an unreconstructed 30° partial dislocation was presented and verified experimentally by high-resolution transmission electron microscopy (HRTEM) by Northrup et al. (1981). Models of the reconstructed 30° partial dislocation were proposed by Marklund (1983), Chelikowsky (1982), and Csányi et al. (2000). In this configuration, the dangling bonds are saturated after the pairs of neighbouring core atoms move closer together to form bonded dimers. The reconstruction breaks the translation symmetry and doubles the period along the dislocation line from b to 2b, where b is the magnitude of the Burgers vector. A defect appears at the boundary between two segments reconstructed in the opposite sense (so-called antiphase defect (Hirsch, 1979) or soliton (Heggie & Jones, 1983)).
The core reconstruction of the 90° partial dislocation was studied for more than 30 years. The driving force for core reconstruction is the same as for the 30° partial, that is the high energy of the unsaturated dangling bonds. A first model was proposed by Hirsch (1979). In this reconstruction there is a displacement that breaks the mirror symmetry normal to the dislocation line, enabling threefold coordinated atoms in the unreconstructed core to come together and bond. In this way two degenerate reconstructions exist. This core reconstruction is shown in figure 2a. The symmetry breaking displacement does not alter the translational symmetry along the dislocation line, which retains the same periodicity as the crystal (Bulatov et al., 2001). Another core reconstruction was proposed by Duesbery et al. (1991). In this structure the mirror symmetry is not broken and the atoms on either side of the dislocation line move towards each other so that each core atom has three nearest neighbours plus two more neighbours at a somewhat greater distance. This reconstruction is known as the quasi-fivefold reconstruction. Simulations, however, indicate that the quasi-fivefold configuration was higher in energy (Bigger et al., 1992). Benetto et al. (1997) proposed a new core reconstruction for the 90° partial dislocation with double the periodicity along the dislocation line (figure 2b). They found also that this reconstruction has a lower potential energy than the single period reconstruction. Further simulations,

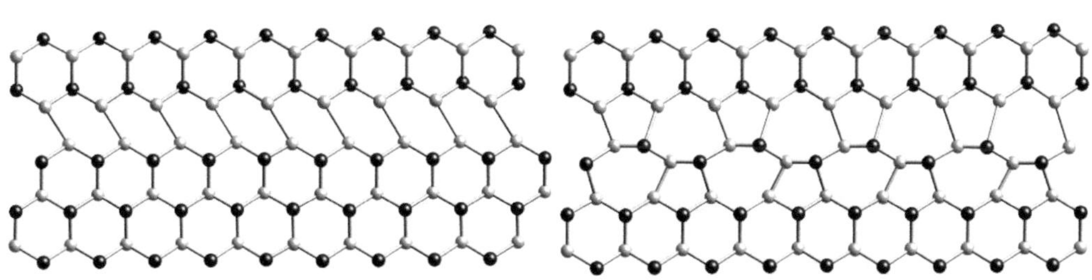

(a) (b)

Fig. 2. Models of the core structure of a single period (a) and a double period reconstruction of a 90° partial dislocation (b) according to Bulatov et al. (2001).

however, have shown that the energy differences between the single and double period structures are very close (Lehto & Öberg, 1998).

3. Electronic properties of dislocations in silicon

Dislocations interfere the translational symmetry of the crystal. As a consequence energy levels in the band gap result. First analyses were done by Read (1954a, b) who concerned with long-range screening and occupation statistics in the presence of the macroscopic band bending due to the dislocation. Based on early experiments of the plastic deformation of heavily doped p-type Ge single crystals (Gallagher, 1952; Pearson et al., 1954) Read concluded that only an acceptor level is introduced by edge dislocation. According to this model the dislocation is negatively charged. The line charge of the dislocation is screened by ionized donor atoms in a cylinder. Free electrons cannot penetrate this space charge cylinder and are scattered by specular reflection at its surface. For the position of the energy level of the neutral dislocation Read (1954a) obtained a value of 0.2 eV below the conduction band. The acceptor model of the dislocation states was not confirmed by measurements on p-type Ge and Si with lower doping levels (Schröter, 1969; Weber et al., 1968). It was concluded that dislocations can act as acceptors and as donors and consequently a partially filled band was attributed to the dislocation, in agreement with theoretical predictions (Schröter & Labusch, 1969). Veth and Lannoo (1984) combined the models of Read and Schröter & Labusch. They carried out a self-consistent calculation of the potential in the vicinity of the dislocation, point to an intraatomic Coulomb term *J*, and treat screeing in the dislocation core as dielectric perturbation. Outside the core, classical screeing by ionized dopant atoms or free carriers take place. The transition between the two screening mechanisms was analyzed. From this analysis follows a parameter-free formula for the total shift of the dislocation level E_e with respect to the edge of the undisturbed valence band

$$E_e = \frac{Jp}{\epsilon} + \frac{2e^2p}{\epsilon a}\left[ln\left(\frac{R}{a}\right) - 0.616\right] \tag{8}$$

with p as the number of excess electrons per atom, ϵ the dielectric constant, *a* the distance between two core atoms, and R as the radius of the screening space-charge cylinder (Read cylinder), given by

$$R = (a \cdot \pi \cdot |N_D - N_A|)^{-1/2} \tag{9}$$

In Eq. (8) N_D and N_A are the concentrations of donors and acceptors, respectively. Veth and Lannoo (1984) pointed out that Eq. (8) is linear with p, which fits the experimental data with Read´s model and corresponds to the line charge model by Labusch & Schröter (1980). There are several problems that have to be solved by any model of the charged dislocation core. One is the electrostatic potential around a charged dislocation. Another is the mobility of the charges on the dislocation line.

Computer simulations result in a number of deep levels related to defects on the dislocation core (Alexander & Teichler, 1991). The energy levels depend strongly on the geometry of the defects. For instance, the structure and resulting energy levels of 30° partial dislocations were studied by Marklund (1979), Northrup et al. (1981), Chelikowsky (1982), and Csányi et al. (2000). Deep levels related to 60° or 90° partial dislocations were summarized by Alexander & Teichler, 1991). All the computer simulations clearly demonstrate that deep levels are caused by core bond reconstruction and reconstruction defects. Most of the models, however, prefer reconstruction defects (antiphase defects, solitons) to explain the electronic properties of dislocations (Marklund, 1979; Heggie & Jones, 1983; Justo & Assali, 2001). Furthermore, electronic band gap calculations in combination with electron energy loss spectroscopy of dislocations in GaN revealed that impurities bonded to the dislocation core may induce electronic levels in the band gap (Bangert et al., 2004). On the other hand, the interaction of core defects with otherwise electronically active centers can also result in inactive complexes (Heggie et al., 1993; Jones et al., 1993).

Besides deep levels related to the dislocation core, shallow levels exist corresponding to more extended states, either states associated with stacking faults between two partial dislocations or states of electrons and holes trapped in the elastic deformation field of the dislocation. Calculations of energy levels related to stacking faults refer to the existence of shallow levels up to 0.1 eV above the valence band edge (Marklund, 1981; Mattheiss & Patel, 1981; Lodge et al., 1989). The shift of point defect levels relative to the silicon band edge may also be caused by the dislocation strain field. The response of the Si band structure to homogeneous elastic stresses has been investigated and is described by the deformation potential Ξ_{ij} (Bardeen & Shockley, 1950), which is written for the conduction band edge as (Keyes, 1960)

$$\Delta E_c = \sum_{i,j} \Xi_{ij} \cdot \varepsilon_{ij} \tag{10}$$

where ε_{ij} denotes the components of the strain tensor. Considering one minimum in the centre of the Brillouin zone and assuming an elastically isotropic material as an approximation, the shift of the conduction band minimum is given by the trace of the strain tensor and one component of the deformation potential tensor, Ξ_d (Schröter & Cerva, 2002):

$$\Delta E_c = \frac{b_e \cdot \Xi_d (1-2\nu)}{2\pi(1-\nu)} \cdot \frac{\sin\Theta}{r} \tag{11}$$

In polar coordinate system Θ means the angle between **r** and b_e, the edge component of the Burgers vector. If Ξ_d is positive, the conduction band edge is lowered in the compressed region of an edge dislocation and increases in its tensile region. The behavior is reverse for negative values of Ξ_d. In addition, the strain field results also in an effective shift of the point defect level. If the strain-induced shift of the point defect level is described by the deformation potential Ξ_{pd}, one obtaines a position-depending shift $\Delta E_{c,pd}$ of the conduction band edge and point defect level (Schröter & Cerva, 2002)

$$\Delta E_{c,pd} = -\frac{b(\Xi_d - \Xi_{pd})(1-2\nu)}{2\pi(1-\nu)} \cdot \frac{\sin\Theta}{r} \tag{12}$$

Analogous models for 60° and screw dislocations in p- and n-type elemental semiconductors (Si, Ge) have been proposed by Shikin & Shikina (1995).

The electrical activity of dislocations in silicon and germanium was studied by numerous methods where mostly plastic deformation was applied to produce defined dislocation arrangements (for instance, Schröter & Cerva, 2002; Alexander & Teichler, 2000). Hall effect measurement was primarily applied to verify the electrical activity of dislocations and to propose first models (Gallagher, 1952; Read, 1954a, b; Schröter & Labusch, 1969). Electron paramagnetic resonance (EPR) spectroscopy provides substantial information about the structure and, in combination with other techniques, electronic core defects (Kisielowski-

Kemmerich, 1990; Alexander, 1991; Alexander & Teichler, 1991, 2000). Plastic deformation introduces a variety of EPR-active defects in Si. Some of them denoted as Si-K1, Si-K2, Si-Y, and Si-R have been identified to be associated with the dislocation core, others, namely Si-K3, Si-K4, and Si-K5 with deformation-induced point defect clusters (Alexander & Teichler, 1991). Si-K6 and Si-K7 are ascribed to impurity atoms in the dislocation core. It was concluded that all EPR active centers attributed to dislocations belong to vacancies introduced into the core of the 30° partials forming screw dislocations. There is hitherto no satisfying explanation, why paramagnetic centers are not observable for 60° dislocations (and therefore 90° partials). EPR requires defined charge states of defects which can be different for 60° and screw dislocations.

Properties of deep levels generated by lattice defects are also investigated by deep level transient spectroscopy (DLTS) introduced by Lang (1974). The method probes changes of the capacity of the space charge region of a diode caused by reloading of deep levels. For point defects, emission and capture rate are linearly dependent on the occupation ratio of the defect level so that capacitance transients are exponentially dependent on time during capture and emission. The analysis of the DLTS-line variations with correlation frequency and filling pulse duration is then straightforward and yields the level of the defect (ionization enthalpy and entropy, its electron or hole capture cross section, and its concentration (Schröder & Cerva, 2002)). For dislocations, line charge fluctuations modify the electron emission resulting in a non-exponential transient and gives rise to a broadening of the corresponding DLTS line (Figielski, 1990). Some important features such as the C-line in n-type silicon, the F-line in p-type Si as well as B- and D-line in plastically deformed Si were analyzed in detail (for a review see e.g. Schröter & Cerva, 2002). The interaction of dislocations especially with metal impurities was also intensively studied with DLTS (Seibt et al., 2009a).

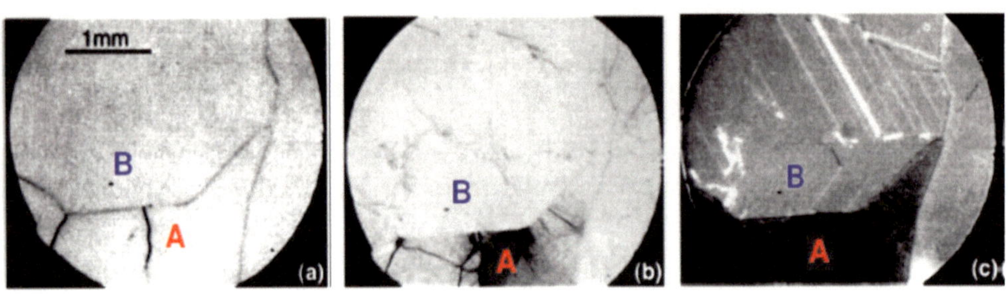

Fig. 3. Temperature dependence of the EBIC contrast of defects in multi-crystalline silicon. Measurements at 300K (a), 80K (b), and 30K (c).

Several techniques have been applied to analyze with spatial resolution the recombination activity of dislocations such as scanning deep level transient spectroscopy (SDLTS, Breitenstein & Wosinski, 1983), photoluminescence, light beam induced current (LBIC), and electron beam induced current (EBIC). EBIC and LBIC are unique among the electrical characterization methods with respect to a spatial resolution, sufficient to measure individual dislocations. In EBIC, for instance, the variation of the current at a Schottky contact resulting from excess electrons and holes generated locally by the electron beam is measured, when the specimen area of interest is scanned. The values of the current at the dislocation I_{dis} and away from it, I_0, are used to define the contrast $C_{dis} = (I_0 - I_{dis})/I_0$ of single dislocations. The measurement involves the dependence of C_{dis} on the temperature and the beam current of the electron probe. The temperature dependence of the defect contrast, $C_{dis}(T)$, is illustrated in figure 3 for different defects (intra-grain dislocations, grain boundaries, etc.) in multi-crystalline silicon. Furthermore, C_{dis} is proportional to the recombination rate of minority carriers at a dislocation. A theoretical description was derived by Donolato (1979, 1983) and Pasemann (1981). Numerous experimental investigations showed that dislocations in different Si materials often exhibit very different EBIC contrast behavior $C_{dis}(T)$ which is caused by different concentrations of deep intrinsic core defects and impurities. Different models were presented to explain the contrast behavior (Schröter and Cerva, 2002). A quantitative explanation of the experimental results was proposed by Kveder et al. (2001) which differs from earlier model (Wilshaw & Booker, 1985) by including electronic transitions between one-dimensional bands and deep localized states due to overlapping of their wave functions. Taking these transitions into account the dislocation recombination activity is properly described.

In 1976, Drozdov et al. (1976) proved lines in the photoluminescence spectra of deformed n- and p-type Si associated with dislocations. The lines are denoted as D1 - D4 (figure 4). The maximum position of the lines were measured at T = 4.2K as D1 = 0.812 eV, D2 = 0.875 eV, D3 = 0.934 eV, and D4 = 1.000 eV. The relative intensity of D1 to D4 depend on the dislocation density and distribution and can vary in different samples. The polarization of the D-lines emission and their response to uniaxial stress has been utilized to establish their relations to dislocations. Lines D1 and D2, on the one hand, and lines D3 and D4, on the other, show similar shifts by applying uniaxial stress and therefore have been grouped as pairs (Drozdov et al. 1977; Sauer et al., 1985). Polarization measurements were carried out to determine the electric field vector $\mathfrak{E}$ of the luminescent light. Using three different registration directions ($[21\bar{1}], [1\bar{1}1], [011]$), the $\mathfrak{E}$ vector within the primary slip plane, along $[1\bar{1}1]$, with a polarization of about 30% was found for D1/D2 (Weber, 1994). D3 and D4 exhibit an $\mathfrak{E}$ vector within the primary glide plane roughly along [011], i.e. along to the main Burgers vector, and with a polarization of about 20%. These findings strongly point to the dislocations as radiative centres for D3/D4. For D1/D2 the situation is more complex. The energy positions of D3 and D4 depend on the distance between partial dislocations suggesting that both originate from recombination processes at straight segments of 60° dislocations (Sauer et al., 1986, 1994). In addition, photoluminescence measurements on

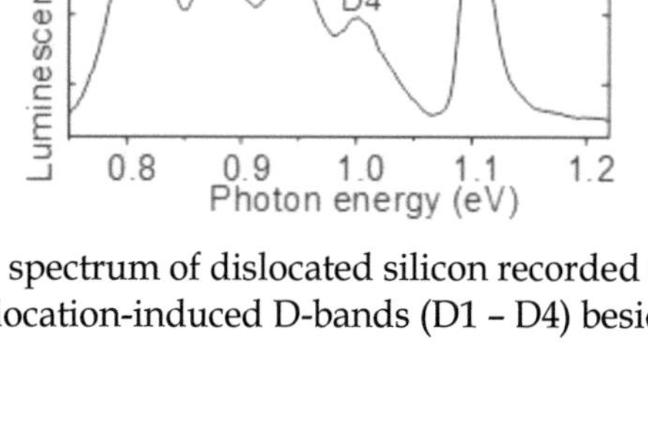

Fig. 4. Photoluminescence spectrum of dislocated silicon recorded at 80K. The spectrum shows the presence of dislocation-induced D-bands (D1 - D4) besides the band-band luminescence (BB).

dislocations in epitaxially grown SiGe layers refer to D3 as a phonon assisted replica of D4 (Weber & Alonso, 1990).
The origin of the D1 and D2 lines is still not understood. There are investigations referring that both lines are related to impurity atoms in the dislocation core (Higgs et al., 1993), dislocation jogs (Watson et al., 1998), or segments of dislocations (Lomer dislocations) appearing due to dislocation reactions, multi-vacancy and/or self-interstitial clusters trapped in the core (Jones et al., 2000).

4. Grain boundaries

Crystallization and recrystallization are typical processes to produce multi-crystalline silicon as the mostly applied material in solar cell manufacture. Multi-crystalline silicon, or in general polycrystalline materials, consists of numerous (single crystalline) grains with different crystallographic orientations separated by grain boundaries. The geometry of a grain boundary is macroscopically characterized by five degrees of freedom: three angles define the crystallographic orientation of both crystals with respect to each other, while two parameters describe the inclination of the grain boundary plane. To fully characterize the boundary geometry on a microscopic level, three additional parameters are required to define the atomic-scale relative translation of the two grains. Depending on the misorientation, grain boundaries are of the tilt type, when the rotation axis lies in the boundary plane, or of the twist type, when the rotation axis is normal to the boundary plane. A general grain boundary may have tilt and twist components.
Based on previous consideration of Burgers, Bragg, and Frank (see Amelinckx, 1982), first models of grain boundaries have been proposed by Shockley & Read (1949), Read & Shockley (1950), van der Merwe (1949), and Cottrell (1953). Besides a classification into tilt and twist boundaries, grain boundaries may be divided by their angle of misorientation Θ_{GB} into low-angle ($\Theta_{GB} < 5°$) and large-angle grain boundaries. More comprehensive definitions distinguish between singular, vicinal or general interfaces (Baluffi & Sutton, 1996), or between general (or random) and special grain boundaries (Chadwick & Smith, 1976). Special grain boundaries exhibit a periodic structure, while general grain boundaries show no appearent periodicity.
Numerous investigations have been carried out about the structure of grain boundaries in silicon. From these investigations it is concluded that (Seager, 1985):

1. Silicon grain boundaries are primarily composed of regular defects: perfect dislocations, partial dislocations, and stacking faults. There are no evidences for distinct amorphous phases at the grain boundary. This is true for all silicon materials grown by different techniques.
2. Low-angle ($\Theta_{GB} < 5°$) tilt and twist boundaries are not composed of regular arrays of perfect dislocations. Instead, several types of dislocations are present in the same boundary; some may be dissociated into partial dislocations forming a stacking fault in between. Most of these low-angle grain boundaries are reconstructed such that no dangling bonds remain.
3. Large-angle grain boundaries are usually composed of distinct facets. These facets with lengths of one or more nanometers are subsections of the boundaries where bonding rearrangements have occurred that are of a few known low-energy configurations. These configurations can usually be predicted using the concepts of the coincidence site lattice (CSL) theory (Gleiter & Chalmers, 1972; Chadwick & Smith, 1976; Sutton &

Baluffi, 1995). The arrangement of these facets is not always a simple, repetitive one and the average boundary interface angle can actually vary substantially over macroscopic distances.

4. Even simple first-order twin boundaries can display this irregular faceted structure at their interfaces. Dislocations frequently terminate at coherent twin boundaries, and the resulting intersection points disturb the atomic arrangements on the boundary plane.

The interaction of intragranular dislocations with grain boundaries is an important issue because grain boundaries are effective obstacles to dislocation motion. Dislocations coming upon a boundary generally do not have the same Burgers vector and slip plane to glide into the next grain. Most commonly, the elastic interaction between dislocations and grain boundaries is repulsive and consequently the dislocations pile up at the boundary. Dislocations, however, may also transmitted directly across the grain boundary if the slip planes on both sides intersect along a line that lies in the boundary plane. For pure screw dislocations, the Burgers vector remains unchanged. In contrast, the transmission of dislocations with an edge component requires the formation of a residual grain boundary dislocation with a Burgers vector equal to the difference of the Burgers vectors of the incoming and outgoing lattice dislocations. A dislocation may alternatively be absorbed by the boundary without emission of a dislocation in the adjacent grain. In this case, the lattice dislocation fully dissociates.

Another important issue related to grain boundaries is the diffusion of impurities. It is generally known that diffusion at grain boundaries is orders of magnitude faster compared to volume diffusion, and it plays a major role in processes that involve material transport, such as recrystallization, grain growth, grain boundary segregation, etc. Based on previous analyses, Queisser et al. (1961) measured the phosphorous diffusion on a particular grain boundary suggesting an enrichment of phosphorous near the boundary dislocations. More recent investigations support the enhanced diffusion at grain boundaries but measurements of the activation energy are quite different (Schimpf et al. 1994). Values of the activation energy ranging from 1.4 eV to 2.9 eV were reported indicating the effect of the grain boundary structure as well as the interaction with other impurities segregated at the boundary on the diffusion. There is a number of other investigations dealing with the diffusion of different elements into polycrystalline silicon. All these investigations show a different behavior for various elements. For instance, an enhanced diffusion was proved for boron and titanium (Corcoran & King, 1990), while the diffusion of Al is suppressed. Other elements tend to diffuse out (Salman et al., 2007).

In order to overcome the difficulties arising from the analyses of polycrystalline materials specific grain boundaries were of growing interest to study their structure and properties (for instance, Bourret & Bacmann, 1987; Thibault-Desseaux et al., 1989). The realization of the so-called bicrystals requires, however, a Czochralski growth process allowing only the formation of specific grain boundaries such as $\Sigma = 9(122)$, $\Sigma = 13(510)$, and $\Sigma = 25(710)$ (Aubert & Bacmann, 1987).

A first model of the electrical activity of grain boundaries in Ge was proposed by Taylor et al. (1952). Based on measurements they concluded that the grain boundary acts as a potential barrier due to surface states. The center zone with a high density of states (assumed as broken bonds) and a space charge on either side represents a double Schottky barrier. The current across the grain boundary, I, is then given by

$$I = e\mu[(n_B - n_A) \cdot exp(\mp eV_{AB}/kT)]/[1 - exp(\mp eV_{AB}/kT)] \,, \tag{13}$$

where μ is the carrier mobility, E is the electric field at the top of the barrier, n_B the carrier density at the barrier top, n_A the carrier density on the bottom of the barrier, and V_{AB} the voltage measured across the barrier. The negative and positive signs are taken for electron current and hole current, respectively. Using the Richardson equation for thermoionic emission, Mueller (1961) write the zero-bias conductance G_0 of a grain boundary as

$$G_0 = (1 - \tfrac{\gamma}{2}) \cdot (e^2 N_c \overline{\upsilon}/4kT^2) \cdot e^{c/k} \cdot T \cdot e^{-\phi_0/kT}\,, \tag{14}$$

with γ the capture rate, e the electron charge, $\overline{\upsilon}$ the average thermal velocity, N_c the effective number of states, and ϕ_0 the barrier height at equilibrium. The model of Taylor et al. (1952) was developed further by Mataré (1984) and was successfully applied to interpret the electronic properties of grain boundaries in bicrystals (Broniatowski, 1985; Bourgoin et al. 1987). Seager (1985) proposed another model by integrating tunneling and thermoionic emission currents resulting in

$$G_0 = \frac{eA^*T}{k} \cdot \left\{ \frac{1}{kT} \int_0^{\phi_0} \frac{exp\left[\left(\frac{2}{E_0}\right)\left[[\phi_0(\phi_0 - E)]^{\frac{1}{2}} - E \cdot ln\left(\frac{\left(\phi_0^{\frac{1}{2}} + (\phi_0 - E)^{\frac{1}{2}}\right)}{E^{\frac{1}{2}}}\right)\right]\right]}{1 + exp[(E + \varsigma)/kT)]} dE + exp\left(\frac{-(\phi_0 + \varsigma)}{kT}\right) \right\} \tag{15}$$

with A* as an effective Richardson constant, ζ = E_C-E_F, $E_0 = (\hbar^2 e^2 N_d / 4 m_t \epsilon \epsilon_0)$, N_d the dopant concentration, and m_t as the tunneling mass. The second term in brackets of Eq. (15) is the standard thermoionic emission, while the first term describes the thermoionic field emission contributions to G_0.

If a dc bias is applied to the grain boundary, the band diagram is modified. Using simplifying assumptions (pinning of the Fermi level at the grain boundary, Mueller, 1961), the energy density of grain boundary states with respect to the applied voltage is given by Seager and Pike (1979) as

$$\overline{N}_T(E) = \left(\frac{\epsilon \epsilon_0 N_d}{2e^2}\right)^{1/2} \cdot \left[\phi_B^{-1/2} + \left(1 + \frac{e}{\phi_B'}\right)(\phi_B + eV)^{-1/2}\right] \tag{16}$$

for eV > kT. In Eq. (16) $\phi_B' = \partial \phi_B / \partial eV$ and ϕ_B is the barrier height given by

$$\phi_B = \phi_0 - kTln(eJ/G_0 kT) \tag{17}$$

Models describing especially the minority carrier transport and recombination processes on grain boundaries under optical illumination were presented, for instance, by Fossum & Sundaresan (1982) and Joshi (1987). Assuming a Gaussian distribution of interface states (other distributions were also discussed, see Joshi, 1987), the electron *n(0)* and hole concentrations *p(0)* at the grain boundary are obtained as

$$n(0) = N_d \cdot \exp(e\phi_B/kT) \tag{18}$$

and

$$p(0) = \frac{n_i^2}{N_d} \exp\left(\frac{\Delta E_F}{kT}\right) exp\left(\frac{e\phi_B}{kT}\right) \tag{19}$$

where n_i is the intrinsic carrier concentration and ΔE_F the separation of the quasi-Fermi levels at the grain boundary. ΔE_F is a function of the illumination level. Using the Shockley-Read-Hall theory, Joshi (1987) calculated the steady-state recombination current density at a grain boundary assuming a single interface energy level in the energy gap exists:

$$J_r(0) = e\sigma_n\sigma_c\bar{v}n_i^2[exp(\Delta E_F/kT) - 1] \cdot \int_{E_v(0)}^{E_c(0)} \frac{n_{gs}(E)}{\sigma_n n(0)+\sigma_n n_i\beta+\sigma_c p(0)+\sigma_c n_i\beta^{-1}} dE, \quad (20)$$

where σ_c and σ_n are the Coulomb and neutral capture cross-sections for a recombination center, respectively, n_{gs} the energy distribution of the states and $\beta = \exp[(E-E_i)/kT]$, where E_i means the energy position of the mean value of the interface states distribution.

The increasing importance of multi-crystalline silicon in the production of solar cells results in a huge number of publications related to the analyses of grain boundaries. The analyses of trap levels on model grain boundaries were extensively investigated by Broniatowski (1985). Numerous measurements on individual grain boundaries in multi-crystalline silicon were presented. Recent results about the electrical activity of grain boundaries obtained by EBIC methods were published, for instance, by Chen et al. (2010), Sekiguchi et al. (2011), or Pandelov et al. (2002). These papers refer to numerous others published previously. Caused by the high local resolution EBIC methods were also utilized to study the segregation of dopands or metallic impurities on grain boundaries (e.g. Seibt et al., 2009). The passivation of interface states on grain boundaries by hydrogen was studied as well (Rinio et al. 2006; Chen et al., 2005).

Furthermore, luminescence-based techniques are widely applied in the characterization of grain boundaries in solar cell materials. Cathodoluminescence (Vernon-Parry et al., 2005) and photoluminescence (Mchedlidze et al., 2010; Dreckschmidt & Möller, 2011) are useful tools to characterize defects at grain boundaries and different multi-crystalline bulk and thin film materials. The high sensitivity of the band-to-band emission of silicon to recombination activity (Würfel, 1982) results in the development of micro-photoluminescence spectroscopy used to study individual defects as well as to characterize the quality of whole solar cell wafers. The method allows the characterization of impurities (metal precipitates) and their effect on the recombination behavior of extended defects (Gundel et al. 2009, 2010). Recently, the D-lines appearing in the photoluminescence spectrum of dislocated silicon were used as well. First results using photoluminescence (Schmid, 2011) and cathodoluminescence were reported (Lee et al., 2009; Sekiguchi et al. 2010). Another approach to study electrically active defects in multicrystalline materials is the so-called dark lock-in thermography (DLIT, Breitenstein et al., 2010).

5. Characterization of individual dislocations

A fundamental problem in studying dislocations is the realization of defined arrangements of these defects. Some of the methods need a higher concentration of the defects to attain the detection limit (such as EPR). In contrast, other methods, as electron microscopy, require only a few or individual dislocations to obtain reasonable results. The dominant method to produce defined dislocation arrangements is plastic deformation. Plastic deformation, however, result also in a large number of point defects and defect reactions making it sometimes difficult to interpret experimental data (Alexander et al., 1983; Alexander & Teichler, 1991). In order to avoid interactions between dislocations or between dislocations and other defects, methods are required allowing the realization and analyses of only a few dislocations or, in the ideal case, of an individual dislocation. First attempts can be traced back into the 1970th. Eremenko et al. (see Shikin & Shikina, 1995) measured the current-voltage characteristics of a 60° dislocation. Similar experiments were also done by Milshtein (1979) a few years later. Their measurements demonstrated a diode behavior of the dislocation. The dislocation was assumed to pass the whole specimen and metallic tips were used as contacts which, however, were singnificantly larger than the dislocation diameter. Another approach to realize defined dislocation arrangements was the application of silicon and germanium bicrystals (Thibault-Desseaux et al., 1989). As pointed out before, only specific orientations of grain boundaries are viable by the Czochralski growth process. Another method to realize defined dislocation arrangements in a reproducible way is semiconductor wafer direct bonding. For wafer bonding commercially available wafers are used making it possible to realize any grain boundary. Especially small-angle grain boundaries having rotation angles below 1° are of interest allowing to separate dislocations by a few hundred nanometers. Such distances are large enough to analyze only a small number or individual dislocations.

The principle of semiconductor wafer direct bonding originally developed to produce silicon on insulator (SOI) substrates and three-dimensional micro-electromechanical systems (MEMS) was comprehensively described elsewhere (Tong & Gösele, 1998). If the native oxide is removed from the wafers, two Si surfaces are brought into contact (hydrophobic wafer bonding). A subsequent annealing transforms the original adhesion forces into Si-Si bonds via the interface. Crystal defects (dislocations) are generated forming a two-dimensional network (or grain boundary) in order to match both crystal lattices. The structure of the dislocation network depends on the surface orientation of both wafers. Screw dislocation networks, networks dominated by 60°, and interactions between both types of networks were realized and studied in detail (for instance, Reiche (2008)). The mesh size of the network or, the dislocation distance, is reproducibly adjusted by controlling the tilt and twist misorientation angles which can be calculated using Frank´s formula (Amelinckx, 1982). Dislocation distances of more than 100 nm are obtained by using misorientation angles below 0.1°. Note that misorientation angles down to 0.005° were realized using aligned wafer bonding processes (Wilhelm et al., 2008).

Properties of dislocation networks formed by semiconductor wafer direct bonding were described in numerous publications (for reviews see, e.g. Kittler et al., 2007; Kittler & Reiche, 2009). The dislocation networks may be considered as model structures resulting in a lot of new information about the structure and properties of dislocations. The electrical properties of bonded hydrophobic silicon wafers were studied for the first time by Bengtsson et al. (1992) using capacitance-voltage measurements. More recent EBIC analyses proved barrier heights generally smaller than 100 meV for different types of bonded hydrophobic wafers (Kittler & Reiche, 2009). The concentration of deep levels along the interface was determined to be a few 10^5 per cm.

The luminescence properties of dislocation networks were also studied. Figure 5a shows the luminescence spectra of different bonded samples. The spectra are obtained from samples having different misorientation. Detailed photoluminescence and cathodoluminescence measurements provide direct evidence that the wavelength of light emitted from the dislocation network could be tailored to some extent by misorientation of the wafers during the bonding procedure. D1 or D3 lines have the largest intensity in the spectra due to the variation of the twist angle from 8.2° to 9°. Thus the luminescence spectrum can be tailored by the misorientation angles in a controlled manner and the dominance of either D1 or D3 radiation can be attained. Further investigations refer that screw dislocations dominantly effect the intensity of the D1 line. The photoluminescence spectra of three different dislocation networks are presented in figure 5b. The corresponding electron microscope images are shown in figure 5c. The dislocation network DN#1 is dominated by 60° dislocations running in the image parallel with a distance in between of about 30 nm. The network of the 60° dislocations is superposed by an additional network of screw dislocations having distances of more than 2 μm (not shown in the image). The other networks in figure 5c (DN#2, DN#3) are characterized by more or less hexagonal meshes caused by the interaction of two networks of 60° dislocations and screw dislocations, both with nearly the same dislocation distances therein. The photoluminescence spectra recorded at low temperature (80K) and room temperature show the presence of the D1-line around

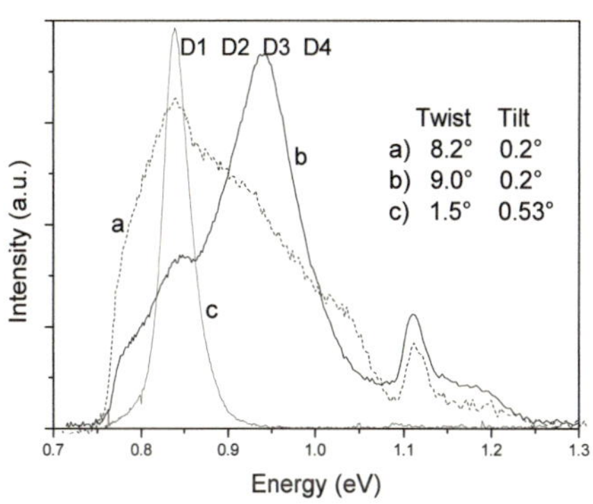

(a)

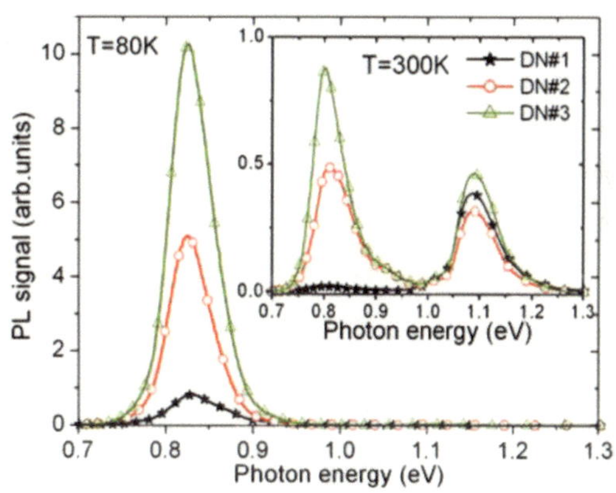

(b)

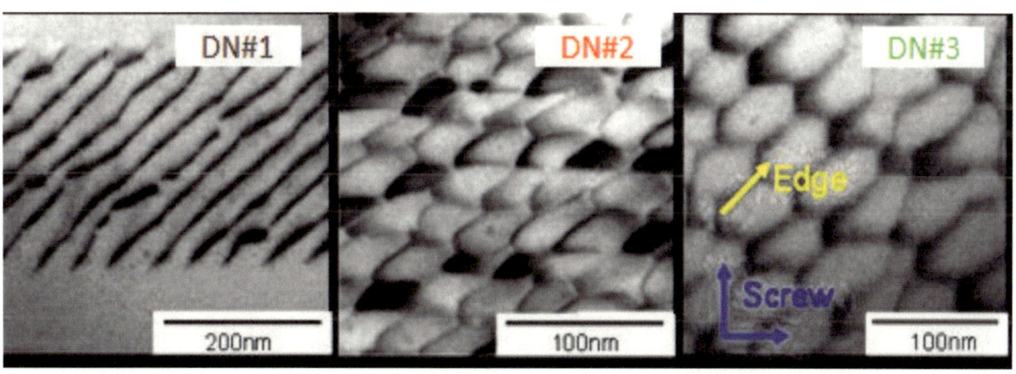

(c)

Fig. 5. The impact of the misorientation and dislocation structure on the luminescence spectra of dislocation networks. (a) The effect of misorientation (tilt and twist components). Cathodoluminescence spectra recorded at 80K. Photoluminescence spectra measured at 80K and room temperature (b) of three dislocation networks shown in (c).

(figure 5b). The spectra clearly prove the different intensity behavior depending on the dislocation structure and distance of the screw dislocations. The intensity of the D1-line is lowest in the spectrum of sample DN#1, characterized by the largest distance of screw dislocations, and increases as the distance of the screw dislocations decreases. The distance of screw dislocations in these particular samples is 15 nm (DN#2) and 32 nm (DN#3). Note that significant intensities of the D1-line are measured for DN#2 and DN#3 which are considerably stronger than that of the band-to-band-luminescence even at room temperature. According to these results it is suggested that radiative recombination is mainly caused by screw dislocations while 60° dislocations attribute preferentially to the non-radiative recombination.

A combination of wafer bonding with preparation methods to separate individual dislocations or a small number of dislocations allows the measurement of their electronic properties by elimination of interactions in between. As shown before, twist angles between two bonded Si wafers below 0.1° result in dislocation distances of more than 100 nm. Using photolithography and etching techniques, individual dislocations can be separated and measured. Typical structures applied were diodes and metal-oxide-semiconductor field-effect transistors (MOSFETs) (Reiche et al. 2010, 2011). The presented data clearly showed an indirect behavior of the drain current on the number of dislocations in the channel. The fact that the highest current is obtained if only a few dislocations are present allows the conclusion that electrically active centers in the dislocation core of the straight dislocation segments are responsible for the electron transport while dislocation nodes and dislocation segments oriented orthogonal to the channel direction act as "scattering centers" and reduce the carrier transport.

The single-electron tunneling on dislocations was recently studied by Ishikawa et al. (2006) on nMOSFETs prepared on dislocation networks produced by wafer bonding of SOI wafers. Measurements were done using a back gate contact (oxide thickness 400nm). Low-temperature measurements (T=15K) proved oscillations in the drain current - gate voltage curves indicating single-electron tunneling (Coulomb blockade oscillations). The lateral size of the Coulomb islands was estimated to be about 20 nm which agreed with the dislocation distance. From this Ishikawa et al. (2006) concluded that Coulomb islands are related to the dislocation nodes in the screw dislocation network. Very recent measurements at T = 4K by the authors proved also the existence of Coulomb blockade oscillations. Using nMOSFETs and applying a front side gate contact (gate oxide thickness 6 nm) lateral sizes of the Coulomb island of about 6 nm were extracted which do not correspond to dislocation nodes. Furthermore, a different behavior is observed for screw and mixed dislocations resulting from the reaction of screw and 60° dislocations. The single-electron tunneling was proved for one set (screw dislocations), while the other shows a more two-dimensional characteristics indicated by a staircase structure.

6. Acknowledgment

We would like to thank T. Arguirov, A. Hähnel, T. Mchedlidze, R. Scholz, W. Seifert, and O. Vyvenko for supporting this work. Parts of this work were financially supported by the German Federal Ministry of Education and Research in the framework of the SiliconLight project (contract no. 13N9734) and the SiGe-TE project (contract no. 03X3541B).

7. References

Alexander, H. (1986). Dislocations in Covalent Crystals, in: *Dislocation in Solids*, Vol. 7, F.R.N. Nabarro, pp. 113-234, North-Holland, Amsterdam

Alexander, H. (1991). Dislocations in Semiconductors, in: *Polycrystalline Semiconductors II*, J.H. Werner and H.P. Strunk, Springer Proc. In Physics, Vol. 54, Springer, Berlin, pp. 2-12

Alexander, H. & Teichler, H. (1991). Dislocations, in: *Materials Science and Technology, Vol. 4. Electronic Structure and Properties of Semiconductor*, W. Schröter, VCH, Weinheim, pp. 249-319

Alexander, H. & Teichler, H. (2000). Dislocations, in: *Handbook of Semiconductor Technology*, K.A: Jackson and W. Schröter, Wiley-VCH, Weinheim, pp. 291-376

Alexander, H., Kisielowski-Kemmerich, C., and Weber, E.R. (1983). Investigations of Well Defined Dislocations in Silicon, *Physica B*, Vol. 116, pp. 583-593

Amelinckx, S. (1982). Dislocations in Particular Structures, in: *Dislocation in Solids*, Vol. 2, F.R.N. Nabarro, pp. 67-460, North-Holland, Amsterdam

Aubert, J.J. & Bacmann, J.J. (1987). Czochralski Growth of Silicon Bicrystals, *Revue Phys. Appl.*, Vol. 22, No. 7, pp. 515-518

Baluffi, R.W. & Sutton, A.P. (1996). Why Should We Interested in the Atomic Structure of Interfaces?, *Mat. Sci. Forum*, Vol. 207-209, pp. 1-12

Bangert, U., Harvey, A.J., Jones, E., Fall, C.J., Blumenau, A.T., Briddon, R., Schreck, M., and Hörmann, F. (2004). Dislocation-Induced Electronic States and Point Defect Atmospheres Evidenced by Electron Energy Loss Imaging, *New J. Phys.*, Vol. 6, pp. 184-189

Bardeen, J. & Shockley, W. (1950). Deformation Potentials and Mobilities in Non-Polar Crystals, Phys. Rev., Vol. 80, No. 1, pp. 72-80

Benetto, J., Nunes, R.W., and Vanderbilt, D. (1997). Period-Double Structure for the 90° Partial Dislocation in Silicon, *Phys. Rev. Lett.*, Vol. 79, No. 2, pp. 245-248

Bengtsson, S., Andersson, G.I., Andersson, M.O., and Engström, O. (1992). The Bonded Unipolar Silicon-Silicon Junction, *J. Appl. Phys.*, Vol. 72, No. 1, pp. 124-140

Bigger, J.R.K., McInnes, D.A., Sutton, A.P., Payne, M.C., Stich, I., King-Smith, R.D., Bird, D.M., and Clarke, L.J. (1992). Atomic and Electronic Structures of the 90° Partial Dislocation in Silicon, *Phys. Rev. Lett.*, Vol. 69, No. 15, pp. 2224-2227

Bourgoin, J.C., Mauger, A., and Lannoo, M. (1987). Electronic Properties of Grain Boundaries in Semiconductors, *Revue Phys. Appl.*, Vol. 22, No. 7, pp. 579-583

Bourret, A. & Bacmann, J.J. (1987). Atomic Structure of Grain Boundaries in Semiconductors, *Revue Phys. Appl.*, Vol. 22, No. 7, pp. 563-568

Breitenstein, O. & Wosinski, T. (1983). Scanning-DLTS Investigation of the EL 2 Level in Plastically Deformed GaAs, *Phys. Stat. Sol.* (a), Vol. 77, K107-K110

Breitenstein, O., Bauer, J., Altermatt, P.P., and Ramspeck, K. (2010). Influence of Defects on Solar Cell Characteristics, *Solid State Phenom.*, Vol. 156-158, pp. 1-10

Broniatowski, A. (1985). Electronic States at Grain Bounaries in Semiconductors, in: *Polycrystalline Semiconductors, Physical Properties and Applications*, G. Harbecke, pp. 95-117, Springer, Berlin

Bulatov, V.V., Yip, S., and Argon, A.S. (1995). Atomic Modes of Dislocation Mobility in Silicon, *Phil. Mag.* A, Vol. 72, No. 2, pp. 453-496

Bulatov, V.V., Justo, J.F., Cai, W., Yip, S., Argon, A.S., Lenosky, T., de Koning, M., and Diaz de la Rubia, T. (2001). Parameter-Free Modelling of Dislocation Motion: The Case of Silicon, *Phil. Mag.*, Vol. 81, No. 5, pp. 1257-1281

Bulatov, V.V. & Cai, W. (2006). *Computer Simulations of Dislocations*, Oxford Univ. Press, Oxford

Burgers, W.G. & Burgers, J.M. (1935). First report on viscosity and plasticity. *Kon. Nederl. Akad. v. Wet., Sect. 1*, Vol. 15, No. 3

Burgers, J.M. (1939). Some Considerations on the Fields of Stress Connected with Dislocations in a Regular Crystal Lattice. *Kon. Nederl. Akad. v. Wet.* Vol. 42, pp. 293-325

Burgers, J.M. (1940). Geometrical Considerations Concerning the Structural Irregularities to be Assumed in a Crystal, *Proc. Phys. Soc. London,* Vol. 52, pp. 23-33

Chadwick, G.A. & Smith, D.A. (1976). *Grain Boundary Structure and Properties*, Academic Press, London

Chelikowsky, J.R. (1982). 30° Partial Dislocations in Silicon: Absence of Electrically Active States, *Phys. Rev. Lett.*, Vol. 49, No. 21, pp. 1569-1572

Chen, J., Yang, D., Xi, Z., and Sekiguchi, T. (2005). Electron-Beam-Induced Current Study of Hydrogen Passivation on Grain Boundaries in Multicrystalline Silicon: Influence of GB Character and Impurity Contamination, *Physica* B, Vol. 364, No. 1, pp. 162-169

Chen, J., Chen, B., Lee, W., Fukuzawa, M., Yamada, M., and Sekiguchi, T. (2010). Grain Boundaries in Multicrystalline Si, *Solid State Phenom.*, Vol. 156-158, pp. 19-26

Cockayne, D.J.H., Ray, I.L.F. and Whelan, M.J. (1969). Investigations of Dislocation Strain Fields Using Weak Beams, *Phil. Mag.* Vol. 20, pp. 1265-1270

Corcoran, Y.L. & King, A.H. (1990). Grain Boundary Diffusion and Growth of Titanium Silicide Layers on Silicon, *J. Electron. Mat.*, Vol. 19, No. 11, pp. 11771183

Cottrell, A.H. (1953). *Dislocations and Plastic Flow in Crystals*, Clarendon, Oxford

Csányi, G., Engeness, T.D., Ismail-Beigi, S., and Arias, T.A. (2000). New Physics of the 30° Partial Dislocation in Silicon Revealed Through ab initio Calculations, *J. Phys.: Condens. Mater.* Vol. 12, pp. 10029-10037

Dehlinger, U. & Kochendörfer, A. (1940). Eigenbewegungen in Kristallgittern. *Z. Phys.* Vol. 116, pp. 576-585

Donolato, C. (1979). Contrast and Resolution of SEM Charge-Collection Images of Dislocations, *Appl. Phys. Lett.*, Vol. 34, No. 1, pp. 80-81

Donolato, C. (1983). Quantitative Evaluation of the EBIC Contrast of Dislocations, *J. Physique Coll.*, Vol. 44, No. 9, pp. C4-269 - 275

Dreckschmidt, F. & Möller, H.-J. (2011). Defect Luminescence at Grain Boundaries in Multicrystalline Silicon, *Phys. Stat. Sol.* (c) Vol. 8, No. 4, pp. 1356-1360

Drozdov, N.A., Patrin, A.A., and Tkachev, V.D. (1976). Recombination Radiation on Dislocations in Silicon, *Pisma Zh. Eksp. Teor. Fiz.*, Vol. 23, No. 11, pp. 651-653, *Sov. Phys. JETP Lett.* Vol. 23, pp. 597-599

Drozdov, N.A., Patrin, A.A., and Tkachev, V.D., On the Nature of the Dislocation Luminescence in Silicon, *Phys. Stat. Sol.* (b), Vol. 83, No. 2, pp. K137-K139

Duesbery, M.S. & Richardson, G.Y. (1991). The Dislocation Core in Crystalline Materials, *Rev. Solid State Mater. Sci.* Vol 17, No. 1, pp. 1-46

Duesbery, M.S., Joos, B., and Michel, D.J. (1991). Dislocation Core Studies in Empirical Silicon Models, *Phys. Rev.* B, Vol. 43, No. 6, pp. 5143-5146

Duesbery, M.S. & Joós, B. (1996). Dislocation Motion in Silicon: The Shuffle-Glide Controversy, *Phil Mag. Lett.*, Vol. 74, No. 4, pp. 253-258

Eshelby, J.D. (1949). Edge Dislocations in Anisotropic Materials. *Phil. Mag*. Vol. 7, No. 40, pp. 903-912

Figielski, T. (1990). Electron Emission from Extended Defects: DLTS Signal in Case of Dislocation Traps, *Phys. Stat. Sol.* (a), Vol. 121, No. 1, pp. 187-193

Fossum, J.G. & Sundaresan, R. (1982). Analysis of Minority-Carrier Transport in Polysilicon Devices, *IEEE Trans. Electron. Dev.*, Vol. ED-29, No. 8, pp. 1185-1197

Frank, F.C. (1951). Crystal Dislocations - Elementary Concepts and Definitions, *Phil. Mag.* Vol. 42, No. 331, pp. 809-819

Friedel, J. (1979). Dislocations - an Introduction, in: *Dislocation in Solids,* Vol. 1, F.R.N. Nabarro, pp. 1-32, North-Holland, Amsterdam

Gallagher, C.J. (1952). Plastic Deformation of Germanium and Silicon, *Phys. Rev.* Vol. 88, No. 4, pp. 721-722

George, A. & Yip, S. (2001).Preface to the Viewpoint Set on: Dislocation Mobility in Silicon, *Scripta Mat.*, Vol. 45, pp. 1233-1238

Gleiter, H. & Chalmers, B. (1972). *High-Angle Grain Boundaries*, Pergamon Press, Oxford

Gomez, A., Cockayne, D.J.H., Hirsch, P.B., and Vitek, V. (1974). Dissociation of Near-Screw Dislocations in Germanium and Silicon, *Phil. Mag.* Vol. 31, pp. 105-113

Gomze, A.M. & Hirsch, P.B. (1977). On the Mobility of Dislocations in Germanium and Silicon, *Phil. Mag*. Vol. 36, No.1, pp. 169-179

Gundel, P. Schubert, M.C., Kwapil, W., Schön, J., Reiche, M., Savin, H., Yli-Koski, M., Sans, J.A., Martinez-Criado, G., Seifert, W., Warta, W., and Weber, E.R. (2009). Micro-Photoluminescence Spectroscopy on Metal Precipitates in Silicon, *Phys. Stat. Sol. RRL*, pp. 1-3

Gundel, P., Schubert, M.C., Heinz, F.D., Kwapil, W., Warta, W., Martinez-Criado, G., Reiche, M., and Weber, E.R. (2010). Impact of Stress on the Recombination at Metal Precipitates in Silicon, *J. Appl. Phys.* 108, pp. 103707-1 - 5

Heggie, M. & Jones, R. (1982). Glide of Partial Dislocations in Silicon, *J. Physique, Coll.*, Vol. 43, No. 10, pp. 45-50

Heggie, M. & Jones R. (1983). Solitons and the Electrical and Mobility Properties of Dislocations in Silicon, *Phil. Mag.*, Vol. 48, No. 4, pp. 365-377

Heggie, M.I., Jones, R., and Umerski, A. (1993). Ab Initio Energy Calculations of Impurity Pinning in Silicon, *Phys. Stat. Sol.* (a), Vol. 138, pp. 383-387

Higgs, V., Lightowlers, E.C., Fitzgerald, E.A., Xie, Y.H., and Silverman, P.J. (1993). Characterization of Compositionally Graded $Si_{1-x}Ge_x$ Alloy Layers by Photoluminescence Spectroscopy and by Cathodoluminescence Spectroscopy and Imaging, *J. Appl. Phys.*, Vol. 73, No. 4, pp. 1952-1956

Hirsch, P.B. (1979). Recent Results on the Structure of Dislocations in Tetrahedrally Coordinated Semiconductors, *J. Physique, Coll.*, Vol. 40, No. 6, pp. C6-27 - 32

Hirsch, P.B. (1985). Dislocations in Semiconductors, *Mat. Sci. Technol.*, Vol. 1, No. 9, pp. 666-677

Hirth, J.P. & Lothe, J. (1982). *Theory of Dislocations*, Wiley Interscience, New York

Hornstra, J. (1958). Dislocations in the Diamond Lattice, *J. Phys. Chem. Solids*, Vol. 5, pp. 129-141

Ishikawa, Y., Yamamoto, C., and Tabe, M. (2006). Single-Electron Tunneling in a Silicon-On-Insulator Layer Embedding an Artificial Dislocation Network, *Appl. Phys. Lett.*, Vol. 88, pp. 073112-1 - 3

Jones, R., Umerski, A., Stich, P., Heggie, M.I., and Öberg, S. (1993). Density Functional Calculations of the Structure and Properties of Impurities and Dislocations in Semiconductors, *Phys. Stat. Sol.* (a), Vol. 138, pp. 369-381

Jones, R., Coomer, B.J., Goss, J.P., Öberg, S., and Briddon, P.R. (2000). Intrinsic Defects and the D1 to D4 Optical Bands Detected in Plastically Deformed Si, *Phys. Stat. Sol.* (b), Vol. 222, No. 1, pp. 133-140

Joshi, D.P. (1987). Grain Boundary Recombination in Polycrystalline Silicon Under Optical Illumination, *Phys. Stat. Sol.* (a), Vol. 108, No. 2, pp. 213-218

Kisielowski-Kemmerich, C. (1990). Vacancies and Their Complexes in the Core of Screw Dislocations: Models which Account for ESR Investigations of Deformed Silicon, *Phys. Stat. Sol.* (b), Vol. 161, pp. 11-42

Keyes, R.W. (1960). The Effects of Elastic Deformation on the Electrical Conductivity of Semiconductors, *Solid State Phys.*, Vol. 11, pp. 149-221

Kittler, M., Yu, X., Mchedlidze, T., Arguirov, T., Vyvenko, O.F., Seifert, W., Reiche, M., Wilhelm, T., Seibt, M., Voß, O., Wolff, A., and Fritzsche, W. (2007). Regular Dislocation Networks in Silicon as a Tool for Nanostructure Devices Used in Optics, Biology, and Electronics, *Small*, Vol. 3, No. 6, pp. 964-973

Kittler, M. & Reiche, M. (2009). Dislocations as Active Components in Novel Silicon Devices, *Adv. Eng. Mater.* Vol. 11, No. 4, 249-258

Kochendörfer, A. (1938). Theorie der Kristallplastizität, *Z. Phys.* Vol. 108, No. 3-4, pp. 244-264

Kveder, V., Kittler, M., and Schröter, W. (2001). Recombination Activity of Contaminated Dislocations in Silicon: A Model Describing Electron-Beam-Induced Current Contrast Behavior, *Phys. Rev.* B, Vol. 63, pp. 115208-1 - 11

Kveder, V. & Kittler, M. (2008). Dislocations in Silicon and D-Band Luminescence for Infrared Light Emitters, *Mat. Sci. Forum*, Vol. 590, pp. 29-56

Labusch, R. & Schröter, W. (1980). Electrical Properties of Dislocations in Semiconductors, in: *Dislocation in Solids*, Vol. 5, F.R.N. Nabarro, pp. 127-191, North-Holland, Amsterdam

Lang, D.V. (1974). Deep-Level Transient Spectroscopy: A New Method to Characterize Traps in Semiconductors, *J. Appl. Phys.*, Vol. 45, No. 7, pp. 3023-3032

Lee, W., Chen, J., Chen, B., Chang, J., and Sekiguchi, T. (2009). Cathodoluminescence Study of Dislocation-Related Luminescence From Small-Angle Grain Boundaries in Multicrystalline Silicon, *Appl. Phys. Lett.*, Vol. 94, pp. 112103-1 - 3

Lehto, N. & Öberg, S. (1998) Effects of Dislocation Interactions: Application to the Period-Double Core of the 90° Partial in Silicon, *Phys. Rev. Lett.*, Vol. 80, No. 25, pp. 5568-5571

Li, C., Meng, Q., Zhong, K., and Wang, C. (2008). Computer Simulation of the 60° Dislocation Interaction with Vacancy Cluster in Silicon, *Phys. Rev.* B, Vol. 77, pp. 045211-1 – 045211-5

Lodge, K.W., Lapiccirella, A., Battistoni, C., Tomassini, N., and Altmann, S.L. (1989). The 90° Partial Dislocation in Silicon: Geometry and Electronic Structure, *Phil. Mag.* A, Vol. 60, No. 5, pp. 643-651

Love, A.E.H. (1927). *A Treatise on the Mathematical Theory of Elasticity*, Cambridge University Press, Cambridge, p. 221

Marklund, S. (1979). Electron States Associated with Partial Dislocations in Silicon, *Phys. Stat. Sol.* (b), Vol. 92, pp. 83-89

Marklund, S. (1981). Energy Levels of Intrinsic and Extrinsic Stacking Faults in Silicon, *Phys. Stat. Sol.* (b), Vol. 108, pp. 97-102

Marklund, S. (1983). Structure and Energy Levels of Dislocations in Silicon, *J. Physique, Coll.*, Vol 44, No. 9, C4-25 - 35

Mataré, H.F. (1984). Carrier Transport at Grain Boundaries in Semiconductors, *J. Appl. Phys.*, Vol. 56, No. 10, pp. 2605-2631

Mattheiss, L.F. & Patel, J.R. (1981). Electronic Stacking Fault States in Silicon, *Phys. Rev.* B, Vol. 23, No. 10, 5384-5396

Mchedlidze, T., Arguirov, T., Kouteva-Argoirova, S., and Kittler, M. (2010). Characterization of Thin Film Photovoltaic Material Using Photoluminescence and Raman Spectroscopy, *Solid-State Phenom.* Vol. 156-158, pp. 419-424

Milshtein, S. (1979). Application of Dislocation-Induced Potentials in Si and Ge, *J. Physique, Coll.*, Vol. 40, No. 6, pp. C6-207 - 211

Mueller, R.K. (1961). Current Flow Across Grain Boundaries in n-Type Germanium I, *J. Appl. Phys.*, Vol. 32, No. 4, pp. 635-639

Nabarro, F.R.N. (1947). Dislocations in a Simple Cubic Lattice. *Proc. Phys. Soc. London*, Vol. 59, pp. 256-272

Northrup, J.E., Cohen, M.L., Chelikowsky, J.R., Spence, J., and Olsen, A. (1981). Electronic Structure of the Unreconstructed 30° Partial Dislocation in Silicon, *Phys. Rev.* B, Vol. 24, No. 8, pp. 4623-4628

Orowan, E. (1934). Zur Kristallplastizität III. *Z. Phys.* Vol. 89, pp. 634-659

Pandelov, S., Seifert, W., Kittler, M., and Reif, J. (2002). Analysis of Local Electrical Properties of Grain Boundaries in Si by Electron-Beam-Induced-Current Techniques, *J. Phys.: Condens. Matter*, Vol. 14, pp. 13161-13168

Pasemann, L. (1981). A Contribution to the Theory of the EBIC Contrast of Lattice Defects in Semiconductors, *Ultramicroscopy*, Vol. 6, pp. 237-250

Pearson, G.L., Read, W.T., and Morin, F.J. (1954). Dislocations in Plastically Deformed Germanium. *Phys. Rev.* Vol. 93, No. 4, pp. 666-667

Peierls, R. (1940). The Size of a Dislocation. *Proc. Phys. Soc. London*, Vol. 52, pp. 34-37

Polanyi, M. (1934). Über eine Art Gitterstörung, die einen Kristall plastisch machen könnte. *Z. Phys.* Vol. 89, pp. 660-664

Queisser, H.J., Hubner, K., and Shockley, W. (1961). Diffusion Along Small-Angle Grain Boundaries in Silicon, *Phys. Rev.* Vol. 123, No. 4, pp. 1245-1254

Ray, I.L.F. & Cockayne, D.J.H. (1971). The Dissociation of Dislocations in Silicon, *Proc. R. Soc. London*, A, Vol. 325, pp. 543-554

Read, W.T. (1954a). Theory of Dislocations in Germanium, *Phil. Mag.* Vol. 45, No. 367, pp. 775-796

Read, W.T. (1954b). Statistics of the Occupation of Dislocation Acceptor Centres, *Phil. Mag.* Vol. 45, No. 370, pp. 1119-1128

Read, W.T. & Shockley, W. (1950). Dislocation Models of Crystal Grain Boundaries, *Phys. Rev.*, Vol. 78, No. 3, pp. 275-289

Reiche, M. (2008). Dislocation Networks Formed by Silicon Wafer Direct Bonding, *Mater. Sci. Forum*, Vol. 590, pp. 57-78

Reiche, M., Kittler, M., Buca, D., Hähnel, A., Zhao, Q.-T., Mantl, S., and Gösele, U. (2010). Dislocation-Based Si-Nanodevices, *Jpn. J. Appl. Phys.*, Vol. 49, pp. 04DJ02-1 - 5

Reiche, M., Kittler, M., Scholz, R., Hähnel, A., and Arguirov, T. (2011). Structure and Properties of Dislocations in Interfaces of Bonded Silicon Wafers, *J. Phys. Conf. Ser.*, Vol. 281, pp. 012017-1 - 10

Rinio, M., Kaes, M., Hahn, G., and Borchert, D. (2006). Hydrogen Passivation of Extended Defects in Multicrystalline Silicon Solar Cells, *Proc. 21st Europ. Photovolt. Solar Energy Conf.*, Dresden

Salman, F., Arnold, J., Zhang, P., Chai, G., Stevie, F.A., and Chow, L. (2007). Redistribution of Implanted Species in Polycrystalline Silicon Films on Silicon Substrate, *Defect & Diff. Forum*, Vol. 264, pp. 7-12

Sauer, R., Weber, J., Stolz, J., Weber, E.R., Küsters, K.-H., and Alexander, H. (1985). Dislocation-Related Photoluminescence in Silicon, *Appl. Phys. A*, Vol. 36, No. 1, pp. 1-13

Schimpf, K., Palm, J., and Alexander, H. (1994). Enhanced Diffusion of Phosphorous at Grain Boundaries in Multicrystalline Silicon, *Cryst. Res. Technol.* Vol. 29, No. 8, pp. 1123-1129

Schmid, R.P., Mankovics, D., Arguirov, T., Mchedlidze, T., and Kittler, M. (2011). Novel Imaging Techniques for Dislocation-Related D1-Photoluminescence of Multicrystalline Si Wafers - Two Different Approaches, *Phys. Stat. Sol. (c)*, Vol. 8, No. 4, pp. 1297-1301

Schröter, W. (1969). Trägerbeweglichkeit in verformtem Germanium, *Phys. Stat. Sol.*, Vol. 31, No. 1, pp. 177-186

Schröter, W. & Labusch, R. (1969). Electrical Properties of Dislocations in Ge and Si. *Phys. Stat. Sol.* Vol. 36, pp. 539-550

Schröter, W. & Cerva, H. (2002). Interaction of Point Defects with Dislocations in Silicon and Germanium: Electrical and Optical Effects, *Solid State Phenom.* Vol. 85-86, pp. 67-144

Seager, C.H. (1985). Grain Boundaries in Polycrystalline Silicon, *Ann. Rev. Mater. Sci.*, Vol. 15, pp. 271-302

Seager, C.H. & Pike, G.E. (1979). Grain Boundary States and Varistor Behavior in Silicon Bicrystals, *Appl. Phys. Lett.*, Vol. 35, No. 9, pp. 709-711

Seibt, M., Khalil, R., Kveder, V., and Schröter, W. (2009a). Electronic States at Dislocations and Metal Silicide Precipitates in Crystalline Silicon and Their Role in Solar Cell Materials, *Appl. Phys.* A, Vol. 96, pp. 235-253

Seibt, M., Abdelbarey, D., Kveder, V., Rudolf, C., Saring, P., Stolze, L., and Voß, O. (2009b). Structure, Chemistry and Electrical Properties of Extended Defects in Crystalline Silicon for Photovoltaics, *Phys. Stat. Sol.* (c), Vol. 6, No. 8, pp. 1847-1855

Seitz, F. (1952). The Plasticity of Silicon and Germanium, *Phys. Rev.* Vol. 88, No. 4, pp. 722-724

Sekiguchi, T., Chen, J., Lee, W., and Onodera, H. (2011). Electrical and Optical Activities of Small Angle Grain Boundaries in Multicrystalline Si, *Phys. Stat. Sol.* (c), Vol. 8, No. 4, pp. 1347-1350

Shikin, V.B. & Shikina, Y.V. (1995). Charged Dislocations in Semiconductor Crystals, *Usp. Fiz. Nauk*, Vol. 165, No. 8, pp. 887-917, *Physics-Uspekhi*, Vol. 38, No. 8, pp. 845-875

Shockley, W. (1953). Dislocations and Edge States in the Diamond Crystal Structure, *Phys. Rev.*, Vol. 91, p. 228

Shockley, W. & Read, W.T. (1949). Quantitative Predictions from Dislocation Models of Crystal Grain Boundaries, *Phys. Rev.*, Vol. 75, p. 692

Spence, J.C.H. (2007). Experimental studies of Dislocation Core Defects, in: *Dislocation in Solids*, Vol. 13, F.R.N. Nabarro and J.P. Hirth, pp. 419-452, Elsevier, Amsterdam

Sutton, A. & Baluffi, R.W. (1995). *Interfaces in Crystalline Materials*, Oxford University Press, Oxford

Taylor, G.I. (1934a). The Mechanism of Plastic Deformation of Crystals. Part I.-Theoretical. *Proc. R. Soc. London*, Vol. 145, No. 855, pp. 362-387

Taylor, G.I. (1934b). The Mechanism of Plastic Deformation of Crystals. Part II. - Comparison with Observations. *Proc. R. Soc. London*, Vol. 145, No. 855, pp. 388-404

Taylor, W.E., Odell, N.H., and Fan, H.Y. (1952). Grain Boundary Barriers in Germanium, *Phys. Rev.*, Vol. 88, No. 4, pp. 867-875

Thibault-Desseaux, J., Putaux, J.L., Bourret, A., and Kirchner, H.O.K. (1989). Dislocations Stopped by the $\Sigma = 9(122)$ Grain Boundary in Si. An HREM Study of Thermal Activation, *J. Physique*, Vol. 50, pp. 2525-2540

Tong, Q.Y. & Gösele, U. (1998). *Semiconductor Wafer Bonding. Science and Technology*, Wiley, New York

Van der Merwe, J.H. (1950), On the Stress and Energies Associated with Inter-Crystalline Boundaries, *Proc. Phys. Soc. London*, A Vol. 63, pp. 616-637

Vernon-Parry, K.D., Davies, G., and Galloway, S. (2005). Electronic and Structural Properties of Grain Boundaries in Electron-Irradiated Edge-Defined Film-Fed Growth Silicon, *Semicond. Sci. Technol.*, Vol. 20, pp. 171-174

Veth, H. & Lannoo, M. (1984). The Electronic Properties of Charged Dislocations in Semiconductors, *Phil. Mag.* B, Vol. 50, No. 1, pp. 93-102

Volterra, V. (1907). Sur l`équilibre des corps élastiques multiplement connexes, *Ann. Sci. Ecole Norm. Super.*, Vol. 24, No. 3, pp. 401-517

Watson, G.P., Benton, J.L., Xie, Y.H., and Fitzgerald, E.A. (1998). Influence of Misfit Dislocation Interactions on Photoluminescence Spectra of SiGe on Patterned Si, *J. Appl. Phys.*, Vol. 83, No. 7, pp. 3773-3776

Weber, H., Schröter, W., and Haasen, P. (1968). Elektronenzustände an Versetzungen in Silizium, *Helv. Phys. Acta*, Vol. 41, pp. 1255-1258

Weber, J. & Alonso, M.I. (1990). Detection of Dislocation-Related Photoluminescence Bands in Si-Ge Alloys Grown by Liquid Phase Epitaxy, in: *Defect Control in Semiconductors*, K. Sumino, Vol. 2, pp. 1453-1457, North-Holland, Amsterdam

Weber, J. (1994). Correlation of Structural and Electronic Properties from Dislocations in Semiconductors, *Solid State Phenom.*, Vol. 37-38, pp. 13-24

Wilhelm, T, Mchedlidze, T., Yu, X., Arguirov, T., Kittler, M. and Reiche, M. (2008). Regular Dislocation Networks in Silicon. Part I: *Structure, Solid State Phenom.*, Vol. 131-133, pp. 571-578

Wilshaw, P.R. & Booker, G.R. (1985). New Results and an Interpretation for SEM EBIC Contrast Arising from Individual Dislocations in Silicon., *Inst. Phys. Conf. Ser.*, Vol. 76, pp. 329-336

Würfel, P. (1982). The Chemical Potential of Radiation, *J. Phys. C: Solid State Phys.*, Vol. 15, pp. 3967-3985

5

High Mass Molecular Ion Implantation

Bill Chang and Michael Ameen
Axcelis Technologies, Inc., Beverly
USA

1. Introduction

Semiconductor device manufacturing is facing stringent challenges in advanced COMS process technology nodes. Ion implantation technology has always been a good solution of last resort since it's got a much wider latitude and stronger flexibility to accommodate new challenges than any other process steps in device fabrication. It is not unusual that people utilize ion implantation not just for doping the silicon substrate, but also for compensating the shortfalls of other process steps. In the past decade, the process window, typically large enough for ion implant engineers to maneuver has gotten narrow, so narrow to a degree so that itself started to create problems which none other process steps can resolve, or compensate. These problems include dopant atoms activation, co-implant species of choice, pre-amorphization implant species of choice, implant damage control, runaway low-energy implant cost. High mass molecular (HMM) ion implantation is investigated in response to all these ion implant related problems.

Ion implantation is a process whereby energetic ions impinge on a target, penetrating below the target surface and giving rise to a controlled, predictable, ion distribution. Here we will focus on Si technology; hence the target will be mostly Si. Implanted ions are typically dopants, such as Boron, Phosphorus, Arsenic, Indium and Antimony. Table 1 shows these commonly used dopant elements in the periodic table of the elements. However, the scaling of device features into the sub-100nm regime has added species such as Ge, C, N, and Xe to this list. Implantation energies cover a wide range from 0.2 keV to >3 MeV; doses range from 1×10^{11} cm^2 to more than 1×10^{16} cm^2; incident angles cover normal incidence (a tilt angle of 0°) to 60°.

The industry has been using BF_2^+, as the molecular form of Boron, to implant in order to attain higher throughput for low-energy applications. This species has the disadvantage of co-implanting fluorine, which retards boron activation and increases contact resistance, both undesirable consequences for doping process (Foad, 2005). HMM implants have recently been introduced as an alternative. As the molecular structure shown in Fig. 1, Octadecaborane ($B_{18}H_{22}$), which has 18 effective dopant atoms in one molecule, has been proven a viable replacement for boron in poly-doping and BF_2 for ultra-shallow junction (USJ) formation.

Besides the advantage of higher productivity, HMM implant process advantages have been noticed and explored. Due to its heavy mass, HMM ion implant can eliminate the use of pre-amorphization implant (PAI). We can use the HMM ions that contains either dopant or co-implant species to replace PAI (Ameen, 2008). Implant damage control is also possible by the use of HMM ion implantation, due to germanium PAI elimination.

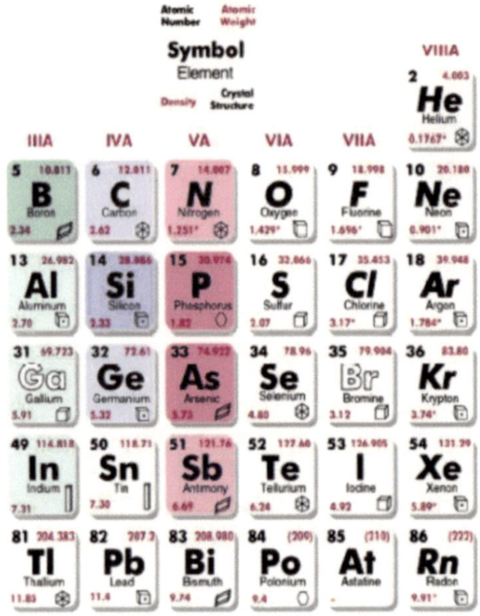

Table 1. Atomic dopant species that are commonly used in ion implantation.

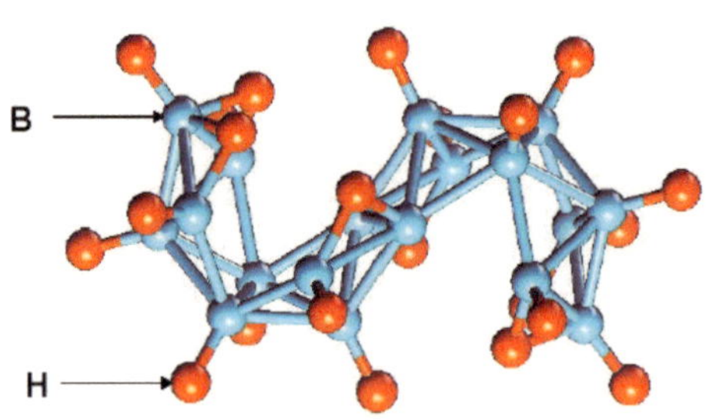

Fig. 1. Structure of Octadecaborane ($B_{18}H_{22}$) molecule.

Attempts have been made to implant molecular carbon ($C_{16}H_{10}$), of which the molecular structure is shown in Fig. 2, to replace Ge-PAI plus monomer carbon for advanced logic manufacturing. The latter mentioned monatomic implants are nowadays popular co-implants for USJ formation in the metal-oxide-semiconductor field-effect transistor (MOSFET) source and drain extension (SDE) doping process (Pawlak, 2006). Carbon is a standard co-implant in the MOSFET SDE implant sequence due to its capability of reducing dopant transient enhanced diffusion (TED). This is achieved through trapping of crystalline interstitial defects by the carbon atoms that are incorporated in the lattice substitutional sites (Carroll, 1998). Unless the silicon substrate is amorphized, the carbon implant could not be incorporated in the silicon lattice sites when undergone thermal annealing. In this case Ge-PAI is required, because carbon mass is too light to cause self-amorphization under most conditions. The use of molecular carbon opens up the possibility of replacing the traditionally used Ge-PAI, which is also known to leave residual damage leading to junction leakage.

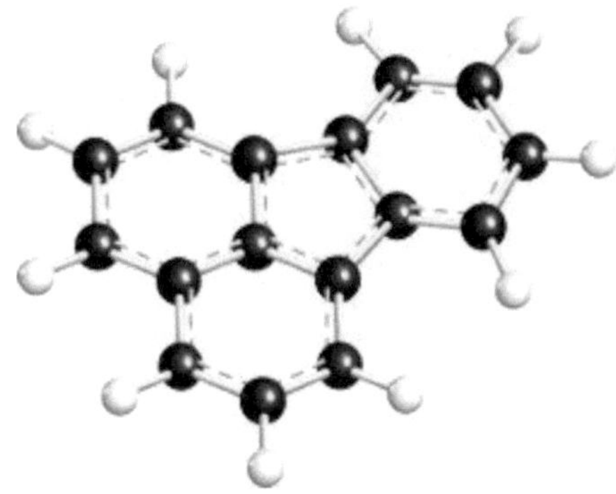

Fig. 2. Structure of $C_{16}H_{10}$ molecule

2. Overview

Semiconductor devices have become omnipresent due to their amazingly miniature in sizes, ever expanding functionalities in time, inexpensive manufacturing cost, and etc. Most of these reasons have to do with one historical event; the advent of commercial ion implanters. Impurity doping process is a major manufacturing step that needs to be repeated over and over for many times for the semiconductor material going from basic substrate to electrically functioning devices. Forty years ago, doping of semiconductor had been predominantly a thermal process, where the impurity is introduced at the substrate surface, and within a closed chamber at an elevated temperature, such as a furnace, the dopant atoms are allowed to diffuse into the substrate under a thermal equilibrium process. The speed, or the distance of impurity atom diffusion is dependent on the surface impurity concentration and process temperature. Usually, this temperature of operation is in the 1000 degree Celsius region. The atom diffusion energy is no greater than several eV's. This makes the doping process long and expensive.

Due to the advent of commercial ion implanters, the impurity doping process has shifted from predominantly thermally enhanced in nature to predominantly kinetically driven in nature. The impurity atoms are now being stripped of or bestowed with electrons in a part of the implanter called the "ion source", where they become ions to be accelerated in high electrical fields. Once the charge particles, or ions have gained the desired energies, they are collimated and then impinging into the substrate at high initial speeds. All of these actions are performed at room temperature. Although the process temperature for ion implantation is relatively low, the dopant ions acquire energies in the keV range. Therefore, the process time is less than one one-thousandth of that of a thermal process. Thus, the productivity is higher, and the cost is significantly lower too.

These advantages provide the semiconductor manufacturers with motivation to quickly adopt ion implantation in the process flow. They also give the process engineers and device engineers a lot of freedom to utilize the technique without having to wary of process constraints and tradeoffs too much. However, in the past decade, the process window, typically large enough for ion implant engineers to maneuver has gotten narrow, so narrow to a degree so that itself started to create problems, which none other process steps can resolve or compensate. These problems include insufficient dopant activation, co-implant species of choice, pre-amorphization implant species of choice, implant damage control, runaway low-energy implant cost

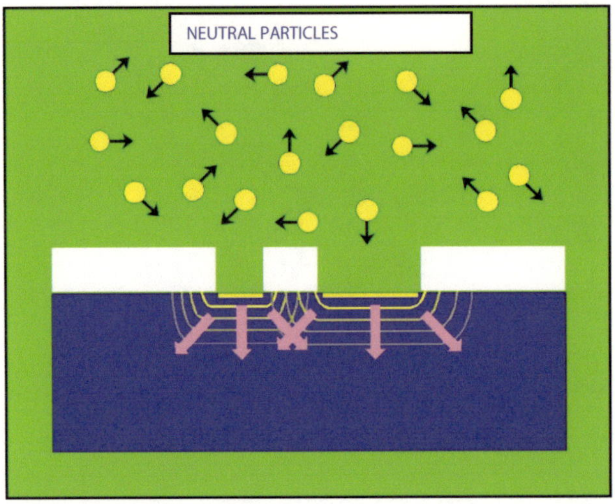

Fig. 3. a) Diffusion process is in thermodynamic equilibrium and energies are thermal (~eV) and random (isotropic)

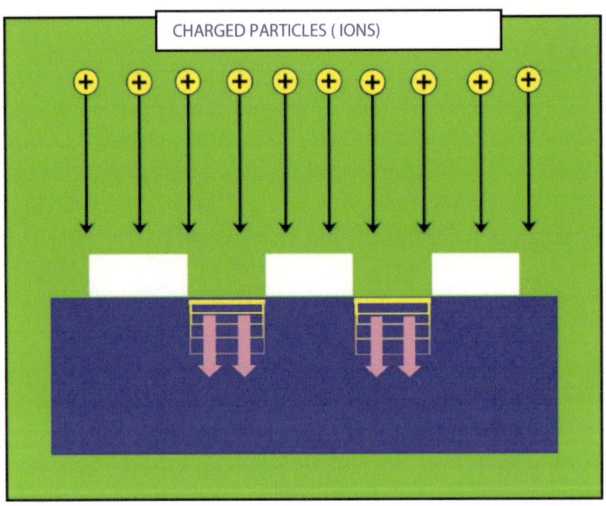

Fig. 3. b) Ion implantation is a process in which energetic, charged particles (atoms or molecules) are accelerated into the near surface of a target substrate at depths from ~10nm to ~1000nm (1 micron)

As time progresses, the process issue and cost issue are still the driving forces that motivate us to look at high mass molecular ion implantation, as oppose to monatomic implantation. However, among these two, the aspect of process requirements usually plays a dominant role in tool selection for semiconductor manufacturing. One obvious reason is that if people can quickly translate process benefits to device performance improvement, or geometry scaling down (in other words, device real estate saving), the cost it associated can be readily justified. In this chapter, we will also address the productivity and cost issue. However, the aspect of cost can hardly be the primary factor for choosing a process. We would like to make sure the production method we choose today can be extended to cover for the future needs. Only by taking the whole picture into consideration, then one can start to appreciate molecular ion implantation being a production method for now and the future.

3. The process issues of implant damage

An ion implant process is basically putting the dopant atoms into the silicon substrate by bombarding the silicon wafers with very energetic ions. This process would inevitable result in crystal damage. The implant damage can take many different forms, such as non-equilibrium excess of vacant lattice sites (vacancies) and self-interstitial atoms (interstitials), vacancy clusters, interstitial clusters, dopant-interstitial and dopant-vacancy clusters, and locally amorphized regions of the crystalline silicon target. Iso-valent ions such as Si, or Ge are sometimes implanted to intentionally take advantage of this collateral damage. The annealing of this damage, and the electrical activation of the implanted dopants, requires that the implanted target receive a subsequent heat treatment. The as-implanted defect configurations evolve during post-implant thermal processing, giving rise to transient enhanced dopant diffusion (TED), and the formation of relatively stable dislocation arrays, which if present in active device regions can lead to degradation of electrical performance. An understanding of all these phenomena is therefore crucial to the design of the implant recipe and the post-implant thermal treatment.

In advanced CMOS processing, this amorphous layer plays important roles for several purposes. The top three are, 1) dopant channeling prevention; 2) dopant activation enhancement; 3) end-of-range (EOR) defect reduction. In other words, they represent the properties of controlled junction depth; higher conductivities; and lower junction leakage currents in the CMOS device respectively.

Achieving an implant profile without appreciable channeling is of practical importance to avoid that slight differences in beam orientation across the wafer result in radically different implant profiles. There are three different methods to prevent implant from channeling. As shown in fig. 4 a) the first choice is by tilting the wafer, which is the easiest way to achieve if it serves the purpose. However, only at high energies, where the critical angles are relatively small, this method can be effective. At low energies, the tradeoff between the amount of angle being tilted and the compromise it incurs to implant profile starts to become significant. If a low tilt angle is not sufficient for preventing implant channeling, we may have to go to a higher tilt angle. On the other hand, the implant shadowing effect, which is caused by device surface topology blocking the incident beam at an angle, starts to get intolerable. Therefore, merely by tilting the wafer plane away from beam incident angle might not be effective. The second alternative is to use sacrificial oxide to prevent implant channeling. This is shown in fig. 4 b). Since ion implantation may also introduce metal contamination to the wafer, it has been a common practice to use a thin layer of sacrificial oxide, from 100Å to 200Å thick to block the elemental contamination from penetrating the wafer surface. Once the implant process is done, this layer of sacrificial oxide would be stripped of from a wet bench using buffered oxide etching solution. However, due to advanced devices are very sensitive to "substrate loss", or so to speak "dopant loss", people have begun to move away from using sacrificial oxide. Finally, the most inconvenient method for preventing implant channeling is, as depicted in fig. 4 c), by inserting a pre-amorphization implant before dopant implant. Usually, this implant species of choice is non-electrically active, or iso-valent atom, such as germanium or silicon. It is indeed an effective way to prevent implant channeling. The drawback is that it adds an additional implant step to the process.

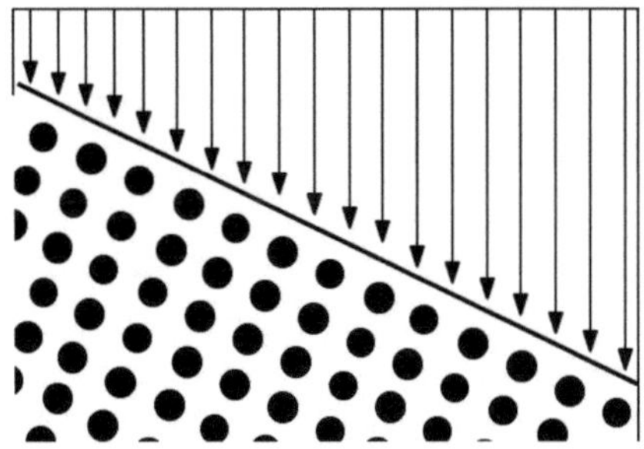

Fig. 4. a) Tilting wafer off the channeling plane.

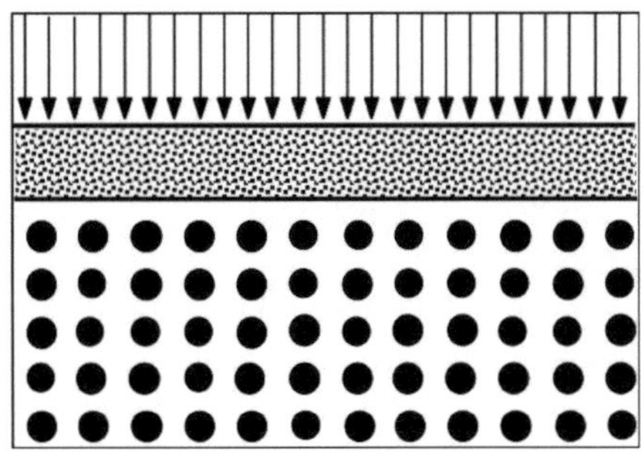

Fig. 4. b) Thin sacrificial oxide for randomizing the direction of incident ions.

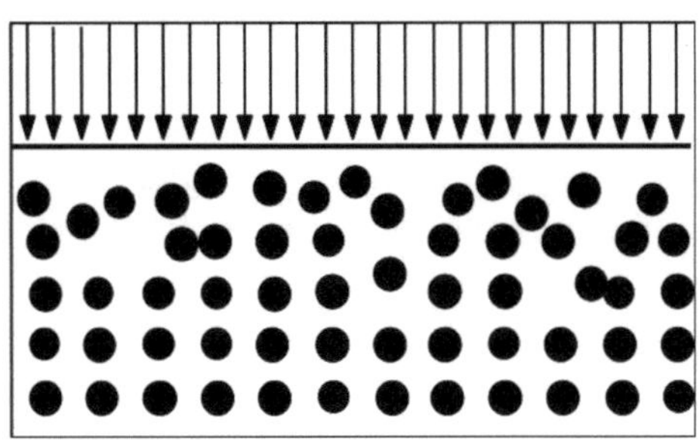

Fig. 4. c) Pre-amorphization implant to randomize the lattice atoms, thus destroy the crystal channels.

4. The process issues of thermal annealing

After ion implantation, the substrate needs to be treated with thermal processes. This is because the silicon substrate is damaged by ion bombardment, and needs to be "annealed", which is a thermal treatment to recover its crystalline structure. Meanwhile dopant atoms can be incorporated into the crystal lattice and become electrically active. As depicted in fig. 5, these two goals should be achieved simultaneously. Since this thermal treatment can also cause dopant diffusion, there would be some dopant redistribution.

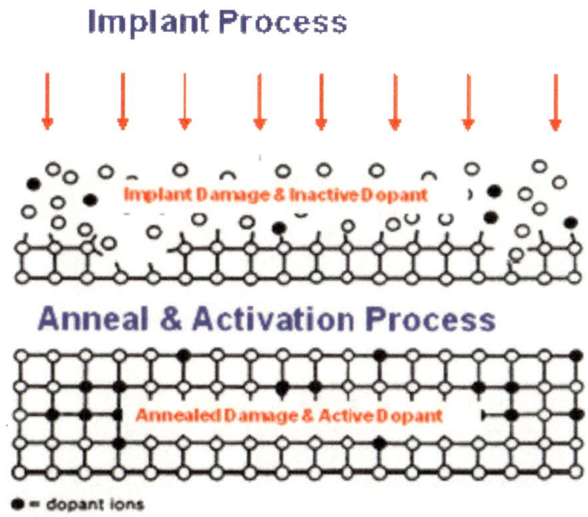

Fig. 5. The implant damage and inactive dopant atoms left in the silicon substrate need a post implant anneal to active the dopant and recover the crystalline structure.

From the logic manufacturing side, when the technology moved beyond 0.25um (deep sub-micron) era, the requirement of SDE is demanding USJ formation. This requirement is in response to the potential short channel effect (SCE) associated with device shrinkage. The geometry of device structure has to be tightly controlled now. In short, the "as implanted" dopant profile and dopant redistribution during anneal need to be well managed. For shallow junctions, dopant concentration levels can be very high. These implanted atoms tend to form high density crystal defects. The thermal budget for implant anneal has been greatly reduced for advanced logic devices due to the concern of excessive dopant re-distribution when the device is undergone high temperature thermal anneal. However, if the thermal budget is insufficient, the crystal defects could not be totally removed, and would lead to adverse effects on device performance, such as high device leakage currents.

It has been known for some time that boron diffusion can be enhanced by damage introduced by the implant process. For example, fig. 6 shows the enhanced diffusion of a boron marker produced by molecular beam epitaxy on a silicon substrate, which was subsequently damaged by $1x10^{14}$ cm^{-2} silicon implants at various energies and then subjected to a $950^oC/30s$ anneal. The enhancement scales linearly with the projected range of the implant which is approximately where the damage induced excess interstitials are initially located (Agarwal, 1997; Gossmann, 2000).

The phenomenon of transient enhanced diffusion (TED) after ion implantation increases the challenge of forming ultra-shallow junctions (Agarwal, 1997, 1999a, 1999b). Ion implantation leads to the displacement of silicon atoms from their lattice positions, creating pairs of vacancies and interstitials. During the initial stage of post-implantation annealing most of the vacancies and interstitials recombine leaving behind a net excess of interstitials approximately equal to the implanted ion dose; this is also referred to as the "+1" approximation (Giles, 1991). These excess interstitials quickly coalesce into extended defects, such as {311}'s (Eaglasham 1994; Stolk, 1997), or more stable dislocation loops. While these extended defects have lower free energy than individual interstitials (Eaglasham 1994; Rafferty, 1996), they are still metastable and dissolve with continued annealing. As they dissolve, they release excess interstitials into the lattice. Since boron diffuses by an interstitial mechanism (Gossmann, 1997) its diffusivity is enhanced by the excess interstitials with the time averaged diffusivity enhancement equal to the time averaged interstitial supersaturation . Both the interstitial supersaturation and the diffusivity enhancement end soon after the defects have dissolved. This phenomenon is depicted in fig. 7.

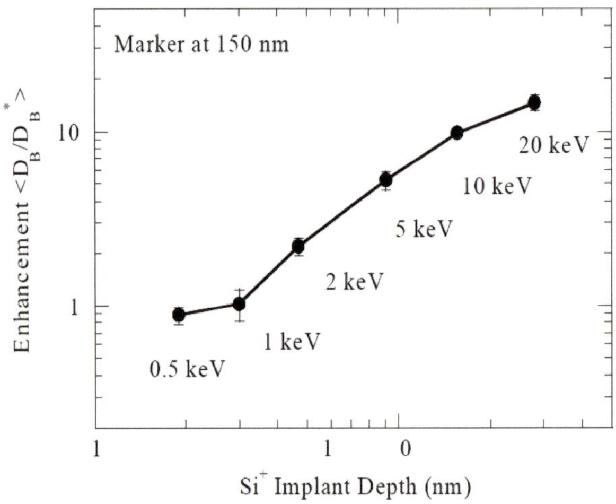

Fig. 6. Enhancement in diffusion of a boron marker layer, grown by molecular beam epitaxy during a 950°C/30s anneal, following implantation of $1x10^{14}$ cm^{-2} Si at various energies (Agarwal, 1997; Gossmann, 2000).

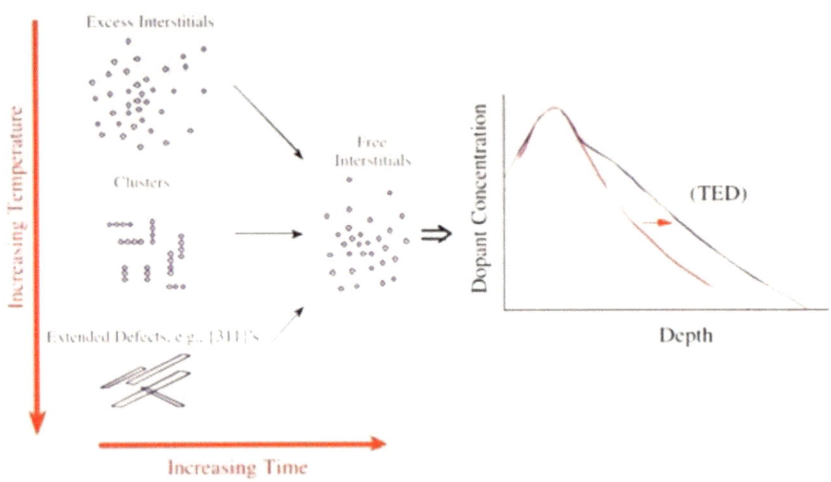

Fig. 7. Boron diffuses by an interstitial mechanism; its diffusivity is enhanced by the excess interstitials.

The increase in junction depth, Δx_j, due to TED to be expressed as (Gossmann, 1998; Rafferty, 1996)

$$\Delta x_j^2 \propto N \cdot R_p \cdot \exp[-(-1.4eV)/kT] \qquad (1)$$

where N is the number of interstitials trapped in the defects (approximately equal to the implanted dose) and R_p is the projected ion range (where the excess interstitials are initially located). The linear dependence on Rp has been demonstrated experimentally, as shown in fig. 6. The activation energy of Δx_j^2 is negative because the interstitial supersaturation due to the presence of the extended defects is larger at lower temperatures. This implies that the final junction will be deeper if the defects are annealed out at a lower temperature than at a higher temperature. This is a key reason why junction anneals are done in a rapid thermal annealing (RTA) rather than in a conventional furnace with a ramp-up rate of a few degrees per minute. An RTA spends significantly less time during the temperature ramp-up at lower temperatures where the diffusivity enhancement is larger.
Since the increase in junction depth due to TED depends on the implant dose (Eq. 1), it is possible that for a high dose implant some damage will remain after a fast ramp-up, allowing TED to continue during the ramp down (Agarwal, 1999). As the ramp-up rate is increased, the temperature at which TED runs out is pushed up until the TED is pushed over to the ramp-down side of the anneal (Agarwal, 2000). This is illustrated in fig. 8.

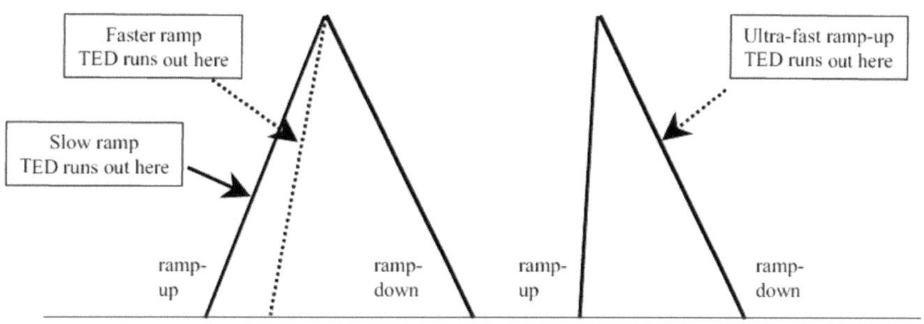

Fig. 8. Schematic illustration of TED continuing during ramp down of a spike anneal that is sufficiently fast (Agarwal, 2000).

In the sub-keV regime, there is more than one way to arrive at the same junction properties. It is very important to minimize the dose first, before reducing the energy further. The dependence of the sheet resistance and junction depth data on the different implant and annealing parameters is summarized in fig. 9. Increasing the ramp-up rate leads to a more shallow junction with higher resistivity. The same is also true when a smaller dose or energy is used. Modifying the implant parameters first helps avoid the risk of poor process repeatability which necessarily accompanies the use of higher ramp-up rates.
As the advanced logic manufacturers manage the implant and anneal together in an effort to meet the process requirements, the treadmill of device scaling is relentlessly pushing the implant dose higher and energy lower. The conventional USJ scaling is inevitably hitting the limits. The USJ formation for SDE is key for 65nm technology node and beyond (Foad, 2005). The obstacles include boron TED, low boron solubility limit in silicon, and most of all, post-anneal residual implant damage. For high dose applications, not all implant damage can be removed by the anneal process due to insufficient thermal budgets from "spike" RTA or ms laser spike anneal (LSA) processes. If this damage is in the wrong place, increased device leakage and catastrophic p-n junction shorts are probable. This scenario is depicted in fig. 10. Engineering the type, extent, and location of post-anneal residual implant damage is one of the primary objects of Front End of Line (FEOL) process integration.

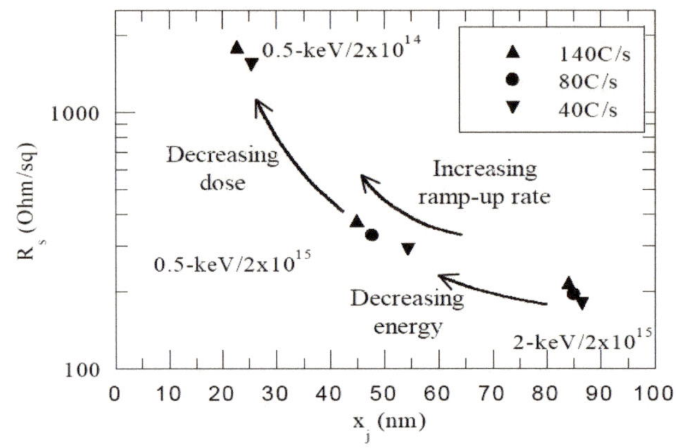

Fig. 9. Sheet resistance vs. junction depth as a function of ramp rate, implantation dose and implantation energy. Note the similarity between increasing the ramp-up rate or reducing the energy and dose (Agarwal, 1999, 2000).

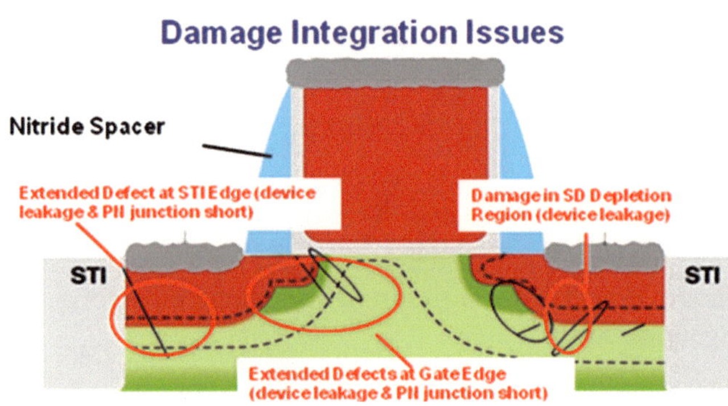

Fig. 10. When the EOR defect damage is in the wrong place, increased device leakage and catastrophic p-n junction shorts are probable.

5. Molecular implants

Molecular implants have long been considered by the IC manufactures as alternatives to atomic implants for low-energy applications (Jacobson, 2001). The major benefit of using molecular species implants is wafer throughput improvement due to higher effect beam currents when implanting at low energy. A molecular ion dissociates into its constituent atoms at the wafer surface. The constituent atoms then continue with a fraction of the total energy. This phenomenon can be utilized to gain wafer throughput in the sub-5.0keV range as implanters in general can deliver higher molecular beam currents at higher extraction voltages, and still provide equivalent processes to the low-energy monatomic implants.

A well-known and long-used example of this in production environments is BF_2^+ implantation as a means of delivering a lower effective energy boron as the molecular type of p-type dopant. More recent experimentation with molecular n-type dopants has

demonstrated that As_2 and P_2 can provide production-worthy beam current and throughput improvements with comparable process results (Chang, 2003).
The formation of aggressive n-type junctions has not posed as severe a challenge as p-type junctions in the past, due to the much larger atomic mass (75 amu for As, versus 11 amu for B) and lower diffusivity in Si. Arsenic dimer implant requires twice the ion energy of the monatomic implant. However, the effective fluence of a dimer implant is two times that of a monatomic implant, since both atoms in the dimer ion contribute to the total dopant dose. Therefore, it requires only half the dose of a monatomic implant. These conditions can be expressed by equations (2) and (3).

$$E_{eff} = E_{extraction} / 2 \tag{2}$$

$$I_{eff} = 2 \times I_{measured} \tag{3}$$

Since ion implanters can in general produce more I_{eff} (molecular) beam current than I_{eff} (atomic) beam current at $E_{extraction}$ under these operating conditions, a significant throughput advantage may in many cases be realized.

5.1 High mass molecular implants

In recent years significant advances have been made in the development of high mass molecular (HMM) beam sources for dopant implantations into silicon. The driver for the development of these sources has been the need for very low energy implants. Energy is partitioned between the atoms of a molecule in direct proportion to their mass. For example, the widely used molecular ion BF_2^+ with atomic mass ~49 having a single boron atom of mass ~11 results in the implantation of boron at an energy that is ~11/49 of the molecular ion energy, e.g. a 10 keV BF_2 implant, for example, is energetically equivalent to a 2.24 keV B implant.
A much more dramatic example of this energy partitioning may be achieved with decaborane ($B_{10}H_{14}$) (Jacobson, 2001) where a 10keV implant is equivalent to a ~1 keV implant. Recently, another large boron containing molecule, Octadecaborane ($B_{18}H_{22}$) has also been identified as a useful molecule for this application (Perel, 2001). It is important to note that with these molecules, one milliampere of ion beam current is equivalent to 10 (for decaborane) or 18 milliamperes (for octadecaborane) of boron current. For this reason the molecular beam obviates many of the space charge limitations associated with the ultra-low energy Boron beams. Conventional ion sources are not suitable for decaborane or octadecaborane implantation since the high arc chamber temperature causes disassociation of the molecule. Ionization chamber temperatures below 300°C are required and a different approach to electron impact ionization of the molecule is required. Figure 11 shows a commercially available octadecaborane ion source (Jacobson, 2005). Also, the ionization process results in a distribution of ions of the form $B_{10}H_x$ or $B_{18}H_x$ with the result that the mass resolved spectrum consists of a typically up to 10 peaks, all containing the same boron content but with varying hydrogen content. As a result, the acceptance of the mass resolving system must be increased to allow for maximum utilization of the available molecular ion current (Perel, 2001). Figure 12 gives a typical mass resolved spectrum obtained from a decaborane source (Jacobson, 2005).

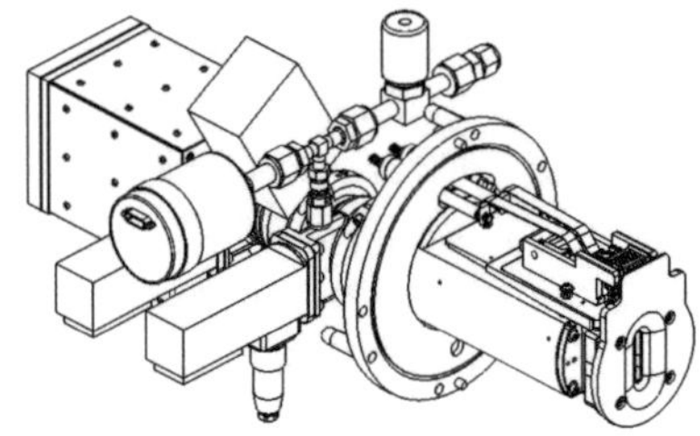

Fig. 11. Ion Source Suitable of Decaborane or Octadecaborane Ion Beam Generation (Jacobson, 2005).

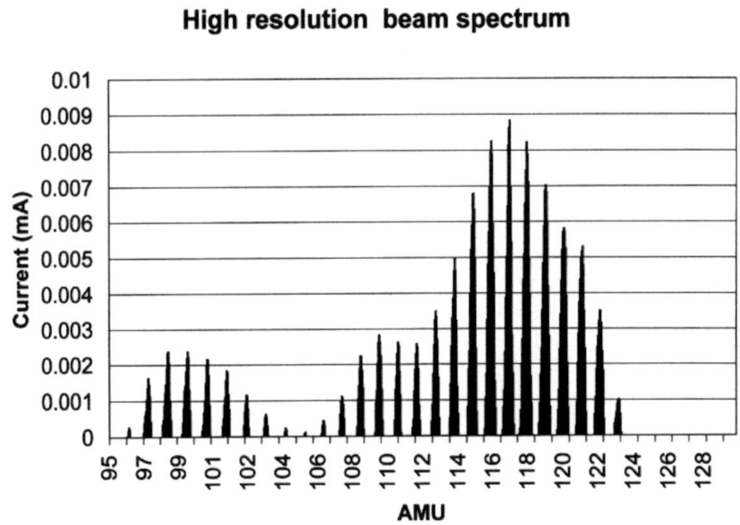

Fig. 12. Typical mass resolved spectrum obtained from a decaborane source.

5.2 High mass molecular implant application for DRAM

The aggressive scaling of DRAM puts severe constraints on the gate formation. Single work function polysilicon gate for PMOS with buried channel will suffer serious short channel effect as the scale shrinkage continues. Meanwhile, its high leakage is not tolerable for the requirements of low power high performance devices. The high leakage comes from the fact that the buried channel is away from the surface; hence, the gate can't control the channel as effectively as surface channel. As the dual work function poly gate shows the advantage of easiness of Vt control and resistance to short channel effects, Surface-channel PMOS with P+ poly gate will take substitution of buried-channel PMOS with N+ poly gate for advanced devices inevitably. Figure 13 shows the channel current flowing underneath the surface in a buried-channel PMOS device of the left, and on the surface in a surface-channel PMOS device on right.

Octadecaborane ($B_{18}H_{22}$) implant technology was evaluated for p+ poly gate doping process in a 72nm node stack DRAM device. For DRAM manufacturing, the 7x-nm-class is about the technology node where the device performance requires dual-poly gate structure for tuning the PMOS and NMOS work functions separately. Since the gate poly is in-situly dosed with n-type dopant during CVD polysilicon deposition, the PMOS gate poly needs to be doped heavily with p-type dopant afterwards, in order to counter dope the gate and transform it from originally n-type to p-type poly. Therefore, it requires low energy (< 5keV) and high dose boron implant (> 5×10^{15} /cm^3). The evaluation criteria were to improve the productivity of the process, which was initially built with conventional atomic boron implantation (^{11}B), while maintaining process equivalency. Before implanting into device wafers, process matching to conventional boron implant was done using both crystalline silicon and poly-silicon on Si wafers (Chang, 2008). For the crystalline silicon wafers, the R_s of blanket $B_{18}H_X^+$ implants were compared to that of atomic boron. For the poly-Si silicon wafers, SIMS dopant profiles were compared. For the device wafers, boron penetration, gate depletion, and final yield were compared. In addition, $B_{18}H_{22}$ implant splits of various energies and doses have been studied for their sensitivities to the electrical performance of the p-MOSFET in the 72nm node stack DRAM devices. In this study, we have demonstrated that $B_{18}H_{22}$ can provide up to 5× wafer throughput advantage over conventional atomic boron process due to much higher effective beam currents. Besides the significant productivity improvement, $B_{18}H_{22}$ implant device characteristics were well matched to the baseline atomic boron process.

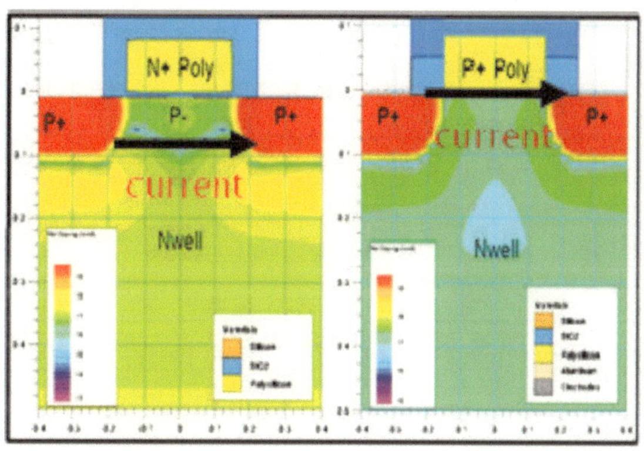

Fig. 13. The channel current flowing underneath the surface in a buried-channel PMOS device of the left, and on the surface in a surface-channel PMOS device on right.

In a BF_2^+ implant, the extraction energy is 49/11 times the desired Boron energy. Under the same principle, a $B_{18}H_{22}$ implant extraction energy is 210/11 times the desired Boron energy. These conditions can be expressed by equations (4) and (15).

$$E_{eff} = E_{extraction} \times \left(\frac{11}{210}\right) \tag{4}$$

$$I_{eff} = 18 \times I_{extraction} \tag{5}$$

Since ion implanters can in general produce more I_{eff} (molecular) than I_{eff} (atomic) at $E_{extraction}$ under these operating conditions, a beam current and thus throughput advantage may be realized. For example, a 2keV boron implant can be run using over 2.5mA of $B_{18}H_{22}^+$ beam current, or 45mA of effective boron current.

In this study, we used Axcelis' OptimaHD Imax implanter for molecular boron implants. The Imax was developed for ionizing, transporting and implanting molecular species such as $C_{16}H_{10}$ and $B_{18}H_{22}$. Figure 14 shows the R_s of B18 implant versus POR boron implant for the P+ gate poly process. The B18-implanted wafers require higher doses to match the POR R_s. The slightly under-dosing of the $B_{18}H_{22}$ implant in this case could be caused by a difference in dose retention between B18 and monomer boron. For low-energy implants, as dose increases, the fraction of dopant loss increases due to the sputtering, where near surface atoms leave the target during implantation due to recoil collisions. This phenomenon is depicted in fig. 15. While a detailed comparison of B18 and B has not been carried out, the retained dose of B18 as a function of energy has been reported (Harris, 2006). From the dose sensitivity test, a dose trim factor of 1.17 (17% higher dose) was determined for the P+ gate poly process, which has a lower target Rs.

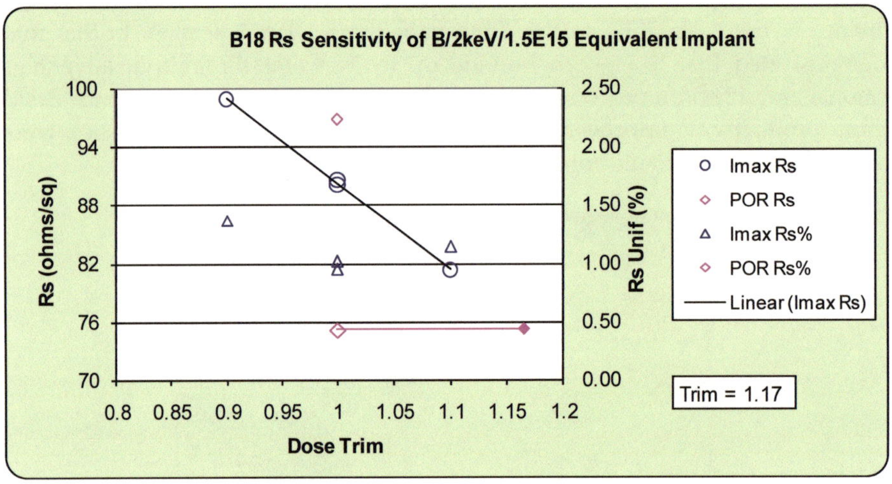

Fig. 14. P+ poly process R_s matching for the recipe of B/2keV/1.5×10^{15}cm^{-2}

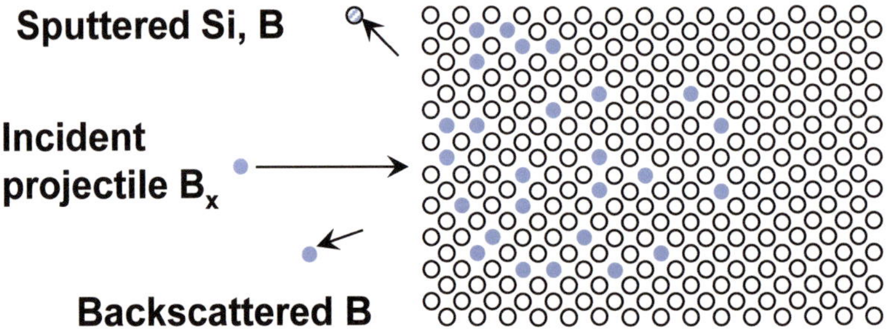

Fig. 15. For low-energy implants, as dose increases, the fraction of dopant loss increases due to the sputtering, where near surface atoms leave the target during implantation due to recoil collisions

In this test, wafers of poly implant conditions were subject to secondary ion mass spectrometry (SIMS) profile analysis. Figures 16 and 17 show the implant profiles of as-implanted and annealed implants from TPOR and Imax. The poly thickness is 90nmin this case. The annealing condition is RTP for a 20s soak at 965C. The implant dose for B18 has been adjusted to account for dopant loss. Meanwhile, the split conditions were designed for a process window check. Table 2 shows the comparison of the accumulated doses in SIMS.

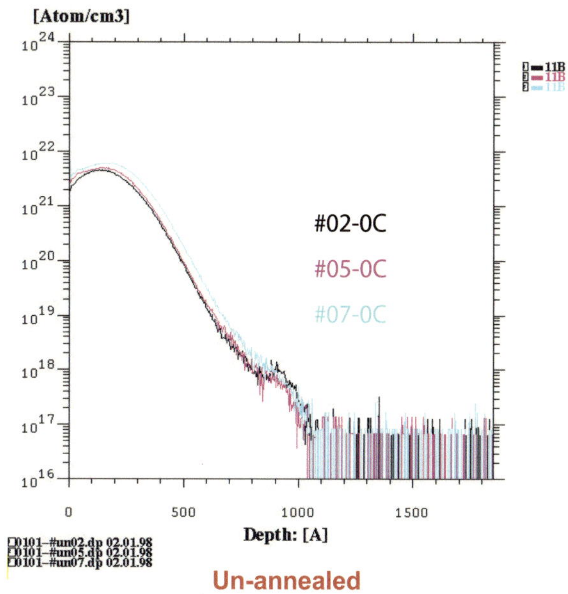

Fig. 16. As implanted SIMS profiles for B and B18 implants.

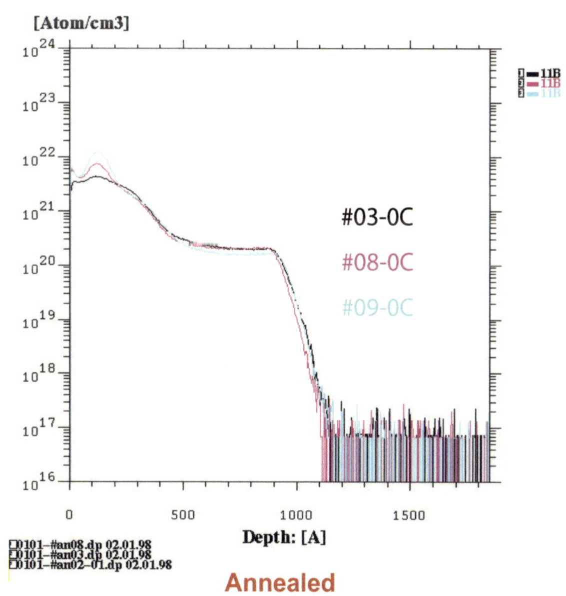

Fig. 17. Annealed SIMS profiles for B and B18 implants.

Remark	Sample no.	Dose (atom/cm^2)
POR (B11/4keV)	#2	1.202E+16
	#3 (anneal)	1.294E+16
IMAX (B18H22/76.8keV)	#5	1.382E+16
	#8 (anneal)	1.554E+16
IMAX (B18H22/80keV)	#7	1.828E+16
	#9 (anneal)	2.041E+16

Table 2. Accumulated SIMS dose for all samples.

Figure 17 shows that B18 implants seems to get a near surface bump as their signature. This could be due to the hydrogen effect. Since for every B18 ion implanted into the wafer, 22 hydrogen atoms would also be implanted. And hydrogen would enhance boron out diffusion. In some literatures, the possibility of hydrogen induced boron pile up in the surface has been discussed (Berry, 2008). Nevertheless, B and B18 implant profiles are matched at the oxide interface for as-implanted and annealed samples. Since the dopant concentrations match at the critical depth of the profile, we can view the SIMS profiles as matched in this case. Therefore, the implant matrix for the product wafers is to split the dose at target, ±10%, ±20% and ±30% for the P+ poly doping recipe. Device PMOS Vth does have a trend corresponding to different dosages. As the dosage gets high, the Vth gets high too. However, the biggest deviation is less than 10mV, we can say that the device results are all meeting the specification (Chang, 2008).

5.3 Molecular implant applications for advanced logic

As device scaling continues previously acceptable implant technologies for p-MOSFET SDE are struggling to meet advanced device requirements. There are three metrics that must be simultaneously achieved; those are device leakage, p-type dopant activation and junction depth control. In order to meet all of these goals, we found that molecular carbon implant is particularly well suited for USJ formation of the p-MOSFET SDE.

Due to preserving device geometry is of primary importance, junction depth control is the first thing to consider. Recent years, people have started to use carbon implant to suppress boron TED. The reason is that when carbon concentration is high enough (above $1\times10^{19}cm^{-3}$), it would create an interstitial "under-saturation" region (Carroll, 1998) (Moroz, 2005). Therefore, boron dopant atoms would less likely to be "kicked-out" by the excessive interstitials in the lattice, and implant profile remains stable during annealing. In order to incorporate carbon into silicon, the implant layer needs to be fully amorphized before annealing. Therefore, germanium pre-amorphization implant (Ge-PAI) was inserted in the process flow. Although it is a common practice to use Ge-PAI now, we all know that Ge-PAI is problematic due to it results in elevated end-of-range (EOR) defect damages, which have been identified as the leakage source for the devices. In the light of this concern, we put the constraints on Ge-PAI usage, so that it would not impact the junction quality. However, the trade-off between limiting Ge-PAI dosage and excessive residual implant damage may lead to an insufficient amorphous layer for carbon incorporation.

The other way to get around of this problem would be to increase the carbon implant dose, so that it reaches the critical dose for the formation of amorphous layer. However, carbon also leaves behind point defects (Mirabella, 2002), and causes device leakage. Although the effect of these point defects left behind by carbon implant are still under investigation, the increase in sheet resistance is observable. This is due to carbon diffuses predominantly by a "kick-out" mechanism. If carbon concentration is too high, it would unavoidably compete with boron dopant atoms for occupying lattice sites, and kick the already electrically active boron atoms out of the lattice sites. Therefore, the use of carbon should be evaluated of its pro's and con's. If we go beyond a certain dosage of carbon, the benefits of activation improvement and diffusion suppression would be compromised by the excessive implant damage and dopant deactivation.

Since High Mass Molecular (HMM) implants have been known to create an amorphous layer as effectively as the heavy ion species (Krull, 2006), implanting molecular carbon is a potential technique to replace the process steps of Ge PAI plus monomer carbon implant. $C_{16}H_{10}$ is shown to be a consistently self-amorphizing method for introducing carbon into the extension region.

In a preliminary study, we used Axcelis' OptimaHD Imax implanter for molecular carbon implants. We proved that a single implant of $C_{16}H_{10}$ can effectively replace a two step Ge + C implant sequence. As logic device technologies advanced into the 40nm node, USJ requirements became very stringent. The x_j target of p-MOSFET SDE implant is very aggressive, less than 20nm per ITRS roadmap (ITRS 2005). In order to meet these requirements, both the implant and anneal of p-type species need to be considered simultaneously because their interaction is essential to the desired outcome. The process of record (POR) for Ge +C in this case is a Ge/12keV/1×10^{15}cm^{-2}+ C/2.5keV/1×10^{15}cm^{-2} implant sequence. We compared the B/400eV/1×10^{15}cm^{-2} implant Rs-Xj results with the presence of the Ge + C, against $C_{16}H_{10}$ implant of the equivalent carbon dose and energy. Figure 18 shows an XTEM image of a $C_{16}H_{10}$ implant at 2.5keV per carbon atom, with1×10^{15} cm^{-2} dose. The amorphous layer is around 12.9nm, whereas, the projected range of this carbon implant is at 10.2nm, according to SRIM. This result is in line with the data previously published (Mirabella, 2002), and sufficient for the purposes of this study.

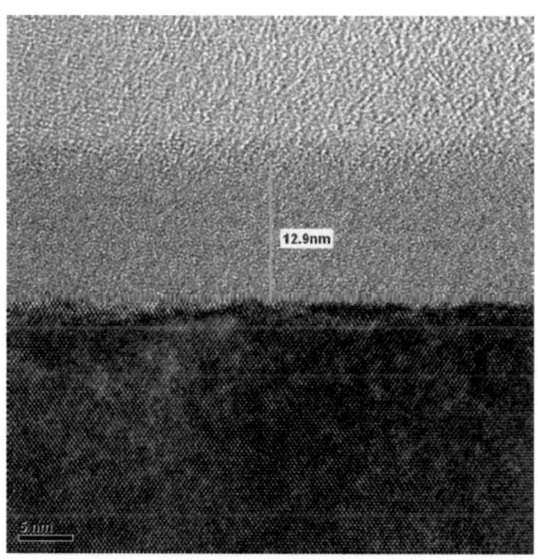

Fig. 18. XTEM image of a $C_{16}H_{10}$ implant at 2.5keV per carbon atom, with 1×10^{15}cm^{-2} dose

For the case of laser spike annealing (LSA) only, a comparison of POR co-implant against $C_{16}H_{10}$ implant effect on the boron SDE implant is made in figure 19. The R_s vs. X_j of the two implants indicate that if LSA only was used, it is easy to achieve the advanced logic process target. The Rs of the boron SDE implant with the one step $C_{16}H_{10}$ implant is comparable to that of the Ge + C co-implant's. However, one can see that monatomic co-implants may still be insufficient for suppressing the boron diffusion above 15nm deep in the substrate. Although the amorphous layer created by Ge/12keV/1×10^{15}cm^{-2} is around 20nm, the total defects it creates could provide a lot of interstitials in the deeper region. If one pays attention to the boron profile, one can see the characteristic signal of the amorphous layer and crystalline layer interface at around 20nm deep. The carbon atoms would segregate at this interface, and influence the subsequent boron diffusion. However, one can argue that the tail region of the annealed boron profile for the Ge + C co-implanted case, being slightly higher at around 15nm is beyond the p-n junction. No matter how the defect damage is distributed, we would still expect that the one step $C_{16}H_{10}$ implant should cause much less implant damage and easier to be annealed. Frontier Semiconductor provides a metrology system that measures the non-contact sheet resistance, and leakage current, called RsL. The RsL leakage current measurement for Ge + C co-implanted USJ shows an average of 28 uA/cm^2 in this case. And the RsL leakage current measurement for Ge + C co-implanted USJ shows an average of 0.7 uA/cm^2 in this case. This is only one fourth of the leakage current from POR.

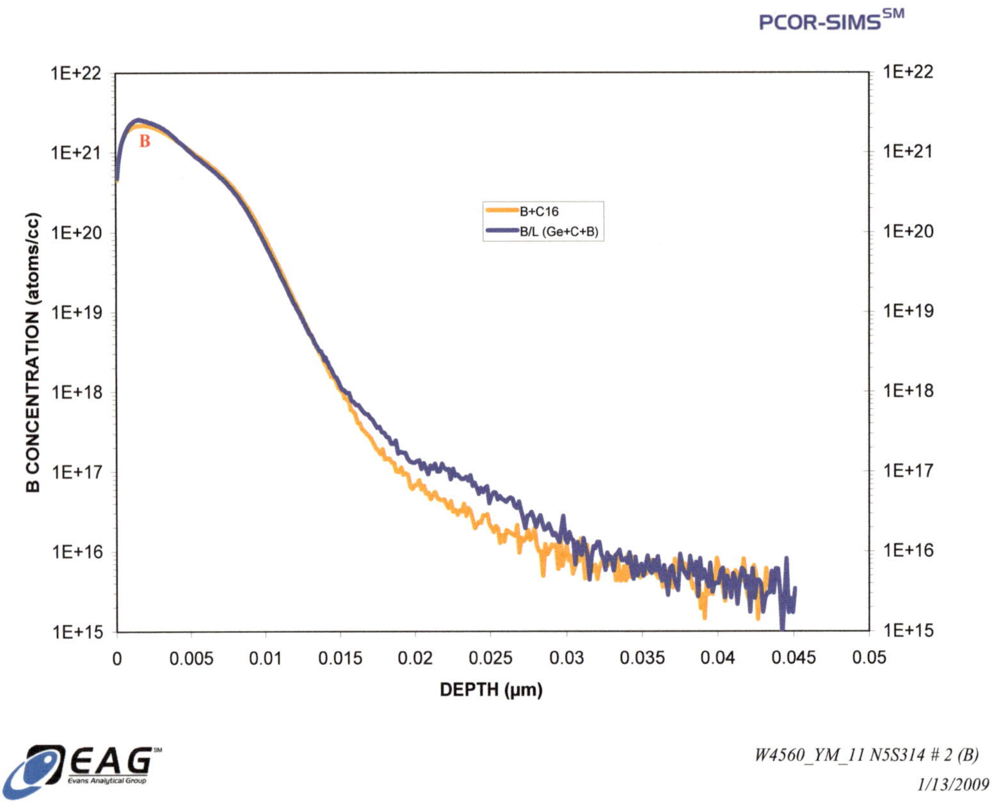

Fig. 19. Comparison of the B/400eV/1×10^{15}cm^{-2} LSA annealed dopant profile with the presence of the Ge + C, and $C_{16}H_{10}$ implant. The POR is a Ge/12keV/1×10^{15}cm^{-2}+ C/2.5keV/1×10^{15}cm^{-2} implant sequence, and $C_{16}H_{10}$ implant is of the equivalent carbon dose and energy.

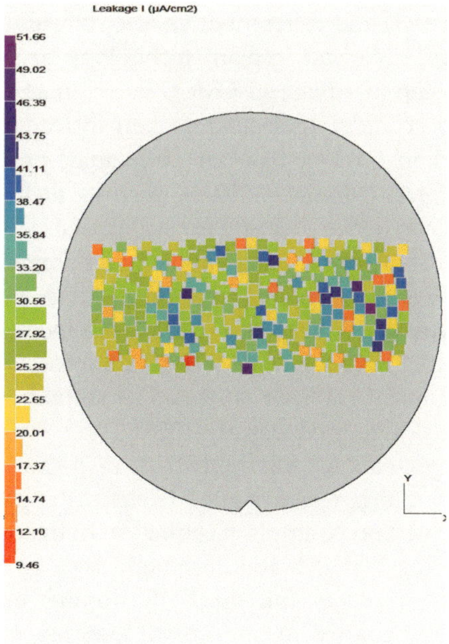

Fig. 20. RsL leakage current measurement for Ge + C co-implanted USJ shows an average of 28 uA/cm^2 in this case.

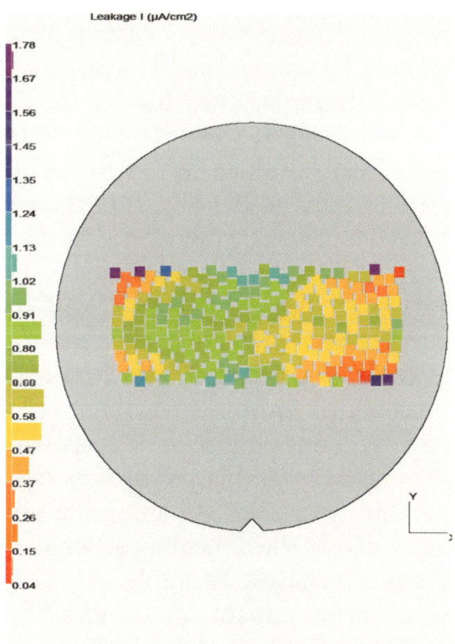

Fig. 21. RsL leakage current measurement for $C_{16}H_{10}$ co-implanted USJ shows an average of 0.7 uA/cm^2 in this case.

We also investigated the combination of $C_{16}H_{10}$, and $B_{18}H_{22}$ implants for USJ formation in a p-MOSFET SDE doping process for a 40nm logic device. We studied the split condition of

various energies, beam currents, and different advanced annealing schemes. The objective of this study is to use molecular carbon implant technology to supersede monomer carbon implants as a new process step in advanced CMOS device manufacturing. There are several reasons for the industry to consider molecular carbon instead of monomer carbon. First, conventional monomer carbon implant has poor implanter productivity. Secondly, carbon implants may have side effects (Mirabella, 2002), such as their competition with electrical dopant for substitutional silicon lattice sites and formation of excessive point defects, and incur penalties as well as benefits. Therefore, its adoption requires complicated integration schemes.
The purpose of this study was on developing the future USJF. Since the annealing program could be altered and the thermal budget be reduced, the focus was put on the interaction between implant and anneal. There are three different annealing programs involved in this study. The first one is a millisecond laser anneal. The second and the third programs are with spike RTP with the peak temperatures at <1000°C and >1000°C, and followed by laser anneal. We denote them as anneal "A" and anneal "B" respectively. In the blanket wafer test part, an implant and anneal matrix was designed to study the possibility of using $C_{16}H_{10}$, to replace the 2-step Ge-PAI + carbon co-implant sequence. In the device wafer test, we use the p-MOSFET of 40nm node logic, which requires high dose and low energy BF_2 implant, along with three other co-implants for the SDE doping process. In this study, the productivity of $B_{18}H_{22}$ for low energy boron implant was also evaluated. We first focus on the process matching of $B_{18}H_{22}$ to the recipe of 3keV BF_2^+ in the process of record (POR). There is also a 2-step Ge-PAI + carbon co-implant sequence precedes the BF_2 SDE implant. In the subsequent annealing process, both RTP spike and LSA annealing are applied in this case. Since there is fluorine in the BF_2 implant, which is known to affect the boron doping profile during anneal, the $B_{18}H_{22}^+$ energy may need some adjustment to reflect the difference in the boron diffusion profile from the influence of fluorine.
If the conventional co-implants were replaced by $C_{16}H_{10}$, the R_s could be further improved when millisecond laser anneal was applied. This offers the process solution to the LSA only scheme. We expect lower device leakage since Ge-PAI was eliminated. In this case, a light RTP spike anneal was applied to remove the implant damage. Although the molecular carbon implant appears to have the process equivalency as the conventional co-implants, it has lost the process advantages in R_s reduction as shown in the LSA only case. Figure 22 shows the 350eV boron post anneal dopant profile of different annealing schemes. As expected, the xj increases in accordance to RTP temperatures. LSA offers diffusionless anneal, and it only shifts the profile for no more than 2nm deeper, and gets the best sheet resistance. If the spike RTP was added prior to LSA, the profile would shift from 5 to 7nm for "A" annealing scheme and "B" annealing scheme respectively. In figure 23, the 350eV boron implant of the 2-step co-implant is compared against the $C_{16}H_{10}$ co-implant. The xj of these two implant schemes all shift 5nm after "A" annealing scheme. We can conclude that, even with the light spike RTP added in the annealing scheme, molecular carbon co-implant would behave the same as the monatomic co-implants. The reason is that millisecond anneal, although can activate boron dopant atoms effectively, it doesn't remove the excessive interstitials resulted from implant damage due to limited thermal annealing. When a spike RTP in the 1000°C regime was applied, the implant induced EOR defect damage would resolve and release the interstitials, which allows the boron TED to run out its course, due to sufficient thermal energy. Therefore, the self-amorphization property of the molecular $C_{16}H_{10}$ implant may not bring process benefits to p-type USJ formation, unless a diffusionless annealing scheme is employed.

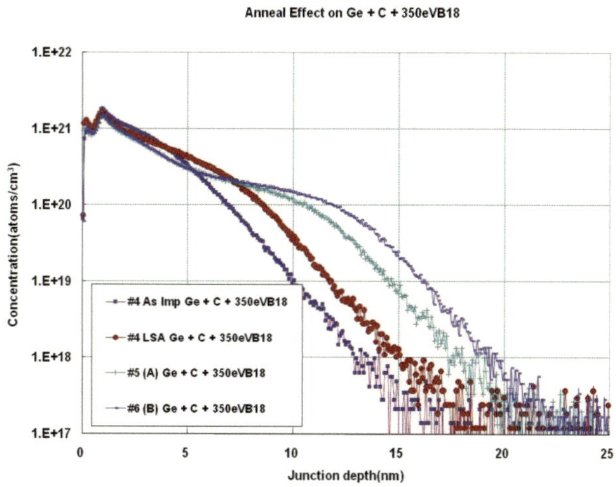

Fig. 22. The 350eV boron post anneal dopant profile of different annealing schemes. LSA is a millisecond laser anneal. Anneal "A" is a <1000°C spike RTP followed by a millisecond anneal; and anneal "B" is a >1000°C spike RTP followed by a millisecond anneal.

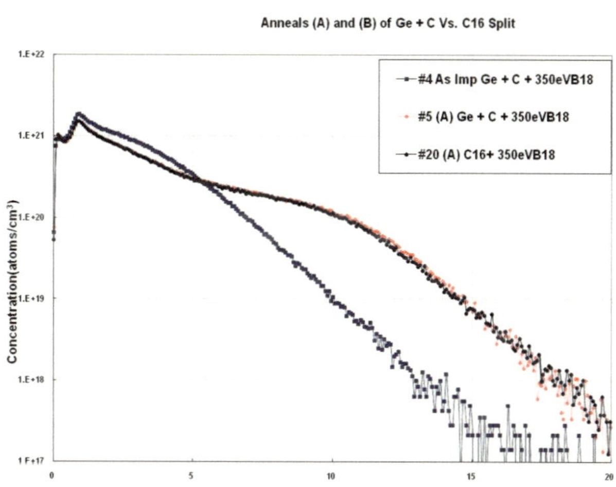

Fig. 23. The 350eV boron post anneal dopant profile of different annealing schemes, and of the Ge-PAI plus mono-atomic carbon 2-step co-implant versus the $C_{16}H_{10}$ co-implant.

Figure 24 shows the overlap capacitance of $C_{16}H_{10}$ plus $B_{18}H_{22}$ implanting into the SDE region of a 40nm logic device. In comparison to the POR implant matrix shown on the left side, $B_{18}H_{22}$ direct replacement of BF_2 as the boron dopant in the POR appears to have a much higher C_{ov}. This indicates that $B_{18}H_{22}$ diffuses faster than BF_2 in the RTP plus LSA annealing scheme. The main reason should be due to the presence of fluorine in the BF_2 implant, which also plays a role in boron TED suppression. However, if the $C_{16}H_{10}$ implant is employed instead of the conventional co-implants, the C_{ov} is restored. In short, molecular implants can at least be shown to have process equivalency even if the annealing scheme is not in favor of molecular implants. On the other hand, $C_{16}H_{10}$ has been shown as a valid replacement for current POR co-implants for PMOS SDE. It not only can be easily integrated into the existing process nodes for Ge-PAI replacement, but also allows a smooth transition to a smaller thermal-budget or diffusionless annealing scheme in the future.

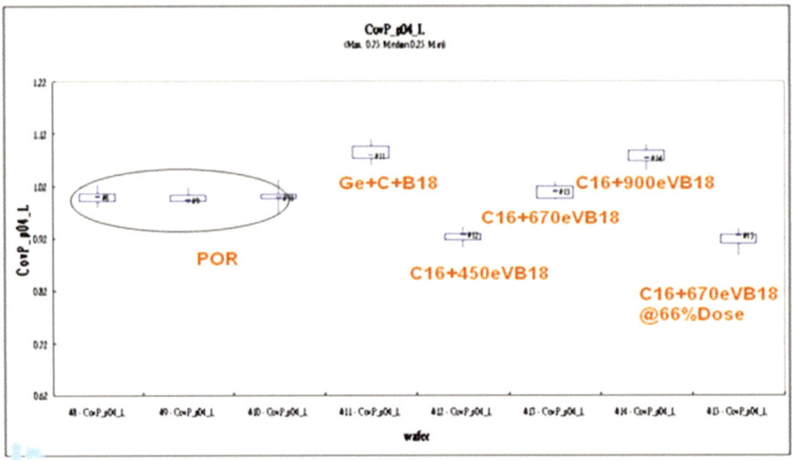

Fig. 24. Overlap capacitance of $C_{16}H_{10}$ plus $B_{18}H_{22}$. implanting into the SDE region of a 40nm logic device, where the POR is mono-atomic Ge-PAI plus C, followed by a BF2 SDE implant.

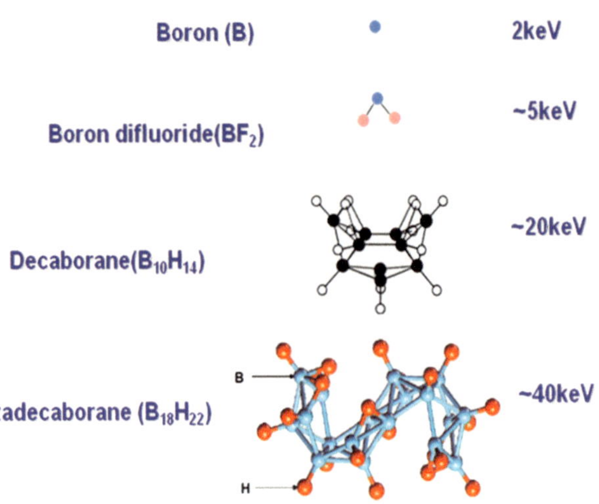

Fig. 25. The 2keV equivalent boron energy beams of BF_2, $B_{10}H_{14}$, and $B_{18}H_{22}$, could be extracted from the source and travel in the ion implanter beam line at an energy of 5keV, 20keV and ~40keV respectively.

6. Conclusion

Owing to the advent of High Mass Molecular implant technology, semiconductor manufacturing fabs now have an opportunity to leap forward in making great productivity and process improvement by utilizing its unique properties of effective beam transportation in the ion implanter beam line, and self-amorphization during process. The amorphous layer could be formed at a relatively lower dose for the HMM implant to avoid the side effects of excessive implant damage. By proper tuning the molecular carbon implant, we can show process equivalency to conventional co-implant scheme, and utilize it for the 40nm p-MOSFET device. Should the annealing scheme be flexible, and the carbon implant only sees the LSA as the post-implant anneal, the adverse effects of carbon implant, such as its competition against boron dopant for activation, etc, can be nullified. Figure 25 shows the industrial trend in the past 40 years for choosing the primary p-type dopant. This trend is in response to the demand for low energy boron implant. Fortunately, people always find a production worthy solution when the request becomes imminent. Even though using BF_3 as the source material is problematic, the industrial people are still clinging to it due to the benefit of higher productivity than monatomic boron for low energy operation. It is authors' believe that as long as Moore's law still holds, both productivity and process issues will compel fab engineers to migrate to the next generation of p-type dopants. It would be only natural for such evolution to take place. Just as sometimes in the past, we migrated from monatomic boron to small molecular BF_2. This time we are just going from BF_2, the smaller molecule to $B_{10}H_{14}$ or $B_{18}H_{22}$, bigger molecules. For the last transition, the productivity improvement was noticeable, but not awesome, due to either boron or BF_2, only has one dopant atom in it. But for this transition, the productivity would get improved from ten to twenty times, due to the HMM $B_{10}H_{14}$ or $B_{18}H_{22}$ ions contain that many more dopant atoms. This would be more than just an evolutionary change. It is so significant a leap for the ion implantation technology so that it should to be deemed as a revolutionary change for the silicon manufacturers to make.

7. Acknowledgment

The author would like to thank Hans Cheng of TSMC, Phoenix Kuo of UMC, and Jay Huang of Rexchip for their many years of collaboration with us in the field of HMM ion implantation for DRAM and advanced logic device manufacturing. Most of all, I would like to extend my appreciation to P.S. Chen and S.H. Yang of NTC, who actually adopted $B_{18}H_{22}$ ion implantation in their 70nm DRAM production for the P+ poly implant process.

8. References

Agarwal, A., et al., (1997), *Reduction of transient diffusion from 1-5 keV Si+ ion implantation due to surface annihilation of interstitials, Appl. Phys.* Lett. 71, 3141.

Agarwal, A., Fiory, A.T., Gossmann, H.-J., Rafferty C.S., Frisella, P., Hebb, J., (1999) *Ultra-shallow junction formation by spike annealing in a lamp-based or hot-walled RTP system: effect of ramp-up rate,* Mater. Sci. in Semicond. Proc. 1: 237.

Agarwal, A., Gossmann, H.-J.,and Fiory, A.T., (1999) *Ultra-shallow junctions by ion implantation and rapid thermal annealing: spike-anneals, ramp rate effects,* Mater. Res. Soc. Symp. Proc. 568, 19.

Agarwal, A., Gossmann,H.-J., Fiory, A. T., Venezia, V. C. and Jacobson, D. C., (2000) *Ultra-shallow junction formation using ion implantation and rapid thermal annealing: physical and practical limits,* ECS PV 2000-9.

Agarwal, A.,et al., (1997) *Boron-enhanced diffusion of boron: the limiting factor for ultra-shallow junctions,* IEDM Tech. Digest, 467.

Ameen, M. S., et al, (2008) *Properties of ultra-low energy boron implants using octadecaborane,* J. Vac. Sci. Technol. B 26(1), p. 373,

Berry, I. L., Waldfried, C., Han, K., Luo, S., Sonnemans, R., Ameen, M., *Effects of resist strip and clean on USJ performance,* 8th International Workshop on Junction Technology.

Carroll, M. S., et al., (1998) *Complete suppression of boron transient-enhanced diffusion and oxidation-enhanced diffusion in silicon using localized substitutional carbon incorporation,* Appl. Phys. Lett., 73(25), 3695

Chang, B., et al., (2002) *Arsenic dimer implants for shallow extensions in 0.13mm devices,* 14th Int'l Conf. on Ion Impl. Tech. Proceedings, B. Brown et al, Eds. IEEE Press, Piscataway, NJ, 111.

Chang, B., et al., (2008) *Octadecaborane Implant Technology for 72nm node Stack DRAM p+ Poly Gate Doping Process,* 8th International Workshop on Junction Technology.

Foad, M., Graoui, H., (2005) *Extending existing fab equipment for 65nm node ultra-shallow junction formation, Semiconductor Manufacturing,* Vol. 6, Issue 8, August 2005

Eaglesham, D. J., Stolk, P. A., Gossmann, H.-J., and Poate, J. M., (1994) *Implantation and transient B diffusion in Si: the source of the interstitials,* Appl. Phys. Lett. 65, 2305

Giles, M. D., (1991) *Transient Phosphorus diffusion below the amorphization threshold,* J. Electrochem. Soc. 138..

Gossmann, H.-J., (1998) *Dopant and point defects during Silicon Processing,* SemiconductorSilicon, H. R. Huff, U. Goselle, and H. Tsuya ed, ECS Proc. Vol. 98-1, 884.

Gossmann, H.-J., Rafferty, C. S., and Keys, P., (2000) *Junctions for deep sub-100nm NMOS: how far will ion implant take us?* Mater. Res. Soc. Symp. Proc. 610, B1.2.1.

Gossmann, H.-J., (1997) *The interstitial fraction of diffusivity of common dopants in Si,* Appl. Phys. Lett., 71: (26), 3862.

Harris H., et. al. (2006) *Dose retention effects in atomic boron and clusterboron (B18H22) implant processes,* Proc. XVI Int'l. Conf. on Ion Implantation Technology, p. 155

ITRS 2007 Front End Process Update, http:/www.itrs.net/Links/2006Update/FinalToPost/07_FEP2006Update.pdf.

Jacobson D. C., et al., (2005) *Using boron cluster ion implantation to fabricate ultra-shallow junctions,* Ext. Abs. of 5th Inter. Workshop on Junction Technol. S2-1.

Jacobson, D. C., et al, (2001) *Decaborane, an alternative approach to ultra low energy ion implantation,* 2000 Conf. on Ion Implantation Technology Proceedings, H. Ryssel, L. Frey, J. Gyulai, H. Glawischnig, Eds, IEEE Press, Piscataway, N.J., 300.

Jacobson, D. C., (2005) private communication, SemEquip Co. Krieger, G., et al., (1991) *Moderately doped NMOS (M-LDD)-hot electron and current drive optimization,* IEEE Trans. Elec. Dev. 38, 121.

Krull, W.A., et al., (2006) *Simplifying the 45nm SDE process with clusterboron and clustercarbon implantation,* International Conference of Ion Implantation Technology. P. 142

Lindhard, J., (1965) *Influence of crystal lattice on motion of energetic charged particles,* Kgl. Dan. Vid. Selks. Mat. Fys.Medd. 34, 1,.

Moroz, V., Foad, M., Graoui, H., Nouri, F., Pramanic, D., Felch, S., (2005) *Ultra-shallow junction for the 65nm node based on defect and stress engineering,* Matter. Res. Soc. Proc. Vol.864. E3.5.1

Mirabella, S., et al., (2002) *Interaction between self-interstitials and substitutional C in silicon: interstitial trapping and C clustering mechanism,* Phys. Rev. B65 045209

Perel, A., et al., (2001) *Decaborane ion implantation,* 2000 Conf. on Ion Implantation Technology Proceedings, H. Ryssel, L. Frey, J. Gyulai, H. Glawischnig, Eds, IEEE Press, Piscataway, N.J., 304.

Rafferty, C. S., Gilmer, G. H., Jaraiz, M., Eaglesham, D. J., and Gossmann, H.-J., (1996) *Simulation of cluster evaporation and transient enhanced diffusion in silicon,* Appl. Phys. Lett. 68: 2395.

Stolk, P. A., et al., (1997) *Physical mechanisms of transient enhanced dopant diffusion in ion-implanted silicon,* J. Appl. Phys. 81: (9), 6031.

6

Infrared Spectroscopic Ellipsometry for Ion-Implanted Silicon Wafers

Bincheng Li[1] and Xianming Liu[1, 2]
[1]*Institute of Optics and Electronics, Chinese Academy of Sciences, Chengdu*
[2]*The Key Laboratory for Optoelectronic Technologies and Systems of Ministry of Education, Chongqing University, Chongqing*
China

1. Introduction

Spectroscopic ellipsometry (SE) is an optical technique that measures changes in the reflectance and phase differences between the parallel (*Rp*) and perpendicular (*Rs*) components of a polarized light beam upon reflection from a surface. As a non-contact and non-destructive optical method, SE is widely used to determine the thickness and optical constants (refractive index n and extinction coefficient k) of films or layered structures, as well as other properties related to optical constants.

Spectroscopic ellipsometry was firstly applied to analyze the optical properties of implanted silicon wafer back in 1979. From then on the SE technique has been widely used in characterization of mono-, micro-, and poly- crystalline, and amorphous silicon wafers, and in implantation and annealing process monitoring. Up to now, most researches were focused on the visible spectral range as implantation induced lattice damage altered the optical properties of implanted silicon wafers in the visible range. However, when the implanted silicon wafers were thermally annealed, their optical properties in the visible range were restored to that of monocrystalline silicon. Visible SE can no longer distinguish between silicon wafers implanted with different dose or energy. On the other hand, Infrared spectroscopic ellipsometry (IRSE) could be a sensitive characterization technique for implanted silicon wafers, as in the infrared spectral range, the optical properties of the implanted silicon wafers are functions of the activated impurity concentration, which is a function of the implantation dose or energy.

In this chapter, the principle of ellipsometry is introduced briefly in section 2. The application of visible spectroscopic ellipsometry to characterize silicon wafers is reviewed in section 3. In section 4, IRSE spectra of implanted silicon wafers with and without thermal annealing are analyzed comparatively. An equivalent optical model is established to fit the IRSE spectra for implanted and annealed silicon wafers. Finally, a summary is presented in section 5.

2. Principle of ellipsometry

Spectroscopic ellipsometry is a sensitive optical technique for determining optical and structural properties of layered surfaces and thin films. In this section, a brief introduction to the principle of ellipsometry measurement is presented.

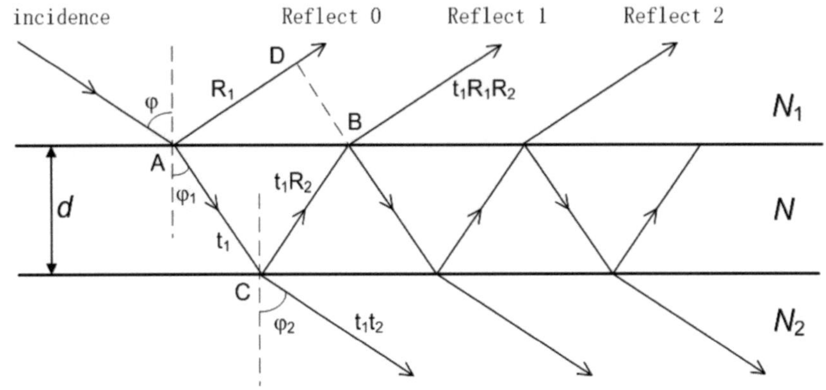

Fig. 1. Reflection and transmission of a transparent plate

When linearly polarized light of a known orientation is reflected or transmitted at oblique incidence from an interface, the reflected or transmitted light becomes elliptically polarized. The orientation and shape of the ellipse depend on the direction of the polarization of the incident light, the angle of incidence, and the optical properties of the interface. Based on the electric field components polarized parallelly (*p*) or perpendicularly (*s*) to the plane of incidence, the polarized light can be classified as *p*- and *s*- polarized light.

Taking a transparent plate for example, the incident light was divided into reflected light and refracted light at the surface of a sample when a monochromatic beam with wavelength λ was incident as show in Fig. 1. Every encounter with the interface of the beam would have a decomposition, therefore, the total reflected light is the summation of the multiple reflection and refraction of light 0, 1, 2, ... , n.

Here, assume d is the plate thickness and $N=n+ik$ is the complex refractive index of the plate, with n the refractive index and k the extinction coefficient. The complex refractive indexes of the medium on both sides of the plate are N_1 and N_2, respectively. The optical path difference l_0 between the reflected beam 0 and the reflected beam 1 is:

$$l_0 = N \cdot (\overline{AC} + \overline{AB}) - N_1 \cdot \overline{AD} \tag{1}$$

where

$$\overline{AC} = \overline{CB} = \frac{d}{\cos\varphi_1} \tag{2}$$

$$\overline{AD} = 2d\tan\varphi_1 \sin\varphi_1 = \frac{2dn\sin^2\varphi_1}{\cos\varphi_1} \tag{3}$$

and

$$l_0 = 2dN\cos\varphi_1 \tag{4}$$

The corresponding phase difference is

$$\delta = \frac{2\pi l_0}{\lambda} = \frac{4\pi d}{\lambda} \cdot \sqrt{N^2 - \sin^2\varphi_1} \tag{5}$$

According to Fresnel reflection formula, the reflective index of *p*-component and *s*-component at two surfaces of the sample can be deduced as follows:

$$r_{1p} = \frac{N\cos\varphi - N_1\cos\varphi_1}{N\cos\varphi + N_1\cos\varphi_1} \tag{6}$$

$$r_{1s} = \frac{N_1\cos\varphi - N\cos\varphi_1}{N_1\cos\varphi + N\cos\varphi_1} \tag{7}$$

$$r_{2p} = \frac{N_2\cos\varphi_1 - N\cos\varphi_2}{N_2\cos\varphi_1 + N\cos\varphi_2} \tag{8}$$

$$r_{2s} = \frac{N\cos\varphi_1 - N_2\cos\varphi_2}{N\cos\varphi_1 + N_2\cos\varphi_2} \tag{9}$$

Assume that the amplitudes of *p*- and *s*-components of the incident beam and the reflected beam are E_{ip} , E_{is} , E_{rp} , E_{rs}, respectively, while

$$R_p = \frac{E_{rp}}{E_{ip}};\ R_s = \frac{E_{rs}}{E_{is}} \tag{10}$$

The ratio of the total reflectance of *p*-component and *s*-component is:

$$\frac{R_p}{R_s} = \frac{E_{rp}/E_{ip}}{E_{rs}/E_{is}} = \frac{(E_p/E_s)_r}{(E_p/E_s)_i} e^{i(\beta_r - \beta_i)} \tag{11}$$

Where $\beta_i = (\beta_p - \beta_s)_i$ is the phase difference of *p*- and *s*-components in the incident light, and $\beta_r = (\beta_p - \beta_s)_r$ is the phase difference of *p*- and *s*-components in the reflected light. Define

$$\tan\Psi = \frac{(E_p/E_s)_r}{(E_p/E_s)_i} \tag{12}$$

$$\Delta = \beta_r - \beta_i \tag{13}$$

then

$$\frac{R_p}{R_s} = \tan\Psi \cdot e^{i\Delta} \tag{14}$$

where $\tan\Psi$ means the attenuation of the relative amplitudes of *p*- and *s*-components after being reflected on each surface (or interface) while Δ means the phase difference change through this process, both of which can be measured by an ellipsometer directly.

Taking multiple reflection and refraction on the surface of the sample into account and combining Fresnel formula, the total reflection coefficients of *p*- and *s*-components are

$$R_p = \frac{r_{1p} + r_{2p}\exp(-i\delta)}{1 + r_{1p}r_{2p}\exp(-i\delta)} \tag{15}$$

$$R_s = \frac{r_{1s} + r_{2s}\exp(-i\delta)}{1 + r_{1s}r_{2s}\exp(-i\delta)} \tag{16}$$

Ellipsometric equation can be represented by

$$\begin{aligned} \tan\Psi \cdot e^{i\Delta} &= \frac{R_p}{R_s} = \frac{r_{1p} + r_{2p}\exp(-i\delta)}{1 + r_{1p}r_{2p}\exp(-i\delta)} \cdot \frac{1 + r_{1s}r_{2s}\exp(-i\delta)}{r_{1s} + r_{2s}\exp(-i\delta)} \\ &= f(N_1, N_2, N, \varphi, d, \lambda) \end{aligned} \tag{17}$$

which shows the relationship between the variation of polarization state Ψ, Δ and the thickness *d*, the complex refractive index *N*. Here both Ψ and Δ are called ellipsometric parameters, each of which has an angular value. Since N_1, N_2, λ and φ are known parameters, and Ψ and Δ can be measured experimentally, *N* and *d* can be determined by a least-square fitting calculation. This is the foundation of calculating the sample's thickness and complex refractive index by the ellipsometric data.

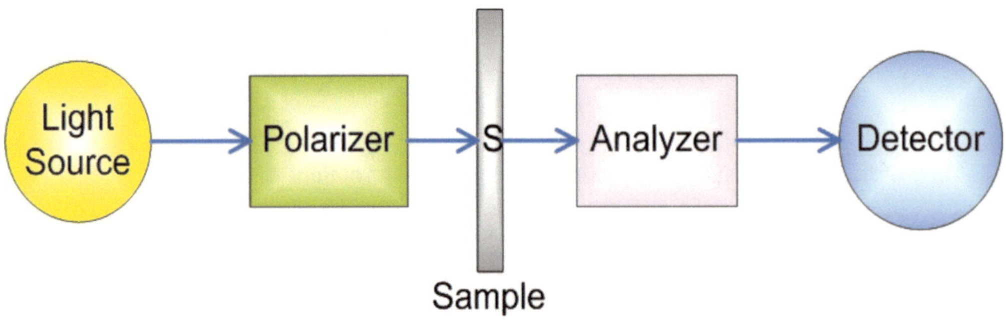

Fig. 2. Schematic diagram of a general spectroscopic ellipsometric arrangement

The changes of the polarization direction of the reflected or transmitted light can be measured with a spectroscopic ellipsometer, by which the relative amplitude and phase change introduced by the interface can be calculated. The schematic diagram of a general spectroscopic ellipsometric arrangement is shown in Fig. 2.

A collimated monochromatic or quasi-monochromatic beam from a light source passes through a variable angle polarizes to produce light of known controlled polarization. The light interacts with the sample under study and its polarization is modified. The modified state of polarization at the output of the system is analyzed by a variable polarization analyzer and can be measured by a photo-detector. Then, an equal optical model that can accurately represent the true physical structure of the sample under study should be developed to interpret the measured data. Finally, the optical and structural parameters of the sample can be obtained by solving an inverse problem.

As spectroscopic ellipsometry is a powerful and versatile non-contact and non-destructive technique for the investigation of the optical properties and structural parameters of thin films or layered structures, it has found wide applications in many different fields, from semiconductor physics to microelectronics and biology, from basic research to industrial applications, etc.. In the following sections, some typical applications of spectroscopic ellipsometry on the measurements of semiconductor silicon are presented.

3. Visible and near infrared ellipsometry

Spectroscopic ellipsometry is sensitive to the optical properties of the film deposited on a substrate. Therefore the physical or chemical properties which link to the optical properties of the sample can be characterized by spectroscopic ellipsometry.

In integrated circuits (IC) manufacturing, silicon wafers were doped by ion implantation to improve their electrical properties. Implantation induced damage altered the optical properties of the silicon wafers in the visible spectral range. Therefore, most spectroscopic ellipsometry measurements on implanted silicon wafers were carried out in the visible spectral range. In the past few decades, spectroscopic ellipsometry was applied to characterize the damage depth profiles, the surface and interface, and the structural changes of the silicon wafers.

As the SE measured damage is introduced by ion implantation, implantation parameters, such as the implantation dose and energy, can therefore be obtained indirectly. In visible spectral range, spectroscopic ellipsometry has been widely used for *in-situ* monitoring of ion implantation dose. As early as the 1980s, spectroscopic ellipsometry was applied to investigate the As^+ and P^+ ion implanted silicon wafers at wavelengths in the range of 325-400 nm. Results have shown that the ellipsometric spectra in this spectral range were very sensitive to the lattice damages introduced by ion implantation. Both implantation dose and energy could be determined by visible SE. In several research papers the monitoring of implantation energy by visible spectroscopic ellipsometry were reported. Spectroscopic ellipsometry could also be used in thin film deposition processes, such as the growth of microcrystalline silicon. The measurements were carried out in the 2.5-5eV (248-496nm) spectral range.

In our work, five groups (G1 to G5) of silicon wafers with properties shown in Table 1 were prepared under different process conditions. All substrates were prepared from the same batch of <111> oriented p-type Czochralski crystalline silicon wafers, (525±20μm, 8-13Ω·cm), with the front surface chemically-mechanically polished. These samples were implanted with As+ ions. G1 and G2 were implanted with implantation doses from 1×10^{11} to $1\times10^{16}/cm^2$ at the same implantation energy of 100keV. G3 and G4 were implanted with implantation energies from 20 to 140keV at implantation dose of $1\times10^{15}/cm^2$. All wafers of G5 were implanted with $1\times10^{15}/cm^2$ at 100keV. After implantation, G1 and G3 were then annealed in a rapid thermal annealing (RTA) system for 30s at 1100°C in an inert nitrogen atmosphere, and G2 and G4 were not annealed. Wafers in G5 were annealed at various temperatures ranging from 500 to 1100°C for 30s.

Wafer group	Implantation dose (As^+/cm^2)	Implantation energy (keV)	Annealing temperature (℃)
G1	1×10^{11} - 1×10^{16}	100	--
G2	1×10^{11} - 1×10^{16}	100	1100
G3	1×10^{15}	20 - 140	--
G4	1×10^{15}	20 - 140	1100
G5	1×10^{15}	100	500 - 1100

Table 1. List of wafer groups prepared with different implantation and annealing conditions

The SE measurements were performed with a generalized ellipsometer at room temperature. The spectral range from 270 to 2000 nm was covered using a rotating-analyzer ellipsometer with automated compensator function (VASE®, J. A. Woollam). All measurements were carried out at an angle of incidence of 75°.

The ellipsometric spectra of ion implanted wafers with different implantation doses without annealing were plot in Fig. 3. The optical properties in the visible spectral range were altered by ion implantation for wafers with implantation dose higher than $10^{14}cm^{-2}$. However, when these highly implanted wafers were annealed at high temperature, the visible ellipsometry (300-800nm) could no longer distinguish the wafers with different implantation doses, even for wafers implanted with a high dose. The visible ellipsometric spectra of ion implanted wafers were close to that of monocrystalline silicon, as presented in Fig. 4.

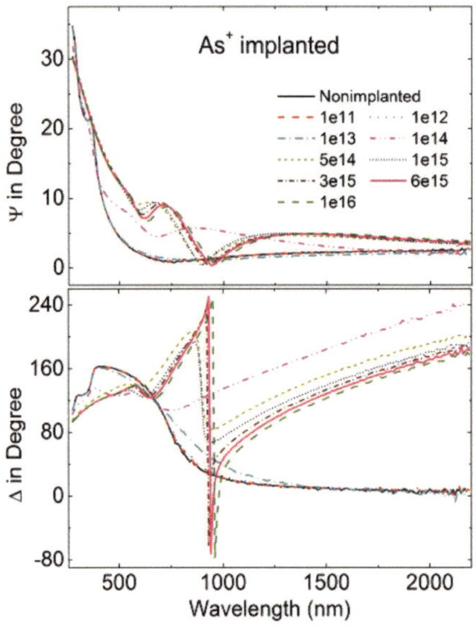

Fig. 3. Ellipsometric spectra of wafers G1 in visible and near infrared range at 75°.

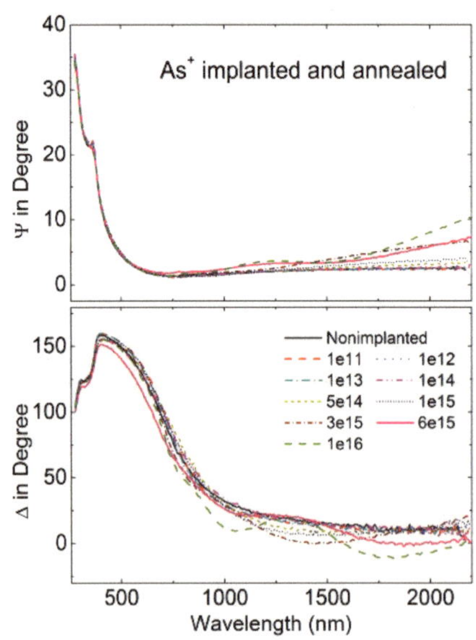

Fig. 4. Ellipsometric spectra of wafers G2 in visible and near infrared range at 75°.

Figure 5 presented the ellipsometric spectra for wafers implanted with different implantation energies. The implanted layer influenced the visible ellipsometric spectra, and

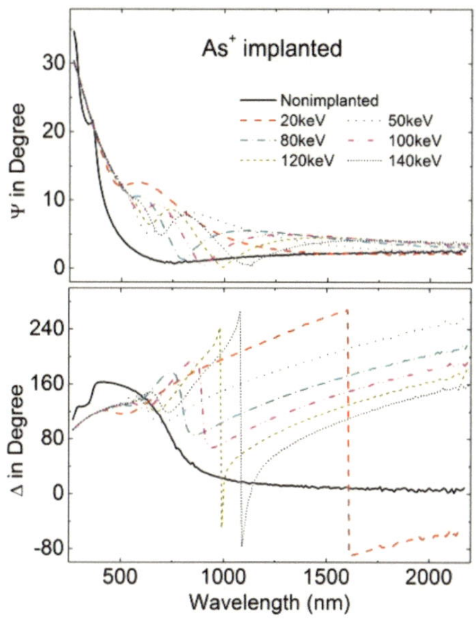

Fig. 5. Ellipsometric spectra of wafers G3 in visible and near infrared range at 75°.

the layer thickness was determined by implantation energy. Similar to that in Fig. 4, the visible ellipsometric spectra for annealed wafers with different implantation energies were close to that of non-implanted monocrystalline silicon, as shown in Fig. 6. Ellipsometric spectra for wafers annealed at different temperature were plot in Fig. 7. Spectra of wafers

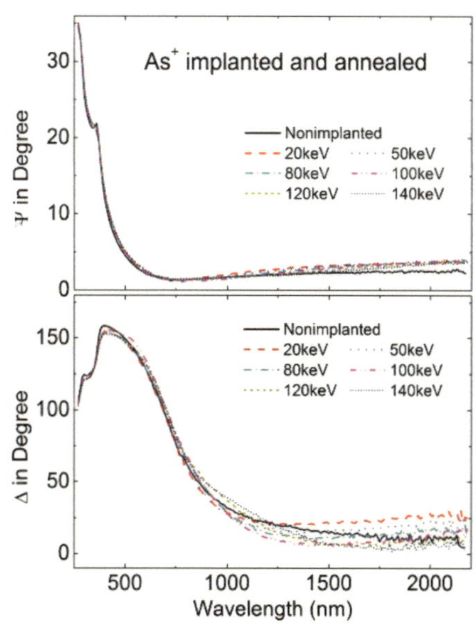

Fig. 6. Ellipsometric spectra of wafers G4 in visible and near infrared range at 75°.

prepared with the same implantation conditions but without thermal annealing were shown for comparison. When the implanted wafers were thermally annealed, the temperature of 600°C was considered to be a threshold, above this annealing temperature the damaged material was reconstructed and returned to its original crystal structure. Thus the annealing temperature of 1100°C in our process made the implantation induced structural damage almost totally recrystallized.

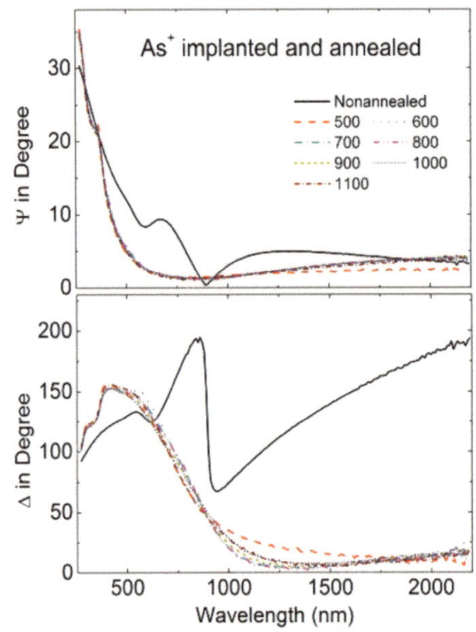

Fig. 7. Ellipsometric spectra of wafers G5 in visible and near infrared range at 75°. For comparison, the spectra of a non-annealed reference wafer were shown by the solid lines.

4. Infrared spectroscopic ellipsometry

The results reported in section 3 indicated that the visible SE is not a sensitive method to investigate the ion implantation induced effects of ion-implanted silicon wafers which were completely annealed. In practice, however, during the ICs processing, ion implantation is the predominant doping method to alter the electrical properties. It is always followed by thermal annealing to recrystallize the damaged material and active the impurities. Evaluating the distribution of the implanted ions and the activated impurities after ion implantation and thermal annealing is essential for the semiconductor device design, simulation and fabrication. From section 3 it is noticed that the effects of implantation dose and energy on the SE parameters begin to be evident from the near infrared range, as shown in Fig. 4. It is of interest to extend the SE measurement into the infrared spectral range.

The infrared spectroscopic ellipsometry measurements were carried out in the 300-5000cm^{-1} (2-30μm) spectral range by a rotating-polarizer, rotating-compensator, Fourier-transform based variable angle spectroscopic ellipsometer (IR-VASE®, J. A. Woollam Co.) at room temperature. The angle of incidence was set to 75° to keep in consistence with the SE measurements performed in the visible spectral range. The same wafers used in Section 3 were comparatively studied in the infrared spectral range, with the ellipsometric spectra of wafers G1-G5 were presented in Fig. 8 to Fig. 12.

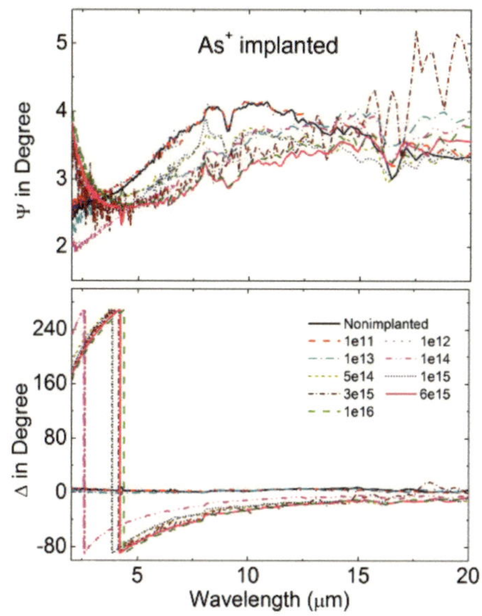

Fig. 8. Ellipsometric spectra of wafers G1 in the infrared range at 75°.

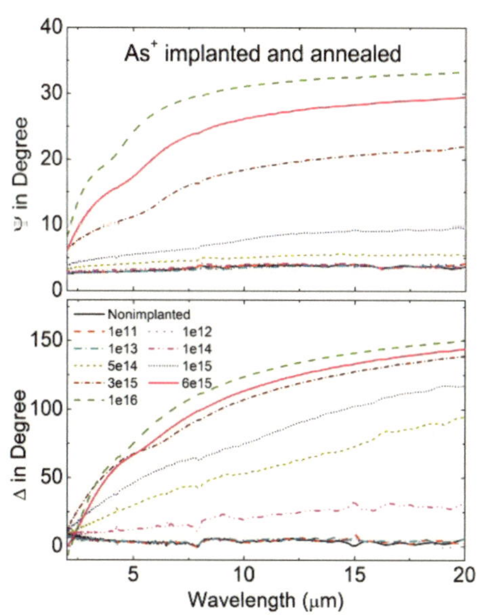

Fig. 9. Ellipsometric spectra of wafers G2 in the infrared range at 75°.

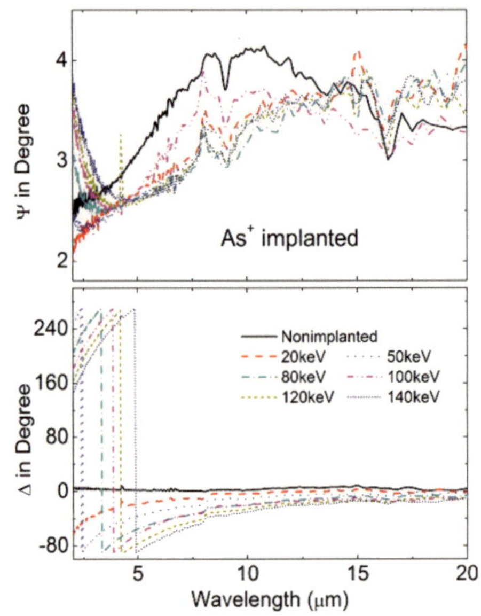

Fig. 10. Ellipsometric spectra of wafers G3 in the infrared range at 75°.

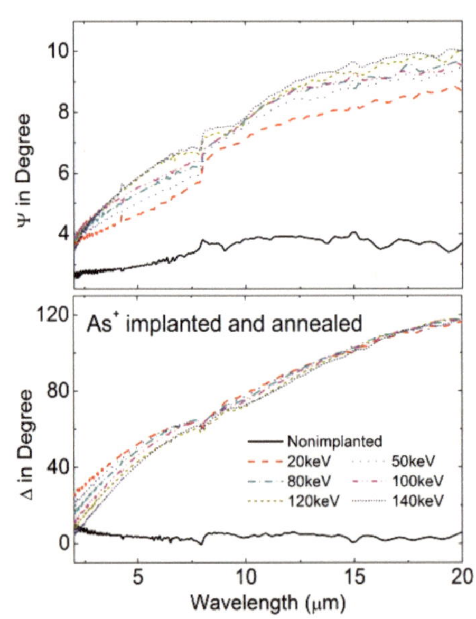

Fig. 11. Ellipsometric spectra of wafers G4 in the infrared range at 75°.

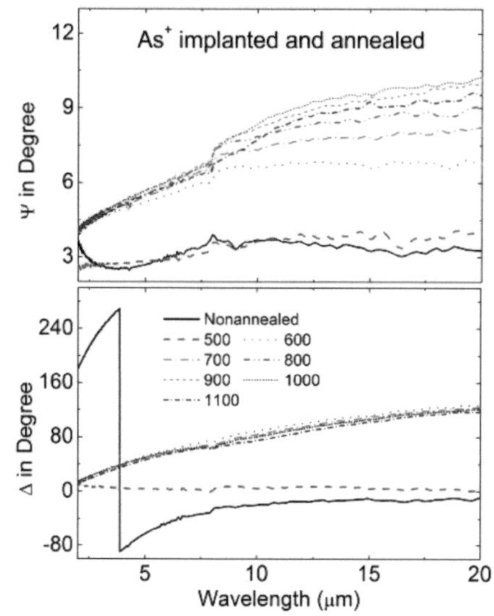

Fig. 12. Ellipsometric spectra of wafers G5 in the infrared range at 75°. For comparison, the spectra of a non-annealed reference wafer were shown by the solid lines.

In opposition to the results obtained in the visible spectral range, as presented in section 3, the IRSE is not a sensitive method for implanted wafers without thermal annealing. The value of ellipsometric parameter Ψ ranged only 2-4°, as shown in Fig. 8, Fig. 10, and Fig. 12. The results indicated that ion implantation alone introduced no significant change to the optical properties of damaged crystal structure in the infrared range. However, once the implanted wafers were thermally annealed at high temperature, the IRSE could effectively distinguish the wafers with different implantation doses, especially for wafers implanted with a high dose, as presented in Fig. 9. On the other hand, the IRSE could not clearly distinguish the wafers implanted with different energies, as shown in Fig. 11. In the following, the infrared ellipsometric spectra for implanted and annealed wafers were analyzed in details.
In the infrared range, different absorption processes exist in a silicon wafer, such as free carrier absorption, impurity absorption and Reststrahlen absorption, as shown in Fig. 13.

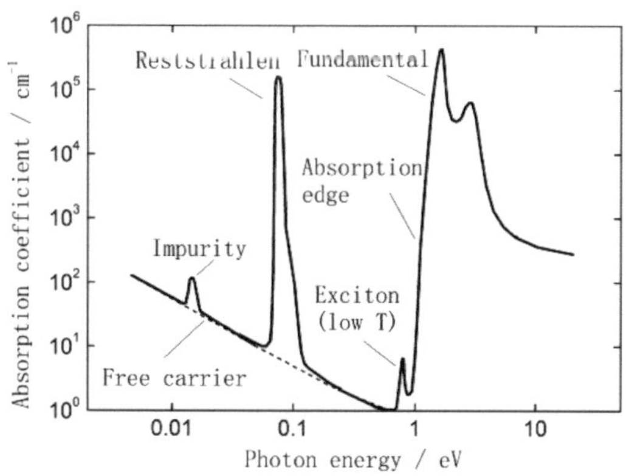

Fig. 13. Absorption coefficient plotted as a function of the photon energy in a silicon wafer, illustrating various possible absorption processes

At room temperature for silicon, the impurity absorption is too weak to be observed. The influence of the Reststrahlen absorption process on the optical properties of implanted silicon wafer is at least two orders of magnitude lower than the influence of the free carrier absorption, thus is negligible. Therefore, the free carrier absorption dominates the optical properties of the implanted layer in the infrared range, which can be described by a classical Drude model:

$$\tilde{\varepsilon}(E = \hbar\omega) = \frac{-\hbar^2}{\varepsilon_0 \rho\left(\tau \cdot E^2 + i\hbar E\right)} \tag{18}$$

with

$$\begin{aligned} \rho &= \frac{1}{eN\mu}, \\ \mu &= e\tau/m^*, \\ m &= m^*/m_0 \end{aligned} \tag{19}$$

where ε is the complex dielectric constant, ε_0 is the vacuum dielectric constant, N is the carrier concentration, e is the electronic charge, m^* is the ratio of the optical carrier effective mass to the electron rest mass, E is the energy of the incident photons, ρ is the resistivity, and τ is the mean scattering time of the free carriers. The parameters Ψ and Δ can be implicitly expressed as a function of the dielectric function ε and the thickness of the wafer under study.

In order to simulate the optical properties of the ion implanted silicon wafer, the atoms distribution was calculated. The calculation was performed with the Monte Carlo simulation by software package TRIM. Figure 14 shows the simulation results as well as the corresponding Gaussian fit for an As^+ implanted silicon wafer.

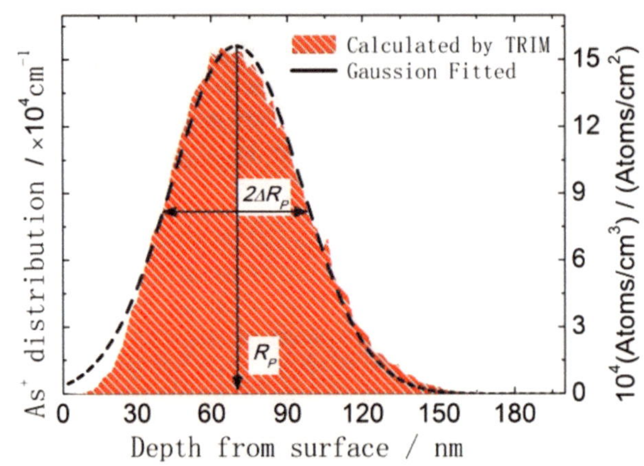

Fig. 14. Calculated As^+ ion distribution and corresponding Gaussian fitted result.

The As^+ ion distribution can be expressed with a Gaussian function:

$$N = N_{max} \exp[-\frac{1}{2}(\frac{d - R_p}{\Delta R_p})^2], \tag{20}$$

where $N_{\max}$ is the maximum carrier concentration, d is the depth, R_p is the range, and ΔR_p is the standard deviation of the Gaussian function. The Gaussian fitted curve is plot in Fig. 14. The fitted values are given in Table 2.

Fitted parameter	Fitted value	Fitting error
R_p (nm)	69.78909	0.2599
ΔR_p (nm)	25.8049	0.32895
A (corresponding N_{max})	15.63882	0.15216

Table 2. As^+ ion distribution parameters fitted with a Gaussian function.

For the evaluation of IRSE data of these implanted wafers, the optical model with an ion-implanted layer described by 30 sub-layers, and a single-crystalline silicon substrate layer is employed to describe the structure of the implanted wafer. Although both the m^* and τ are functions of the doping concentration, it is reasonable to consider them as fixed values in the ion-implanted layer in each fitting. The optical properties of the ion implanted layer can be expressed by the Drude model, while the optical properties of the substrate silicon in the infrared range can be taken from literature (Palik, 1998). Then, the ion distribution parameters and the physical properties of the implanted layer can be fitted. In the multi-parameter fitting program, a mean square error (MSE) was minimized. The MSE was defined as:

$$MSE^2 = \frac{1}{2n-p}\sum_{i=1}^{n}\left[\left(\tan\Psi_i^{\text{mod}} - \tan\Psi_i^{\text{exp}}\right)^2 + \left(\cos\Delta_i^{\text{mod}} - \cos\Delta_i^{\text{exp}}\right)^2\right] \quad (21)$$

Here, *mod* and *exp* represent the data calculated from the theoretical model and the experimental data, respectively. n is the number of measured ψ and Δ pairs being included in the fitting and p is the number of fitting parameters. In this multi-parameter fitting, six parameters are set as free parameters to minimize the MSE, that is, N_{max}, R_p, ΔR_p, m, μ and the implanted layer thickness l.

To reduce the number of iteration and improve the computation efficiency in the multi-parameter fitting procedure, it is important to give a reasonable initial set of values for the free parameters in the fitting. Here, R_p and ΔR_p values fitted in Table 2 are set as the initial values. For N_{max}, the initial value is $\frac{dose}{\sqrt{2\pi}\cdot\Delta R_p}$. As shown in Fig. 9, only the annealed wafers in G2 with As^+ ion implantation dose higher than 1×10^{14} cm^{-2} can be distinguished by infrared ellipsometric spectra, which are fitted with above model. The optimized fitted parameters for the ion-implanted layers of wafers implanted with different doses are listed in Table 3, in which the ρ and τ values are calculated with the fitted values. The IRSE fitted results for wafers in Table 3 are shown in Fig. 15 by the solid lines. The units of the horizontal abscissa are wave number, in accordance with the measurement conditions of the infrared ellipsometer. Good agreements between the experimental SE data and the best fits are observed in this spectral range.

From Table 3, it is observed that the impurities were activated by the rapid thermal annealing, which resulted in the redistribution. For implanted wafers with higher doses, more impurities were activated and the impurities diffused farther. Therefore, the N_{max}, R_p and ΔR_p of the implanted layer increased with the increasing implantation dose, especially for wafers with high implantation doses. Meanwhile, the mean scattering time of the free carriers decreased due to the increasing impurity concentration.

Dose (cm^{-2})	1×10^{14}	5×10^{14}	1×10^{15}	3×10^{15}	6×10^{15}	1×10^{16}
l (nm)	118.1	152.2	166.5	171.9	249.6	252.2
R_p (nm)	60.4	66.1	62.3	62.9	85.9	97.5
ΔR_p (nm)	65.6	64.8	66.1	73.6	99.2	104.8
N_{max} (cm^{-3})	8.3×10^{18}	2.8×10^{19}	9.9×10^{19}	3.2×10^{20}	7.3×10^{20}	1.5×10^{21}
μ ($cm^2/V\cdot s$)	166.6	58.4	26.6	20.1	11	7.2
m	0.36	0.37	0.67	0.78	1.43	1.95
ρ ($10^{-3}\Omega\cdot cm$)	4.51	3.82	2.37	0.97	0.78	0.58
τ (fs)	34.06	12.15	10.1	8.96	8.95	7.99
MSE	0.386	0.316	0.257	0.21	0.302	0.276

Table 3. Best-fit model parameters of the ion-implanted layer. ρ and τ are calculated with the fitted values.

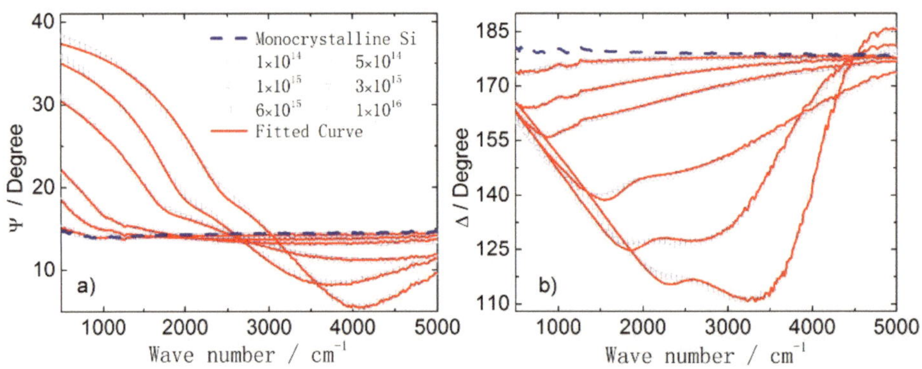

Fig. 15. The measured and best-fitted infrared ellipsometric spectra Ψ and Δ for silicon wafers As^+ implanted with different doses.

5. Conclusion

In this chapter the application of spectroscopic ellipsometry to silicon characterization and processes monitoring has been reviewed. The comparative studies on the infrared spectroscopic ellipsometry for implanted silicon wafers with and without thermal annealing have been presented. Several conclusions can be summarized as follows:

- For implanted but non-annealed silicon wafers, the optical properties in the visible spectral range are determined by ion implantation induced lattice damages.
- For implanted and annealed silicon wafers, the optical properties in the visible spectral range are close to that of monocrystalline silicon, as the lattice damages are recovered by thermal annealing.
- In infrared spectral range, the optical properties of the implanted and annealed silicon wafers are functions of the activated impurities concentration, which is determined by the implantation dose, the implantation energy and the annealing temperature.
- The optical properties of the implanted and annealed silicon wafers in the infrared spectral range can be described with a Drude free-carrier absorption equation.

Therefore, the infrared ellipsometric spectra can be analyzed with the corresponding model to better characterize the implanted and annealed silicon wafers.

6. Reference

Palik, E D. (1998). *Handbook of Optical Constants of Solids.* Academic, ISBN 978-012-5444-23-1, San Diego, USA

Watanabe, K; Miyao, M; Takemoto, I & Hashimoto, N. (1979). Ellipsometric study of silicon implanted with boron ions in low doses. *Applied Physics Letters,* Vol. 34, No. 8, pp. 518-519, ISSN 0003-6951

Fried, M; Polgar, O; Lohner, T; Strehlke, S & Levy-Clement, C. (1998). Comparative study of the oxidation of thin porous silicon layers studied by reflectometry, spectroscopic ellipsometry and secondary ion mass spectroscopy. *Journal of Luminescence,* Vol. 80, No. 1-4, pp. 147-152, ISSN 0022-2313

Cabarrocas, P R I; Hamma, S; Hadjadj, A; Bertomeu, J & Andreu, J. (1996). New features of the layer-by-layer deposition of microcrystalline silicon films revealed by spectroscopic ellipsometry and high resolution transmission electron microscopy. *Applied Physics Letters,* Vol. 69, No. 4, pp. 529-531, ISSN 0003-6951

Drevillon, B; Godet, C & Kumar, S. (1987). Insitu Spectroscopic Ellipsometry Investigation of the Nucleation of Microcrystalline Silicon. *Applied Physics Letters,* Vol. 50, No. 23, pp. 1651-1653, ISSN 0003-6951

Kang, T D; Lee, H; Park, S J; Jang, J & Lee, S. (2002). Microcrystalline silicon thin films studied using spectroscopic ellipsometry. *Journal of Applied Physics,* Vol. 92, No. 5, pp. 2467-2474, ISSN 0021-8979

Nelson, F J; Kamineni, V K; Zhang, T; Comfort, E S; Lee, J U & Diebold, A C. (2010). Optical properties of large-area polycrystalline chemical vapor deposited graphene by spectroscopic ellipsometry. *Applied Physics Letters,* Vol. 97, No. 25, pp. 253110-3, ISSN 0003-6951

Fujiwara, H; Koh, J; Wronski, C R & Collins, R W. (1997). Application of real time spectroscopic ellipsometry for high resolution depth profiling of compositionally graded amorphous silicon alloy thin films. *Applied Physics Letters,* Vol. 70, No. 16, pp. 2150-2152, ISSN 0003-6951

Lohner, T; Khanh, N Q; Petrik, P; Fried, M; Kotai, E & Gyulai, J. (1997). Ion implantation induced damage accumulation studied by Rutherford Backscattering Spectrometry and Spectroscopic Ellipsometry. *Materials Science Applications of Ion Beam Techniques,* Vol. 248, No. pp. 229-232, ISSN 0255-5476

Podraza, N J; Li, J; Wronski, C R; Dickey, E C & Collins, R W. (2009). Analysis of controlled mixed-phase, amorphous plus microcrystalline. silicon thin films by real time spectroscopic ellipsometry. *Journal of Vacuum Science & Technology A,* Vol. 27, No. 6, pp. 1255-1259, ISSN 0734-2101

Petrik, P; Polgar, O; Lohner, T; Fried, M; Khanh, N Q & Gyulai, J. (1998). Ion implantation-caused damage depth profiles in single-crystalline silicon studied by Spectroscopic Ellipsometry and Rutherford Backscattering Spectrometry. *Vacuum,* Vol. 50, No. 3-4, pp. 293-297, ISSN 0042-207X

Petrik, P; Lohner, T; Fried, M; Khánh, N Q; Polgár, O & Gyulai, J. (1999). Comparative study of ion implantation caused damage depth profiles in polycrystalline and single crystalline silicon studied by spectroscopic ellipsometry and Rutherford ackscattering spectrometry. *Nuclear Instruments and Methods in Physics Research B,* Vol. 147, No. pp. 84-89, ISSN 0168-583X

Fried, M; Lohner, T; Aarnink, W A M; Hanekamp, L J & van Silfhout, A. (1992). Nondestructive determination of damage depth profiles in ion-implanted semiconductors by spectroscopic ellipsometry using different optical models. *Journal of Applied Physics,* Vol. 71, No. 6, pp. 2835-2843, ISSN 0021-8979

Yamada, T; Harada, N; Kitahara, K & Moritani, A. (2003). Study of mechanism of plasma surface modifications in Si by spectroscopic ellipsometry. *Surface and Coatings Technology,* Vol. 173, No. pp. 854-857, ISSN 0257-8972

McMarr, P J; Vedam, K & Narayan, J. (1986). Spectroscopic ellipsometry: A new tool for nondestructive depth profiling and characterization of interfaces. *Journal of Applied Physics,* Vol. 59, No. 3, pp. 694-701, ISSN 0021-8979

Hikino, S-i & Adachi, S. (2004). Structural changes in ion-implanted and rapid thermally annealed Si(100) wafers studied by spectroscopic ellipsometry. *Journal of Physics D: Applied Physics,* Vol. 37, No. pp. 1617–1623, ISSN 0022-3727

Jans, J C; Hollering, R W J & Lifka, H. (1991). Characterization of Annealed High-Dose Oxygen-Implanted Silicon by Spectroscopic Ellipsometry and Reflectometry. *Journal of Applied Physics,* Vol. 70, No. 11, pp. 6643-6646, ISSN 0021-8979

Petrik, P; Lohner, T; Fried, M; Khanh, N Q; Polgar, O & Gyulai, J. (1999). Comparative study of ion implantation caused damage depth profiles in polycrystalline and single crystalline silicon studied by spectroscopic ellipsometry and Rutherford backscattering spectrometry. *Nuclear Instruments & Methods in Physics Research Section B-Beam Interactions with Materials and Atoms,* Vol. 147, No. 1-4, pp. 84-89, ISSN 0168-583X

Petrik, P; Cayrel, F; Fried, M; Polgar, O; Lohner, I; Vincent, L; Alquier, D & Gyulai, J. (2004). Depth distribution of disorder and cavities in high dose helium implanted silicon characterized by spectroscopic ellipsometry. *Thin Solid Films,* Vol. 455-56, No. pp. 344-348, ISSN 0040-6090

Tsunoda, K; Adachi, S & Takahashi, M. (2002). Spectroscopic ellipsometry study of ion-implanted Si(100) wafers. *Journal of Applied Physics,* Vol. 91, No. 5, pp. 2936-2941, ISSN 0021-8979

Cortot, J P & Ged, P. (1982). Analysis of Arsenic and Phosphorus Ion-Implanted Silicon by Spectroscopic Ellipsometry. *Applied Physics Letters,* Vol. 41, No. 1, pp. 93-95, ISSN 0003-6951

Lioudakis, E; Christofides, C & Othonos, A. (2006). Study of the annealing kinetic effect and implantation energy on phosphorus-implanted silicon wafers using spectroscopic ellipsometry. *Journal of Applied Physics,* Vol. 99, No. 12, pp. 123514-6, ISSN 0021-8979

Hu, Y Z; Li, M; Conrad, K; Andrews, J W; Irene, E A; Denker, M; Ray, M & Mcguire, G. (1992). Insitu Spectroscopic Ellipsometry Studies of Hydrogen-Ion Bombardment of Crystalline Silicon. *Journal of Vacuum Science & Technology B,* Vol. 10, No. 3, pp. 1111-1117, ISSN 1071-1023

Kasap, S & Capper, P. (2006). *Springer Handbook of Electronic and Photonic Materials.* Springer Science, 978-038-7260-59-4, Heidelberg

Tiwald, T E; Thompson, D W & Woollam, J A. (1998). Optical determination of shallow carrier profiles using Fourier transform infrared ellipsometry *Journal of Vacuum Science & Technology B: Microelectronics and Nanometer Structures,* Vol. 16, No. 1, pp. 312-315, ISSN 1071-1023

Ziegler, J F; Biersack, J P & Littmark, U. (1985). *The Stopping Range of Ions in Solids.* Pergamon, ISBN 978-008-0216-03-4, Tarrytown, NY

Tiwald, T; Thompson, D; Woollam, J; Paulson, W & Hance, R. (1998). Application of IR variable angle spectroscopic ellipsometry to the determination of free carrier concentration depth profiles. *Thin Solid Films,* Vol. 313-314, No. pp. 661-666, ISSN 0040-6090

7

Silicon Nanocrystals

Hong Yu, Jie-qiong Zeng and Zheng-rong Qiu
Southeast University
China

1. Introduction

Silicon has many advantages over other semiconductor materials: low cost, nontoxicity, practically unlimited availability, and decades of experience in purification, growth and device fabrication. It is used for almost all modern electronic devices. However, the indirect energy gap in bulk crystalline Si makes it unable to emit light efficiently and thus unsuitable for optoelectronic applications. For example, lasers, photodetectors are not constructed from silicon. On the other hand, although silicon is widely used for solar cell fabrication, the efficiency can not exceed the Schockley and Queisser limit in single band gap device, because of its inability to absorb photons with energy less than the band gap and thermalisation of photon energy exceeding the band gap. One approach for tackling this disadvantage is to use tandem cells, which can implement the increasing of the number of band gaps (Conibeer et al., 2006; Cho et al., 2008). Moreover, the band gap in silicon is too small to interact effectively with the visible spectrum. If the gap could be adjusted, silicon would be used for either electronic or optical application. In 1990, it was firstly observed experimentally by Canham (Canham, 1990), that photoluminescence (PL) occurs in the visible range at room temperature in porous silicon (PS). Since then the silicon clusters or silicon quantum dots (Si-QDs) or silicon nanocrystals (Si-NCs) have attracted much of research interest, and many of theoretical models, computations, and experimental results on band structures, PL and other electronic properties have been reported during the last decades (Öğüt et al., 1997; Fang & Ruden, 1997; Wolkin et al., 1999; Wilcoxon et al., 1999; Soni et al., 1999; Vasiliev et al., 2001; Garoufalis & Zdetsis, 2001; Carrier et al., 2002; Nishida, 2004; Biteen et al., 2004; Tanner et al., 2006). The results from these reports show that in low-dimension silicon structures, such as silicon nanocrystals or silicon quantum dots, electronic and optical properties can be quite different from those of silicon bulk counterpart, for instance, free-standing Si-NCs show strong luminescence, the color of which depends on the size of the Si-NCs, and the gap and energy increase when their size is reduced. Therefore, the energy gap can be tuned as a function of the size of quantum dots.

We are especially interested in the theoretical study on the band gap and the optical spectrum with respect to the size of the Si-NCs or Si-QDs and surface terminations and reconstructions.

The effective mass approximation (EMA) is used by Chu-Wei Jiang and M. Green to calculate the conduction band structure of a three-dimensional silicon quantum dot superlattice with the dots embedded in a matrix of silicon dioxide, silicon nitride, or silicon carbide(Jiang & Green, 2006), and later the EMA is only of partial use in determining the absolute confined energy levels for small Si-NCs, because it has been found a decreasingly

accurate prediction of the confined energy level by the EMA as the Si-QDs size decreases(Conibeer et al., 2008).
Time dependent density functional theory (TDDFT) has been performed by Aristides D. Zdetsis and C. S. Garoufalis over the last ten years (Garoufalis & Zdetsis, 2001; Zdetsis & Garoufalis, 2005). In their calculations the Si dangling bonds on the surface of the Si-NCs are passivated by hydrogen and oxygen. In the DFT method, they have used the hybrid nonlocal exchange correlation functional of Becke and Lee, Yang and Parr, which includes partially exact Hartree–Fock exchange (B3LYP). Their results are in excellent agreement with accurate recent and earlier experimental data. It is found that the diameter of the smallest oxygen-free nanocrystal that could emit PL in the visible region of the spectrum is around 22 Å, whereas the largest diameter falls in the range of 84– 85 Å. The high level and the resulting high accuracy of their calculations have led to the resolution of existing experimental and theoretical discrepancies. Their results also clarify unambiguously and confirm earlier predictions about the role of oxygen on the gap size. More recently, they report accurate high level calculations of the optical gap and absorption spectrum of ultra small Si-NCs of 1nm, with hydrogen and oxygen passivation (with and without surface reconstruction) (Garoufalis & Zdetsis, 2009). They show that some of the details of the absorption and emission properties of the 1 nm Si nanoparticles can be efficiently described in the framework of TDDFT/B3LYP, by considering the effect of surface reconstruction and the geometry relaxation of the excited state. Additionally, they have examined the effect of oxygen contamination on the optical properties of 1nm nanoparticles and its possible contribution to their experimentally observed absorption and emission properties.
By performing the same method, TDDFT, the optical absorption of small Si-NCs embedded in silicon dioxide is studied systematically by Koponen (Koponen et al., 2009). They have found that the oxide-embedded Si-NCs exhibit absorption spectra that differ significantly from the spectra of the hydrogen-passivated Si-NCs. In particular, the minimum absorption energy is found to decrease when the Si-NCs are exposed to dioxide coating. The absorption energy of the oxide-embedded Si-NCs remains approximately constant for core sizes down to 17 atoms, whereas the absorption energy of the hydrogen-passivated Si-NCs increases with decreasing crystal size. They suggest a different mechanism for producing the lowest-energy excitations in these two cases.
Wang, et al, generate and optimize geometries and electronic structures of hydrogenated silicon nanoclusters, which include the T_d and I_h symmetries by using the semi-empirical AM1 and PM3 methods, the density functional theory DFT/ B3LYP method with the 6-31G(d) and LANL2DZ basis sets from the Gaussian 03 package, and the local density functional approximation (LDA), which is implemented in the SIESTA package(Wang et al., 2008). The calculated energy gap is found to be decreasing while the diameter of silicon nanocluster increases. By comparing different calculated results, they conclude that the calculated energy gap by B3LYP/6-31G(d)//LDA/SIESTA is close to that from experiment. For investigation of the optical properties of Si-NCs as a function of surface passivation, they carry out a B3LYP/6-31G(d)//LDA/SIESTA calculation of the Si_{35} and Si_{47} core clusters with full alkyl-, OH-, NH_2-, CH_2NH_2-, OCH_3-, SH-, C_3H_6SH-, and CN- passivations. In conclusion, the alkyl passivant affects the calculated optical gaps weakly, and the electron-withdrawing passivants generate a red-shift in the energy gap of silicon nanoclusters. A size-dependent effect is also observed for these passivated Si nanoclusters.
The optical absorption spectra of Si_nH_m nanoclusters up to 250 atoms are computed using a linear response theory within the time-dependent local density approximation (TDLDA)

(Vasiliev et al., 2001). The TDLDA formalism allows the electronic screening and correlation effects, which determine exciton binding energies, to be naturally incorporated within an ab initio framework. They find the calculated excitation energies and optical absorption gaps to be in good agreement with experiment in the limit of both small and large clusters. The TDLDA absorption spectra exhibit substantial blueshifts with respect to the spectra obtained within the time-independent local density approximation.

2. Structure of silicon quantum dots

Typically, the size of Si-QDs is less than ten nanometers which is close to the exciton Bohr radius of bulk silicon. Owing to the extreme small dimensions, silicon quantum dots exhibit strong quantum confinement which causes the band gaps to widen, the electronic states to become discrete, and the oscillator strength of the smallest electronic transitions to increase. Generally, at ideal conditions, we consider that the interior of the dot has the structure of crystalline silicon while the surface of the dot is passivated with specific atoms depending on the surrounding environment of the dot, such as hydrogen, oxygen and so on.

2.1 Physical characterization

Direct physical evidence of the crystallinity of Si-QDs has been obtained from high resolution TEM, see Fig. 1a and b (Conibeer et al., 2006). Crystal planes are apparent in many of the darker areas in these HRTEM images. The darker areas are denser material in a less dense matrix, which are attributed to Si-NCs in a SiO_2 matrix.

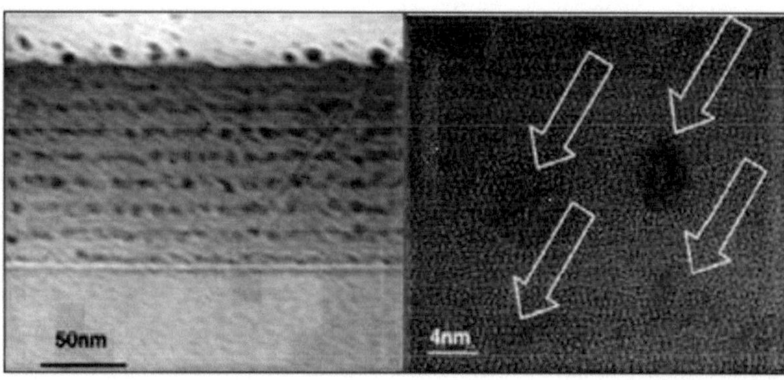

(a) (b)

Fig. 1. Cross-sectional TEM (a) HRTEM (b) images of Si QDs in oxide. (a) Shows the layered structure and (b) shows individual nanocrystals in which crystal planes can be seen

Scientists also have studied about the nanostructures of Si QDs in other dielectrics such as silicon nitride and silicon carbide, see Fig. 2a and b (Conibeer et al., 2006, 2008). Results from HRTEM images are very promising, showing crystalline nanocrystals in the nitride matrix and carbide matrix.

From the introduction stated above, we can see that it is reasonable to suppose that the interior of the dot has the structure of crystalline silicon while the surface of the dot is passivated with specific atoms at ideal conditions.

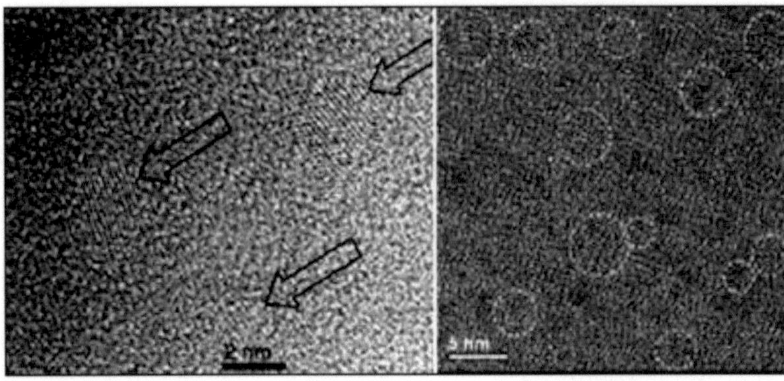

(a) (b)

Fig. 2. HRTEM images of Si-QDs in (a) silicon nitride and (b) silicon carbide

2.2 Ideal structure

Lots of experimental researches have been made on the electronic and optical properties of Si-QDs. However, several factors contribute to making the interpretation of measurements a difficult task. For instance, samples show a strong dispersion in the QD size that is difficult to be determined. In addition, Si-NCs synthesized by different techniques often show different properties in size, shape and the interface structure (Guerra et al., 2009). For the reasons stated above, the majority of experimental work give diverse results. Therefore, theoretical model calculations for some ideal structures have been considered very necessary to investigate the properties of Si-QDs. Generally, passivated-surface silicon nanoclusters are the ideal theoretical structure for us to study Si-QDs. In this section, we will introduce several ideal structures of Si-QDs in theoretical simulation.

2.2.1 Hydrogen-passivated silicon quantum dots

Hydrogen is often used as the passivating agent for the silicon nanocluster surface in most of the theoretical calculations and computations. It is generally accepted that hydrogen-passivated silicon nanocluster (Si_nH_m cluster) is the simplest structure to represent Si-QDs in a vacuum environment and can reproduce most of the experimental results in despite of neglecting some interface effects.

In Fig. 3, some idealized hydrogen-passivated quantum dots of silicon are illustrated. The interior of the dot consists of silicon atoms in the diamond structure; the surface of the dot is hydrogen-passivated. We can see that hydrogen atoms remove all dangling bonds on the surface.

2.2.2 Oxidized silicon quantum dots

Si-QDs are believed to be the luminescence centres in PS. Experimental measurements on PS samples have indicated that a large PL redshift is observed as soon as the freshly etched samples (oxygen-free PS) are transferred from Ar to a pure oxygen atmosphere or to air; however, no redshift at all is detected when the samples are kept in pure hydrogen atmosphere or in vacuum (Wolkin et al., 1999). Therefore, it is obviously that the chemistry of oxygen at the surface has played an important role when PS is exposed to air and has to be considered in theoretical models for this problem.

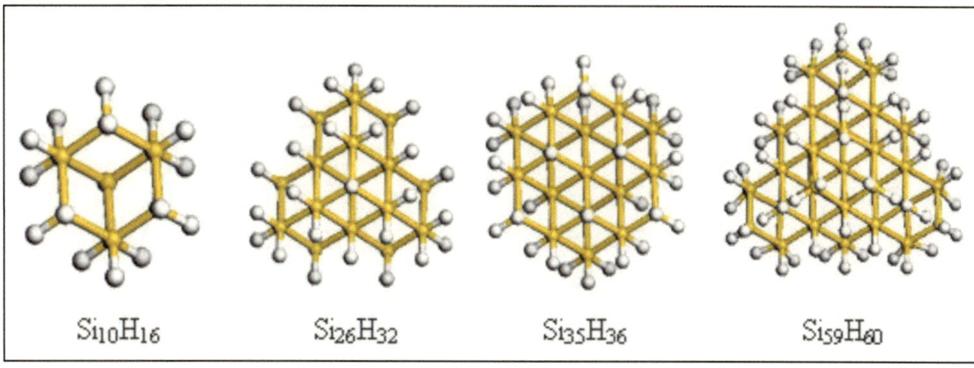

Fig. 3. Structures of Si_nH_m clusters represented in terms of ball-and-stick models. Yellow balls denote Si atoms and the white balls denote surface H atoms

There are totally four possible oxygen passivations on the surface of Si-QDs: (i) double-bonded, (ii) backbonded, (iii) bridge-bonded and (iiii) inserted oxygen configurations. The ball and stick representations of these oxygen passivation configurations are illustrated in Fig. 4. Compared with corresponding hydrogen-passivated silicon clusters, oxidized clusters are built up by substituting H or Si atoms on the surface with oxygen. From (b) to (e), they are double-bonded, backbonded, bridge-bonded and inserted oxygen configurations, respectively. It should be noted that the backbonded oxygen configuration is different in geometry from the bridge-bonded oxygen configuration, in that the former can preserve hydrogen coverage on a Si nanocrystal and retain the number of Si atoms when it is oxidized, while in the latter, one oxygen atom replaces a surface SiH_2 dihydride on the initial dot causing a decrease of the number of H and Si atoms. In another word, for the backbonded oxygen configuration, the oxygen atom is situated between the nearest-neighbor Si atoms while in the bridge-bonded oxygen configuration it is situated between the second nearest-neighbor Si atoms, referring Fig. 4(c) and (d).

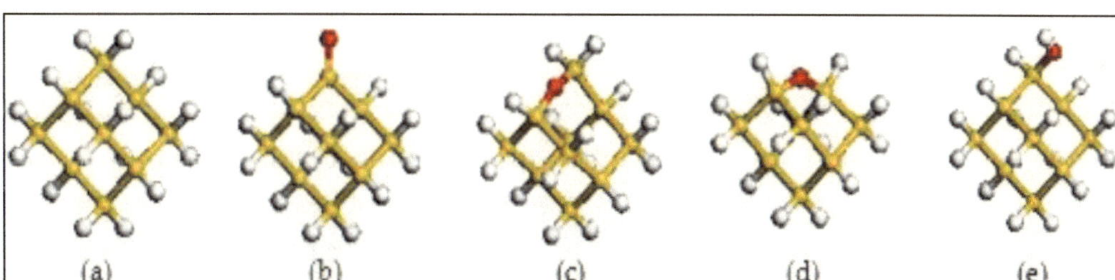

Fig. 4. Ball and stick representations of four possible oxygen passivation configurations, the yellow balls represent Si atoms, red balls represent O atoms, and the white balls, H atoms: (a) initial hydrogen-passivated silicon cluster, (b) double-bonded oxygen passivation configuration (c) backbonded (d) bridge-bonded (e) inserted oxygen configuration

Actually there may exist other oxygen-contamination on the surface of Si-QDs when oxidized. Zdetsis and Garoufalis (Zdetsis & Garoufalis, 2005) proposed a structure that can maintain the Td symmetry of the nanocrystals, considering only double-bonded oxygen configuration, see Fig. 5.

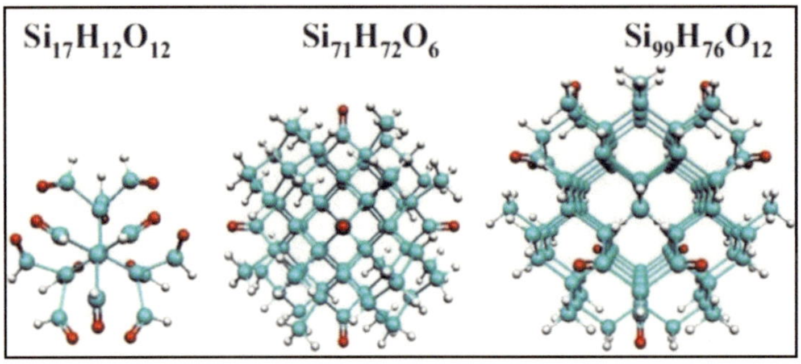

Fig. 5. Structure of oxidized nanocrystals with Td symmetry: The blue spheres are Si atoms, the red spheres are O atoms, and the white spheres are H atoms

2.2.3 Silicon quantum dots embedded in different matrixes

Si-QDs are often embedded in a dielectric matrix based on the fabrication of Si-QDs. Therefore, it is necessary to build an appropriate structure to simulate Si-QDs in such kind of environment.

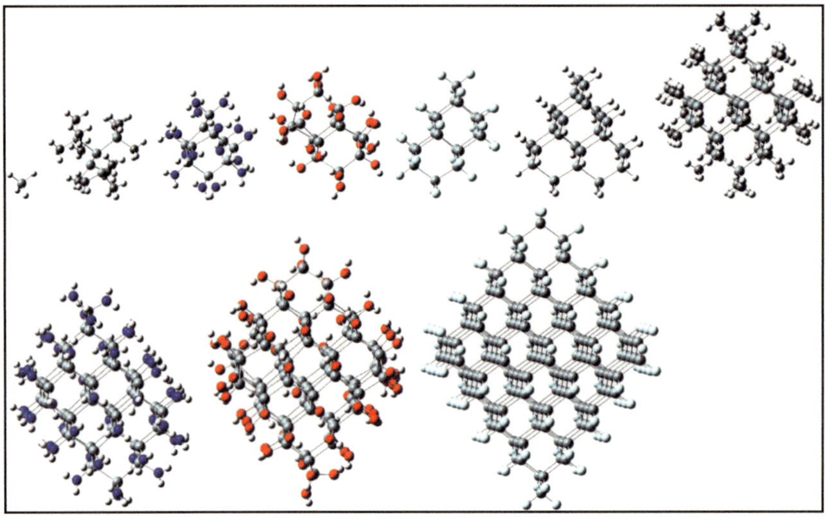

Fig. 6. Structures of Si-QDs embedded in an ionic environment, SiO_2, Si_3N_4, SiC matrix and a co-valent environment (left to right; top to bottom): SiH_4, $Si_5(CH_3)_{12}$, $Si_{10}(NH_2)_{16}$, $Si_{14}(OH)_{20}$, $Si_{18}F_{24}$, $Si_{26}H_{32}$, $Si_{35}(CH_3)_{36}$; $Si_{53}(NH_2)_{48}$, $Si_{84}(OH)_{64}$, $Si_{165}F_{100}$

König et al (König et al., 2009) have used F, OH, NH_2, CH_3 and H groups as the passivating agent for the Si clusters surface to simulate Si-QDs embedded in an ionic environment, SiO_2, Si_3N_4, SiC matrix and a co-valent environment, respectively, see Fig. 6. Apparently, these structures are simply built by replacing all the hydrogen atoms on the surface of hydrogen-passivated Si clusters with F, OH, NH_2 and CH_3 groups, respectively.

However, there are many more complex structures when Si-QDs embedded in a SiO_2 matrix. For instance, Koponen et al (Koponen et al., 2009) put forward some structures in which Si clusters are embedded in one or two neutral SiO_2 shells and the outermost layer of the cluster is hydrogen passivated in order to get rid of dangling bonds and to better mimic the effect of bulk silicon oxide. One of these structures is presented in Fig. 7(a). In addition, Guerra et al (Guerra et al., 2009) obtained a crystalline embedded structure from a β-cristobalite (BC) matrix by removing all the oxygen atoms included in a cut-off sphere, whose radius determines the size of the Si-QDs. The final optimized structure of the Si_{32} in β-cristobalite matrix is illustrated in Fig. 7(b).

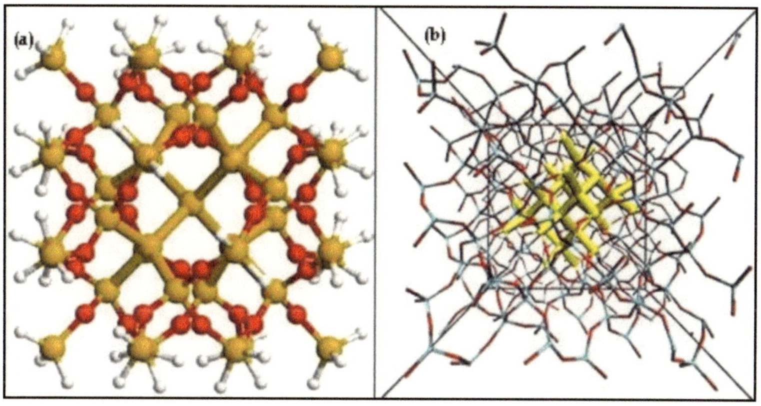

Fig. 7. (a) A stick and ball representation of $Si_{17}@Si_{24}O_{36}H_{60}$ (Si clusters embedded in one neutral SiO_2 shell), (b) The final optimized structure of the Si_{32} in β-cristobalite matrix

From the introduction stated above, we can clearly see that different structures should be used for simulating Si-QDs in different environment. Besides the structures presented above, there are many other structures used to simulate Si-QDs, we'll not elaborate here. It is crucial to choose an appropriate structure for a certain problem, otherwise the results may be inaccuracy and make no sense.

3. Method of calculation

Local-density approximations (LDA) are a class of approximations to the exchange correlation energy functional in DFT that depends solely upon the value of the electronic density at each point in space (and not, for example, derivatives of the density or the Kohn-Sham orbits). It was used to study on the band gap and PL of Si-QDs in early researches because of its simplicity and low computation cost. However, it is well known that time-independent LDA calculations typically underestimate the experimental photo-absorption gaps of Si-QDs (Vasiliev et al., 2001). TDLDA is then developed by Vasiliev as a natural extension of the ground state density-functional formalism and LDA, designed to include the proper representation of excited states.

The computation software we used is CASTEP module of Material Studio. CASTEP is an ab initio quantum mechanical program employing DFT to simulate the properties of solids, interfaces, and surfaces for a wide range of materials classes such as ceramics, semiconductors, and metals. First principle calculations allow researchers to investigate the nature and origin of the electronic, optical, and structural properties of a system without the need for any experimental input. It allows to choosing local, gradient-corrected, and non-local functionals for approximating exchange and correlation effect, and non-local functional include screened exchange, HF, B3LYP and PBE0. The screened exchange LDA (sX-LDA) is used to calculate the band structures of silicon quantum dots by considering the computational cost and accuracy. The structures with hydrogen and oxygen passivations are partially shown in Fig. 3 and Fig.4, respectively.

4. Energy gaps of silicon quantum dots

Researches of Si-QDs can be divided into two groups: experimental measurements and theoretical simulations. The results of experiments are coincide with that of simulations to some extent, in spite of some discrepancies attributing to the measurement errors in experiments and the idealized structures in simulations. In this section, some results from other researchers are introduced first. Our computational results are then illustrated in the figures.

4.1 Experimental results

Very Large amounts of works have been done on light emission from Si-QDs because they are believed to be the luminescence centres in PS which had been observed having visible PL and electroluminescence at room temperature. Generally, the properties of Si-QDs are studied by measuring their PL in experiments. As we mentioned above, it is so hard to precisely control the purity and size of the sample that the results of different experiments may be differ from each other to some extent.

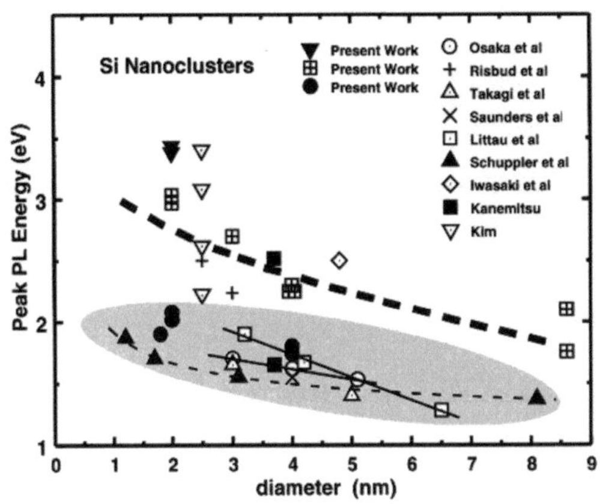

Fig. 8. Summary of experimental data on peak PL energy versus Si- QD size

Wilcoxon et al (Wilcoxon et al., 1999) had summarized most of the existing experimental results for PL peak energies as a function of the diameter of the Si-QDs, see Fig. 8. We note that there are differences in the dependence on size reported by different authors. As explained by Wilcoxon et al, all PL peaks of SiO_2 capped Si-QDs or Si-QDs embedded in glass matrices fall into the shaded region of the diagram while the experimental data for the oxygen-free samples (for the same value of the diameter) fall above this shaded region. So we can clearly see that the interface situation plays a significant role in the luminescence from Si-QDs.

In addition, Wolkin et al (Wolkin et al., 1999) obtained a similar conclusion by comparing the PL of PS samples in different atmosphere, see Fig. 9. They first examined the freshly etched samples (oxygen-free PS) with different porosities (equal to different sizes of Si-QDs) and emitting throughout the visible spectrum. Then the oxygen-free PS samples were exposed to air and their PL energies were measured again. We can clearly see from Fig. 9 that after exposure to air, a redshift of the PL is observed, which can be as large as 1 eV for blue luminescent samples that contain QDs smaller than 2 nm. Besides, authors pointed out that a PL redshift was also observed as soon as the oxygen-free PS samples were transferred from Ar to a pure oxygen atmosphere while no redshift at all was detected when the samples were kept in pure hydrogen atmosphere or in vacuum. Moreover, they found that there is an upper limit of the emission energy (2.1 eV) which is independent of size. So the conclusion is that the surface passivation plays an important role, especially the chemistry of oxygen at the surface can cause an evident change on the PL of Si-QDs and an upper limit of energy gap.

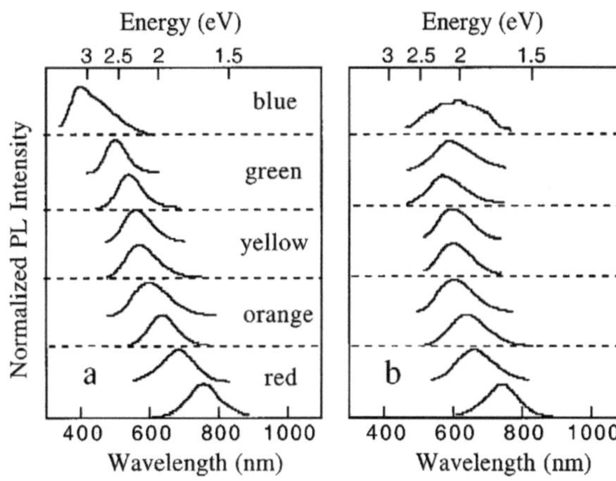

Fig. 9. Room temperature PL spectra from PS samples with different porosities kept under Ar atmosphere (a) and after exposure to air (b)

On the whole, despite some inevitable differences, the experimental results show that for ultra pure Si-QDs as their sizes decrease, there is a considerable blue-shift in the peak PL energy which is due to the increase of band gap. However, if the Si-QD is not pure enough, much more complex results will be obtained due to the influence of impurity contamination and complicate interface situation.

4.2 Theoretical calculation results

As we have mentioned above, a number of computations based on TDDFT, TDLDA, and other methods have been performed to simulate the properties of Si-QDs with different structures and passivations. In this section, we are going to introduce our calculation results and some theoretical results from other reports.

4.2.1 Hydrogen-passivated silicon quantum dots

Zdetsis and Garoufalis calculated the band gaps of hydrogen-passivated Si-QDs using TDDFT, the results are shown in Table 1 (Zdetsis & Garoufalis, 2005). As it can be easily seen in Table 1, with the size of Si-QDs decreases, the band gaps increases.

We have calculated the energy gaps of silicon QDs of different sizes from 0.6nm to 2nm in diameter by using the CASTEP computation program, which employ the plane-wave peseudopotentials method based on DFT. Some of the structures of Si-QDs we used are shown in Fig. 3. In the computations, screening exchange LDA (sX-LDA)/CA-PZ is used as the type of exchange-correlation potential. As comparison, LDA/CA-PZ is selected as the type of DFT exchange-correlation potential and calculations are performed again. The results are shown in Fig. 10. It can be seen that LDA/CA-PZ suffer from the well-known underestimation of the energy gap while sX-LDA/CA-PZ can obtain a more accurate result which is well consistent with the results from other literatures (Vasiliev et al., 2001; Conibeer et al., 2006).

No. of Si atoms	Total No. of atoms	$TDDFT^b$ (eV)
5	17	6.66
17	353	5.03
29	65	4.53
35	71	4.42
47	107	4.04
71	155	3.64
99	199	3.39
147	247	3.19

Table 1. The band gaps for small Si nanoclusters

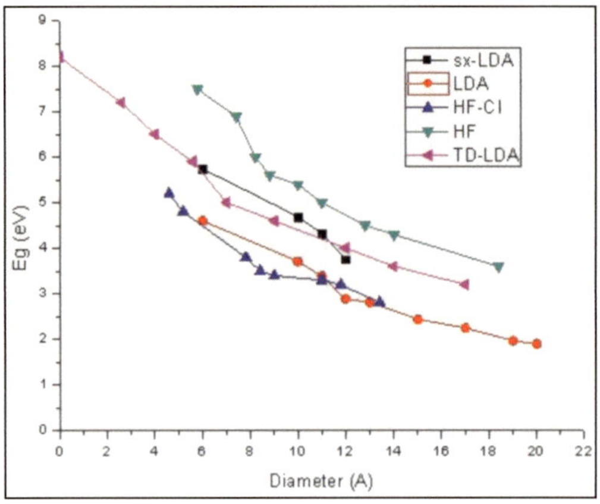

Fig. 10. Energy gaps of silicon QDs as a function of the QD diameter with H-termination from our work (■) & (●), and from the literatures [(▲) & (◀) from Vasiliev et al; (▼) from Conibeer et al]

4.2.2 Oxidized silicon quantum dots

Luppi and Ossicini (Luppi & Ossicini, 2005) carried out density-functional theory calculations on oxidized Si-QDs using local density approximation, considering three kinds of oxygen configurations: double-bonded, backbonded and bridge-bonded. The structures' schematic diagram has been shown in Fig. 4. They pointed out that the multiple presence of silanonelike Si=O bonds can be a reliable model for explaining the PL redshift observed in oxidized PS samples. The same conclusion has been reached by Zdetsis and Garoufalis (Zdetsis & Garoufalis, 2005) using TDDFT. Some of their calculated structures have been shown in Fig. 5.

However, a model based on a backbonded oxygen configuration has been proposed by Nishida to explain the observed PL redshift (Nishida, 2006). Nishida carried out self-consistent calculations using the extended Hückel-type non-orthogonal tight-binding method (EHNTB) for three different oxygen configurations (double-bonded, backbonded and inserted). The results are illustrated in Fig. 11. The author pointed out that the energy gaps calculated for Si dots double-bonded to oxygen are gradually decrease from 2.2 eV to about 1.7 eV with increasing dot size and the inserted oxygen configuration does not cause a significant energy-gap redshift even in the smallest Si-QD. At last he found out that the energy gaps calculated for the Si-QDs backbonded to oxygen coincide well with luminescence redshifts observed in PS.

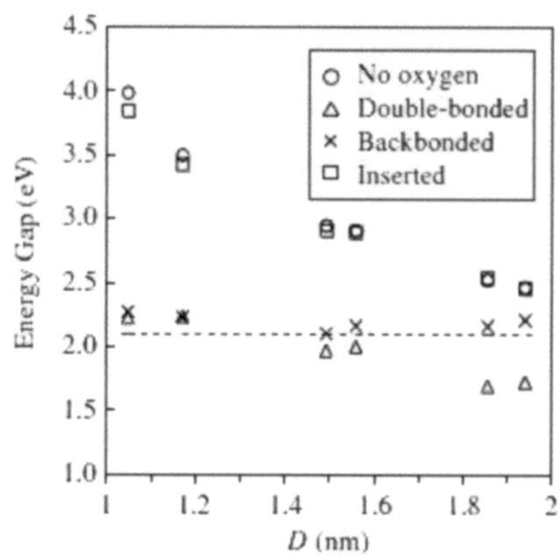

Fig. 11. Calculated energy gaps as a function of the diameter (D) of the Si -QDs studied. The dashed line shows the oxidation-induced peak energy (at ~2.1 eV) in PL spectra observed in PS

Obviously, the two conclusions stated above contradict each other. For this reason, we have performed calculations using the CASTEP computation program for all four kinds of oxygen configurations. The results we obtained based on LDA method are shown in Fig. 12, from which we can safely come to a conclusion that it is the formation of a Si=O double bond which is responsible for the PL redshifts in PS and other three types of oxygen configurations only cause a few decrease in energy gap.

The discrepancy between the above results may be due to the difference of the oxygen geometries, initial silicon clusters and the models based on different theories. So it is clear that different structures or models may obtain quite different results.

4.2.3 Silicon quantum dots embedded in different matrixes

König et al performed a systematic analysis of Si-QDs embedded in different environment by non-periodic spatial space density-functional–Hartree–Fock (DF–HF) computations of Si clusters comprising 1–165 Si atoms corresponding to a spherical QD diameter of d_{QD}=3.4–18.5Å (König et al., 2009). Some of the structures they used have been shown in Fig. 6 and their simulation results are illustrated in Fig. 13. Obviously, it shows that different interface termination causes different energy gaps. However, we can see that the entire trends of different interface terminations are the same. It is that the energy gap increase with the size decreasing.

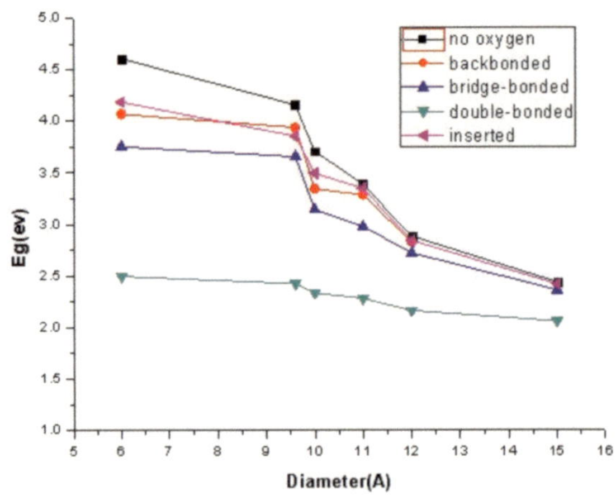

Fig. 12. Calculated energy gaps as a function of the diameter (D) of the Si-QDs

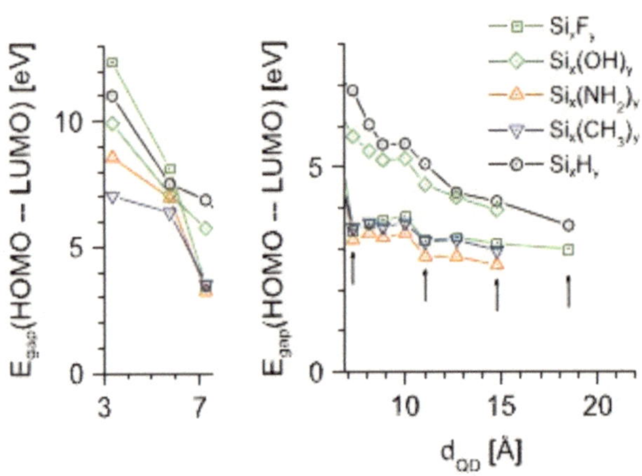

Fig. 13. Band gaps as a function of the diameter of Si-QDs with different interface terminations. For clarity, the graph is split into two sub-graphs with different energy scale. The points for the $Si_{10}X_{16}$ clusters (d_{QD} = 7.3 Å) are shown in both graphs

There are many more complicate calculations simulating Si-QDs in specific environments, we'll not list them one by one here. From the results above, we know that many factors will influence the properties of Si-QDs, other than the size, the interface termination also have a significant impact, so we should make an comprehensive considerations to study the properties of Si-QDs.

5. Conclusion

The band gap of Si-NCs or Si-QDs is size-dependent, and the energy of PL can be tuned from the near infrared to the ultraviolet by changing the size of nanocrystals. The surface passivation also plays an important role in determining the band gap. The third generation photovoltaic (PV) solar cell using tandem cells, which is based on Si quantum dots nanostructures, is proposed by Martin Green group (Conibeer et al., 2006, 2008). A number of theoretical models, computations results on PL have been reported, and much of research effort on the properties of Si-QDs has been performing currently. Based on CASTEP quantum mechanical program with choice of sX-LDA/CA-PZ, the band gap of Si-QDs with H-passivation and O-passivation is calculated. The results show that band gap increases when the size of quantum dots decreases for both H-passivation and O-passivation structures. The computation method sX-LDA/CA-PZ is comparable with TDLDA in the computational accuracy. Further research on the band gap and PL of Si-QDs by choosing other non-local functional as the exchange correlation is in progress.

6. Acknowlegment

This work is partially supported by National High-Tech Research and Development Program of China under Grant No 2007AA04Z301.

7. Reference

Biteen, J. S.; Lewis, N. S., and et al. Size-dependen oxygen-related electronic states in silicon nanocrystals. *Applied Physics Letters*, Vol. 84 (2004), pp. 5389-5391

Canham, L. T.. Silicon quantum wire array fabrication by electrochemical and chemical dissolution of wafers. *Applied Physics Letters* , Vol.57 (1990), pp.1046-1048

Carrier, P. ; Lewis, L. J. and et al. Optical properties of structurally relaxed Si/SiO_2 superlattices : the role of bonding at interfaces. *Physical Review B*, Vol. 65, (2002), pp.165339-1—165339-11

Cho, E. C., Park, S. W., and et al. Silicon quantum dot/crystalline silicon solar cells. *Nanotechnology*, Vol.19, (2008), pp.1-5

Conibeer, G. ; Green, M. and et al. Silicon nanostructures for third generation photovoltaic solar cells. *Thin Solid Films*, Vol. 511-512 (2006), pp.654-662

Conibeer, G. ; Green, M. and et al. Silicon quantum dot nanostructures for tandem photovoltaic cells. *Thin Solid Films*, Vol. 516(2008), pp.6748-6756

Fang, T. N. & Ruden, P. P.. Electronic structure model for n- and p- type silicon quantum dots. Superlattices and Microstructures, Vol. 22 (1997), pp.590-596

Garoufalis, C. S. & Zdetsis, A. D.. High level Ab Initio calculations of the optical gap of small silicon quantum dots. *Physical Review Letters*, Vol. 87 (2001), pp.276402-1—276402-4

Garoufalis, C. S. & Zdetsis, A. D.. Optical properties of ultra small nanoparticals:potential role of surface reconstruction and oxygen contamination. *Journal of Math. Chem.*, Vol. 46 (2009), pp.952-961

Guerra, R.; Degoli, E. and et al. Size, oxidation, and strain in small Si/SiO2 nanocrystals. *Physical Review B*, Vol. 80 (2009), pp.155332-1—155332-5

Jiang, C. W. & Green, M. A.. Silicon quantum dot superlattices: modeling of energy bands, densities of states, and mobilities for silicon tandem solar cell applications. *Journal of Applied Physics*, Vol. 99(2006), pp.114902-1—114902-7

König, D.; Rudd, J. and et al. Impact of interface on the effective band gap of Si quantum dots. *Solar Energy Materials & Solar Cells*, Vol. 93 (2009), pp.753- 758

Koponen, L.; Tunturivuori, L. and et al. Effect of the surrounding oxide on the photoabsorption spectra of Si nanocrystals. *Physical Review B*, Vol. 79 (2009), pp.235332-1—235332-6

Luppi, M. & Ossicini, S.. Ab initio study on oxidized silicon clusters and silicon nanocrystals embedded in SiO2: Beyond the quantum confinement effect. *Physical Review B*, Vol. 71 (2005), pp.035340-1 – 035340-15

Nishida, M.. Electronic state calculations of Si quantum dots: Oxidation effects. *Physics Review B.*, Vol. 69 (2004), pp.165324-1 – 165324-5

Nishida, M.. Calculations of the electronic structure of silicon quantum dots: oxidation-induced redshifts in the energy gap. *Semicond. Sci. Technol.*, Vol. 21 (2006), pp.443–449

Öğüt, S. & Chelikowsky, J. R.. Quantum confinement and optical gaps in Si nanocrystals. *Physical Review Letters*, Vol. 79 (1997), pp.1770-1773

Soni, R. K.; Fonseca, L. F. and et al. Size-dependent optical properties of silicon nanocrystals. *Journal of Luminescence*, vol. 83-84 (1999), pp.187-191

Tanner, M. G.; Hasko, D. G. and et al. Investigation of silicon isolated double quantum-dot energy levels for quantum computation. Microelectronic Engineering, Vol. 83 (2006), pp.1818-1822

Vasiliev, I.; Chelikowsky, J. and et al. Ab initio absorption spectra and optical gaps in nanocrystalline silicon. *Physical Review Letters*, Vol. 86 (2001), pp.1813-1816

Wang, B. C.; Chou, Y. M. and et al. Structural and optical properties of passivated silicon nanoclusters with different shapes: a theoretical investigation. *Journal of Phys. Chem. A*, Vol. 112 (2008), pp.6351-6357

Wilcoxon, J.; Samara, G. and et al. Optical and electronic properties of Si nanoclusters synthesized in inverse micelles. *Physical Review B*, Vol. 60 (1999), pp.2704-2714

Wolkin, M.; Jorne, J. and et al. Electronic states and luminescence in porous silicon quantum dots: the role of oxygen. *Physical Review Letters*, Vol. 82 (1999), pp.197-200

Zdetsis, A. & Garoufalis, C.. Real space ab initio calculations of exitation energies in small silicon quantum dots, In : *Quantum Dots: Fundamentals, Applications, and Frontiers*, B. A. Joyce et al. (2005), pp.317–332

8

Defect Related Luminescence in Silicon Dioxide Network: A Review

Roushdey Salh
Institute of Physics, Faculty of Science and Technology, Umeå University, Umeå Sweden

1. Introduction

The discovery of strong luminescence at room temperature from silicon cluster has attracted an enormous attention in recent years due to its potential applications in Si-based optoelectronic devices, especially blue and UV luminescence devices. Several studies have addressed the question of augmentation and stabilization of luminescence emission with different treatments, such as electron irradiation, thermal treatments and ion implantations. Even though observed visible light can be explained by the quantum confinement (QC) effect, many experimental results obey to the QC model supported surface state. Theoretically, the effects of quantum confinement on band gap of silicon indicate that it should be possible to obtain a blue light or shorter wavelength photons from the porous silicon if the silicon crystal size can be reduced to a certain value, e.g. to the diameter of 2-5 nm. However, it is rather difficult in practice to obtain a blue/white luminescence even when nanocrystals with diameters smaller than 2.0 nm are present. Some authors attributed this difficulty to the formation of surface Si–O bonds, which lead to mid-band gap trapped electrons and hole exciton states and thereby remove the dependency of the luminescence on cluster size. But in this case a luminescence emission was obtained in the red and yellow regions. It was found that some post-treatments such as anodic oxidation and oxidation in air could reduce the effective size of the Si nanocrystallites, thus causing a blue shift of the luminescence spectrum. If the red-yellow emission of the PS could be combined with a blue/green emission by modification of the PS surface such as oxidation or metal coating, it would be possible to obtain white light. This would mean an economical route to silicon based LEDs. Silicon dioxide, SiO_2, is widely distributed in the environment, and is present in the form of sand on all beaches and deserts. It is the starting material for the production of silicate glasses and ceramics. It may occur in crystalline or amorphous form, and is found naturally in impure forms such as sandstone, silica sand or quartz. Its specific gravity and melting point depend on the crystalline structure. Silica is known to occur in many crystalline phases or modes (α-quartz, β-quartz, α-cristobalite, β-cristobalite, α-tridymite, β-tridymite, γ-tridymite, coesite, melanophlogopite, keatite, fasriges, stishovite, chalcedon, agate, moganite, and others) besides some amorphous phases (opal, hyalite, sintered pearl, lechateierite, natural silica glass) [Fanderlik 1991]. Silicon dioxide is not only one of the most abundant materials on earth as mentioned, but also a critical material component of considerable technological importance. Today's modern electronics greatly depends on silicon dioxide for the manufacture of semiconductors and microelectronic devices, besides it is the basic material of which the most of common communication optical fibers are presently made. In its pure amorphous form, it is drawn into fibers for use in optical communications. Its high melting temperature (≈1700 °C) and its chemical stability make it an excellent thermal and electrical insulator and hence the principal dielectric in silicon electronic technology, particularly in metal-oxide-semiconductor (MOS) devices [Nicollian and Brews 2002], where it is used as a gate oxide. High quality SiO_2 films are obtained by thermal oxidation of silicon; thermally oxidized SiO_2 forms smooth and low-defect interfaces with Si.

From a fundamental point of view, silicon dioxide is equally useful as a model amorphous material because of its chemical simplicity. It is not surprising that much research effort has been expended to understand the properties of this material. Despite hundreds of publications available, only in a few cases have an unequivocal identification of the atomic structure of the defects and an explanation of their production mechanisms been attained [Griscom 1977, Skuja 1998]. Most of the available information comes from theoretical models and other experimental investigations of naturally or synthetically prepared specimens, but many questions still remain unsolved. Accordingly, many of the optical features characterizing the as-grown material (that is the material prior to any treatment) are not attributable to a specific defect having a known structure. Also, the identification of the precursor of radiation-induced defects and the exact mechanism of their generation are still debated. As the number of device components on a chip continues to increase, where the performance and reliability of these devices are controlled by electrically active defects, the precise control of oxidation and its understanding on an atomic scale of silica has been a subject of great interest and importance in the modern Si technologies, especially those related to thin oxidize layers with high quality. It has been observed that irradiation, mechanical stress, temperature change and the presence of impurities can effect the formation of defects and/or the transformation of existing defects to other defects in SiO_2 [Mitchell and Denure 1973, Trukhin et al. 1998, Fitting et al. 2004]. Irradiation of SiO_2 induces various types of structural defects, which cause degradation in SiO_2-based electronic and optical devices. Charge trapping at these defect sites degrades the current/voltage performance of metal-oxide semiconductor (MOS) devices, for example, and ultimately leads to device failure when such a device is exposed to high-energy radiation in space-based applications. Such and other problems are caused or at least initiated by the creation or transformation of luminescent point defects. Therefore a physical understanding of the structure and formation mechanisms of these defects is becoming increasingly important. But because of the nature of the chemical bond in silicon dioxide, a reliable guess of the detailed structure of particular luminescent defects has not been possible until now. A point defect embedded in a solid matrix (crystal or amorphous) is strongly interacting with its neighboring atoms. Then the study of the electronic states of the luminescent defects is a many-body particles system problem.

A half century of spectroscopy examinations of point defects in crystalline and glassy silica have led to identification and characterization of about a dozen intrinsic defect types (those constructed of silicon and/or oxygen atoms or ions) and around ten times as many defect varieties involving the presence of impurities. Luminescent defects in SiO_2 could be electrically charged or uncharged, diamagnetic or paramagnetic. Many such defects have been associated with optical absorption or emission bands, which may pose problems for long-path-length optical systems (fiber optics). Luminescent defects in amorphous silicon dioxide (a-SiO_2) have been defect investigated by numerous spectroscopic techniques, including optical absorption spectroscopy (OA), electron spin resonance (ESR) or the often so-called electron paramagnetic resonance (EPR), photoluminescence spectroscopy (PL), photoluminescence excitation spectroscopy (PLE), cathodoluminescence spectroscopy (CL), infrared spectroscopy (IR), Raman spectroscopy and many more. All of these experimental approaches have yielded valuable insights, but have not given an explicit model for defect types in silica network. ESR spectra have afforded some detailed atomic-scale structure characterizations, especially of two fundamental classes of defects: a dangling electron on Si atom denoted by E'-center and the oxygen deficient center (ODC). These defects have been known in *a*-SiO_2 since 1956 [Weeks 1956].

2. Structures, defects and properties of silicon dioxide

2.1 Structure of pure silicon dioxide

The basic structural unit of vitreous SiO_2 and silicate glasses is the SiO_4 tetrahedron [Mozzi and Warren 1969, Bell and Dean 1972, Gerber and Himmel 1986]. A tetrahedron is a polyhedron composed of four triangular faces, three of which meet at each vertex. A regular tetrahedron is one in which the four triangles are regular, or "equilateral," and is one of the platonic solids (convex polyhedron). There are four oxygen atoms, one located at each apex of a regular tetrahedron and a single silicon atom is located at the center of the tetrahedron, Fig. 2.1. This silicon atom has a valence charge of 4 in sp^3 orbitals, meaning that it is looking to acquire four electrons through sharing with other atoms to complete its outermost energy shell, known as the valence shell. An oxygen atom has two electrons in its outermost shell that are available to bond with the silicon atom. If four oxygen atoms surround one silicon atom, where each oxygen atom offers one electron, then the silicon atom's outermost shell will be complete and stable. The resulting arrangement comprises a silicate molecule. One electron remains, allowing those oxygen atoms to search for another silicon atom to share an electron and form another tetrahedron. Tetrahedrons are linked together through oxygen bonds and the arrangement of these links between the basic tetrahedral units determines the classification of the silicate. When the tetrahedra are not linked together, as each exists in isolation, the material is classified as a Nesosilicate. If groups of two tetrahedra are linked together, the material is then classified as a Sorosilicate. If all of the tetrahedra link back onto each other to form a closed ring, then the material becomes a Cyclosilicate.

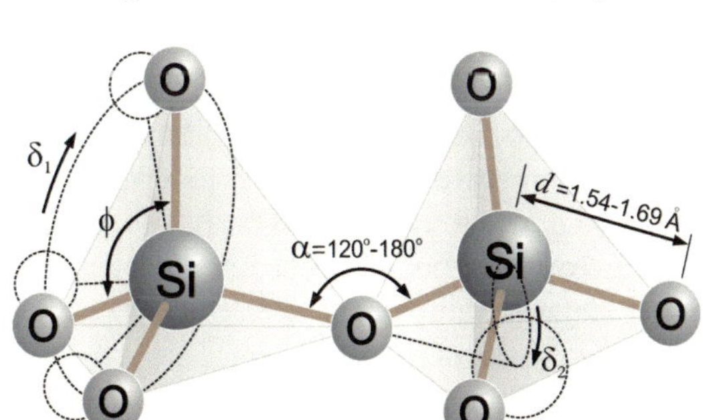

Fig. 2.1 Three-dimensional schematic of a pure fragment of the regular silica structure. The structure is defined by several parameters; the Si-O bond length (*d*), the tetrahedral angle (φ), the inter-tetrahedral bond angle (α), and the bond torsion angles (δ_1, δ_2), [Henderson and Baker 2002].

The bonding in silicon dioxide is complex. The Si-O bond energy is very high (4.5 eV) compared to the Si-Si bond (2.3 eV) and has an approximately 50:50 ionic:covalent nature and lone-pair oxygen orbitals are present [Fanderlik 1991, Edwards et al. 2000]. The covalency favors tetrahedral coordination of Si by O and maintains the O−Si−O bond angle (φ) very close to the ideal tetrahedral angle of 109.5°. Some of the literature indicates that the Si-O bond length (d) in various modifications of silicates may vary in the range of 0.154 nm to 0.169 nm [Slater 1965, Weber and Cromer 1965], while the mean value specified for d is 0.162±0.005 nm. In most cases one can treat the SiO_4 tetrahedra as rigid units, which can be linked together by their corners to form pairs, rings, chains, sheets, or frameworks. In this case the inter-tetrahedral Si−O−Si bond angle (α) can be defined. Measurements have been shown that the angle (α) varies with the different tetrahedra of the polyhedron, according to respective form of silicas (for vitreous SiO_2 from 120° to 180° and for quartz between 146° and 155°) [Fanderlik 1991, Nicholas et al. 1991].

Fig. 2.2 shows how the energy of a Si_2O_7 pair of tetrahedra varies as a function of Si−O bond length and the Si−O−Si bond angle [van Santen et al. 1991]. It is demonstrating that Si−O−Si bonds prefer to be linked rather than straight. However, the gently-sloping energy valley extending to 180°, which implies large variations in Si−O−Si bond angle (α) could occur in a structure with relatively slight energy penalty in order to accommodate changes in temperature, pressure, bulk composition, or local atomic environment. Silica does not produce any separate molecules (O=Si=O) because double bonds do not form. The parameters, such as the inter-tetrahedral bond angle (α) and the bond torsion angles (δ_1, δ_2), that define the way in which individual tetrahedra are linked together are highly variable [Mozzi and Warren 1969]. This variation in parameters distinguishes a glass from its corresponding crystalline analogue. To avoid the formation of a structural entity with an unsaturated electron cloud at the silicon atom in the case of a single bond, the $[SiO_4\}^{-4}$ tetrahedra join to produce a polymer with various spatial arrangements. As a result of the alternative possibilities of structural arrangement, there exist a number of modifications of crystalline silica. In other words, the different types of silicate structure arise from the ways in which these tetrahedra are arranged. They may exist as separate unlinked entities, as linked finite arrays, as infinite one-dimensional chains, as infinite two-dimensional sheets or as infinite three-dimensional networks.

2.2 Intrinsic defects in silicon dioxide

The presence of defects in the silica matrix can dramatically change its structural, electrical, and optical properties. Many parameters, such as manufacturing processes, irradiation, mechanical stress, change of temperature, and the presence of impurities may cause the formation of defects and/or lead to the transformation of the existing defects to other types of defect. Commonly, defects can be grouped according to their structure and size as point defects, dislocations (linear defects), and plane defects [Stevens-Kalceff and Philips 1995, Kofstad 1988]. The following types of point defects can be considered as intrinsic or extrinsic. Intrinsic point defects involve atoms of the host matrix only, i.e. vacancies (the host atoms are missing, Schottky defects or Frenkel pairs) and self-interstitials (additional host atoms at an interstitial position). Extrinsic point defects involve atoms chemically different from the host crystal, such as those used for electrical doping. Defects in a perfect silica glass could include oxygen or silicon vacancies and their interstitials, Si−Si or O−O homobonds or under-coordinated silicons or oxygens.

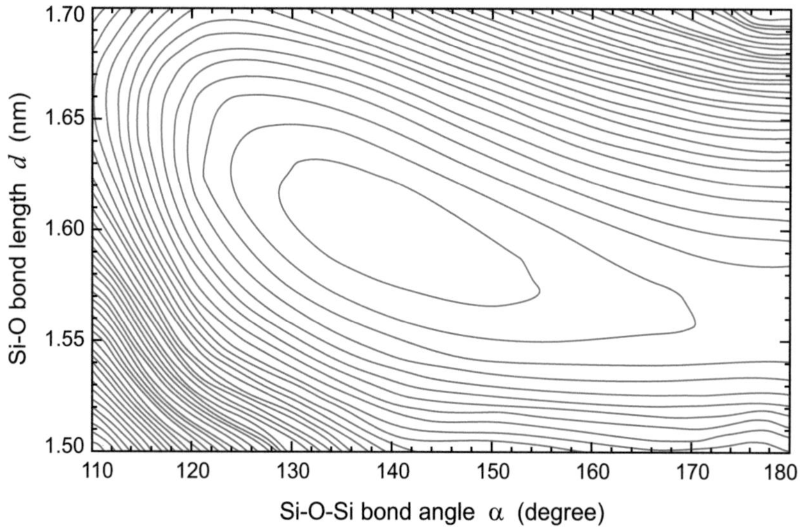

Fig. 2.2 Total energy as a function of Si–O bond length and Si–O–Si bond angle. Contour lines in units of kJ/mol, [van Santen et al. 1991].

Due to the high Si–O bond energy (4.5 eV), the crystalline quartz is resistant to chemical weathering (it is only soluble in hydrofluoric acid and in hot alkalis) and to corrosion [Lamkin et al. 1992, Lide 2004]. It is hard (Mohs' hardness 7), brittle, and has a very high melting point at around 1710°C [Lide 2004]. Due to its wide energy band gap of about 9 eV, it is optically transparent and shows low electrical conductivity [Fan et al. 1998].

Generally, a homobond is electrostatically neutral although both Si–Si and O–O bonds may become positively charged by trapping holes. Irrespective of their electrical charges, defects can be divided into two classes: diamagnetic and paramagnetic. As a general role, all stable paramagnetic defects have optical absorption bands associated with them, since they represent half-occupied energy transitions to the valence band and electron transitions to the conduction band are both possible. Diamagnetic defects may have absorption bands associated with electron transitions to the conduction band. The confirmed examples of diamagnetic defects in *a*-SiO_2 have electron absorption bands in the ultraviolet or vacuum ultraviolet spectral regions, implying that the uppermost filled levels of these states lie below the middle of the 9 eV band gap [Griscom 1977].

A variety of defect structures are known to exist in silica materials and were one of the major subjects of extensive experimental and theoretical studies [Stevens-Kalceff 2000, Song et al. 2001, Lu et al. 2002]. Many aspects regarding the nature of the defects and their correlated properties are still controversial and not yet completely understood. Quite a lot of defect types have been discussed in the literature and many reproduction models have been proposed for each one. In this part we will review the main defects in the silica network but whether any of these models is correct remains an open question of considerable interest.

2.2.1 *E´*-center

Probably the best known paramagnetic defect in all forms of SiO_2 is the *E´*-center which was first detected in late fifties using electron paramagnetic resonance (EPR) spectroscopy [Weeks 1956, Weeks and Nelson 1960, Griscom et al. 1974, Gobsch et al. 1978]. It is associated with the 5.85 eV absorption band in quartz and silica glass and no associated emission band has been observed where its nonradiative mechanism has been reported by some authors [Pacchioni et al. 1998a , Kajihara K. et al. 2003]. From studies of the hyperfine structure in the EPR spectrum it is known that *E*´-center can comprise an unpaired electron in a dangling tetrahedral (*sp*³) orbital of a single silicon atom which is bonded to just three oxygens in the glass network [Griscom 1979a, Isoya et al. 1981]. This generic *E*´-center is shown in Fig. 2.3, which is often denoted by ≡Si•, where the three parallel lines represent three oxygen separate bonds to one silicon atom and the dot denotes the unpaired electron.

Fig. 2.3 Generic *E*´-center. The large atom is silicon, the smaller ones are oxygens

Previous EPR studies on irradiated *a*-SiO_2 have demonstrated that there are several distinguishable variants of the *E*´-center in terms of their *g* values but in common all have the structure ≡Si• [Griscom 1990a]. These *E*´-center variants are also distinguished by virtue of different annealing kinetics depending on both the character of the irradiation and the water contents in dry or wet oxidized SiO_2, as shown in Fig. 2.4 [Griscom et al. 1983, Griscom 1984, Griscom 1985].
Four main types of *E*´-centers, labeled E'_{α} , E'_{β} , E'_{γ} and E'_{S} have been identified in vitreous silica depending on their spectroscopic signatures [Skuja 1998]. Several models have been suggested based on different precursors for each of these defects where some of these types are associated with hydrogen atoms. Optically stimulated electron emission technique (OSEE) shows that each one of these types of *E*´-centers has a distinguishable absorption band in the range of 5.7 eV [Zatsepin et al. 2000], see Fig. 2.5.

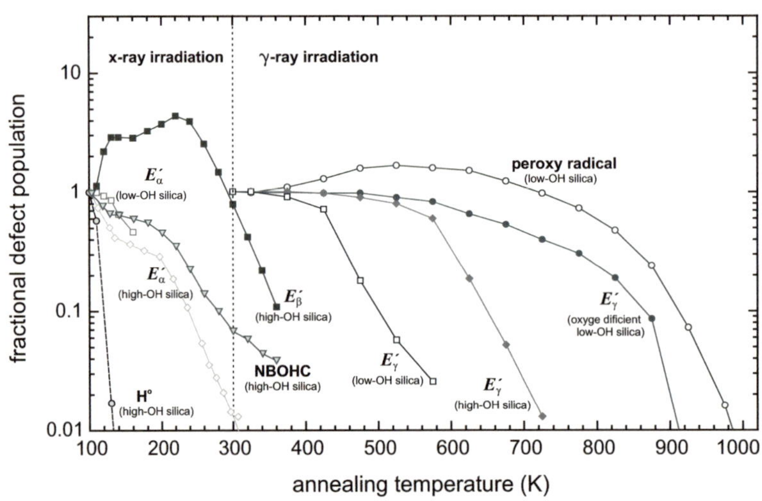

Fig. 2.4 Normalized isochronal anneal curves for radiation-induced defect centers (*E*´, NBOHC and peroxy radical) in high-purity silica (low-OH silica: <5 ppm OH and high-OH silica: 1200 ppm OH), [Griscom 1984].

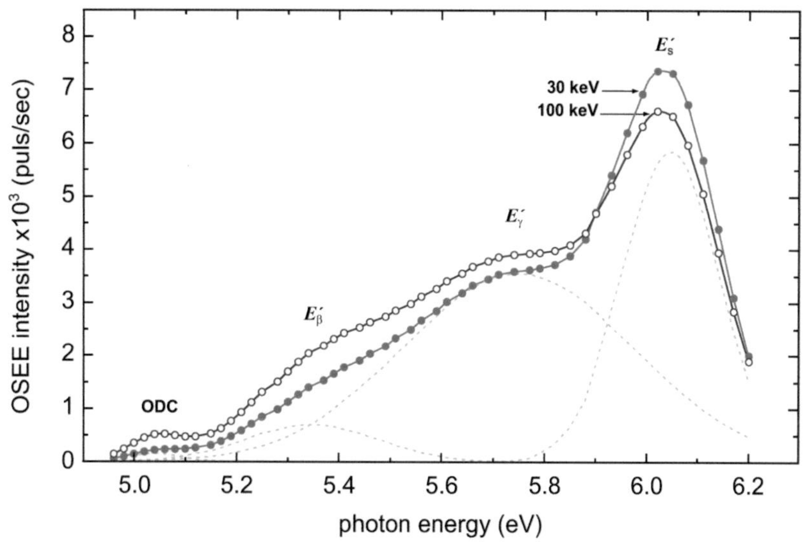

Fig. 2.5 OSEE spectra of glassy SiO_2 irradiated by Fe^+ ions of two different energies, 30 and 100 keV, the absorption bands of E'_β, E'_γ and E'_S-centers are detected, besides a very weak absorption band associated with oxygen deficient centers (ODC), [Zatsepin et al. 2000].

It was inferred that the E'_α variant in silica initially observed by Griscom [Griscom 1984], is a defect which tends to anneal in times on the order of minutes up to hours above 100 K. It was suggested that this center is created by a radiolytic process which moves an oxygen atom from an undisturbed network site ≡Si−O−Si≡) into a neighboring position which must be chemically bonded, since insufficient energy can be transferred from an X-ray generated compton electron to result in a net breakage of bonds [Uchino et al. 2001]. Fig. 2.6 illustrates one of the conceivable ways in which such a process could come about. The oxygen-oxygen (peroxyl ≡Si−O−O) bond suggested to be formed in Fig. 2.6 should be a relatively stable entity according to recent theoretical calculations [Griscom 1979a]. Still, less exotic mechanisms for E'_α production, not inconsistent with the data, might be the momentary rupture of strained oxygen bonds ≡Si··O−Si≡). Here •O−Si≡ is the notation for the non-bridging oxygen hole center (NBOHC), and is in fact seen by electron spin resonance (ESP) in X-ray irradiated silicas in numbers comparable to the E'_α-center.

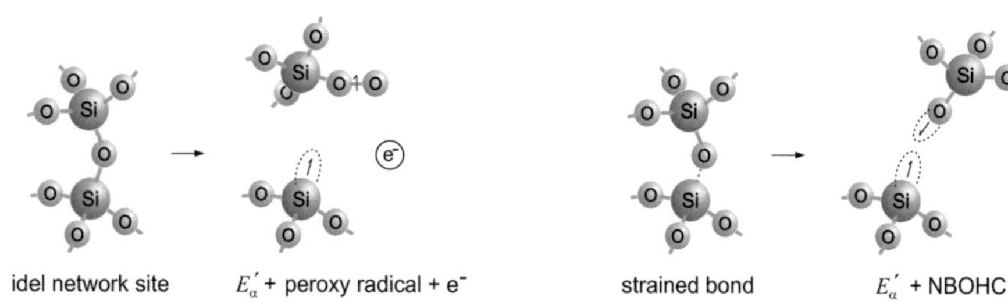

Fig. 2.6 Schematic models for the E'_α-center in pure a-SiO_2, the arrow denotes the unpaired spin and dashed balloons represent their orbital.

E'_β in silica network (E'_2 in quartz) features a proton trapped in the oxygen vacancy and the silicon atom containing the unpaired spin relaxed outwards [Griscom 1991, Weeks 1963], i.e.

the interaction of the unpaired spin associated with a long-bond silicon with the hydrogen atom is weak enough to not saturate each other. Two possible formation reactions of E'_{β} are shown in Fig. 2.7.

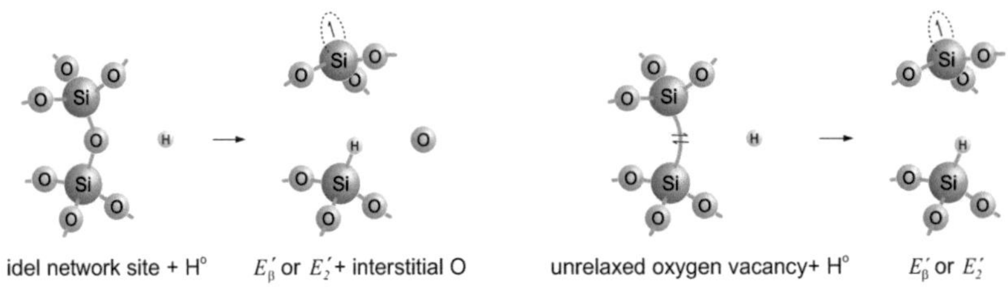

Fig. 2.7 Schematic models for the E'_{β}-center in pure *a*-SiO_2. The arrow denotes the unpaired spin and dashed balloons represent their orbital. The E'_{β}-center is considered to be the closest analog for E'_{2}-center in quartz.

E'_{γ}-center is the closest analog of the E'_{1}-center in α-quartz [Griscom 1980, Boero et al. 1997]. According to current theoretical calculations [Feigl et al. 1974, Yip and Fowler 1975, Mysovsky et al. 2004], E'_{γ} is suggested to consist of a positively charged single oxygen vacancy composed of a nearly planar ≡Si^+ unit and a singly occupied dangling bond ≡Si•, namely, ≡Si+ •Si≡ [Uchino et al. 2000b, Agnello et al. 2002]. An unrelaxed oxygen monovacancy (≡Si··Si≡) or an unperturbed SiO_2 fragment (≡Si−O−Si≡) is assumed to be the precursor of this defect as shown in Fig. 2.8. There is no indication that hydrogen is involved in this defect [Feigl et al. 1974]. E'_{γ} is stable for years at room temperature [Griscom 1984].

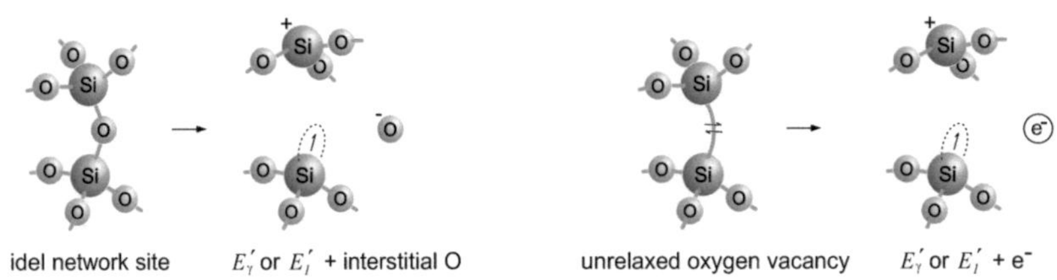

Fig. 2.8 Schematic models for the E'_{γ}-center in pure α-SiO_2. The arrow denotes the unpaired spin and dashed balloons represent their orbital. The E'_{γ}-center is considered to be the closest analog for E'_{1}-center in quartz.

Relaxing of the Si atom with the unpaired spin towards oxygen vacancy results in the E'_{4}-center. It is in fact the most reliably characterized of these defects depending on the experimental and theoretical analysis [Isoya et al. 1981]. E'_{4}-center consists of a hydrogen substituting for an oxygen atom in α-quartz [Mysovsky et al. 2004]. This center, Fig. 2.9, is observed in crystalline silicon dioxide (α-quartz) but there is no evidence of existence of E'_{4}-center in silica glass [Griscom and Friebele 1986]. Some other authors [Rudra et al. 1985, Majid and Miyagawa 1993, Snyder and Fowler 1993] suggested that the E'_{2} and E'_{4} are in fact the same defect, but with long-bond silicon relaxed through the plane of its three oxygen neighbors such that the unpaired spin points away from the vacancy. But this configuration is predicted to be slightly lower in energy than the E'_{4} configuration. In surface center studies, several variants of surface E'-centers were found [Bobyshev and Radtsig 1988]. Two of them are depicted in Fig. 2.10, E'_{S}(*1*) which seems like the normal E′-center but with a constant isotropic hyperfine splitting, and the second is E'_{S}(2) which has a dangling silicon bond with a neighboring hydroxyl (OH) group [Skuja 1998].

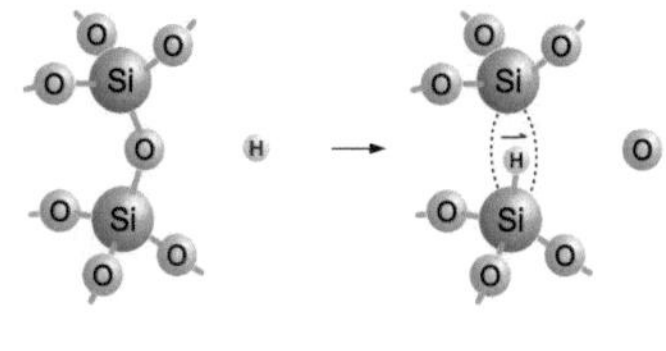

Fig. 2.9 Schematic model for E'_4-center in crystalline SiO_2. The arrow denotes the unpaired spin which interacts with a hydrogen atom to relax the oxygen vacancy by forming Si–Si bond.

Fig. 2.10 Schematic models for surface E'-center (E'_S) in pure a-SiO_2. The arrow denotes the unpaired spin and dashed balloons present their orbital. Hydrogen saturates the dangling oxygen bond

2.2.2 Oxygen-deficiency center (ODC)

It should be mentioned first that all E'-center types are also considered as oxygen deficiency centers but in this subsection a review of a different (non-paramagnetic) kind of oxygen deficiency center will be given. This defect center is entitled simply by a neutral oxygen vacancy which is often denoted ODC and indicated generally as ≡Si–Si≡.

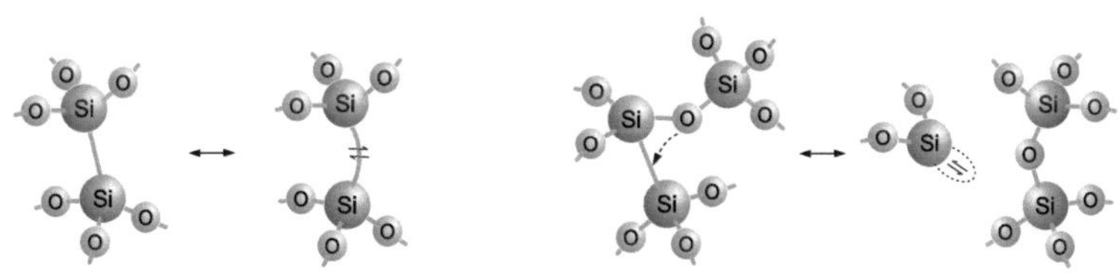

Fig. 2.11 Schematic illustration of the transformation between ODC(I) and ODC(II) visualizing two possible models for the ODC(II), the unrelaxed oxygen vacancy and the twofold coordinated silicon.

It is diamagnetic and can be directly investigated by photoluminescence (PL) or cathodoluminescence (CL) spectroscopy. The literature mostly describes two models for the ODCs: neutral oxygen vacancy ODC(I) and the twofold coordinated silicon ODC(II) denoted as =Si••. The ODC(I) represents one of the essential defects in all silicon dioxide modifications in a form of simple oxygen vacancies; here two Si atoms could relax and make a silicon silicon bonding (relaxed oxygen vacancy ≡Si–Si≡) or stay in unstable interaction condition and form an unrelaxed oxygen vacancy (≡Si···Si≡) which each one of them could be a precursor for the other under some undeclared circumstances, see Fig. 2.11, and both are considered as a key role in many defect-type generations and transformations in the silica matrix, as shown in Figs. 2.7 and 2.8. The 7.6 eV absorption band in irradiated and as grown a-SiO_2 has been ascribed to the neutral oxygen vacancy ODC(I) [Imai et. al 1988,

Hosono et al. 1991]. The ODC(I) can also be converted to ≡Si−H groups in thermal reaction with hydrogen molecules according to the visualized reaction shown in Fig. 2.12.

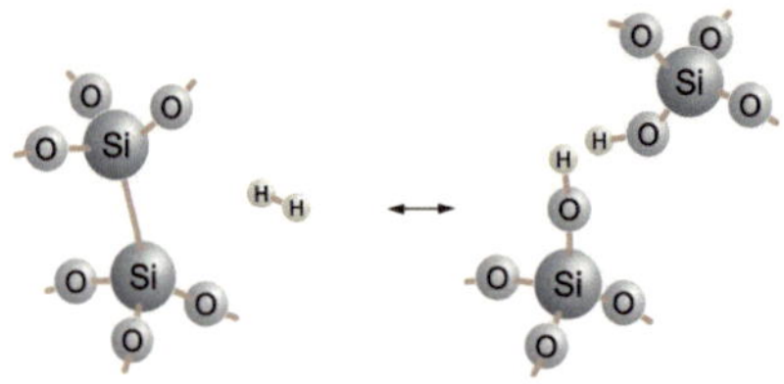

Fig. 2.12 Schematic illustration of the ODC(I) conversion to silanol groups in thermal reaction with hydrogen molecules.

In addition, two photoluminescence (PL) bands, 4.4 eV (decay constant τ=4 ns) and 2.7 eV (decay constant τ=10.4 ms) have been observed under excitation of the 5 eV, 6.9 eV or 7.6 eV bands, indicating the interaction of ODC(II) with ODC(I) [Nishikawa et al. 1994, Seol et al. 1996]. Based on their lifetimes, the 4.4 eV and 2.7 eV bands have been ascribed to singlet-singlet ($S_1 \rightarrow S_o$) and triplet-singlet ($T_1 \rightarrow S_o$) transitions at the site of oxygen-deficient type defects, respectively [Skuja 1998]. The interconversion between the ODC(I) and ODC(II) in an energy diagram is given in Fig. 2.13. The origin of ODC(II) associated with the optical absorption band at ~5 eV is one of the most controversial issues in the defect research field of a-SiO2 [Skuja et al. 1984, Griscom 1991, Skuja 1992a, Skuja 1998]. The first model hypothesis suggested for ODC(II) was a neutral diamagnetic oxygen vacancy [Arnold 1973], later two other models have been proposed for ODC(II): twofold coordinated silicon [Skuja et al. 1984, Skuja 1992a] and the unrelaxed oxygen vacancy [Imai et al. 1988] as shown in Fig. 2.11.
The oxygen vacancy model was further supported by the finding that two-photon photobleaching of SiODC(II) by KrF laser ($\hbar\omega$=5 eV) generates E'-centers [Imai et al. 1988]. But the origin of the ODC(II) is still a matter of controversy.

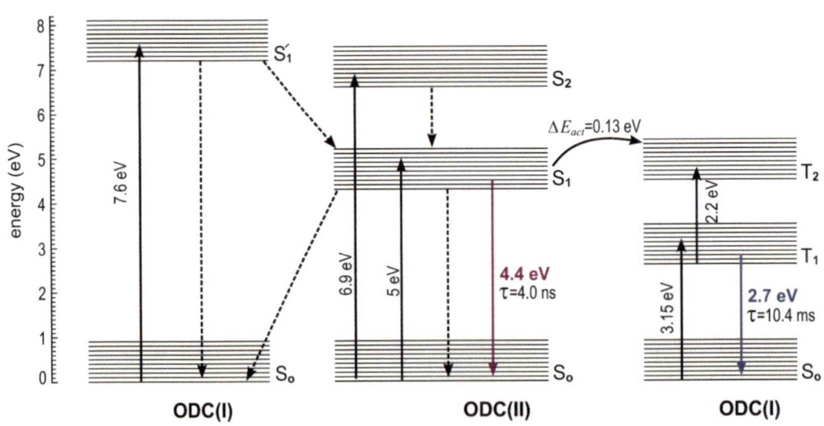

Fig. 2.13 Diagram of the energy levels proposed for the optical transitions at the site of two oxygen-deficient centers: ODC(I) and ODC(II). Transformation between the two states is assumed by the excitation at 7.6 eV. Solid and dotted arrows represent radiative and non-radiative electronic transitions, respectively. ΔE_{act} is the thermal activation energy for singlet-triplet conversion, τ are the radiative decay times, [Skuja 1998 and Nishikawa 2001].

2.2.3 The non-bridging oxygen hole center (NBOHC)

This center can be visualized as the oxygen part of a broken bond (Figs. 2.6 and 2.15). It is electrically neutral and paramagnetic and represents the simplest elementary oxygen-related intrinsic defect in silica. It is well characterized both by EPR and by optical spectroscopies like photoluminescence (PL) and cathodeluminescence (CL).

The main optical characteristics of NBOHC are shown in Fig. 2.14, it has an absorption band at 4.8 eV with FWHM=1.07 eV, oscillator strength f=0.05; an asymmetric absorption band at 1.97 eV, FWHM=0.17 eV, $f=1.5\times10^{-4}$; a photoluminescence band excited in these two absorption bands, at 1.91 eV , FWHM=0.17 eV, decay constant around 20 μs. Out of these three characteristics, the most unique fingerprint of this center is the 1.9 eV luminescent band in the red region of the visible light spectra.

It has been postulated that the NBOHC arises when hydrogen atoms are liberated radiolytically from one member of a pair of OH groups in wet silica (high OH group) according to Fig.2.15 [Stapelbrok et al. 1979].

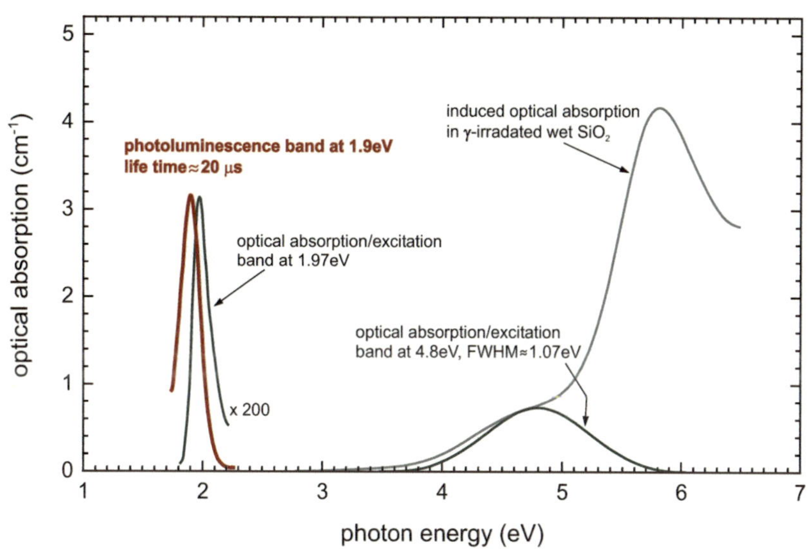

Fig. 2.14 Optical absorption and luminescence spectra of γ-irradiated wet silica illustrating the main optical properties of NBOHC: the absorption/excitation bands at 4.8 eV and 1.97 eV, and the photoluminescence band at 1.9 eV, [Pacchioni et al. 2000].

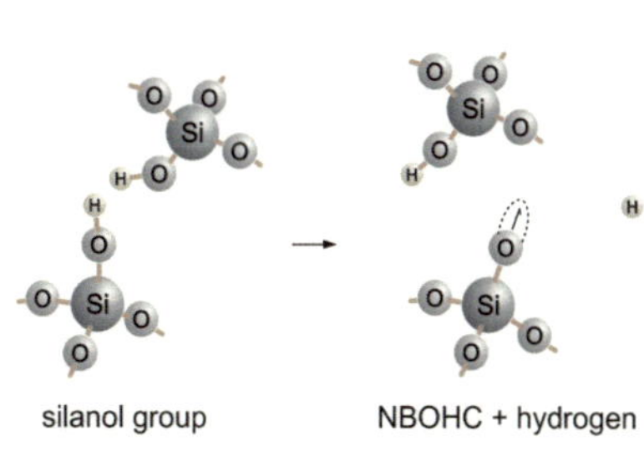

Fig. 2.15 A model of atomic structure of the non-bridging oxygen hole center (NBOHC) showing the possible generating processes of NBOHC in wet silica.

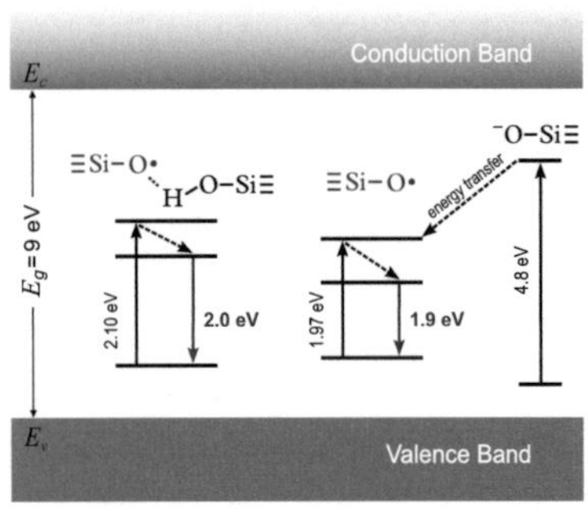

Fig. 2.16 Energy band diagram of different NBOHC energy states, [Munekuni et al. 1990].

However this reaction is not the only way of creating NBOHC. Oxygen dangling bonds may be created as well in wet and in dry silica (negligible amounts of OH groups) by rupturing of the strained Si−O bonds (≡Si···O−Si≡) in the silica network (Fig. 2.6). Particularly there are no spectroscopic distinctions which have been established between the centers formed by these two precursors, but on the other hand some authors [Munekuni et al. 1990] proposed some differences in their emission energies, see Fig. 2.16.
If softer irradiation (X-ray) was used, the centers were created only in groups of Si−O−R (R: alkali ion). This behavior provides evidence that the centers are created in reactions similar to that visualized in Fig. 2.15, and they were attributed to NBOHC [Skuja 1994a, Skuja et al. 2006]. On silica surfaces, the same red luminescence band can be created by adding O atoms to surface *E*´-centers [Streletsky et al. 1982]. Another generic oxygen hole center is the self-trapped hole (STH), which exists in two different variants. STH_1 comprises a hole trapped on a normal bridging oxygen in the network (≡Si−°O−Si≡), while the STH_2 is suggested to consist of a hole delocalized over two bridging oxygens [Griscom 1991, Griscom 2000].

2.2.4 Peroxy bridge (POL)

In oxygen-excess silica, some of the excess oxygen is expected to form "wrong" oxygen-oxygen bonds, called peroxy bridges or peroxy linkages (≡Si−O−O−Si≡). Calculations of atomic oxygen diffusion in SiO_2 suggested that POL structure is the lowest energy configuration for an oxygen interstitial [O`Reilly and Robertson 1983]. However, a definitive spectroscopic confirmation of their presence in silica is still absent. The experimental evidence is only indirect, but it is thought to be responsible for the exclusive (without the accompanying Si−H groups) generation of Si−OH groups during H_2 treatment of oxygen rich silica [Imai et al. 1987], as shown in Fig. 2.17. This reaction is accompanied by an increase of VUV optical absorption for $h\nu$>7 eV indicating that the POL could possibly absorb in this region. POL was initially suggested to be the main precursor of peroxy radical defects, Fig. 2.18, as we will show in the following subsection [Friebele et al. 1979]. The calculation put the energy of the POL absorption band at around 6.4-6.8 eV with a small oscillator strength, f=2×10^{-4} [Pacchioni and Ierano 1998b], such absorption would be hard to detect against the background of other bands in vacuum UV.

2.2.5 Peroxy radical (POR)

The Peroxy radical (POR) in silica is a paramagnetic defect with a hole delocalized over anti-bonding π-type orbitals of the O−O bond in the structure illustrated in Figs. 2.17 and 2.18. EPR spectroscopy shows that the POR is the best characterized oxygen excess defect in silica

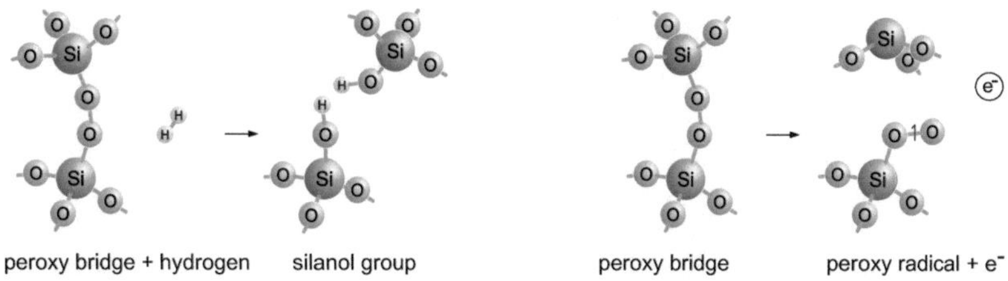

Fig. 2.17 Models presenting the suggested atomic structure of a peroxy bridge (POL) and its role in producing other possible defects in silica matrix.

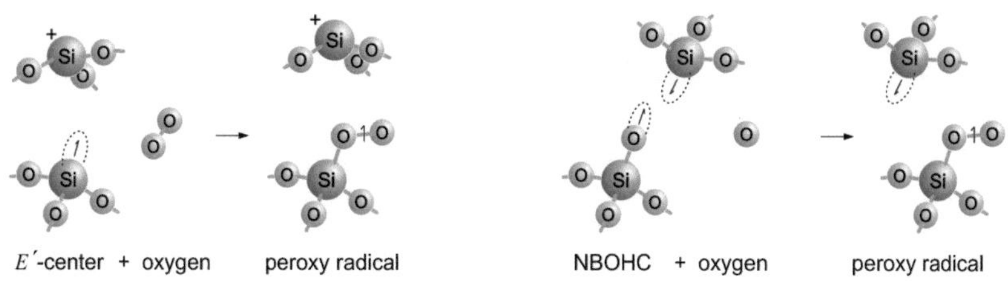

Fig. 2.18 Models presenting some possible generation modes for the peroxy radical (POR) structure in the silica matrix.

[Griscom 1991, Friebele et al. 1979]. However, the optical properties of POR in bulk silica are not accurately known. Good correlations between the isochronal annealing curves of EPR signals of POR and of the optical absorption band at 7.7 eV were reported in γ-irradiated dry silica [Stapelbroek et al. 1979]. The optical absorption spectrum observed for peroxy radicals on the surface of SiO_2 by the diffuse reflectance technique in the region around 5.4 eV with FWHM 1.3 eV and oscillator strength $f\approx0.067$ was calculated [Bobyshev and Radtsig 1988].

2.2.6 The self-trapped exciton (STE)

The electronic excitation of solids produces mainly electrons, holes and excitons. Transient (short living) defects can be created through the combination of the electronic excitation energy of electron-hole pairs and electron-phonon interaction. The conversion of excitation to defects is initiated by self-trapping of excitons, by the trapping of electrons by self-trapped holes or by the consecutive trapping of an electron and hole by a defect. These transient defects can produce either radiative or non-radiative electronic transition, while non-luminescent transient defects disappear by recombination of defect pairs. Self-trapping is a widespread phenomenon in insulators [Hayes and Stoneham 1985, Song and Williams 1993].

The existence of the self-trapped excitons in crystalline SiO_2 is supported by experimental measurements of the optically detected magnetic resonance and transient volume change [Itoh et al. 1988]. The luminescence bands between 2 and 3 eV in the silica spectrum have been ascribed to the STE. Some authors suggested that the STE is the source of the characteristic blue luminescence in crystalline SiO_2, but it has been observed that this luminescence band is removed in quartz by intense electron irradiation (15 keV) at room temperature due to the electron hole recombination as shown in Fig. 2.19 [Griscom 1979b, Trukhin 1978, Trukhin 1980, Barfels 2001]. Almost the same luminescence band can be detected in the emission spectra of amorphous SiO_2 but with much lower intensities than the other characteristic bands. STE perturbed by small distortions due to a structural defect give emissions in the same energy range. For example, Ge implanted quartz exhibits a luminescence band at 2.5 eV close to 2.8 eV in non-implanted quartz [Hayes and Jenkin 1988]. The excitation spectra for STE luminescence in α-quartz show a peak at 8.7 eV, which is ascribed to the first exciton peak. The absorption edge has been determined as 9.3 eV [Itoh et al. 1989], so the exciton binding energy is about 0.6 eV for α-quartz [Bosio and Czaja 1993]. The large energy difference between the band edge absorption (about 9 eV) and luminescence (2.8 eV) points to strong electron-photon coupling. The optical absorption spectra and the excitation spectra for fused silica are similar to those of α-quartz but exhibit modifications due to the amorphous structure [Trukhin 1992].

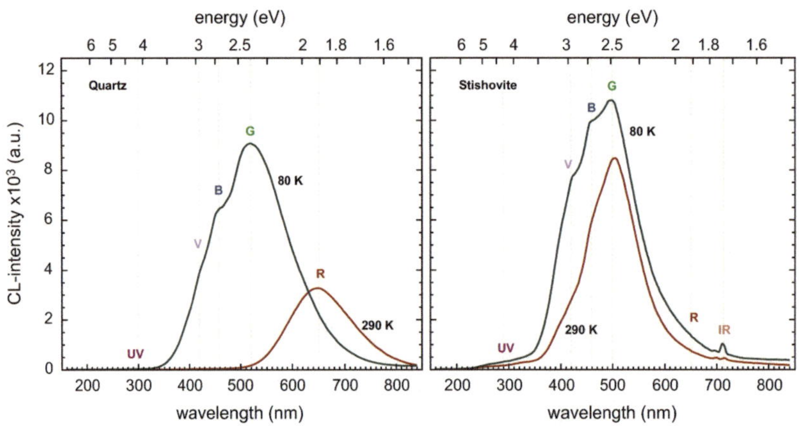

Fig. 2.19 CL-spectra of some crystalline SiO_2 (Quartz and Stishovite) at 290 K and 80 K, [Barfels 2001].

Meanwhile, several models have been proposed implying to clarify the STE. One of the first models considering that an oxygen atom will removed to a peroxy bridge position [Griscom 1979a, Griscom 1979b] and other models are based on the proposition that a threefold coordinated silicon explains the transient absorption at 5.2 eV (*E´*-center) [Trukhin 1992, Trukhin 1994], see Fig.2.20. All of these suggested models are based on the idea that the silicon-oxygen bond (Si−O) gets ruptured and forms an oxygen-oxygen bond (−O−O−) based on the fact that different local structures of the SiO_2 network provide different distances for oxygen-oxygen bonding. Each oxygen atom bonded to two silicon atoms by two types of Si−O bond, one by long bond ≈1.612 Å and another by short bond ≈1.607 Å, as shown in Fig. 2.20 by dashed and solid bond connections [Hayes et al. 1984, Trukhin 1994]. These models explain different STE luminescence properties of different structures.

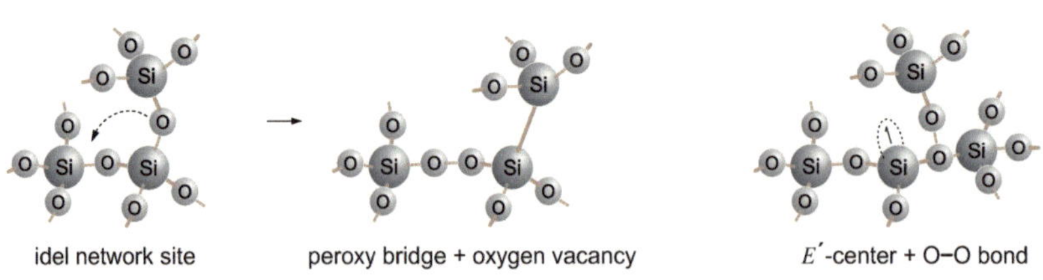

Fig. 2.20 Models of self-trapped exciton (STE) showing a creation of oxygen vacancy, *E´*-center and peroxy bridge due to the decay of a STE associated with an excited Si−O bond in crystalline SiO_2.

2.2.7 Interstitial oxygen

Mostly all variants of manufactured high-purity dry SiO_2 contain natural interstitial oxygen atoms and an additional amount can be generated by ejecting oxygen atoms from their normal sites in the SiO_2 network during the irradiation. Another important point that has to be considered is the molecular oxygen. Principally O_2 molecules can be formed in irradiation processes from the already present oxygen atoms [Morimoto et al. 1992]. The first spectroscopic observation of O_2 molecules in silica was performed by Raman spectroscopy in optical fibers [Carvalho et al. 1985]. O_2 was detected by O–O stretching vibration at wavenumber 1549 cm^{-1}. Using this method the O_2 molecule concentration is found to be in the range of 10^{14}-10^{18} molecules per cm^3 in dry silicon dioxide [Skuja et al. 1998a].

In gas phase, oxygen molecules (O_2) dissolved to two atomic oxygen (2O) for $h\nu$>5.1 eV (λ<242 nm); the same can be expected to occur in silica glass following the photolytic reaction shown in Fig. 2.21. The atomic oxygens appearing as a result of this reaction might be expected to be relatively mobile in silica even at room temperature; then they interact with other oxygen molecules to form ozone (O_3) just as it occurs in the earth's atmosphere [Baulch et al. 1980, Finlayson-Pitts and Pitts 1986]. There are indications that the atomic oxygen becomes mobile at around 400 °C as detected by formation of the interstitial oxygen molecules at this temperature [Skuja et al. 2002, Kajihara et al. 2004]. Depending on the quantum yield for the gas-phase reaction of ozone dissociation, it is believed that O_3 molecules are responsible for both the 4.8 eV absorption and 1.9 eV luminescence in certain oxygen-rich silicas [Awazu and Kawazoe 1990, Griscom 1991, Skuja et al. 1995], which usually are attributed to the NBOHC by many authors [Kajihara et al. 2001, Cannas and Gelardi 2004, Bakos et al. 2004b]. The 1.9 eV luminescence band shifts between 1.8 and 2 eV depending on the excitation wavelength within the 2 eV absorption band [Skuja et al. 1995]. Other bands of possibly related origin have been observed in the 2.0-2.5 eV regions [Munekuni et al. 1990, Skuja 1994a]. O^-_2 and O^-_3 can be other candidates for the 1.8 eV band but the only truly unanimously agreed point is that the 1.8 eV luminescence band is related to excess oxygen in silica [Skuja et al. 1995].

Although the amount of interstitial O, O_2 and O_3 can be negligible in comparison with the whole oxygen content in a silica network, the presence of these interstitial atoms or fragments has to be considered when analyzing a large amount of accumulated defects in the silica matrix.

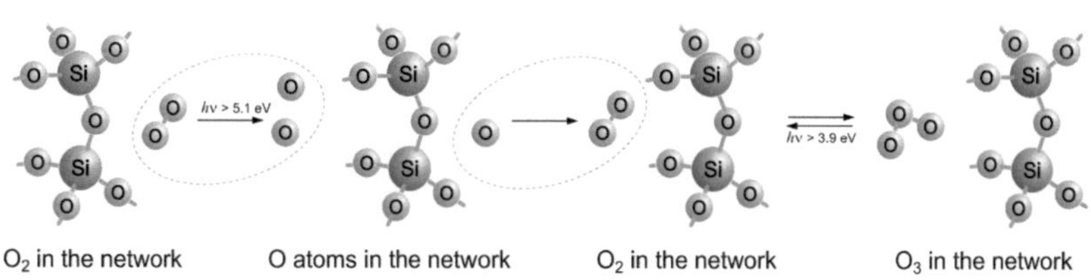

Fig. 2.21 Main interstitial atomic and molecular oxygen as well as ozone model in SiO_2.

2.2.8 Hydrogen-related defects and the state of water in SiO_2

Hydrogen in the form of steam has been used intentionally during thermal oxidation to increase oxidation rates. Very often it is also incorporated unintentionally into SiO_2 layers. Hydrogen is proposed to passivate the silicon or oxygen dangling bonds in SiO_2 network [Cartier et al. 1993, Fair 1997]. This passivation also decreases the number of non-bridging

oxygens which in turn reduces the viscosity of the silica layers substantially [Rafferty 1989]. Some authors considered hydrogen to be an intrinsic defect since it is commonly found in silicon dioxide. Hydrogen and water are ubiquitous impurities in SiO_2. The energy levels of silanol (≡Si–O–H) and hydride (≡Si–H) groups have been calculated by tight-binding calculation [Robertson 1988]. The hydride groups seem to produce a filled *s* state just below the valence band (VB) and an empty σ^* state in the gap just below the conduction band (CB). This group is probably both an electron and a hole trap.
Hydride groups are expected in thermally grown silica, particularly near the interface due to the interaction of interstitial H_2 molecules with Si–O–Si bridges. The H_2 is a byproduct of the fast oxidation of Si by any ambient water.

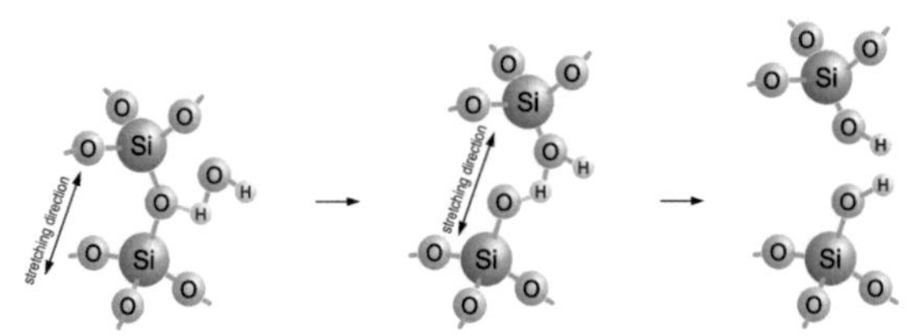

Fig. 2.22 A model of the structural state of water in SiO_2 network and its transformation to silanol groups, [Brunner et al. 1961, Bakos et al. 2004a].

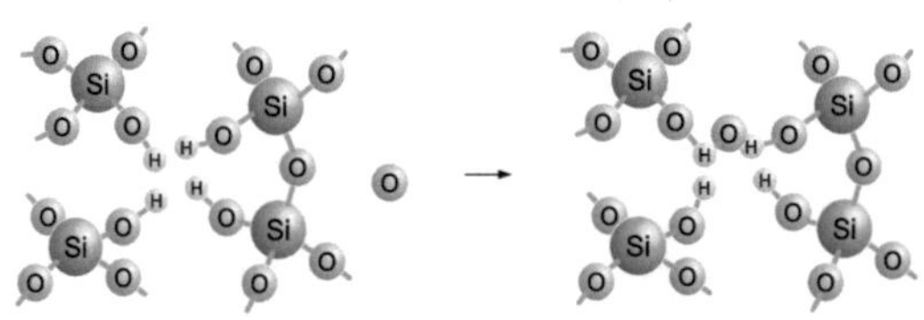

Fig. 2.23 A model of the structural state of water in SiO_2 lattice and its formation, [Nuttall and Weil 1980, McLaren et al. 1983].

It is proposed that water can be incorporated into the SiO_2 lattice according to the reaction illustrated in Fig. 2.22 [Brunner et al. 1961, Bakos et al. 2004a]. This is a preferred stress-induced mechanism and referred to "as hydrolytic weakening" [Griggs and Blacic 1965, Griggs 1967]. Calculations of total energy indicate that this reaction which uses an energy of ~0.1 eV is spontaneous for a Si–O bond stretched beyond 4% of its normal bond length [Heggie 1992]. It has also been suggested that the solubility of water in the quartz lattice depends on the aluminum content [Kronenberg et al. 1986].
Another possible configuration shown in Fig. 2.23, assumes that hydrogen is incorporated in the quartz structure by means of (4H)Si defects where $4H^+$ substitutes for $Si4^+$ [Nuttall and Weil 1980 , McLaren et al. 1983].

3. Fundamental of cathodoluminescence

3.1 Electron beam interaction with matter

When an electron hits a solid surface, it penetrates into the microstructure of the solid and interacts with its atoms. The resulting effects allow the extraction of analytical information

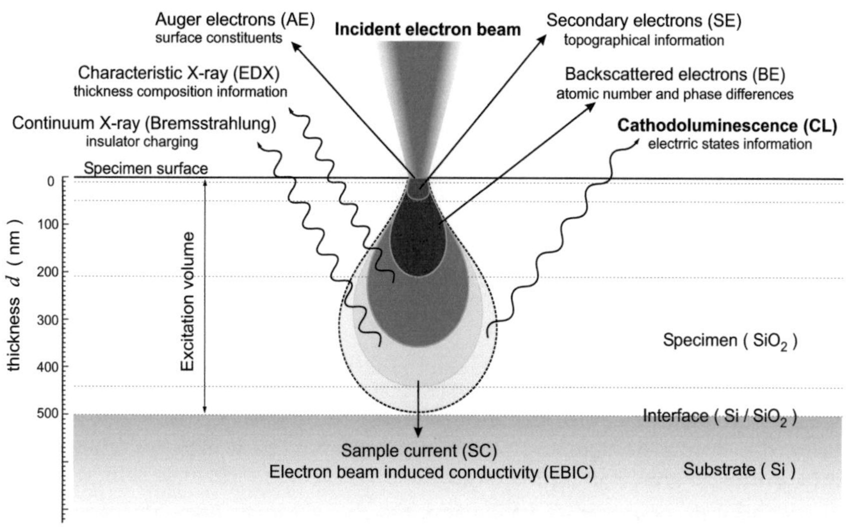

Fig. 3.1 Schematic representation of processes resulting from electron bombardment (modified after Reimer 1998). Note that the emissions come from different depths, e.g. CL and X-rays are emitted from deeper section levels than the secondary electrons.

on various properties of the material. The elementary interaction process involves collision of the electron with an atom, either with the positive atomic nucleus or with the electrons. Since the colliding particles are charged, the interaction is governed essentially by electrostatic Coulomb force. In an electron-nucleus collision, the atomic nucleus hardly changes its position since it is much heavier than the electron. The electron is therefore deflected at the nucleus without significant energy loss (elastic scattering). If the electron collides with the electrons of the atomic shells, then energy is transferred due to the equal masses of the colliding particles (inelastic interaction). Both scattering processes operate simultaneously side by side and provide a wide variety of useful signals. As a consequence of these different scattering events, an electron probe which originally was sharply focused in vacuum spreads over a greater volume after penetrating the solid. The shape and size of this scattering volume depends upon the material as well as the energy and angle of incidence of the probe electrons [Reimer 1998].

The elastic interactions between electrons and atoms or between electrons and the crystal lattice gives rise to a number of effects; the emission of secondary electrons (SE), back-scattering of electrons (BE), Auger electrons (AE), characteristic X-ray (EDX) and cathodoluminescence emission (CL) but mainly most of the electron beam's energy will be absorbed (sample current) and converted into heat (phonons) [Ozawa 1990]. AEs and more generally SEs can only escape from the first few nanometers of the surface. This means that these signals originate from the top of the interaction volume, the lateral size of which is determined by the electron probe size which in turn defines the spatial resolution in AES and SEM. Backscattered electrons are higher in energy and can escape from deeper areas below the surface where the interaction volume is wider laterally, this results in a degraded image resolution for backscattered electron imaging in the SEM relative to that of secondary electron imaging [Reimer 1998]. Analyzing these signals may provide valuable information about the surface and atomic structure of the investigated substance. Two of these signals

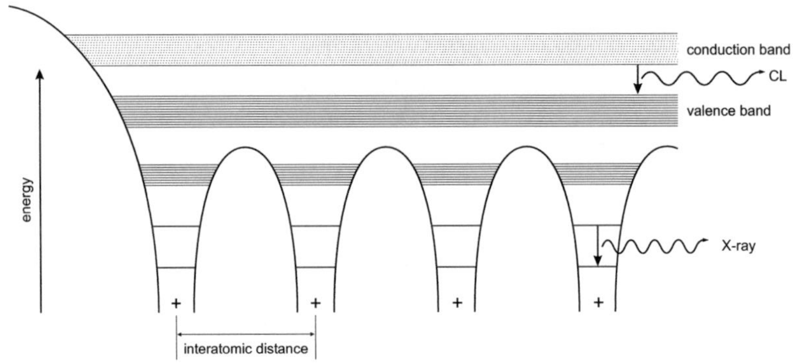

Fig. 3.2 Energy band diagram of one-dimensional lattice, [Yacobi and Holt 1990].

were among our main interest in this study, CL for luminescence defects investigation and EDX used for the atomic element component analysis. Fig. 3.1 illustrates schematically these effects and their major common applications.

The term cathodoluminescence (CL) designates the luminescence induced by energetic electron bombardment. Cathodoluminescence is excited from the region near the surface of the specimen and up to deeper than X-rays excitation volume, see Fig. 3.1. Therefore, the energy of the incident electron beam can be rather low (1-5 keV) to excite significant cathodoluminescence signals of use for the various applications. The CL signal is formed by detecting photons of the ultraviolet (UV), visible and near-infrared (IR) regions of the spectrum. These photons are emitted as the result of electronic transitions between the conduction and valence bands in semiconductors, and transitions between energy levels lying in the band gap of materials with wide band gap (insulators), Fig. 3.2. Many useful signals in these cases are due to transitions that involve impurities and a variety of defects.

X-ray spectrometry is based on the emission of X-rays from the atoms of a solid when struck by particles or waves of sufficiently high energy (electron beam in our case), followed by the relaxation of the atoms to the ground state after ionization. The X-ray spectrum produced by electron excitation consists of two parts, a continuous spectrum of deceleration radiation, known as "bremsstrahlung", and a line spectrum of the characteristic radiation. Characteristic X-rays will only be excited in the volume in which the electron energy exceeds the ionization energy of the inner shell involved. The general concept of X-ray emission-region and emission-process are shown in Figs. 3.1 and 3.2. The electron beam energy is typically in the range 10-20 keV and the energy of the X-rays emitted depends on the material under examination. The X-rays are generated in a region about 1-4 μm in depth, thus EDX is not a surface science technique, but it is a method used for providing sample composition information besides charging properties of insulators. Both CL and EDX techniques are commonly used in combination with a scanning electron microscope (SEM).

3.2 Process of luminescence excitation

Luminescence is produced by electron transitions between different energy levels. It may be produced by a band-band transition, transitions from band to localized state, or even between two localized states in a defect molecule, see Figs. 3.3 and 3.4. Luminescence is considered to form mainly in three steps. First, absorption of the excitation energy and stimulation of the atoms into an excited state, then transformation and transfer of the

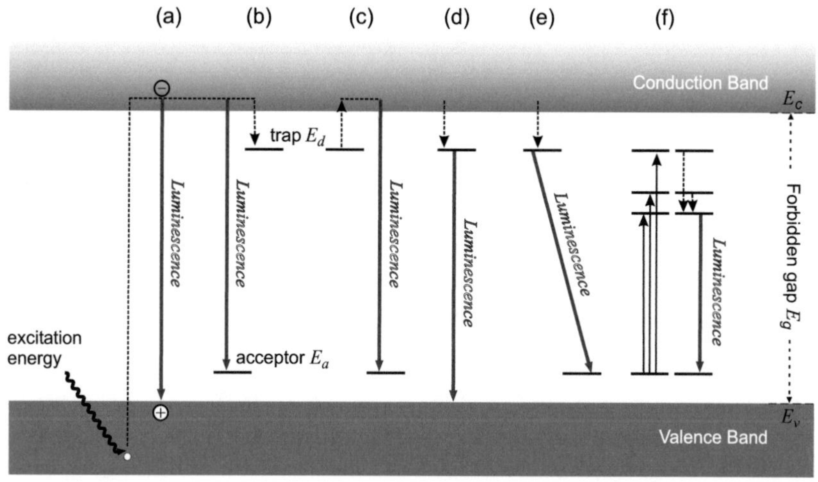

Fig. 3.3 Mechanisms of cathodoluminescence due to recombination processes in semiconductors or insulators: (a) recombination by direct band-to-band transition leads to intrinsic cathodoluminescence ; (b-e) recombination via localized states in the forbidden gap results in extrinsic cathodoluminescence; (f) excitation and recombination within the defect energy levels. E_a=accepter level, E_d=donor level, modified after Krbetschek et al. 1997.

excitation energy, and finally emission of light and relaxation to a non-excited state [Yacobi and Holt 1990]. The first two steps mainly depend on the mode of excitation, whereas the third step depends on the character of specific luminescence centers.

In crystals, complex interaction between atoms leads to the broadening of the energy levels into bands. In insulators like silica, a broad gap, called the forbidden gap, exists between the valence band and the conduction band. A precondition for cathodoluminescence is the existence of activators (impurity ions, lattice defects), which cause and occupy discrete energy levels in this forbidden zone. Luminescence centers can be differentiated by their energy position within the forbidden gap and divided into electron traps near the conduction band (donor level) and recombination sites in the vicinity of the valence band (acceptor level).

CL involves electron radiation induced excitation of an electron from the ground state to an excited state. Recombination of an electron from the conduction band with a hole in the valence band is possible in variety of ways: it may take place by a band-to-band transition or step by step via localized energy states within the forbidden gap. In a band-to-band transition shown in Fig. 3.3a, light is emitted with a photon energy $h\nu=E_g=E_c-E_v$; this type of CL is called intrinsic luminescence. Intrinsic CL is only likely to occur in semiconductors (GaAs and ZnS) in which recombination can take place directly, i.e without changes in momentum. In indirect semiconductors (Si, Ge, and GaP) the special band structure requires a photon to take part in the recombination process so that momentum is conserved.

The extrinsic CL is a recombination that takes place via localized energy states within the forbidden gap. Excitation results in the trapping of an electron or in the recombination with a luminescent or a non-luminescent center (Fig. 3.3b). In the case of a luminescent transition, a photon is emitted. If the atom or the ion is placed in a crystal lattice, non-luminescent transitions are possible due to absorption or emission of lattice vibrations. A trapped electron can be excited again, transits into the conduction band and may recombine with an activator element level resulting in emission of a photon (Fig. 3.3c), or a trapped electron may relax to the valence band and emit a photon (Fig. 3.7d). In the case of a small energy difference between electron trap and activator level, a direct luminescent transition of the electron to the recombination centre is possible (Fig. 3.3e). A More complex process is the excitation of several energy states lying deeper inside the forbidden gap and the subsequent relaxation and luminescence emission of a single activator is also possible (Fig. 3.3f). Such states can be due to point defects (impurities or lattice defects) and act as traps for electrons or holes. The variety of possible transitions and the thermally induced broadening of the energy states make the CL spectra very difficult to interpret.

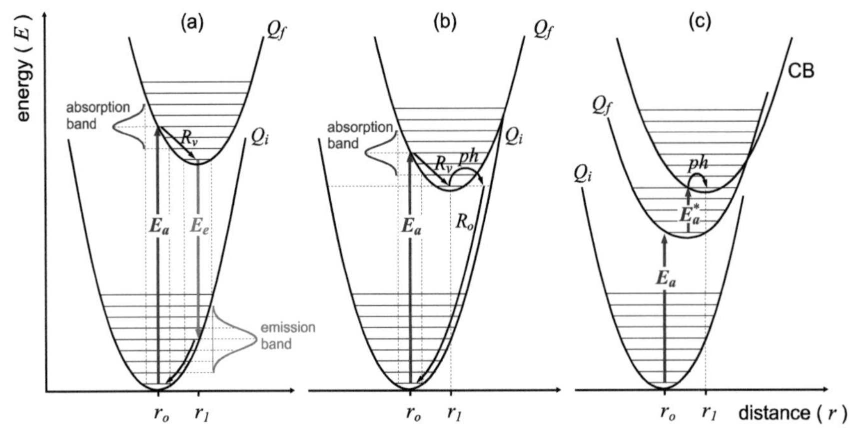

Fig. 3.4 Possible processes of excitation of a luminescence center explained in depending on configuration coordinate diagram assuming (a) radiative transition , (b) non-radiative transition , and (c) thermally assisted excitation into the conduction band, [Krbetschek et al. 1997].

The process of excitation and emission of an individual luminescence center can be described using a configuration coordinate diagram. It shows the potential energy curve of the absorbing center as a function of the relative distance (*r*) of the nucleus to the equilibrium position (electron in the ground state). Fig. 3.4 shows excitation of a defect atom/molecule from the ground state (Q_i) to excited state (Q_f) combined with radiative (Fig. 3.4a) and nonradiative (Fig. 3.4b) relaxation. As a particular case of excitation, thermally assisted excitation (Fig. 3.4c) into the conduction band and the relationship between the configurational coordinate and the band model are also shown.

In optical absorption (E_a) the center is promoted from the ground state to a higher vibration level of the excited state. After relaxation (*R*) to the zero vibration level of the excited state the electron returns with luminescence emission (E_e) or via a nonradiative transition with the emission of photons (ρ) and relaxes to the zero vibrational level (R_o). Since the electron transitions are faster than lattice relaxation, the absorption occurs adiabatically as a vertical transition (the Frank-Condon principle). The energy difference between absorption and emission is called Stokes shift. A special case of the excitation is thermally assisted ionization (Fig. 3.4c). After excitation to a higher energy level, the electron can escape to the conduction band (CB) by phonon absorption and recombine with other activators. This process may form the basis of infrared stimulated luminescence in feldspars [Hütt et al.

1988]. More complex processes may result from charge transfer. The electron is excited by high energy excitation into the charge transfer band and stored in a trap. After a second excitation with lower energy (e.g. thermal, infrared radiation) the electron escapes from the trap to an activator by a radiative or nonradiative transition. The life time τ of the trapped state depends on the depth (energy level) of the trap and the temperature, and can be determined using the Boltzmann relation $\tau = s^{-1} \exp(E/K_B T)$, where s is frequency factor, τ the absolute temperature and E the activation energy, i.e. the energetic depth of the trap.

3.3 Luminescence emission bands in silicon dioxide

The real structure of silica materials significantly influences their luminescence behavior. Moreover, defect contents and implanted elements dramatically changes the cathodoluminescence (CL) properties of silicas. The luminescence of silica is generally weak at room temperature, but it is highly variable depending on the specific conditions of silica formation. Crystalline SiO_2 modifications and amorphous silica show similar main luminescence bands. This is due to the fact that short-range order defects in silicate structure are determined mainly by silicon-silicon and silicon-oxygen interaction rather than by interactions between oxygen atoms [Berry and Vaughan 1985]. Since a strict correlation between the intensities of CL bands and concentrations of specific defects or trace elements does not exist, the interpretation of the origin of luminescence of silica materials is difficult. Furthermore, the band positions of the luminescence emissions can vary depending on the particular SiO_2 structure and the experimental conditions. The CL spectra of silica consist mainly of several emission bands in the blue up to the red spectral region. Investigation of natural and synthetic silica specimens show various emission bands which are ascribed to different intrinsic and extrinsic defects. The CL spectra of silica materials are generally consist of some different broad emission bands (defect's optical bands) instead of sharp spectral lines. Overlapped bands are commonly expected too, and it seems to be difficult to resolve these bands to their real components. Two reasons for the broadening of these transitions have been suggested: the first is due to the different static local environments (inhomogeneities) of defects in disordered glass called inhomogeneous broadening, and the second is due to the coupling of electronic transition to phonons called homogeneous broadening. The homogeneous broadening is the same for all related defect sites throughout the sample and occurs even for defects in otherwise ideal crystal. The main contribution to the defect's optical bands broadening (FWHM) in amorphous SiO_2 which is our main interest in this study, is usually from homogeneous broadening (0.2-1 eV) while the magnitude of inhomogeneous broadening is of the order 0.1 eV [Skuja 2000]. Table 3.1 summarizes the main PL and CL emission bands previously reported for α-SiO_2 and a-SiO_2 and their proposed associations. Further on, cathodoluminescence can be divided into two categories: intrinsic CL, which is characteristic of the host lattice, and extrinsic CL which results from impurities. Larger defects such as dislocations and clusters may also have an effect on the CL signal. Intrinsic luminescence is enhanced by non-stoichiometry (vacancies), structural imperfections (poor ordering, radiation damage, shock damage), and impurities (non-activators), which distort the crystal lattice. The CL properties are mostly the result of luminescence-activating ions such as transition metals, rare-earth elements or actinides. Changes of the crystal field properties i.e., field strength, site symmetry, and coordination number, cause differences in the magnitude of the energy levels of the activator elements and also the splitting of closely spaced levels.

wave length (nm)	energy (eV)	Region
1240	1.00	IR
1181	1.05	IR
1127	1.10	IR
1078	1.15	IR
1033	1.20	IR
992	1.25	IR
954	1.30	IR
918	1.35	IR
886	1.40	IR
855	1.45	IR
827	1.50	IR
800	1.55	IR
775	1.60	IR
751	1.65	Red
729	1.70	Red
708	1.75	Red
689	1.80	Red
671	1.85	Red
653	1.90	Red
636	1.95	Red
620	2.00	Yellow
605	2.05	Yellow
591	2.10	Yellow
577	2.15	Yellow
564	2.20	Yellow
551	2.25	Green
551	2.25	Green
539	2.30	Green
528	2.35	Green
517	2.40	Green
506	2.45	Green
496	2.50	Green

CL, PL emission band (eV)	Specimen	Association	References
0.97	*a*-SiO_2 & α-SiO_2	O_2 , O_2^-	Skuja et al. 1998b, Stevens-Kalceff 2000, Skuja et al. 2000a, Stevens-Kalceff et al. 2002.
1.3 - 1.75	C & Si doped *a*-SiO_2	Si & C nanocluster	Chae et al 1999, Rebohle et al. 2001a, Perez-Rodriguez et al. 2003.
	a-SiO_2 and *a*-SiO_2	Si-nanocluster	Takagahara & Takeda 1992, Shimizu-Iwayama et al. 1993, Kanemitsu et al. 1994, Zunger & Wang 1996, Kim et al 2000a, Kim et al 2000b, Yi et al. 2003, Yu et al. 2004, Wilkinson & Elliman 2004, Imakita et al. 2005, Khriactchev et al. 2004.
	SiO_x	Si-nanocluster (d=5nm)	Ledoux et al. 2002, Yi et al. 2002.
	a-SiO & *a*-SiO_2	Si-nanocluster (d=4-2nm at 1.3-1.6eV)	Zacharias et al. 1998, Zacharias et al. 2002, Zacharias et al. 2003.
	SiO_x	Si-nanocluster (d=10-25nm at 1.4-1.5eV)	Lau et al. 2005.
1.80	SiO_x	Si nanocluster	Cheylan et al. 2001, Sakurai 2003.
	a-SiO_2	ODC, Si nanocluster	Sakurai et al. 1999.
	silica xerogel	NBOHC	Hinic et al. 1999 , Hinic et al. 2003.
1.85 - 1.95	silica fiber	NBOHC	Munekuni et al. 1990.
	a-SiO_2 & α-SiO_2	NBOHC	Skuja 1992b, Cannas & Gelardi 2004, Mohanty et al. 2003.
	SiO_x	NBOHC	Yu et al. 2004.
	a-SiO_2	O_2^-, O_3^-	Skuja et al. 1995.
	a-SiO_2	O_2, O_3	Skuja et al. 2000a, Skuja et al.2000b.
2.0 - 2.2	*a*-SiO_2 & silica gel	hydrogen defects (OH, SiH or SiH_2)	Friebele et al. 1985, Hinic et al. 2003.
	ion doped *a*-SiO_2 & α-SiO_2	H & He implantation + annealing	Morimoto et al. 1996.
	a-SiO_2	trivalent Si center	Mitchell & Denure 1973, Ludwig et al. 1995.
	a-SiO_2	Si nanocluster (d=3nm)	Kim et al 2000a, Ledoux et al. 2002, Yi et al. 2003, Khriactchev et al. 2004.
	a-SiO_2	Si implantatition, Si nanocluster	Shimizu-Iwayama et al. 1997, Im et al. 1999.
	ion doped *a*-SiO_2	Si & C implantation + annealing	Perez-Rodriguez et al. 2003.
	a-SiO_2	STE	Itoh et al. 1990.
2.2-2.5	*a*-SiO_2 & α-SiO_2	STE	Trukhin 1992, Trukhin 1993, Stevens-Kalceff & Wong 2005.
	Ge doped α-SiO_2	STE associated with Ge	Itoh et al. 1988, Luff & Townsend 1990.
	Si doped *a*-SiO_2	Si implantation + annealing	Kim et al. 2000c.
	a-SiO & *a*-SiO_2	Si-nanocluster	Zacharias et al. 2003.
	silica xerogel	organic components or *E'* center	Hinic et al. 2003.
	powder silica	H-related defect, surface Si-H species	Glinka 2000, Glinka et al. 1999.
	SiO_x	H-related defect	Yu et al. 2004.

Table 3.1 (a): Reported CL and PL luminiscent bands in amorphous silicon dioxide (*a*-SiO_2) and crystalline quartz (α-SiO_2) and their proposed associations in IR-Green region.

nm	eV	Region
477	2.60	Blue
468	2.65	
460	2.70	
451	2.75	
443	2.80	
435	2.85	
428	2.90	
421	2.95	
414	3.00	
407	3.05	
400	3.10	Violet
394	3.15	
388	3.20	
382	3.25	
376	3.30	
370	3.50	
365	3.40	
344	3.60	
342	3.65	
335	3.70	UV
361	3.75	
295	4.20	
292	4.25	
289	4.30	
285	4.35	
282	4.40	
279	4.45	
276	4.50	
272	4.55	
270	4.60	
267	4.65	
264	4.70	
261	4.75	
258	4.80	

2.40 - 3.00	ion doped a-SiO_2	Sn^+, Si^+, Ge^+, C^+, Cu^+ implantation	Calestani et al.2005,Perey-Rodriguez et al.2003,Pellegrino et al.2004,Plaksin et al.2004.
	C doped a-SiO_2	crystalline SiC nonoclusters	Rebohle et al. 2001a, Rebohle et al. 2002b.
2.60 - 2.80	α-SiC_2	STE	Trukhin & Plaudis 1979,Griscom 1985,Itoh et al.1988,Tanimura et al.1983,Itoh et al.1990.
	a-SiC_2	STE	Hayes et al. 1984, Stevens-Kalceff 1998, Glinka 2000.
2.7 - 2.75	sol-gel SiO_2	ODC, Triplet-Singlet transition of Si	Griscom 1985, Paleari 2005.
	ion doped a-SiO_2	ODC, Si implantation	Griscom 1990a, Skuja 1992a, Kuzuu & Murahara 1992, Kim et al. 2000a,Kim et al. 2000b.
	Si doped SiO_x	neutral oxygen vacancy	Yu et al. 2004.
	ion doped a-SiO_2	O implantaition, O-related center	Hagni 1987, Chiodini et al. 2000, Anedda et al. 2002.
	a-SiC_2 & α-SiO_2	O_2^-	Stevens-Kalceff 2000.
	a-SiC_2 ,α-SiO_2 & SiO_x	E' center	White et al. 2002.
2.9 - 3.10	a-SiO_2	STE's associated with Ge, Al & H	Khanlary et al. 1993, Yang et al. 1994.
	ion doped a-SiO_2 & α-SiO_2	Al, C, & Ge implantation	Koyama 1980, Stevens-Kaceff 1998, Zhang & Raghavachari 1997, Lee et al. 2000.
	a-SiO_2	-O-O- defects	Nishikawa et al. 1992.
	fused SiO_2 , a-SiO_2	O_2^-	Guzzi et al. 1987, Skuja & Trukhin 1989.
	ion doped a-SiO_2	ODC	Anedda et al. 2003a.
	α-SiC_2	Intrinsic defect	Alonso et al. 1983, Stevens-Kaceff & Philips 1995, Gorton et al. 1996.
3.10 - 3.30	ion doped a-SiO_2 & α-SiO_2	Al^+, H^+, Li^+, Na^+, K^+ & Ge^+ implantation	Qin et al. 1996, Tong et al. 1997, Trukhin et al. 2003a, Trukhin & Poumellec 2003c.
	a-SiO_2	ODC or Al^{3+}-H center	Stevens-Kalceff & Wong 2005.
	α-SiC_2 , a-SiO_2	impurity incorporation during growth or Ge	Skuja & Trukhin 1989, Luff & Townsend 1990.
	Ge doped a-SiO_2	oxygen vacancy related defect	Rebohle et al. 2002a, Rebohle et al. 2002b.
	a-SiO_2	Triplet-Singlet emission of Ge	Hosono et al. 1992, Skuja 1998, Trukhin et al.2003b, Cannizzo et al. 2005.
	SiO	oxygen-related defect, Si nanocluster	Yang et al 2005.
3.70	ion doped a-SiO_2	twofold coordinated Sn, Ge	Skuja 1992a.
	sol-gel a-SiO_2	E'-like surface center	Chiodini et al. 2000.
	sol-gel silica	isolated silanol groups	Yao et al. 2001, Paleari et al. 2005.
	sol-gel silica	OH-related center	Anedda et al. 2003b.
4.20 - 4.40	ion doped a-SiO_2	Si excess defect	Weeks et al. 2005.
	a-SiO_2	ODC, Singlet-Singlet transition	Griscom 1985, Griscom 1991, Skuja 1992a, Boscaino et al. 1996.
	sol-gel a-SiO_2	ODC, Singlet-Singlet transition of Si	Chiodini et al. 2000, Paleari et al. 2005, Cannas et al. 2002.
	α-SiO_2 & a-SiO_2	divalent Ge defect	Hosono et al. 1992, Skuja 1998, Trukhin & Poumellec 2003c.
	a-SiO_2	Si implantation	Magruder et al. 2003.
	Ge doped sol-gel a-SiO_2	GODC, Singlet-Singlet transition of Ge	Agnello et al. 2003, Trukhin et al.2003b, Cannas et al. 2003a, Cannizzo et al. 2005.

Table 3.1 (b): Reported CL and PL luminiscent bands in amorphous silicon dioxide (a-SiO_2) and crystalline quartz (α-SiO_2) and their proposed associations in Blue-UV region.

Therefore the CL spectrum is not a characteristic property of the activator but a property of the mineral. The cathodoluminescence intensity generally increases with the concentration of the activator to a maximum and then decreases. This decrease is referred to as concentration quenching (self-quenching). It can be explained by the transfer of a part of the excitation energy to other activator ions which is more effective than luminescence emission. Some ions (quenchers) show broad and intense charge transfer bands in the absorption spectrum and cause quenching of the luminescence of activators with interfering emission bands.

The quenching by ions with intense charge transfer bands especially influences the visible and the near ultraviolet (UV) spectral ranges, whereas luminescence emissions in the infrared (IR) are more or less unaffected. Apparently, the quenchers cause new closely-spaced energy levels so that the electron can easily return to the ground state with the emission of low-energy photons (IR) or by losing heat energy [Marshall, 1988].

Quenching due to lattice defects may occur if the crystal structure is damaged by mechanical processes, radiation, growth defects or impurities. These lattice defects create new energy levels between the conduction and the valence bands resulting in absorption of the excitation energy, non-luminescent energy transfer or low frequency emission. Another process which may be responsible for lowering the luminescence intensity is thermal quenching. Principally, heating of a sample results in the release of electrons and accordingly in the excitation of thermoluminescence. During irradiation of a sample with high-energy particles (e.g., electrons, ions) a part of the energy can be transformed into heat which influences the energy transfer and non-luminescent transitions. As a consequence, the luminescence intensity decreases during electron bombardment and then stabilizes.

3.4 Cathodoluminescence spectroscopy (CL)

Conventionally, there are several kinds of cathodoluminescence devices that can be installed in SEM. One of them is a spectrometer-type device, which incorporates a parabolic mirror mounted on the spectrometer port designed for high efficiency of light collection, as used in this study. This specially designed mirror is used for collecting the CL. Despite the high generation factors for the electron-hole pairs, the external photon yield is small. Furthermore, the intensity generated inside the specimen is considerably reduced by absorption in the specimen and by the total reflection at the specimen surface.

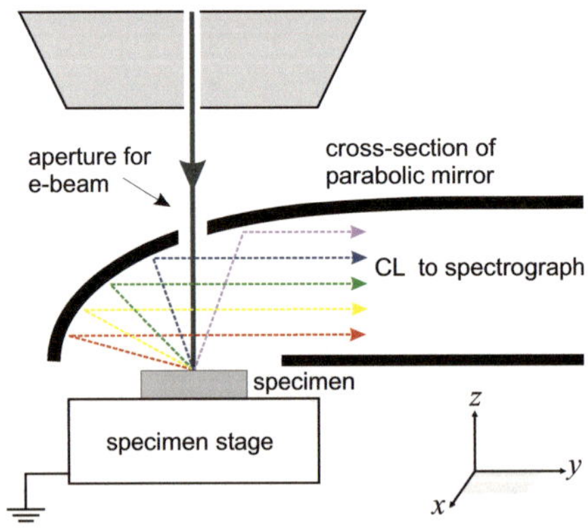

Fig. 3.5 Schematic diagram of the CL collection optics within the SEM (not to scale). The CL photons emitted by the electron-irradiated specimen are collected by a parabolic mirror and directed into the spectrograph by an optical guide.

One of the fundamental problems in CL measuring technique is therefore how to collect the emitted radiation over the largest possible solid angle with high efficiency. Fig. 3.5 shows a configuration using a parabolic mirror in which the primary electrons (PE) impact on the specimen through a hole in the parabolic mirror. Light quanta emitted into a solid angle π can be detected, where the specimen is located almost inside the mirror during the CL measurement for better CL signal collection. A silica fiber optic transmits the CL light to the entrance slit of grating spectrograph of type Spex-270M and it is registered in a single shot technique by a liquid nitrogen cooled charge coupled device CCD camera (Princeton Instruments, EEV 1024×256) with a spectral resolution of about 4 nm.

The CL was excited using a continuous stationary electron beam with energy E_0=10 keV and stable current of ≈0.6 μA with focused beam diameter of ≈1 μm in TV scanning mode both at room temperature (RT) and liquid nitrogen temperature (LNT), working distance fixed at 14 mm where the electron beam was focused on a small area (≈100×100 μm^2) of the specimen surface. The CL excitation and recording use the same parameters (experimental conditions) in order to get comparable spectra collected from different silicon dioxide samples.

The visible light spectrum over the wavelengths 200-800 nm collected by the CL mirror is composed of the real luminescence light and the background light produced by the electron beam. The background spectra is registered separately and subtracted from the measured CL signal to eliminate disturbing light radiation from the thermionic cathode and other sources. The background elimination was done periodically by switching the primary electron beam on and off (beam blinking) in each start of new CL measurement.

The whole experiment was performed in high vacuum. In order to achieve a good high vacuum, some special procedures were needed. Initially, the vacuum chamber was pumped down to 10^{-2} mbar using a rotary pump. Then the chamber was pumped down to about 10^{-6} mbar using a turbomolecular pump. After 24 hours of pumping, the pressure got down to around 10^{-7} mbar and the chamber achieved the HV operation conditions, which is good enough to perform both CL and EDX measurements.

3.5 Energy dispersive X-ray spectroscopy (EDX)

X-ray spectrometry is based on the emission of X-rays from the atoms of a solid when struck by particles or waves of sufficiently high energy (electron beam in our case), followed by the relaxation of the atoms to the ground state after ionization. The X-ray spectrum produced by electron excitation consists of two parts, a continuous spectrum of deceleration radiation, known as "bremsstrahlung", and a line spectrum of the characteristic radiation. The electron beam energy is typically in the range 10-20 keV and the energy of the X-rays emitted depends on the material under examination.

X-ray data are obtained with a wavelength dispersive spectrometer. The X-ray energy dispersive system component is a semiconductor crystal which acts simultaneously as a detector and analyzer [Fuchs et al. 1990]. In a similar way as in a proportional counter tube, X-ray quanta which penetrate the semiconductor detector generate pairs of charge carriers whose number is proportional to the energy of the incident quanta. The spectrum is obtained by evaluating the pulses (current) generated and amplified in the detector system according to their height and sorting them. This takes place in a mutichannel analyzer. Each channel corresponds to a specific pulse height and to a specific energy after suitable calibration. The amount of the current pulse t generated depends on the number of electron-hole pairs created, which in turn depends on the energy of the incoming X-ray. Thus, an X-ray spectrum can be acquired giving information on the elemental composition of the material under examination.

The semiconductor detector is a special silicon diode called a lithium drifted silicon detector Si(Li), which is accommodated in a vacuum chamber and kept at a temperature of about 100 K by means of a cold finger. A beryllium window (about 10 μm thick) closes the evacuated chamber containing the detector and preamplifier. The window protects the detector from condensation and contamination as well as from scattered electrons and light. The entire system can be adjusted both horizontally and vertically.
Elements of low atomic number are difficult to detect by EDX. In all Si(Li) detectors, The absorption of the soft X-rays by the Beryllium window precludes the detection of elements below an atomic number of 11 (Na). In windowless systems, elements with as low atomic number as 4 (Be) have been detected, but the problems involved get progressively worse as the atomic number is reduced.

3.6 Fourier transform infrared spectroscopy (FTIR)

One of the most common methods of analyzing crystalline silica is based on the absorption of infrared light. Infrared radiation spans a part of the electromagnetic spectrum having wavelengths from 0.78 to 1000 μm. It is bounded by the red end of the visible region at high frequencies (≈300 THz) and the microwave region at low frequencies (≈300 GHz). At temperatures above absolute zero, all the atoms in molecules are in continuous vibration with respect to each other. When the frequency of a specific vibration is equal to the frequency of the IR radiation directed at the molecule, the molecule absorbs the radiation.

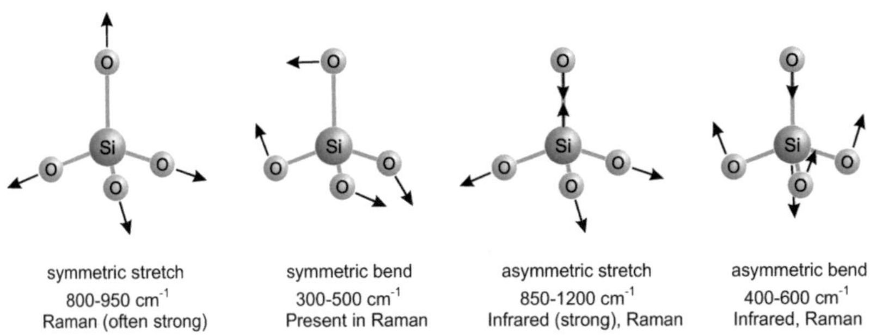

Fig. 3.6 Major vibrational modes for a nonlinear group, [Silverstein et al. 1981].

Different functional groups absorb characteristic frequencies of IR radiation. Thus, IR spectroscopy is an important and popular tool for structural clarification and compound identification in the sample and is even a common spectroscopic technique used for quantitative determination of compounds in mixtures. Infrared radiation is absorbed and the associated energy is converted into different types of motions. The absorption involves discrete, quantized energy levels. However, the individual vibrational motion is usually accompanied by other rotational motions. These combinations lead to the absorption bands, not the discrete lines, commonly observed in the middle IR region. The major types of molecular vibrations are stretching and bending. The various types of vibrations for SiO_2 are illustrated in Fig. 3.6. IR spectra are obtained by detecting changes in transmittance (T) or absorption (A) intensity as a function of frequency. Transmittance is the ratio of radiant power transmitted by the sample (I) to the radiant power incident on the sample (I_o). Absorbance is the logarithm to the base 10 of the reciprocal of the transmittance (T), $A=\log(T^{-1})=-\log(I/I_o)$, The transmittance spectra provide better contrast between intensities of strong and weak bands because transmittance ranges from 0 to 100% *T* whereas absorbance ranges from infinity to zero.

In this study we will focus on the most frequently considered portion of the middle IR region, between 400 and 4000 cm^{-1} (25 to 2.5 μm). The total number of observed absorption bands is generally different from the total number of fundamental vibrations. It is reduced because some modes are not IR active and a single frequency can cause more than one mode of motion to occur.

Absorption bands in the region of 2500-4000 cm^{-1} usually come from stretching vibrations between hydrogen and some other atoms with a mass of 19 or less. The O−H and N−H stretching frequencies fall in the 2500-3700 cm^{-1} region, with various intensities. Hydrogen bonding has a significant influence on the peak shape and intensity, generally causing peak broadening and shifts in absorption to lower frequencies. The C−H stretching bands occur in the region of 2700-3300 cm^{-1} [Settle 1997]. The bands 2000-2300 cm^{-1} and 935-1000 cm^{-1} are associated with Si−H bond-stretching band and Si\textendash H bond-bending type mode, respectively. Another peak at 3675 cm-1 corresponds to the Si−OH bond [Hosono et al. 1999].

The absorption bands at the 1850-2700 cm^{-1} region usually come only from triple bonds and other limited types of functional groups, such as C≡C at 2100-2260 cm^{-1}, C≡N at 2220-2260 cm^{-1}, allenes C=C=C at 1900 to 2000 cm^{-1}, S−H at 2550 to 2600 cm^{-1}, P−H at 2275 to 2440 cm^{-1} [Settle 1997 , Trukhin et al. 1999], Si-C at 800 cm^{-1}, Si−O at 500 cm^{-1} [Prado et al 2003].

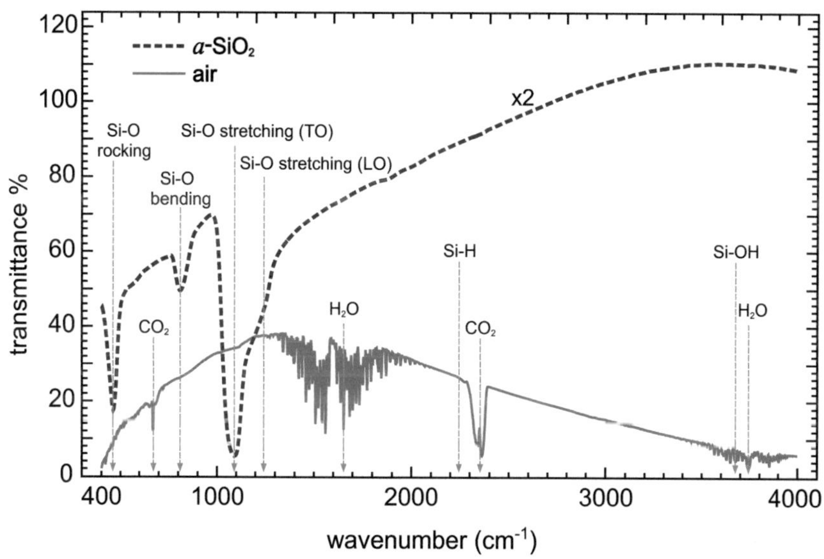

Fig. 3.7 A single-beam IR spectrum of *a*-SiO_2 and air (background), showing different Si−O vibration modes and the contribution of the ambient water and carbon dioxide CO_2.

Moreover, the transmission spectra show peaks at 890-1062 cm^{-1} and 2260 cm^{-1}, corresponding to the Si−O stretching frequency [Tsu et al. 1989, Saito and Ikushima 2002, Suzuki et al. 2003, Prado et al 2003], at 945 cm^{-1} assigned to Si−OH mode, and at 1650 cm^{-1} associated with H−O−H (H_2O) [Gendron-Badou et al. 2003]. Typical FTIR spectra in the range 400-4000 cm^{-1} of air and pure *a*-SiO_2 layer are shown in Fig. 3.7, where some vibrational modes of Si−O and hydrogen incorporated molecules are pointed. Commonly

the transmission peaks at 1100, 471 cm^{-1} are attributed respectively to triply degenerated stretching and bending vibration modes of the tetrahedron, but 800 cm^{-1} corresponds to an inter-terahedral Si−O−Si bending vibration mode [Gendron-Badou et al. 2003].
Most commercial instruments to separate and measure IR radiation use dispersive spectrometers or Fourier transform spectrometers. Fourier transform spectrometers have recently replaced dispersive instruments for most applications due to their superior speed and sensitivity. An IR spectrometer consists of three basic components: radiation source, interferometer (monochromator in dispersive IR system), and detector. Infrared transmittance measurements were performed using a FTIR spectrometer (Nicolet Magna IR 550) in the spectral range 400-4000 cm^{-1}. The spectral resolution was fixed to 4cm^{-1} and the spectra were recorded at room temperature (RT) by means of non-polarized light under normal incidence. Most benchtop FTIR spectrometers are single-beam instruments. Unlike double-beam grating spectrometers, single-beam FTIR does not obtain transmittance or absorbance IR spectra in real time. A typical operating procedure is described as follows: A background spectrum (Fig. 3.7) is first obtained by collecting an interferogram (raw data), followed by processing the data by Fourier transform conversion. This is a response curve of the spectrometer and takes account of the combined performance of source, interferometer, and detector. The background spectrum (air) also includes the contribution from any ambient water (two irregular groups of lines at about 3600 cm^{-1} and about 1600 cm^{-1}) and carbon dioxide (doublet at 2360 cm^{-1} and sharp spike at 667 cm^{-1} present in the optical bench. Next, a single-beam sample spectrum is collected, it contains absorption bands from the sample and the background (air). The background spectra have been eliminated in order to get the real vibrational response of the sample. Reference samples of the silicon substrate have also been used for background division.

4. References

Agnello S. , Boscaino R. , Buscarino, Cannas M. and Gelardi F. M. : *Structural relaxation of E' centers in amorphous silica*, Phys. Rev. B 66 (2002) 113201.

Agnello S. , Boscaino R. , Cannas M. , Cannizzo A. , Gelardi F. M., Grandi S. and Leone M. : *Temperature and excitation energy dependence of decay processes of luminescence in Ge-doped silica*, Phys. Rev. B 68 (2003) 165201.

Alonso P. J., Halliburton L. E. , Kohnke E. E. and Bossoli R .B. : *X-ray-induced luminescence in crystalline SiO_2*, J. Appl. Phys. 54 (1983) 5369.

Anedda A. , Carbonaro C. M. , Clemente F. and Corpino R. : *Ultraviolet excitation fine tuning of luminescence bands of oxygen-deficient centers in silica*, J. Appl. Phys. 92 (2002) 3034.

Anedda A. , Carbonaro C. M. , Clemente F. , Corpino R , Corpino R. and Serpi A. : *Excitation pattern of the blue emission in Ge-doped silica*, J. Non-Cryst. Solids 315 (2003a) 161.

Anedda A. , Carbonaro C. M. , Clemente F. , Corpino R. , Grandi S. , Mustarelli P. and Magistris A. : *OH-dependence of ultraviolet emission in porous silica*, J. Non-Cryst. Solids 322 (2003b) 68.

Arnold G. W. : *Ion-implantation effects in noncrystalline SiO_2*, IEEE Trans. Nucl. Sci. NS 20 (1973) 220.

Awazu K. and Kawazoe H.: *O_2 molecules dissolved in synthetic silica glasses and their photochemical reactions induced by ArF excimer laser radiation*, J. Appl. Phys. 68 (1990) 3584.

Bakos T. , Rashkeev S. N. and Pantelides S. T. : *H_2O and O_2 molecules in amorphous SiO_2:Defect formation and annihilation mechanisum* , Phys. Rev. B 69 (2004a) 195206.

Bakos T. , Rashkeev S. N. and Pantelides S. T.: Optically active defects in SiO_2: *The nonbridging oxygen center and interstitial OH molecule,* Phys. Rev. B 70 (2004b) 75203.

Barfels T. : PhD dissertation , *Kathodolumineszenz amorpher und kristalliner Modifikationen von SiO_2 und GeO_2,* Rostock university (2001).

Baulch D. L. , Cox R. A. , Hampton R. F. , Kerr J. A. , Troe J. and Watson R. T.: *Evaluated Kinetic and Photochemical Data for Atmospheric Chemistry, CODATA Task Group on Chemical Kinetics,* J. Phys. Chem. Ref. Data 9 (1980) 295.

Bell R. J. and Dean P. : *The Structure of Vitreous Silica: Validity of the Random Network Theory,* Philosophical Magazine 25 (1972) 1381.

Berry F. J. and Vaughan D. J. (eds) : *Chemical bonding and spectroscopy in mineral chemistry,* Chapman and Hall, London, (1985).

Bobyshev A. A. and Radtsig V. A. : *Optical absorption spectra of paramagnetic defects in vitreous silica* , Sov. Phys. Chem. Glass 14 (1988) 501.

Boero M. , Pasquarello A. , Sarnthein. and Car R. : *Structure and Hyperfine Parameters of E'_1 Centers in alpha-Quartz and in Vitreous SiO_2,* Phys. Rev. Lett. 78 (1997) 887.

Boscaino R. , Cannas M. , Gelardi F. M. and Leone M. : *Photoluminescence band at 4.4 eV in oxygen-deficient silica: temperature effects,* J. Phys.: Condens. Matter 8 (1996) L545.

Bosio C. and Czaja W. : *The fundamental absorption edge of crystalline and amorphous SiO_2,* Europhys. Lett. 24 (1993) 197.

Brunner G. O. , Wondratschek H. and Laves F. : *Ultrarotuntersuchung über den Einbau von H in natürlichen Quartz,* Z. Electrochem 56 (1961) 735.

Calestani D. , Lazzarini L. , Salviati G. and Zha M. : *Morphological, structural and optical study of quasi-1D SnO_2 nanowires and nanobelts,* Cryst. Res. Technol. 40 (2005) 937.

Cannas M. , Agnello S. , Boscaino R. , Gelardi F. M. , Leone M. and Boizot B. : *Optical properties of oxygen-deficiency related centers in amorphous SiO_2 investigated by synchrotron,* Radiation effects and Defects in Solids 157 (2002) 1045.

Cannas M. , Agnello S. , Boscaino R. , Gelardi F. M. , Grandi S. and Mustarelli P. C.: *Ultraviolet emission lifetime in Si and Ge oxygen deficient centers in silica,* J. Non-Cryst. Solids 322 (2003) 129.

Cannas M. and Gelardi F. M. : *Vacuum ultraviolet excitation of the 1.9 eV emission band related to nonbridging oxygen hole centers in silica,* Phys. Rev. B 69 (2004) 153201.

Cannizzo A. , Agnello S. , Grandi S. , Leone M. , Magistris A. and Radzig V. A.: *Luminescence activity of surface and interior Ge-oxygen deficient centers in silica,* J. Non-Cryst. Solids 351 (2005) 1805.

Cartier E. , Stathis J. H. , and Buchanan D. A. : Passivation and depassivation of silicon dangling bonds at the Si/SiO_2 interface by atomic hydrogen} , Appl. Phys. Lett. (1993) 1510.

Carvalho W. , Dumas P. , Corset J. and Neuman V. : *Raman Spectra and Oxygen-Related Absorption Bands in Pure Silica Core Fibres,* J. Raman Spectrosc. 16 (1985) 330.

Chae K. H. , Son J.H., Chang G. S. , Kim H. B. , Jeong J. Y. , Im S. , Song J. H., Kim K. J., Kim H. K. and Whang C. N. : *Visible photoluminescence in ion beam mixed SiO_2/Si/SiO_2 layers,* NanoStructured Materials 11 (1999) 1239.

Cheylan S. , Elliman R. G. , Gaff K. , Durandet A. : *Luminescence from Si nanocrystals in silica deposited by helicon activated reactive evaporation,* Appl. Phys. Lett. 78 (2001) 1670.

Chiodini N. , Meinardi F. , Morazzoni F. , Paleari A. , Scotti R. and Martino D. Di. : *Ultraviolet photoluminescence of porous silica,* Appl. Phys. Lett. 76 (2000) 3209.

Edwards A. H. , Fowler W. B. and Robertson J. *: Structure and imperfections in amorphous and crystalline silicon dioxide* , R. Devine A. B. , Duraud J.-P. and Dooryhée E. (eds.), John Wiley & Sons Ltd., (2000), p.253.

Fair R. B. : *Physical models of boron diffusion in ultrathin gate oxides,* J. Electrochem. Soc. 144 (1997) 708.

Fan X. D. , Peng J. L. and Bursill L. A. : *Joint Density of States of Wide-Band-Gap Materials by Electron Energy Loss Spectroscopy,* Modern Phys. Lett. B 12 (1998) 541.

Fanderlik I. (ed.) : *Silica Glass and its Application , Glass Science and Technology 11,* ELSEVIER, Amsterdam, (1991).

Feigl F. J. , Fowler W. B. and Yip K. L. : Oxygen vacancy model for the E'_1 center in SiO_2, Solid State Commun. 14 (1974) 225.

Feder R. , Spiller E. and Topalian J. : *X-Ray Lithography,* Polymer Eng. Sci. 17 (1977) 385.

Finlayson -Pitts B. J. and Pitts J. N. : *Jr., Atmospheric Chemistry, Fundamental sand Experimental Techniques,* John Wiley, New York, (1986).

Fitting H.-J. , Ziems T. , von Czarnowski A. and Schmidt B. : *Luminescence center transformation in wet and dry* SiO_2, Radiation Measurements 39 (2004) 649.

Friebele E. J. , Griscom D. L. , Stapelbroek M. and Weeks R. A. : *Fundamental Defect Centers in Glass: The Peroxy Radical in Irradiated, High-Purity, Fused Silica,* Phys. Rev. Lett. 42 (1979) 1346.

Friebele E. J. , Griscom D. L. and Marrone M. J. : The optical absorption and luminescence bands near 2 eV in irradiated and drawn synthetic silica} , J. Non-Cryst. Solids 71 (1985) 133.

Fuchs E. , Oppolzer H. and Rehme H. : *Particle Beam Microanalysis: Fundamental, Methods and Applications,* VCH Verlagsgesellschaft, Weinheim, (1990).

Gerber Th. and Himmel B. : *The Structure of Silica Glass,* J. Non-Cryst. Solids 83 (1986) 324.

Gendron-Badou A. , Coradin T. , Maquet J. , Fröhlich F. and Livage J. : *Spectroscopic characterization of biogenic silica,* J. Non-Cryst. Solids 316 (2003) 331.

Glinka Y. D. , Lin S.-H. and Chen Y.-T. : *The photoluminescence from hydrogen-related species in composites of* SiO_2 *nanoparticles,* Appl. Phys. Lett. 75 (1999) 778.

Glinka Y. D. : *Two-photon-excited luminescence and defect formation in* SiO_2 *nanoparticles induced by 6.4-eV ArF laser light,* Phys. Rev. B 62 (2000) 4733.

Gobsch G. , Haberlandt H. , Weckner H.-J. and Reinhold J. : phys. stat. sol. (b) 90 (1978) 309.

Gorton N. T. , Walker G. , Burley, S. D. : *Experimental analysis of the composite blue CL emission in quartz - is this related to aluminium content?* In: Abstracts SLMS International Conference on Cathodoluminescence, Nancy, Sept (1996) , p.59.

Griggs D. T. and Blacic J. D. : Quartz : Anomalous weakness of synthetic crystals, Science 147 (1965) 292.

Griggs D. T. : *Hydrolytic weakening of quartz and other silicates,* Geophys. J. R. Astron. Soc. 14 (1967) 19.

Griscom D. L. , Friebele E. J. and Sigel G. H. : *Observation and analysis of the primary 29Si hyperfine structure of the E' center in non-crystalline* SiO_2 , Solid State Commun. 15 (1974) 479.

Griscom D. L. : *The electronic structure of* SiO_2*: A review of recent spectroscopic and theoretical advances,* J. Non-Cryst. Solids 24 (1977) 155.

Griscom D. L. : *E' center in glassy SiO_2 : Microwave saturation properties and confirmation of the primary ^{29}Si hyperfine structure*, Phys. Rev. B 20 (1979a) 1823.

Griscom D. L. : Proc. 33rd Frequency Control Symposium (Elrctronic Indstrial Association, Washington, DC, (1979b), p.98.

Griscom D. L. : *E' center in glassy SiO_2: ^{17}O, ^{1}H, and "very weak" ^{29}Si superhyperfine structure*, Phys. Rev. B 22 (1980) 4192.

Griscom D. L. , Stapelbroek M. and Friebele E. J. : *ESR studies of damage processes in X-irradiated high purity a-SiO_2:OH and characterization of the formyl radical defect*, J. Chem. Phys. 78 (1983) 1638.

Griscom D. L. : *Characterization of three E'-center variants in X- and γ-irradiated high purity a-SiO_2*, Nucl. Instrum. and Methods Phys. Res. B 1 (1984) 481.

Griscom D. L. : *Defect structure of glasses : Some outstanding questions in regard to vitreous silica*, J. Non-Cryst. Solids 73 (1985) 51.

Griscom D. L. and Friebele E. J. : *Fundamental radiation-induced defect centers in synthetic fused silicas: Atomic chlorine, delocalized E' centers, and a triplet state*, Phys. Rev. B 34 (1986) 7524.

Griscom D. L. : *Glass Science and Technology*, D. R. Uhlmann and N. J. Kreidl (eds), Academic, London, Vol. 4B (1990a) 151.

Griscom D. L. : *Optical Properties and Structure of Defects in Silica Glass*, J. of the Ceramic Society of Japan 99 (1991) 899.

Griscom D. L. : *The natures of point defects in amorphous silicon dioxide*, Pacchioni et a. (eds.) , *Defects in SiO_2 and related Dielectrics: Science and Technology*, Kluwer Academic Publishers, (2000), p.117.

Guzzi M. , Martini M. , Mattaini M. , Pio F. and Spenolo G. : *Luminescence of fused silica: Observation of the O^-_2 emission band*, Phys. Rev. B 35 (1987) 9407.

Hagni R. D. : *Industrial applications of cathodoluminescence microscopy*. In: Hagni R. D. (ed.): *Process Mineralogy IV: Applications to precious metal deposits, industrial minerals, coal, liberation, mineral processing, agglomeration, metallurgical products, and refractories, with special emphasis on cathodoluminescence microscopy*. TMS, Warrendale, Pennsylvania, (1987), p.37.

Hayes W. , Kane M. J. , Salminen O. , Wood R. L. and Doherty S. P. : *ODMR of recombination centres in crystalline quartz*, J. Phys. C 17 (1984) 2943.

Hayes W. and Jenkin T. J. L. : *Paramagnetic hole centres produced in germanium-doped crystalline quartz by X-irradiation at 4K*, J. Phys. C 18 (1985) L849.

Hayes W. and Stoneham A. M. : *Defects and Defect Processes in Nonmetalic Solids*, John Wiley, New York, (1985).

Hayes W. and Jenkin T. J. L. : *Optically detected magnetic resonance studies of exciton trapping by germanium in quartz*, J. Phys. C 21 (1988) 2391.

Heggie M. I. : *A molecular water pump in quartz dislocations*, Nature 355/23 (1992) 337.

Henderson G. S. and Baker D. R. (eds.) : *Synchrotron Radiation: Earth, Environmental and Material Sciences Applications*, Short Course Series 30, Mineralogical Association of Canada. (2002)159-178. Henderson G. S. , The Geochemical News, number 113, October (2002), pp.13.

Hinic I. , Stanisic G. and Popovic Z. : *Photoluminescence properties of silica aerogel during sintering process*, J. Sol-Gel Science and Technology 14 (1999) 281.

Hinic I. , Stanisic G. and Popovic Z. : *Influence of the synthesis conditions on the photoluminescence of silica gels*, J. Serb. Chem. Soc. 68 (2003) 953.

Hosono H. , Abe Y. , Imagawa H. , Imai H. and Arai K. : *Experimental evidence for the Si-Si bond model of the 7.6-eV band in SiO2 glass*, Phys. Rev. B 44 (1991) 12043.

Hosono H. , Abe Y. , Kinser D. L . , Weeks R. A. , Muta K. and Kawazoe H. : *Nature and origin of the 5-eV band in SiO_2:GeO_2 glasses*, Phy. Rev. B 46 (1992) 11445.

Hosono H. , Mizuguchi M. , Kawazoe H. and Ogawa T. : *Effects of fluorine dimer excimer laser radiation on the optical transmission and defect formation of various types of synthetic SiO_2 glasses*, Appl. Phys. Lett. (1999) 2755.

Im S. , Jeong J. Y. , Oh M. S. , Kim H. B. , Chae K. H. , Whang C. N. and Song J. H. : *Enhancing defect-related photoluminescence by hot implantation into SiO_2 layers*, Appl. Phys. Lett. 47 (1999) 961

Imai H. , Arai K. , Saito T. , Ichimura S. Nonaka H. , Vigroux J. P. , Imagawa H. , Hosono H. and Abe Y. : *UV and VUV optical absorption due to intrinsic and laser induced defects in synthetic silica glasses*, in R. A. B. Devine (ed.), *The physics and Technology of Amorphous SiO_2*, Plenum, New York, (1987), p.153.

Imai H. , Arai K. , Imagawa H. , Hosono H. and Abe Y. : *Two types of oxygen-deficient centers in synthetic silica glass*, Phys. Rev. B 38 (1988) 12772.

Imakita K. , Fujii M. , Yamaguchi Y. , and Hayashi S. : *Interaction between Er ions and shallow impurities in Si nanocrystals within SiO_2*, Phys. Rev. B 71 (2005) 115440.

Isoya J. , Weil J. A. and Hallibruton L. E. : *EPR and ab initio SCF-MO studies of the $Si{\cdot}H-Si$ system in the E'_4 center of α-quartz*, J. Chem. Phys. 74 (1981) 5436.

Itoh C. , Tanimura K. and Itoh N. : *Optical studies of self-trapped excitons in SiO_2*, J. Phys. C : Solid State Phys. 21 (1988) 4693.

Itoh C. , Tanimura K. , Itoh N. and Itoh M. : *Threshold energy for photogeneration of self-trapped excitons in SiO_2*, Phys. Rev. B 39 (1989) 11183.

Itoh C. , Suzuki T. and Itoh N. : *Luminescence and defect formation in undensified and densified amorphous SiO_2*, Phys. Rev. B 41 (1990) 3794.

Kajihara K.,Skuja L.,Hirano M. and Hosono H.: *Formation and decay of the nonbridging oxygen hole centers in SiO_2 glasses induced by F_2 laser irradiation: In situ observation using a pump and probe technique*, Appl.Phys.Lett. 79 (2001) 1757.

Kajihara K. , Ikuta Y. , Hirano M. and Hosono H. : *Effect of F_2 laser power on defect formation in high-purity SiO_2 glass*, J. Non-Cryst. Solids 322 (2003) 73.

Kajihara K. , Skuja L. , Hirano M. and Hosono H. : *Role of Mobile Interstitial Oxygen Atoms in Defect Processes in Oxides: Interconversion between Oxygen-Associated Defects in SiO_2 Glass*, Phys. Rev. Lett. 92 (2004) 15504.

Kanemitsu Y. , Suzuki K. , Kondo M. and Matsumoto H. : *Luminescence from a cubic silicon cluster*, Solid State Commun. 89 (1994) 619.

Khanlary M. R. , Townsend P. D. and Townsend J. E. : *Luminescence spectra of germanosilicate optical fibres.1: radioluminescence and cathodoluminescence*, J. Phys. D 26 (1993) 371.

Khriachtchev L. , Räsänen M. , Novikov S. and Pavesi L. : *Systematic correlation between Raman spectra, photoluminescence intensity, and absorption coefficient of silica layers containing Si nanocrystals*, Appl. Phys. Lett. 85 (2004) 1511.

Kim H. B. , Son J. H. , Whang C. N. , Chae K. H. , Lee W. S. , Im S. , Kim S. O. , Woo J. J. and Song J. H. : *Light-Emitting Properties of Si-Ion-Irradiated SiO_2/Si/SiO_2 Layers*, J. the Korean Phys. Soci. 37 (2000a) 466.

Kim H. B. , Son J. H. , Chae K. H. , Jeong J. Y. , Lee W. S. , Im S. , Song J. H. and Whang C. N. : *Photoluminescence from Si ion irradiated SiO_2 Si SiO_2 films with elevated substrate temperature,* Materials Science and Engineering B 69-70 (2000b) 401.

Kim H. B. , Kim T. G. , Son J. H. , Whang C. N. , Chae K. H. , Lee W. S. , Im S. and Song J. H. , *Effects of Si-dose on defect-related photoluminescence in Si-implanted SiO_2 layers,* J. Appl. Phys. 88 (2000c) 1851.

Kofstad P. : *High Temperature Corrosion,* ELSEVIER, London and New York, (1988).

Koyama H. : *Cathodoluminescence study of SiO_2,* J. Appl. Phys. 51 (1980) 2228.

Krbetschek M. R. , Götze J. , Dietrich A. and Trautmann T. : *Spectral Information from Minerals Relevant for Luminescence Dating,* Radiation Measurements 27 (1997) 695.

Kronenberg A. K. , Kirby S.H. , Aines R. D. and Rossmann G. R. : *Solubility and diffusional uptake of hydrogen in quartz at high water pressure: implications for hydrolytic weakening in the laboratory and within the earth,* Tectonophysics 172 (1986) 255.

Kuzuu N. and Murahara M. : *X-ray-induced absorption bands in type-III fused silicas,* Phys. Rev. B 46 (1992) 14486.

Lamkin M. , Riley F. and Fordham R. : *Oxygen Mobility in Silicon Dioxide and Silicate Glasses: A Review,* J. Eur. Ceram. Soc. 10 (1992) 347.

Lau H. W. , Tan O. K. , Liu Y. , Ng C. Y. , Chen T. P. , Pita K. and Lu D. : *Defect-induced photoluminescence from tetraethylorthosilicate thin films containing mechanically milled silicon nanocrystals,* J. Appl. Phys. 97 (2005) 104307.

Ledoux G. , Gong J. , Huisken F. , Guillois O. and Reynaud C. : *Photoluminescence of size-separated silicon nanocrystals: Confirmation of quantum confinement,* Appl. Phys. Lett. 80 (2002) 4834.

Lee W. S. , Jeong J. Y. , Kim H. B. , Chae K. H. , Whang C. N. , Im S. and Song J. H. : *Violet and orange luminescence from Ge-implanted SiO_2 layers,* Materials Science and Engineering B 69-70 (2000) 474.

Lide D. R. : *CRC Handbook of Chemistry and Physic,* 85th edition, CRC Press, (2004).

Lu Z. Y. , Nicklaw C. J. , Fleetwood D. M. , Schrimpf R. D. and Pantelides S. T. : *Structure, Properties, and Dynamics of Oxygen Vacancies in Amorphous SiO_2,* Phys. Rev. Lett. 89 (2002) 285505.

Ludwig M. H. , Menniger J. , Hummel R. E. : *Cathodoluminescing properties of spark-processed silicon,* J. Phys: Condens. Matter 7 (1995) 9081

Luff B. J. and Townsend P. D. : *Cathodoluminescence of synthetic quartz,* J. Phys. Condensed Matter 2 (1990) 8089.

Magruder R. H. , Weeks R. A. and Weller R. A. : *Luminescence and absorption in type III silica implanted with multi-energy Si, O and Ar. ions,* J. Non-Cryst. Solids 322 (2003) 58.

Majid F. B. and Miyagawa I. : *Detection of Several Types of E'-center by ESR Double Modulation Spectrum Method,* Chem. Phys. Lett. 209 (1993) 496.

Marshall D. J. : *Cathodoluminescence of geological materials,* Allen & Unwin Inc., Winchester Mass., (1988).

McLaren A. C. , Cook R. F. , Hyde S. T. , Tobin R. C. : *The mechanisms of the formation and growth of water bubbles and associated dislocation loops in synthetic quartz,* Phys. Chem. Minerals 9 (1983) 79.

Mitchell J. P. and Denure D. G. : *A study of SiO layers on Si using cathodoluminescence spectra,* Solid State Electron. 16 (1973) 825.

Mohanty T. , Mishra N. C. , Bhat S. V. , Basu P. K. and Kanjilal D . : *Dense electrnic excitation induced defects in fused silica,* J. Phys. D: Appl. Phys. 36 (2003) 3151.

Morimoto Y. , Igarashi T. , Sugahara H. and Nasu S. : *Analysis of gas release from vitreous silica,* J. Non-Cryst. Solids 139 (1992) 35.

Morimoto Y. , Weeks R. A. , Barnes A. V. , Tolk N. H. and Zuhr R. A. : *The effect of ion implantation on luminescence of a silica,* J. Non-Cryst. Solids 196 (1996) 106.

Mozzi R. L. and Warren B. E. : *The Structure of Vitreous Silica,* J. Appl. Crystallogr. 2 (1969) 164.

Munekuni S. , Yamanaka T. , Shimogaichi Y. , Tohmon R. , Ohki Y. , Nagasawa K. and Hama Y. : *Various types of nonbridging oxygen hole center in high-purity silica glass,* J. Appl. Phys. 68 (1990) 1212.

Mysovsky A. S. , Sushko P. V., Mukhopadhyay S. , Edwards A. H. and Shluger A. L.: *Calibration of embedded-cluster method for defect studies in amorphous silica,* Phys. Rev. B 69 (2004) 85202.

Nicholas J. B. , Hopfinger A. J. , Trouw F. R. and Iton E. : *Molecular Modeling of Zeolite Structure. 2. Structure and Dynamics of Silica Sodalite and Silicate Force Field,* J. Am. Chem. Soc. 113 (1991) 4792.

Nicollian E. H. and Brews J. R. : *MOS (Metal Oxide Semiconductor) Physics and Technology,* WILEY-VCK, New York, (2002).

Nishikawa H. , Shiroyama Y. , Nakamura R. , Ohki Y. , Nagasawa K. and Hama Y.: *Photoluminescence from defect centers in high-purity silica glasses observed under 7.9-eV excitation,* Phys. Rev. B 45 (1992) 586.

Nishikawa H. , Watanabe E. , Ito D. and Ohki Y. : *Decay kinetics of the 4.4 eV photoluminescence associated with the two states of oxygen-deficient-type defect in amorphous* SiO_2, Phys. Rev. Lett. 72 (1994) 2101.

Nishikawa H. : *Structure and properties of amorphous silicon dioxide-Isuse on the reliability and novel applications,* Chapter 3 pp.93 in: Hari Singh Nalwa, Silicon-Based Materials and Devices, Vol. 2, (2001).

Nuttall R. H. D. and Weil J. A. : *Two hydrogenic trapped-hole species in α-quartz,* Solid State Commun. 33 (1980) 99.

O'Reilly E. P. and Robertson J. : *Theory of defects in vitreous silicon dioxide,* Phys. Rev. B 27 (1983) 3780.

Ozawa L. : *Cathodoluminescence: Theory and Application* , Kodansha Ltd., Japan, (1990).

Pacchioni G. and Ierano G. : *Optical Absorption and Nonradiative Decay Mechanism of E' Center in Silica,* Phys. Rev. Lett. 81 (1998a) 377.

Pacchioni G. and Ierano G. : *Ab initio theory of optical transitions of point defects in* SiO_2, Phys. Rev. B 57 (1998b) 818.

Pacchioni G. , Skuja L. and Griscom D. L. : *Defects in* SiO_2 *and related dielectrics: Science and technology,* (2000), p.73.

Paleari A. , Chiodini N. , Di Martino D. and Meinardi F. : *Radiative decay of vacuum-ultraviolet excitation of silica synthesized by molecular precursors of Si-Si sites: An indicator of intracenter relaxation of neutral oxygen vacancies,* Phys. Rev. B 71 (2005) 75101.

Pellergrino P. , Perez-Rodriguez A. , Garrido B. , Gonzalez-Varona O. , Morante J. R., Marcinkevicius S. , Galeckas A. and Linnros J. : *Time-resolved analysis of the white photoluminescence from* SiO_2 *films after Si and C coimplantation,* Appl. Phys. Lett. 84 (2004) 25.

Perez-Rodriguez A. , Gonzalez-Varona O. , Garrido B. , Pellegrino P. , Morante J. R., Bonafos C. , Carrada M. and Claverie A. : *White luminescence from Si^+ and C^+ ion-implanted SiO_2 films,* J. Appl. Phys. 94 (2003) 254.

Plaksin O. A. , Takeda Y. , Okubo N. , Amekura H. , Kono K. , Umeda N. , Kishimoto N. : *Electronic transitions in silica glass during heavy-ion implantation,* Thin Solid Films 464-465 (2004) 264.

Prado R. J. , D'Addio T. F. , Fantini M. C. A. , Pereyra I. and Flank A. M. : *Annealing effects of highly homogeneous a-$Si_{1-x}C_x$:H,* J. Non-Cryst. Solids 330 (2003) 196.

Qin G. G. , Lin J. , Duan J. Q. and Yao G. Q. : *A comparative study of ultraviolet emission with peak wavelengths around 350 nm from oxidized porous silicon and that from SiO_2 powder,* Appl. Phys. Lett. 69 (1996) 1689.

Rafferty C. S. : *Stress Effects in Silicon Oxidation Simulation and Experiments,* PhD thesis , Stanford University, (1989).

Rebohle L. , Gebel T. , Fröb H. , Reuther H. and Skorupa W. : *Ion beam processing for Si/C-rich thermally grown SiO_2,* Appl. Surf. Sci. 184 (2001a) 156.

Rebohle L. , Gebel T. , von Borany J. , Skorupa W. , Helm M. , Pacifici D. , Franzo G. and Priolo F. : *Transient behavior of the strong violet electroluminescence of Ge-implanted SiO_2 layers,* Appl. Phys. B 74 (2002a) 53.

Rebohle L. , von Borany J. , Fröb H. , Gebel T. , Helm M. and Skorupa W. : *Ion beam synthesized nanoclusters for silicon-based light emission,* Nucl. Instrum. Methods Phys. Res. B 188 (2002b) 28.

Reimer L. : *Scanning Electron Microscopy: Physics of Image Formation and Microanalysis, Springer series in optical sciences,* Vol. 45, (1998).

Robertson J. : *The Physics and Technology of Amorphous SiO_2,* Roderich A. B. Devine (ed.), National center for Telecommunication Studies, Meyland, France, (1988).

Rudra J. K. , Fowler B. W. and Feigl F. J. : *Model for the E'_2 center in Alpha Quartz* , Phys. Pev. Lett. 55 (1985) 2614.

Saito K. and Ikushima A. J. : *Effects of fluorine on structure, structural relaxation, and absorption edge in silica glass,* J. Appl. Phys. 91 (2002) 4886.

Sakurai Y. , Nagasawa K. , Nishikawa H. and Ohki Y. : *Characteristic red photoluminescence band in oxygen-deficient silica glass,* J. Appl. Phys. 86 (1999) 370.

Sakurai Y. : *Oxygen-related red photoluminescence bands in silica glasses,* J. Non-Cryst. Solids 316 (2003) 389.

Seol K. S. , Ohki Y. , Nishikawa H. , Takiyama M. and Hama Y. : *Effect of implanted ion species on the decay kinetics of 2.7 eV photoluminescence in thermal SiO_2 films,* J. Appl. Phys. 80 (1996) 6444.

Settle F. A. (ed.) : *Handbook of Instrumental Techniques for Analytical Chemistry,* Prentice Hall, (1997).

Shimizu-Iwayama T. , Ohshima M. , Niimi T. , Nakao S. , Saitoh K. , Fujita T. and Itoh N. : *Visible photoluminescence related to Si precipitates in Si^+-implanted SiO_2,* J. Phys.: Condens. Matter 5 (1993) L375.

Shimizu-Iwayama T. , Nakao S. and Saitoh K. : *Optical and structural characterization of implanted nanocrystalline semiconductors,* Nucl. Instrum. Methods Phys. Res. B 121 (1997) 450.

Skuja L. , Streletsky A. N. and Pakovich A. B. : *A new intrinsic defect in amorphous SiO_2: Twofold coordinated silicon,* Solid State Commun. 50 (1984) 1069.

Skuja L. and Trukhin A. : Comment on " Luminescence of fused silica: *Observation of the* O^-_2 *emission band*, Phys. Rev. B 39 (1989) 3909.

Skuja L. : *Isoelectronic series of twofold coordinated Si, Ge, and Sn atoms in glassy* SiO_2*: a luminescence study*, J. Non-Cryst. Solids 149 (1992a) 77.

Skuja L. : *Time-resolved low temperature luminescence of non-bridging oxygen hole centers in silica glass*, Solid State Commun. 84 (1992b) 613.

Skuja L. : *The origin of the 1.9eV luminescence band in glassy* SiO_2, J. Non-Cryst. Solids 179 (1994a) 51.

Skuja L. , Suzuki T. , Tanimura K. : *Site-selective laser-spectroscopy studied of the intrinsic 1.9-eV luminescence center in glassy* SiO_2, Phys. Rev. B 52 (1995) 15208.

Skuja L. : *Optically active oxygen-deficiency-related centers in amorphous silicon dioxide*, J. Non-Cryst. Solids 239 (1998) 16.

Skuja L. , Güttler B. , Schiel D. and Silin A. R. : *Quantitative analysis of the concentration of interstitial* O_2 *molecules in* SiO_2 *glass using luminescence and Raman spectroscopy*, J. Appl. Phys. 83 (1998a) 6106.

Skuja L. , Güttler B. , Schiel D. and Silin A. R. : *Infrared photoluminescence of preexisting or irradiation-induced interstitial oxygen molecules in glassy* SiO_2 *and* α*-quartz*, Phys. Rev B 58 (1998b) 14296.

Skuja L. , *Optical properties of defects in silica*, pp.73-116 in Pacchioni G., Skuja L. and Griscom D. L. (eds) : *Defects in* SiO_2 *and related dielectrics: Science and technology*, (2000).

Skuja L. , Hirano M. and Hosono H. : *Oxygen-Related Intrinsic Defects in Glassy* SiO_2*: Interstitial Ozone Molecules*, Phys. Rev. Lett. 84 (2000a) 302.

Skuja L. , Mizuguchi M. , Hosono H. and Kawazone H. : *The nature of the 4.8 eV optical absorption band induced by vacuum-ultraviolet irradiation of glassy* SiO_2, Nucl. Instrum. Methods Phys. Res. B 166-167 (2000b) 711.

Skuja L. , Kajihara K. , Kinoshita T. , Hirano M., Hosono H. : *The behavior of interstitial oxygen atoms induced by* F_2 *laser irradiation of oxygen-rich glassy* SiO_2, Nucl. Instr. & Methods in Physics Research B 191 (2002) 127.

Skuja L. , Kajihara K. , Hirano M. , Saitoh A. , Hosono H. : *An increased* F_2*-laser damage in 'wet' silica glass due to atomic hydrogen: A new hydrogen-related E'-center*, J. Non-Cryst. Solids 352 (2006) 2297.

Slater J. C. : *Quantum Theory of Molecules and Solids: Symmetry and Energy Bands in Crystals*, McGraw-Hill, New York, Vol. 2, (1965).

Snyder K. C. and Fowler W. B. : *Oxygen vacancy in* α*-quartz: A possible bi- and metastable defect*, Phys Rev. B 48 (1993) 13238.

Song J. , Corrales L. R. , Kresse G. and Jonsson H. : *Migration of O vacancies in alpha-quartz: The effect of excitons and electron holes*, Phys. Rev. B 64 (2001) 134102.

Song K. S. and Williams R. T. : *Self-Trapped Excitons*, Springer Verlag, Berlin, (1993).

Stapelbroek M. , Griscom D. L. , Friebele E. J. and Sigel Jr. G. H. : *Oxygen-associated trapped-hole centers in high-purity fused silicas*, J. Non-Cryst. Solids 32 (1979) 313.

Stevens-Kalceff M. A. and Philips M. R. : *Cathodoluminescence microcharacterization of the defect structure of quartz*, Phys. Rev. B 52 (1995) 3122.

Stevens-Kalceff M. A. : *Cathodoluminescence microcharacterization of the defect structure of irradiated hydratedand anhydrous fused silicon dioxide*, Phys. Rev. B 57 (1998) 5674.

Stevens-Kalceff M. A. : *Electron-Irradiation-Induced Radiolytic Oxygen Generation and Microsegregation in Silicon Dioxide Polymorphs*, Phys. Rev. Lett. 84 (2000) 3137.

Stevens-Kalceff M. A. , Stesmans A. and Wong J. : *Defects induced in fused silica by high fluence ultraviolet laser pulses at 355 nm,* Appl. Phys. Lett. 80 (2002) 758.

Stevens-Kalceff M. A. and Wong J. : *Distribution of defects induced in fused silica by ultraviolet laser pulses before and after treatment with a CO_2 laser,* J. Appl. Phys. 97 (2005) 113519.

Streletsky A. N. , Pakovich A. B. , Gachkovski V. F. , Aristov Yu. I. , Rufov Yu. N. and Butyagin P. Y. : *Khim. Fizika,* Sov. Chem. Phys., No.7 (1982) 938.

Suzuki T. , Skuja L. , Kajihara K. , Hirano M. , Kamiya T. and Hosono H. : *Electronic Structure of Oxygen Dangling Bond in Glassy SiO_2: The Role of Hyperconjugation,* Phys. Rev. Lett. 90 (2003) 186404.

Takagahara T. and Takeda K. : *Theory of the quantum confinement effect on excitons in quantum dots of indirect-gap materials,* Phys. Rev. B 46 (1992) 15578.

Tanimura K. , Tanaka T. and Itoh N. : Creation of Quasistable Lattice Defects by Electronic Excitation in SiO_2, Phys. Rev. Lett. 51 (1983) 423.

Tong S. , Liu X.-N. , Gao T. and Bao X.-M. : *Intense violet-blue photoluminescence in as-deposited amorphous Si:H:O films,* Appl. Phys. Lett. 71 (1997) 698.

Trukhin A. N. : *Investigation of the Photoelectric and Photoluminescent Properties of Crystalline Quartz and Vitreous Silica in the Fundamental Absorption Region,* phys. stat. sol. (b) 86 (1978) 67.

Trukhin A. N. and Plaudis A. E. : *Investigation of intrinsic luminescence of SiO_2* , Sov. Phys. Solid State 21 (1979) 644.

Trukhin A. N. : *Studey of Exitons in SiO_2- Luminescent Centers as Exciton Detectors,* phys. stat. sol. (b) 98 (1980) 541.

Trukhin A. N. : *Excitons in SiO_2: a review,* J. Non-Cryst. Solids 149 (1992) 32.

Trukhin A. N. : *Luminescence of a self-trapped exciton in GeO_2 crystal,* Solid State Commun. 85 (1993) 723.

Trukhin A. N. : Self-*trapped exciton luminescence in α-quartz,* Nucl. Instr. and Meth. in Phys. Res. B 91 (1994) 334.

Trukhin A. N. , Goldberg M. , Jansons J. , Fitting H.-J. and Tale I. A. : *Silicon dioxide thin film luminescence in comparison with bulk silica,* J. Non-Cryst. Solids 223 (1998) 114.

Trukhin A. N. and Fitting H.-J. : *Investigation of optical and radiation properties of oxygen deficient silica glasses,* J. Non-Cryst. Solids 248 (1999) 49.

Trukhin A. N. , Jansons J. , Fitting H.-J. , Barfels T. and Schmidt B. : *Cathodoluminescence decay kinetics in Ge^+, Si^+, O^+ implanted SiO_2 layers,* J. Non-Cryst. Solids 331 (2003a) 91.

Trukhin A. , Poumellec B. and Garapon J. : *Study of the germanium luminecence in silica: from non-controlled impurity to germano-silicate core of telecommunication fiber performs* , J. Non-Cryst. Solids 332 (2003b) 153.

Trukhin A. and Poumellec B. : *Photosensitivity of silica glass with germanium studied by photoinduced of thermally stimulated luminescence with vacuum ultraviolet radiation,* J. Non-Cryst. Solids 324 (2003c) 21.

Tsu D. V. , Lucovsky G. and Davidson B. N. : *Effects of the nearest neighbors and the alloy matrix on SiH stretching vibrations in the amorphous SiO_r:H ($0<r<2$) alloy system,* Phys. Rev. B 40 (1989) 1795.

Uchino T. , Takahashi M. and Yoko T. : *Model of oxygen-deficiency-related defects in SiO_2 glass,* Phys. Rev. B 62 (2000b) 2983.

Uchino T. , Takahashi M. and Yoko T. : *Structure and Generation Mechanism of the Peroxy-Radical Defect in Amorphous Silica,* Phys. Rev. Lett. 86 (2001) 4560.

van Santen R. A. , de Man A. J. M. , Jacobs W. P. J. H. , Teunissen E. H. and Kramer G. J. : *Lattice Relaxation of Zeolites,* Catalysis Letters 9 (1991) 273.

Weeks R. A. : *Paramagnetic Resonance of Lattice Defects in Irradiated Quartz,* J. Appl. Phys. 27 (1956) 1376.

Weeks R. A. and Nelson C. M. : *Trapped electrons in irradiated quartz and silica. I. Optical absorption,* J. Am. Ceram. Soc. 43 (1960) 399.

Weeks R. A. : *Paramagnetic Spectra of E´2 Centers in Crystalline Quartz,* Phys. Rev. 130 (1963) 570.

Weeks R. A. , Magruder R.H. , Gaylon R. and Weller R.A. : *Effects of B and N implantation on optical absorption and photoluminescence and comparison to the effects of implanting Si, Ge, O, and Ar in silica,* J. Non-Cryst. Solids 351 (2005) 1727.

Weber J. T. and Cromer D. T. : *Orbital Radii of Atoms and Ions,* J. Chem. Phys. 42 (1965) 4116.

Wilkinson A. R. and Elliman R. G. : *The effect of annealing environment on the luminescence of silicon nanocrystals in silica,* J. Appl. Phys. 96 (2004) 4018.

White B. D. , Brillson L. J. , Bataiev M. , Brillson L. J. , Fleetwood D. M. , Schrimpf R. D. , Choi B. K. , Fleetwood D. M. and Pantelides S. T. : *Detection of trap activation by ionizing radiation in SiO_2 by spatially localized cathodoluminescence spectroscopy,* J. Appl. Phys. 92 (2002) 5729.

Yacobi B. G. and Holt D. B. : *Cathodoluminescence Microscopy of Inorganic Solids,* Plenum Press, New York and London, (1990).

Yang X. H. , Townsend P. D. and Holgate S. A. : *Cathodoluminescence and depth profiles of tin in float glass* , J. Phys. D 27 (1994) 1757.

Yang X. , X. , Wua L. , Li S. H. , Li H. , Qiu T. , Yang Y. M. , Chu P. K. and Siu G. G. : *Origin of the 370-nm luminescence in Si oxide nanostructures* , Appl. Phys. Lett. 86 (2005) 201906.

Yao B. , Shi H. , Zhang X. and Zhang L. : *Ultraviolet photoluminescence from nonbridging oxygen hole centers in porous silica,* Appl. Phys. Lett. 78 (2001) 174.

Yi L. X. , Heitmann J. , Scholz R. and Zacharias M. : *Si rings, Si clusters, and Si nanocrystals-different states of ultrathin SiO_x layers,* Appl. Phys. Lett. 81 (2002) 4248.

Yi L. X. , Heitmann J. , Scholz R. and Zacharias M. : *Phase separation of thin SiO layers in amorphous SiO/SiO_2 superlattice during annealing,* J. Phys. : Condens. Matter 15 (2003) S2887.

Yip K. L. and Fowler W. B. : *Electronic structure of E'_1 centers in SiO_2* , Phys. Rev. B 11 (1975) 2327.

Yu Z. , Aceves M. , Carrillo J. , Flores F. , Falcony C. , Domínguez C. , Llobera A. and Morales-Acevedo A. : *Photoluminescence in off-stoichiometric silicon oxide compounds,* Superficies y Vacio 17 (2004) 1.

Zacharias M. , Tsybeskov L. , Hirschman K. D. , Fauchet P. M. , Bläsing J. , Kohlert P. and Veit P. : *Nanocrystalline silicon superlattices: fabrication and characterization,* J. Non-Cryst. Solids 227-230 (1998) 1132.

Zacharias M. , Heitmann J. , Scholz R. , Kahler U. , Schmidt M. and Bläsing J. : *Size-controlled highly luminescent silicon nanocrystals: A SiO/SiO_2 superlattice approach,* Appl. Phys. Lett. 80 (2002) 661.

Zacharias M. , Yi L. X. , Heitmann J. , Scholz R. , Reiche M. and Gösele U. : *Size-controlled Si nanocrystals for photonic and electronic applications,* Solis State Phenomena 94 (2003) 95.

Zatsepin A. F. , Biryukov D. Yu. and Kortov V. S. : *Analysis of OSEE Spectra of Irradiated Dielectrics,* Latvian Journal of Physics and Technical Sciences. No.6 (2000) 83.

Zhang B. L. and Raghavachari K. : *Photoabsorption and photoluminescence of divalent defects in silicate and germanosilicate glasses: First-principles calculations,* Phys. Rev. B 55 (1997), 15993.

Zunger A. and Wang L.-W. : *Theory of silicon nanostructures* , Appl. Surf. Sci. 102 (1996) 350.

9

Silicon Nanocluster in Silicon Dioxide: Cathodoluminescence, Energy Dispersive X-Ray Analysis and Infrared Spectroscopy Studies

Roushdey Salh
Institute of Physics, Faculty of Science and Technology, Umeå University, Umeå Sweden

1. Introduction

This chapter is extended to various electronical and optical modifications of amorphous silica (a-SiO_2) layers as they are applied in microelectronics, optoelectronics, as well as in the forthcoming photonics. Scanning electron microscopy (SEM), energy dispersive X-ray analysis (EDX), Fourier transform infrared spectroscopy (FTIR) and cathodoluminescence (CL) have been used to investigate thermally grown pure amorphous silicon dioxide and ion-implanted layers with thickness d_{ox}=100-500 nm. The main luminescent centers in silicon dioxide layers are the red luminescence (650 nm; 1.85 eV) of the non-bridging oxygen hole center (NBOHC; ≡Si–O•), a blue (460 nm; 2.7 eV) and a ultra violet luminescence (290 nm; 4.3 eV) of the oxygen deficient centers (ODC's; ≡Si···Si≡), and a yellow luminescence (570 nm; 2.2 eV) appears especially in hydrogen treated silica indicating water molecules, and on the other hand, in silicon excess samples indicating higher silicon aggregates. A quite different CL dose behavior of the red luminescence is observed in dry and hydrogen-treated samples due to dissociation and re-association of mobile hydrogen and oxygen to radicals of the silica network. Additionally implanted hydrogen diminishes the red luminescence in wet oxide but maintains the blue and the UV bands. Thus hydrogen passivates the NBOHC and keeps the ODC's in active emission states. A model of luminescence center transformation is proposed based on radiolytic dissociation and re-association of mobile oxygen and hydrogen at the centers as well as formation of interstitial H_2 , O_2 , and H_2O molecules.
Non-stoichiometric SiO_x layers produced by direct ion implantation or reactive sputtering are used to investigate whether the different luminescent centers are related to oxygen or to silicon. Oxygen implantation as well as direct silicon implantation led to an oxygen surplus as well as an oxygen deficit, respectively. The related luminescence damages provide direct evidence to the nature of the defects. Oxygen-deficient thin silica layers SiO_x with different stoichiometric degree $1 \leq x \leq 2$, were prepared by thermal evaporation of silicon monoxide in vacuum and in ambient oxygen atmosphere of varying pressure onto crystalline silicon substrates. The chemical composition has been calibrated and determined by FTIR spectroscopy. The CL spectra of the oxygen-deficient layers shows the development of typical silica luminescence bands at the composition threshold $x \leq 1.5$ onwards to x=2. The

green-yellow luminescence (2.15 eV) strongly increases with the annealing temperature up to 1300 °C which is attributed to the formation of small silicon aggregates in the network, from dimers over trimers even to hexamer rings.

Ion implantation doses between 3×10^{16} and 5×10^{16} ions/cm^2 led to an atomic dopant fraction of about 4 at.% at the half depth of the SiO_2 layers. In ion implanted SiO_2 layers additional emission bands are observed. A huge violet band in Ge^+ implanted layers appears at λ=410 nm (3.1 eV). This band corresponds to the Ge-related oxygen deficient center (Ge-ODC). The thermal annealing process of the Ge^+ implanted layers leads first to a strong increase of the violet luminescence due to formation of Ge dimers, trimers or higher aggregates, finally to destruction of the luminescence centers by further growing to Ge nanoclusters. Scanning transmission electron microscopy (STEM) shows the growing in Ge cluster size with increasing annealing temperature up to 1100 °C. As a result of ion implantation, we can state that; group IV elements (C, Si, Ge, Sn, Pb) in SiO_2 increase the intensity of the luminescence in the blue region and group VI elements (O, S, Se) increase the intensity in the red region, confirming the association of the defect centers in the blue and the red region with oxygen deficient centers and oxygen excess centers, respectively. The cathodoluminescence spectra of sulfur-implanted SiO_2 layers and oxygen implanted layers under special conditions show besides the characteristic luminescence bands a multimodal structure beginning in the green region at 500 nm over the yellow-red region and extending to the near IR measured up to 820 nm. The energy step differences of the sublevels amount to an average 120 meV and indicate vibronic-electronic transitions, probably of O_2^- interstitial molecules, which could be demonstrated by a respective configuration coordinate model.

2. CL of stoichiometric and non-stoichiometric SiO_2

2.1 Wet and dry SiO_2 layers

The CL spectra of pure SiO_2 consist of several broad bands and certainly of some overlapped components especially at the region around $\lambda\approx$ 460-620 nm (2.7-2.0 eV). The irradiation response of these luminescence bands indicates that they are associated with different defect centers. Even at specimen temperatures as low as LNT where thermal contributions to the bandwidth are almost minimized, the SiO_2 CL emission bands are still broad because of the degree of coupling between the host lattice and the defects associated with the luminescence emissions [Griscom 1990b].

The initial (1 sec) and the saturated (5 h) CL spectra of pure SiO_2 layers are presented in Fig. 2.1. The main CL emission bands in wet and dry specimens at temperatures between liquid nitrogen (LNT) and room temperature (RT) are the red luminescence R at 650 nm (1.9 eV) associated with the NBOHC [Fitting et al. 2005b], the blue B and ultraviolet UV bands at 460nm (2.7 eV) and 290 nm (4.3 eV) respectively, associated with the Si related oxygen deficient center (Si-ODC) [Skuja 1994b]. Some shoulders can be also seen in the green-yellow G-Y region between 500-580 nm (2.5-2.1 eV). A luminescence band at 500 nm (2.5 eV) in crystalline SiO_2 and another at 560 nm (2.2 eV) in amorphous SiO_2 is often ascribed to the self trapped exciton (STE) [Skuja et al. 1978, Trukhin et al. 1998]. Another CL band which is not often discussed in the literature is easily seen in the yellow Y region at $\lambda\approx$570-580 nm (2.18-2.14 eV) especially at LNT, but it is also expected in RT spectra where the plane between the B band and the R bands can accommodate more than one overlapped emission band.

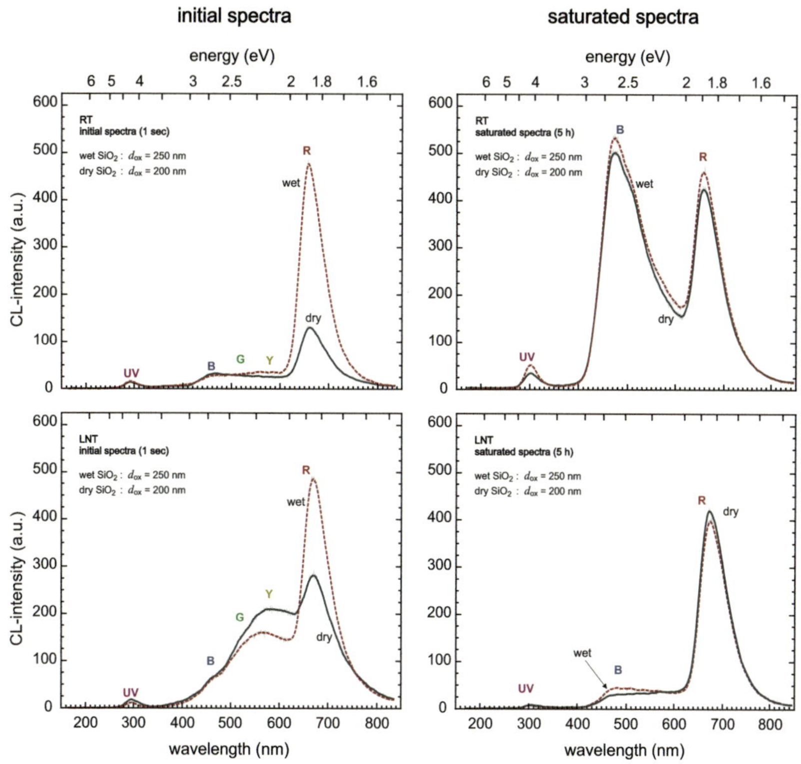

Fig. 2.1 The initial (1sec) and saturated (5h) CL spectra of wet and dry SiO_2 at room temperature (RT) and liquid nitrogen temperature (LNT); electron beam energy E_o=10 keV and current density j_o=5.4 mA/cm^2.

CL bands in this region will be discussed in more detail in the following sections. Based on our basic experimental observations we presume a very thin ice (H_2O) layer to have been produced on the surface of the sample as an effect of low temperatures which could be one of the reasons for the Y band, but under any circumstances one can see that the local intrinsic defects like ODC and NBOHC dominates the CL spectrum at LNT too.

Looking to the 1 sec spectra of both wet and dry SiO_2 in Fig. 2.1, we can state that some of the detected luminescence bands have the same origin and they behave similarly under electron beam irradiation but others are totally different or are formed/transformed by more complex reactions. The UV luminescence always peaks at very low intensities which scarcely change under irradiation. This band was not detected in many crystalline SiO_2 modifications at room temperature even or at liquid nitrogen temperature [Barfels 2001] but it is clearly seen in bulk and thin layers of amorphous SiO_2 [Trukhin et al. 1998, Bakaleinikov et al. 2004]. Absence of the crystalline order seems to be the origin of the UV luminescence band. The blue B luminescence starts with the same intensity in both dry and wet at RT and LNT and reaches the saturation level (5 h) together. The B band grows drastically during the irradiation at room temperature (RT), while it is expected that a high energetic electron beam creates more oxygen vacancy or in other words oxygen deficient centers (ODC).

The main difference between the CL spectra of wet and dry SiO_2 is located in the Y and R region. The Y luminescence is detected with relatively higher intensity in wet SiO_2 at RT and it is more visible at LNT in both dry and wet SiO_2 which is probably associated with some crystalline H_2O molecules on the sample surface. A considerable increase in the R band intensity is clearly seen in the initial spectra in wet SiO_2. This is the main dissimilar point between dry and wet oxide layers. We suspect a direct connection of the Y and R bands with atomic or molecular hydrogen. The saturated spectra of wet and dry SiO_2 have almost the same profile, see Fig. 2.1. That means, the red luminescence is starting from different precursors in dry and wet SiO_2 and these precursors are destroyed or transformed to other similar structural defects in both kinds of layers during the electron irradiation.

2.2 Dose-temperature effect

It is well known that CL spectra of different SiO_2 modifications change during the initial period of excitation. The time evolution of the CL spectrum of wet and dry oxide *a*-SiO_2layers during electron irradiation at room temperature (RT) and liquid nitrogen temperature (LNT) is presented in Fig. 2.2. Comparison of time resolved bands shows clearly the increase of the initial CL intensity of the emission bands as the specimen temperature is reduced to LNT may be because of the reduction in the thermally assisted conversion of STE's to complementary defect pairs (oxygen deficient centers and oxygen excess centers) via nonradiative relaxation processes [Stevens-Kalceff 1998].

In Fig. 2.2 the amplitudes of the main luminescence peaks: red, R (650 nm, 1.85 eV), blue, B (460 nm, 2.7 eV), and UV (290 nm, 4.4 eV) have been recorded as a function of the irradiation time.

The UV band shows slightly increasing behavior for some 10 sec then stabilizing up to the saturation state. Using "track-stop" techniques [Fitting et al. 2004] it was possible to describe this band more comprehensively. The UV band is detected initially in the "track" mode followed by an increase at the beginning of the "stop" measurement, then a maximum (turnaround), decreasing and increasing again to a long term invariance, larger for the wet oxide than for the dry one.

The B band has a time dependence at LNT that differs from that at RT. Whereas it decreases at LNT for both the wet and the dry SiO_2, it increases from very low intensity for RT, i.e. it is fully generated during the irradiation process, probably from precursors like ODC centers [Fitting et al. 2002b]. The long term irradiation shows an increase to saturation even over several hours. The Y band has the same time dependence as the B band both at RT and LNT. It gives us an impression that both defect centers associated with the blue and the yellow luminescence are at least of the same kind, probably oxygen deficiency centers (ODC).

The most obvious difference in the luminescence of wet and dry oxides appears in the R dose dependence, Fig. 2.2 (top). Whereas the R luminescence of the wet oxide starts from an initial intensity, decreases to a minimum and then increases to a saturation, the red band dependence of the dry oxide differs: starting from a much lower intensity, increasing and approaching nearly the same saturation intensity as that of wet oxide. This is the main new finding for the different wet and dry oxides and therefore it will be discussed in more detail in dependence of other ion implantations, especially hydrogen which is the main key difference between the dry and wet oxide silicas. The red (≈1.9 eV) luminescence is generally associated with NBOHC and attributed to the recombination of electrons in the highly localized nonbridging oxygen band gap state, with holes in the valence-band edge [Stevens-Kalceff 1998]. A remarkable difference in the dose behavior of the red peak in dry and wet specimens is found. This can be considered as a well recognized proof that the NBOHC defect structure of SiO_2 is extremely sensitive to hydrogen treatment which can result in the formation of defects and/or the formation of existing defect precursors in the presence of hydrogen atoms [Fitting et al. 2005a]. Other methods than CL have provided direct/indirect evidence for the existence of a number of different defect precursors for the NBOHC. Based on these facts, we may express two interactions where NBOHCs are involved;

$$\text{Dry } SiO_2 : \equiv Si{-}O \cdots Si\equiv \rightarrow \equiv Si{-}O\bullet + \bullet Si\equiv \tag{2.1}$$

$$\text{Wet } SiO_2 : \equiv Si{-}O\bullet + H^{o} \leftrightarrow \equiv Si{-}O{-}H \tag{2.2}$$

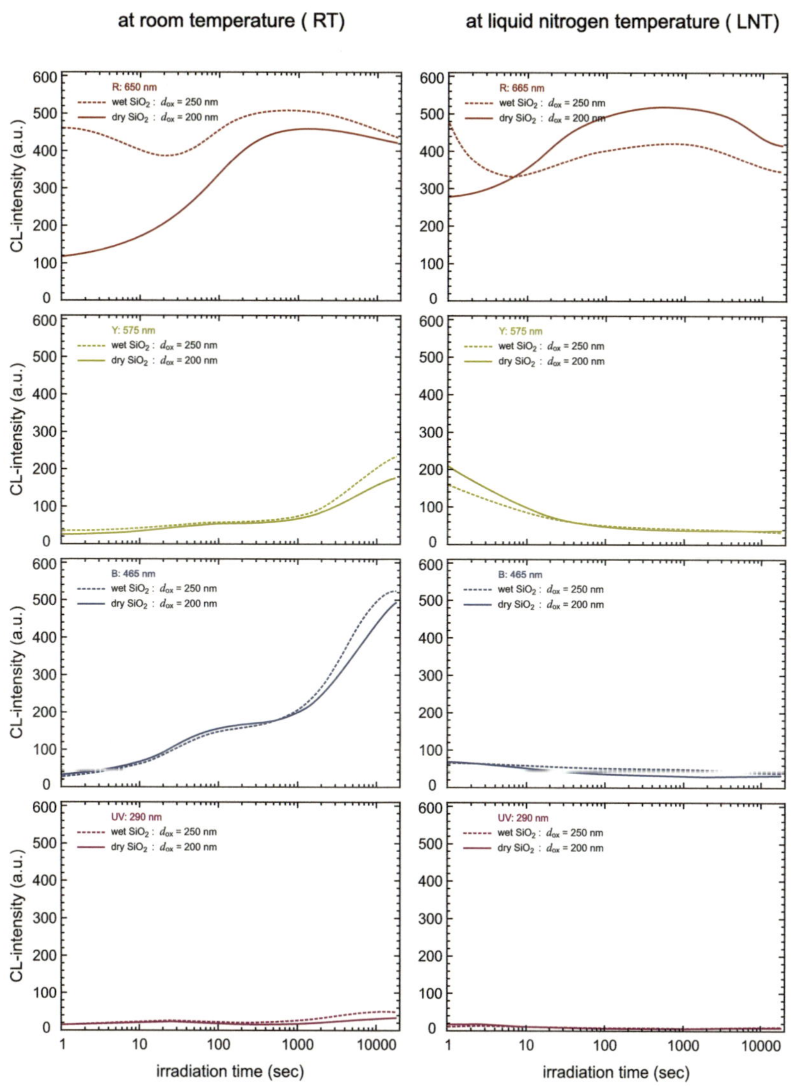

Fig. 2.2 CL dose dependencies of the red (R), the blue (B), the yellow (Y) and the ultraviolet (UV) bands in dry and wet SiO2 at room temperature (RT) and liquid nitrogen temperature (LNT); electron beam energy E_o=10 keV: current density j_o=5.4 mA/cm[2].

The most common production mode for the NBOHC in dry SiO_2 which contains a negligible amount of hydrogen and silanol groups, is by the strained bonds "..." between Si and O atoms, eq. (2.1), here both the *E´*-center and NBOHC can form in dry oxide layers. In wet SiO_2, hydrogen diffuses through the network and simultaneously undergoes reactions with the NBOHC during the first seconds of irradiation (destructive mode of NBOHC) followed by a slow creation mode [Tandon 2004]. This decay process is attributable to the simultaneous recombination of NBOHC with dissociated hydrogenous species, eq. (2.2). In wet SiO_2 layers, NBHOC are likely to react with interstitial molecular hydrogen at room temperature according to eq. (2.3).

$$\equiv Si{-}O\bullet + H_2 \rightarrow \equiv Si{-}O{-}H + H^{o} \quad (2.3)$$

Both of these reactions of eqs. (2.2) and (2.3) reasonably explain the elimination of NBOHC in wet layers. Indeed, infrared (IR) measurements confirm that H^{o} or H_2 react with dangling bonds created by neutron irradiation [Bakos et al. 2004a]. But the most interesting point is that the formation and destruction mechanism of NBOHC has absolutely no dependence on a specimen's temperature because it is affected only by OH groups or hydrogen concentration, as is clearly shown with the red curves in Fig. 2.2.

In the dry SiO_2, the NBOHC defect concentration has the tendency to increase (creation) at both RT and LNT during the beginning of irradiation but it decreases in wet specimens (hydrogenated) due to interaction with the mobile hydrogen [Kajihara et al. 2002]. The intensity difference of NBOHC between the beginning and the end of electron irradiation is much less in wet SiO_2, which confirms the reliability of eqs. (2.1-2.3).

2.3 Lifetime measurement of the resolved CL bands

To identify whether the red bands in wet and dry oxide are due to the same electronic state we have measured their respective lifetimes. In Fig. 2.3 the pulsed luminescence of the red peak R (650 nm, 1.9 eV) is measured, and the red R luminescence decay with its lifetime, τ, is determined from the pulse switching-off decay (bottom). In the pulsed CL response we observe an increase and pumping of the luminescence over about 20 μs after switching on the electron beam. The overall pulse duration is 100 μs.

During this excitation the luminescence again decays very rapidly over about 30 μs to lower intensity. Of course, the first suggestion for such a change could be based on thermal quenching effects of the red luminescence due to electron beam heating of the irradiation spot on the sample. Indeed, the electron beam density with 0.1 A/cm^2 is about 20 times higher than that in the CL dose measurements in Figs. 2.1 and 2.2. However, this very rapid rise, turnaround, and decrease of the red luminescence at the beginning of the electron beam pulse may be discussed also in the context of Fig. 2.2 and in terms of center activation and destruction.

The red luminescence lifetime after switching-off the electron beam has been enlarged in the bottom of Fig. 2.3 showing a mean lifetime of about (4.7±0.2) μs for the wet oxide and (5.3±0.2) μs for the dry silicon dioxide layers, respectively. Thus we may state that the R luminescence lifetime in wet oxide is somewhat smaller than in dry oxide, but of the same order with τ~5 μs. However, the time dependence of the decay is not exponential, but a stretched exponential with (5≤τ≤20μs over longer decay ranges [Trukhin et al. 2003a]. The lifetime of the blue luminescence is measured too, in Fig. 2.4 the B band decay is indicating a rapid component with τ≈(50±8) μs as well as a slow one with τ≈(7.1±0.4) ms for the wet oxide, and τ≈(70±2) μs as well as a slow one with τ≈(7.8±0.8) ms for the dry oxide. Ultimately, the decay kinetics of the UV luminescence was measured by Goldberg [Goldberg 1996] at LNT and RT. The decaying process was found to be occurring over the first 5 ns followed by a hyperbolic one proportion to $t^{-1.5}$ in the μs range.

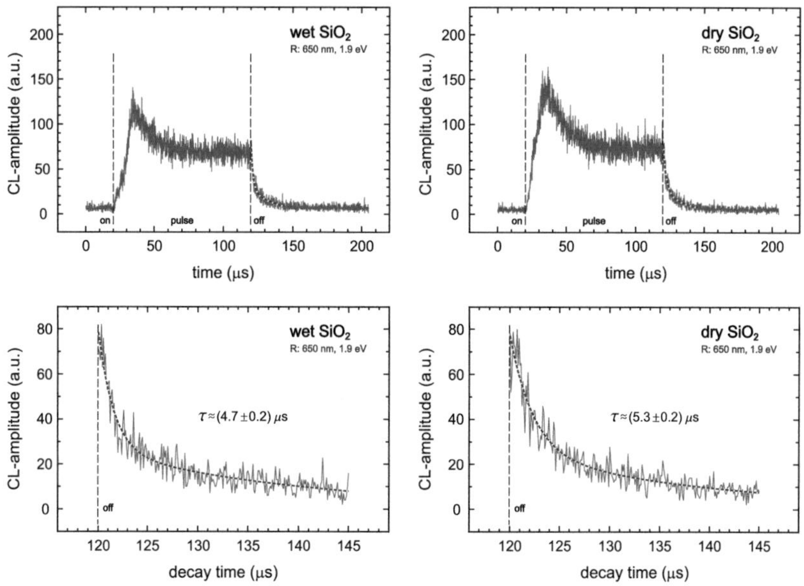

Fig. 2.3 CL pulsed excitation of the red R luminescence: 650 nm, 1.9 eV (top) with its temporary decay and lifetime τ (below) for wet (left) and dry (right) SiO_2 at RT; electron beam energy E_o=5 keV: current density j_o= 0.1 A/cm^2, [Fitting et al. 2005b].

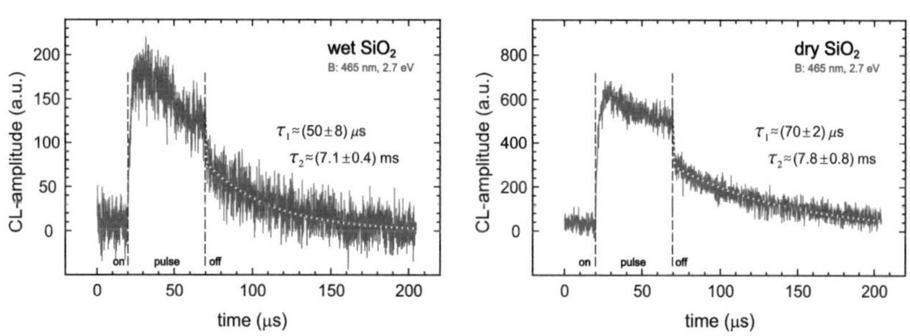

Fig. 2.4 CL pulsed excitation of the blue B luminescence: 465 nm, 2.7 eV with its temporary decay and lifetime τ for wet and dry SiO_2 at RT; E_o=5 keV: j_o= 0.1 A/cm^2.

2.4 Under-stoichiometric silica layers

The presence of intrinsic defects like oxygen vacancies or a presence of hydrogen atoms, plays an important role in numerous circumstances particularly in the growth of the layer structure. Thermal-annealing procedure leads to elimination of hydrogen and production of silicon nanoclusters in different sizes in silica network depending on the applied temperature. At annealing temperatures even below 900 °C, the hydrogen release from the

SiO_x layers permits the formation of Si-O• radicals introducing structural disorder into the layer network [Fitting et al. 2005a]. At higher temperatures, the structure will undergo a network reaction similar to eq. (2.4), then giving rise to the formation of nanostructures.

$$SiOx \rightarrow (x/2)\, SiO_2 + (1-x/2)Si \qquad 1 \leq x \leq 2 \tag{2.4}$$

In many cases cluster formation is affected by various kinds of defects in the Si-O network, which can be either crystalline or amorphous. Fourier transform infrared spectroscopy (FTIR) [Zacharias et al. 2003] and Cathodoluminescence (CL) [Fitting et al. 2002b] has been employed to characterize the silicon cluster growing and to study the effects of ionizing radiation on the structure of luminescent defects in SiO_x systems [Trukhin and Fitting 1999, Trukhin et al. 1999]. The UV (4.3 eV), the blue (2.7 eV) and the red (1.9 eV) luminescence centers have been classified as process-induced centers, which also exist in irradiated SiO_2. Again the yellow luminescence band (2.15 eV) is distinguished more clearly in the SiO_x layer, which we think deserves more attention and has to be studied intensively. In the present section, the annealing temperature dependence of the CL intensities of the 4.3, 2.7, 2.15 and 1.9 eV peaks and their correlation with the stoichiometry x will be demonstrated separately in order to associate the luminescence bands with the nature of different defects.

2.5 Fourier transform Infrared (FTIR) measurement of stoichiometric and under-stoichiometric silica layers

The vibrational spectrum of SiO_2 is full of details. It consists of longitudinal-optical (LO) and transversal-optical (TO) split features at around 460 cm^{-1}, 800 cm^{-1}, and 1100 cm^{-1}. The LO modes are not detected (lower frequencies) or only detected in oblique incidence IR transmission measurements. The highest frequency feature exhibits the largest absorption coefficient, therefore in the literature most attention is paid to this absorption band. In SiO_2 it features a TO absorption peak at around 1090 cm^{-1} with a LO absorption at around 1250cm^{-1}, being visible when applying oblique incident IR radiation. In the present work we focus on the region of the Si–O–Si stretching mode (700-1400 cm^{-1}).

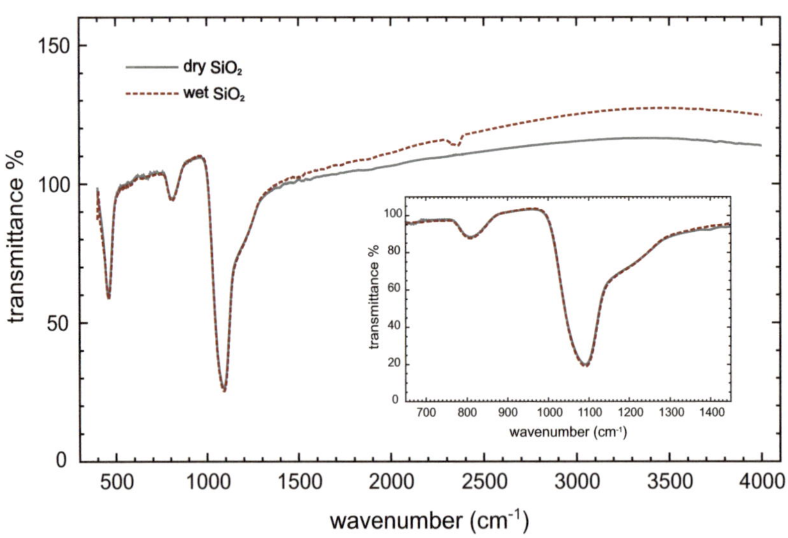

Fig. 2.5 Fourier transform infrared (FTIR) spectra of thermally grown pure dry and wet SiO_2.

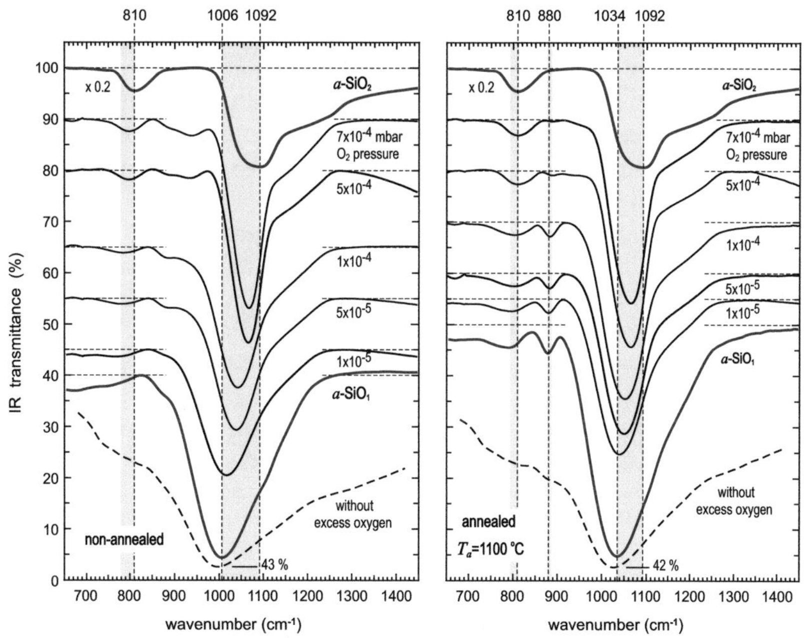

Fig. 2.6 Infrared spectra of SiO_x layers on Si substrate grown by thermal evaporation of SiO in ambient oxygen pressures, the stoichiometric SiO_1 and SiO_2 layers are presented for calibration of non-annealed samples (left) and thermally annealed samples, T_a=1100 °C (right).

At first we present the IR spectra of thermally grown *a*-SiO_2 in Fig. 2.5, then a comparison of the non-annealed and the thermally annealed (T_a=1100 °C) silica layers with different stoichiometry *x*, before they have been irradiated (Fig. 2.6).

Typical IR transmittance spectra of wet and dry silica are presented in Fig. 2.5. Obviously there are no essential differences in spectra of silica samples with different thermal oxidization method (wet or dry). The only bands observed are due to the fundamental SiO_2 vibrational bands. The oxygen to silicon ratio of several SiO_x thin layers deposited by thermal evaporation of silicon monoxide SiO and simultaneous oxidation are studied by FTIR spectroscopy in order to determine the stoichiometric degree *x*.

The infrared properties of the non-annealed and annealed samples at 1100 °C are presented in Fig. 2.6. Various silicon-oxygen related absorption bands can be identified in the wave number region from 700 to 1400 cm^{-1}. Due to the fact that the samples are produced under such conditions that they are totally hydrogen-free we can exclude all hydrogen related IR modes. The band around 810 cm^{-1} is attributed to Si−O textendash Si bond bending motion in SiO_2 [Tsu et al. 1989, Gucsik et al. 2004]. This band position increases when the sample composition approaches the pure stoichiometric silica (SiO_2) structure, while it is not detectable in the sample produced without excess oxygen exposure. The most intense feature in the spectra of Fig. 2.6 appears in the range of 1000-1100 cm^{-1}, which is ascribed to the Si−O−Si stretching vibrations [Lehmann et al. 1984]. The Si−O vibration frequency in thermally annealed samples shifts from 1034 cm_{-1} in the sample without excess of oxygen to higher frequencies at 1092 cm^{-1} in the sample of stoichiometric SiO_2 composition. The transversal-optical (TO) peak positions have been determined by the zero-transition of the first derivative. The increase in wavenumber is due to the fact that the number of Si–Si bonds within the tetrahedral units decreases with the concomitant increase of Si–O bonds. The same fact can explain the creation and the frequency shift of the 810 cm^{-1} bending band. These peak positions can be used for a basic estimation of the stoichiometry of homogeneous SiO_x structures and as a measure for the phase transition from SiO_1 to higher x compositions SiO_x up to SiO_2.

In Fig. 2.7 we show the calibration of the stoichiometry degree x in SiO_x layers as a function of the Si–O–Si TO stretching mode frequency. We should mention that the position of the TO stretching mode as a function of the stoichiometric degree x is always expressed as a linear regressions-type formula, see e.g. [Tsu et al. 1989, Lehmann et al. 1984]. Different peak position determination methods could be the reason for the discrepancy between our data and those presented in the literature where mostly symmetric axis band positions have been used previously. Here we consider the zero transition of the first derivative to obtain the absolute peak position. Earlier a numerous equation in dependence on x was obtained [Lehmann et al. 1984]:

$$\nu_x=(48.8x+976)\ cm^{-1} \tag{2.5}$$

The estimated x values in a linear relation according to Lehmann [Lehmann et al. 1984] are shown as the dashed line in Fig. 2.7. Based on our IR data of the two well-calibrated and fixed SiO_1 and SiO_2 layers we had to slightly modify the Lehmann relation and have obtained the following approach for the growth of x in SiO_x from nearly $x>1$ up to stoichiometric SiO_2:

$$\nu_x=(58x+976)\ cm^{-1} \tag{2.6}$$

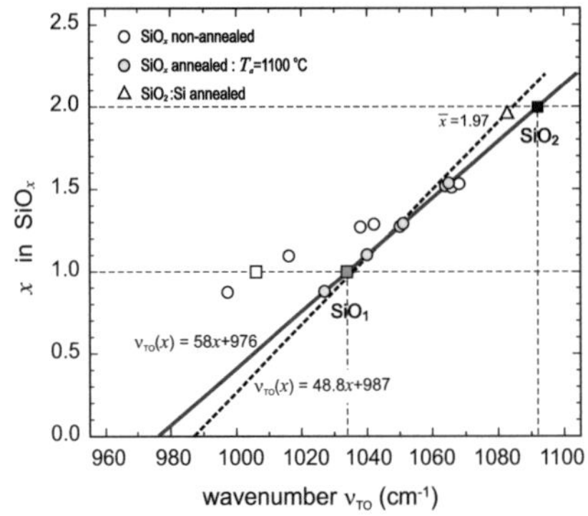

Fig. 2.7 Calibration of stoichiometry degree x in SiO_x layers with the TO stretching mode of IR measurements; dashed line: according to [Lehmann et al. 1984]; solid line: given by the present SiO_1 and SiO_2 data of known stoichiometry, the wavenumbers of the annealed SiO_x samples were placed on this line; Δ SiO_2:Si ion-implanted sample from [Fitting et al. 2002b].

This relation is presented in Fig. 2.7 as a solid line for the thermally annealed samples, where x=1 and x=2 are given as fixed points and the other data were placed on this straight line between SiO_1 and SiO_2 in order to determine their x. Of course, the non-annealed samples will not fit to this straight line because their atomic network is still much more disordered than for the thermally annealed samples leading to higher TO mode softening, see Figs. 2.5 - 2.7. Additionally, a Si^+ ion-implanted sample SiO_2:Si with an Si excess of 4 at.% in the maximum of the implantation profile has been investigated previously, providing a mean stoichiometric degree over the full silica layer [Fitting et al. 2002b]. This value is close to x=2 but there is already a remarkable shift of $-\Delta\nu_{TO}$=10 cm^{-1} as can be seen in Fig. 2.7, supporting the sensitivity of the IR measurement.

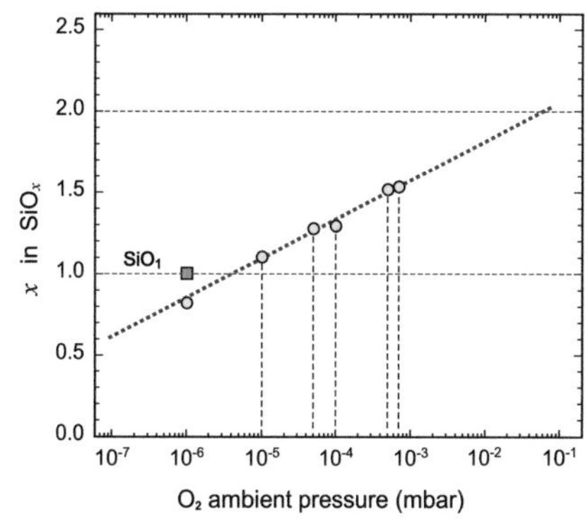

Fig. 2.8 Stoichiometry x of SiO_x layers as grown by thermal evaporation of SiO in oxygen ambient pressure, based on FTIR measurements after thermal annealing of the samples.

Then in Fig. 2.8 the x values obtained by FTIR are correlated with the ambient oxygen pressure during the evaporation of SiO_1. So the stoichiometric transition from SiO_1 to SiO_2 as a function of the ambient oxygen pressure can be estimated. An extrapolation shows that SiO_2 layers can be manufactured roughly at oxygen pressure nearly to 1×10^{-1} mbar.

2.6 CL of under-stoichiometric silica layers

In order to characterize the SiO_x layers by their optical luminescence features, cathodoluminescence (CL) studies of the SiO_x samples were carried out at both room temperature (RT) and liquid nitrogen temperature (LNT). see Fig. 2.9. This experiment was performed using samples that were thermally annealed at T_a=600, 800, 1100, 1300 °C besides the normal non-annealed samples, this way we could follow the structure change/growing under the influence of temperature treatment and the electron beam irradiation. Here we used also the experimental parameters as described before in the previous sections.

Samples with x<1.3 show almost no CL signals and samples with x≈1.3 show quite weak and smoothed CL intensities even when they annealed at high temperatures or cold down to LNT. Thus, if the atomic ratio of silicon atoms is high compared with the ratio of oxygen, i.e. x<1.3, the samples do not exhibit any characteristic CL, indicating deficiency of Si−O bonds to form ordinary silica structures and any prominent silica defects.

Upon reaching x>1.5 , the CL intensities grow considerably but almost only the yellow luminescence Y at 2.15 eV is detectable in the initial CL spectra (1 sec). After longer electron beam irradiation (30 min) the characteristic silica bands UV, B, and R appear too, as can be seen in Fig. 2.9.

Here the yellow Y luminescence band does not occur accidentally in hydrogen rich silica samples, there we have attributed a similar band to hydrogen molecules on interstitial sites in the silica network [Fitting et al. 2005a, Fitting et al. 2005b]. In Fig. 2.9 we see that the CL

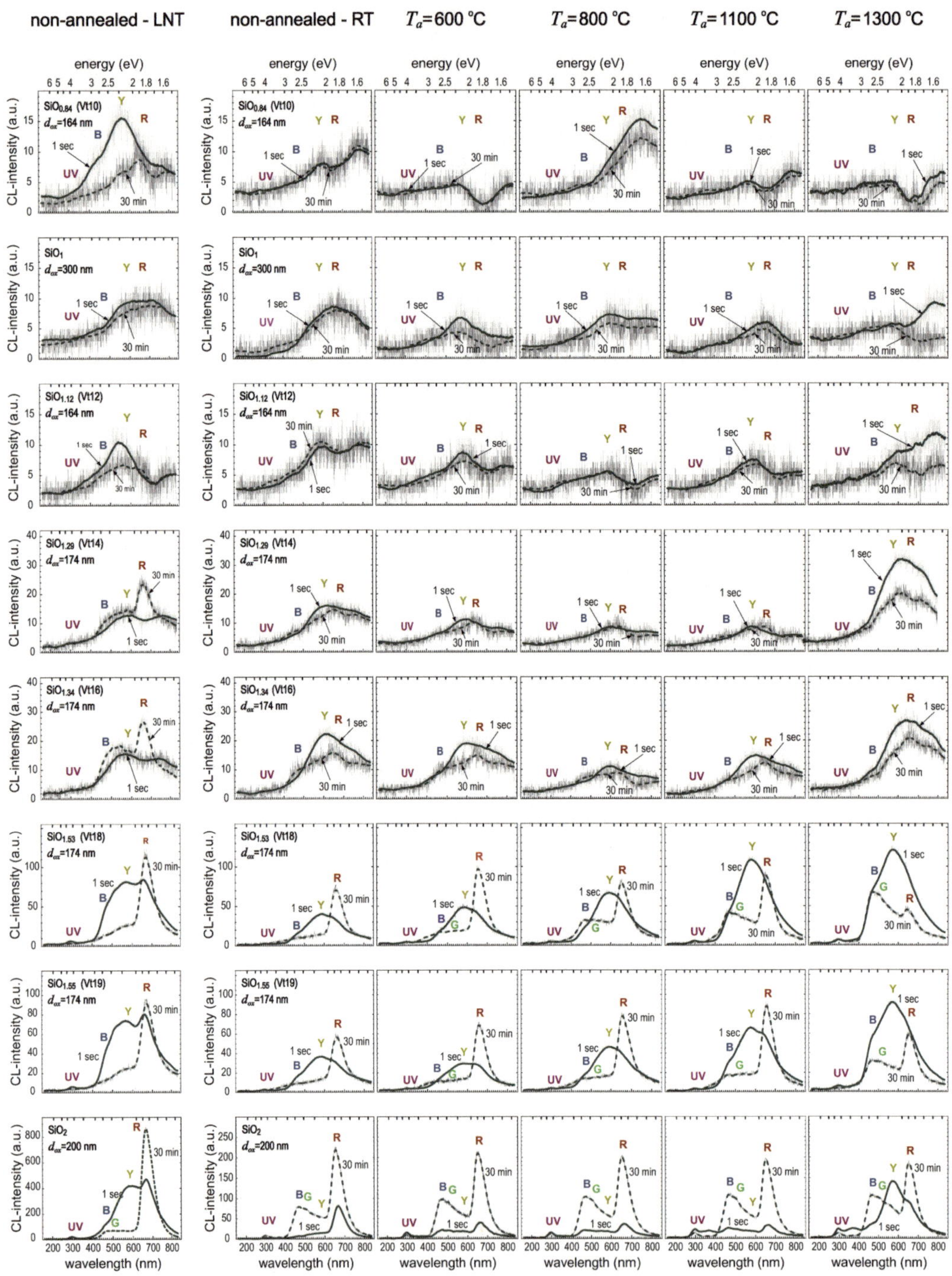

Fig. 2.9 CL spectra of SiO_x layers with thickness d_{ox} grown on Si substrate with different ambient oxygen pressure and having been thermally annealed at different temperatures T_a. The initial and the saturated CL spectra are labeled by "1sec" and "30 min", respectively.

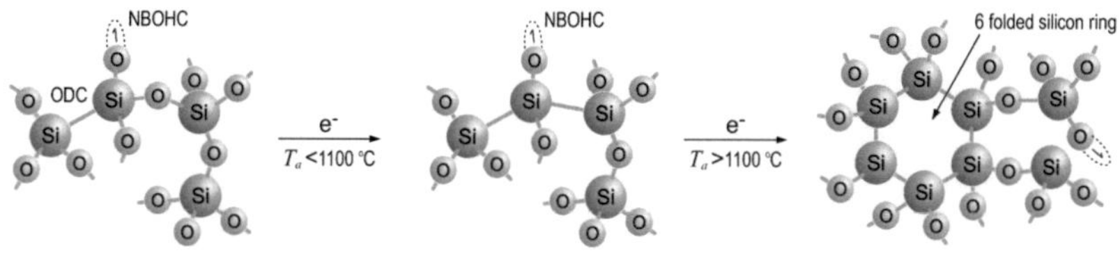

Fig. 2.10 Schematic presentation of the nonbridging oxygen hole center (NBOHC) and oxygen deficiency center (ODC) forming a Si dimer in the SiO_2 network (left) transformed by thermal annealing at temperatures T_a and/or electron beam irradiation to Si trimers (middle) and further on to elementary 6-fold silicon rings (Si hexamers) as a first step of Si nanocrystal formation (right).

spectra of dry and under-stoichiometric SiO_x layers (both are H-free), annealed up to relatively high temperatures in vacuum show a very intense yellow luminescent band at 2.15 eV. After annealing to temperatures T_a>1100 °C the Y luminescence is detectable too, even with higher intensity, which means it is increasing with T_a. We may conclude that the silicon atoms tend to re-arrange themselves in small clusters as chains or rings in the SiO_x network [Nicklaw et al. 2000], and that could be one of the most dominant origins for the Y luminescence in dry silica. In the absence of water there are strained bonds or defects of the form ≡Si−O··Si≡, where O··Si represents a strained (weak) Si−O bond that could be transformed into small membered silicon rings (3, 4-membered) [Ivanda et al. 2003] as well as to nonbridging oxygen hole centers (NBOHC) as an effect of high temperature treatment, as illustrated in Fig. 2.10.

A careful temporal observation of Fig. 2.10 makes it clearer that the cluster can be separated into core and surface components. The core is nucleation of silicon atoms (silicon bonded together) but on the surface both silicon and oxygen have dangling bonds, with the oxygen atom tending toward the surface. These fragments residing on the surface tend to be highly mobile [Schweigert et al. 2002], so we may state that the yellow luminescence is exciting from the surface of the silicon nanoclusters and not from the silicon core.

The red R (1.9 eV) emission is generally associated with the NBOHC and attributed to the recombination of electrons in the highly localized nonbridging oxygen band gap state with holes in the valence-band edge [Stevens Kalceff 1998]. On the other hand, tempering to T_a≈1300 °C leads to a further increase of the yellow band Y and to a strong reduction of the red band R. Probably, at these high temperatures oxygen is released from the SiO_x network diminishing the red luminescence but forming more oxygen deficient centers ODC and Si-rings.

In other words, more oxygen dangling bonds and oxygen deficient centers can be created due to high temperature exposure. Then silicon fragments or rings are produced in the silica network which seems to be the most probable candidates for the yellow Y luminescence center in silica.

2.7 EDX, CL and TEM investigation of Si cluster formation in modified silica

The formation of oxygen deficient centers (ODC) or even higher silicon aggregates by means of electron beam irradiation has been manifested already earlier [Fitting et al. 2002b, Trukhin et al. 1999, Fitting et al. 2005b, Stevens Kalceff 1998]. Even Auger electron spectroscopy (AES) has clearly evidenced that oxygen is dissociated from SiO_2 due to electronic or thermal processes during electron beam excitation, see e.g. [Stevens Kalceff 1998, Cazaux 1986]. Thus the blue B and the red R luminescence bands grow under electron bombardment to a saturation after an irradiation dose of about 3 As/cm^2 [Fitting et al. 2002b], see the bottom row in Fig. 2.9.

In order to demonstrate the ODC increase during electron bombardment we have determined the stoichiometric degree x of SiO_x by means of energy-dispersive X-ray analysis (EDX) during electron irradiation (E_o=10 keV, j_o=2 mA/cm^2) in initially stoichiometric silica layers SiO_2. The results are presented in Fig. 2.11. There we see a decrease of x with irradiation time, as expected, for room temperature (RT) faster than for liquid nitrogen temperature (LNT). The initial values with x>2.04 are somewhat higher than the expected ones x=2 of stoichiometric silica but we suppose an excess of oxygen on interstitial sites within the silica network remaining still from the oxidation process and the thermal diffusion of oxygen through SiO_2 towards the interface at the Si substrate. It is well known that over-stoichiometric silica with x>2 does not exist [Helm and Deal 1993], besides some peroxy bridges ≡Si−O−O−Si≡ or radicals as candidates or precursors for luminescent defects [Pacchioni et al. 2000], however, in very low concentration. This interstitial excess oxygen may form O_2 and O_2^- molecules [Fitting et al. 2005c].

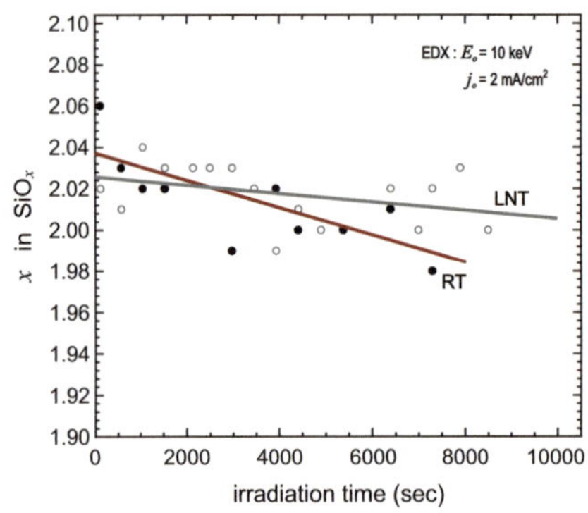

Fig. 2.11 EDX measurement of the oxygen to silicon ratio x in initially stoichiometric silica layers SiO_2 during electron beam irradiation at room temperature (RT) and liquid nitrogen temperature (LNT).

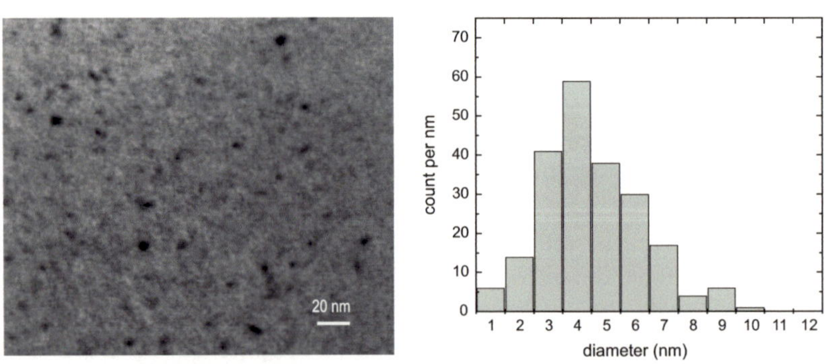

Fig. 2.12 TEM micrograph showing Si nanoclusters embedded in the silica matrix (left) and the related cluster size distribution (right) of a 250 nm SiO_2 layer having been irradiated for 30 min by an electron beam of 5 keV and 2.7 A/cm^2, [Salh et al. 2006].

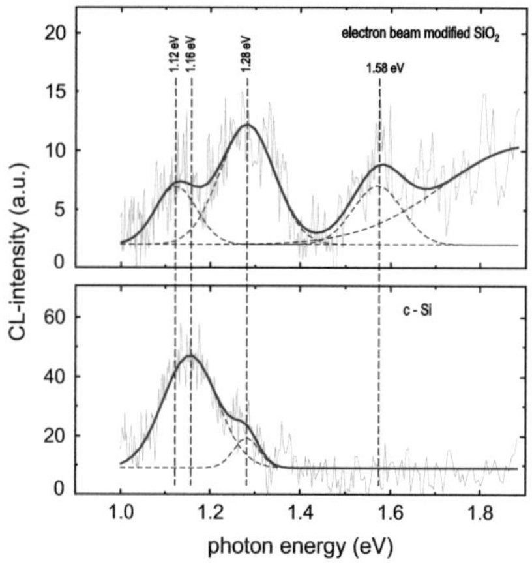

Fig. 2.13 CL spectra of electron beam modified SiO_2 showing in the near IR the fundamental transition of c-Si at hν=1.1 eV, of *a*-Si at 1.3 eV and a transition in Si nanocrystals (quantum dots) at 1.6 eV.

In order to demonstrate the oxygen dissociation and the formation of ODC, and finally, of Si clusters under electron bombardment we have carried out an additional experiment under high dose electron irradiation. Therefore thin 250 nm-thick silica films have been prepared by wet thermal oxidation on silicon wafers. Afterwards the silicon substrate had been removed by mechanical milling and 3 keV Ar^+ ion etching. In this way self-supporting 250 nm thin silica films had been prepared for imaging in a transmission electron microscope (TEM). More details of preparation are given in [Kolesnikova et al. 2005].

These films were modified by heavy electron beam irradiation: beam energy 5 keV, current 20 nA over an area of 3/4 µm² yielding a high current density of 2.7 A/cm². Thus we may assume electronic as well as thermal dissociation [Kolesnikova et al. 2005] of oxygen from the thin SiO_2 layers and more and more the appearance of under-stoichiometric SiO_x. This SiO_x will undergo a phase separation as described by eq.(2.4). After 30 minutes of electron beam irradiation we observe Si cluster formation as presented in Fig. 2.12. The Si clusters embedded in the silica appear as dark spots. Their size distribution is shown in the right part of Fig. 2.12. There we see a most probable cluster diameter of 4 nm and a maximum diameter of 10 nm. As it has already been shown in the context of formation of Ge nanocrystallites in Ge-implanted silica, [Fitting et al. 2002b], such largely extended clusters will diminish the Si-related luminescence. The right size for elementary small luminescent clusters should be searched in intermediate regions, i.e. according to Fig. 2.10 between Si dimers (ODC) and hexamer rings.

Fig. 2.13 shows the CL spectra of pure crystalline Si and the spectra of Si nanoclusters embedded in the host silica. Luminescence bands are observed at around 1.1 eV and 1.3 eV assigned to crystalline and amorphous silicon phases, respectively. Another band at 1.6 eV is also to be seen after heavy electron beam bombardment in the SiO_2 structure. In spite of extensive experimental and theoretical work during the last years, the light-emitting mechanism which explains emission at 1.6 eV has not been fully understood yet. It is believed that the oxygen-related light-emitting centers are positioned at the interface between the Si nanoclusters and the host oxide [Prokes et al. 1998]. A broad CL emission band is characteristic of Si nanoclusters. Although the spectra vary considerably in intensity after longer irradiation, the peak position does not shift significantly, implying a similar mean size for the nanocrystals. No unique relation between the CL or PL emission energies and Si nanocluster sizes has been reported in the literature making the quantitative comparison of the results difficult [Wilkinson and Elliman 2004]. Other authors estimated that 5 nm Si nanoclusters emit at 1.6 eV, while 3 nm Si nanoclusters give PL at 2 eV [Ledoux et al. 2002]. On the contrary it was reported that 4 nm Si nanoclusters emit at 1.3 eV and the 1.6 eV PL correspond to very small sizes of about 1 nm [Iacona et al. 2000].

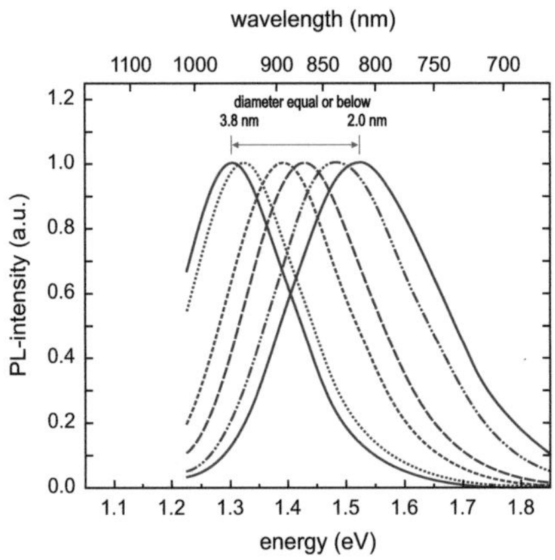

Fig. 2.14 Normalized photoluminescence spectra showing a blue shift correlated with the Si nanocrystal size, [Zacharias et al. 2002].

Recently Si nanocrystals have been fabricated by thermal treatment of SiO-SiO_2 nanolattices, in a way which makes it possible to control not only the size but also the density and the arrangements of the nanocrystals independently of the stoichiometry [Zacharias et al. 2002, Yi et al. 2003, Torchynska 2006]. In this method a strong photoluminescence (PL) and a size dependent shift of the PL position are shown as a proof of size control. A strong blue shift from 960 nm (1.3 ev) to 810 nm (1.5 eV) with decreasing nanocrystal size was observed with respective cluster sizes 3.8 nm and 2 nm, respectively, see Fig. 2.14.

3. Hydrogen implanted SiO_2

Direct hydrogen implantation or exposure of SiO_2 to water vapor results in the appearance of various OH bands, for example water molecules are known to form silanol (Si−O−H) groups in the oxide [Bakos et al. 2004b], but the relative concentration of silanol to interstitial water depends on the way the oxide was manufactured and subsequently treated, raising questions about the most stable form of water in the oxide and the role of hydrogen in this reaction. Hydrogen plays an essential role in the thermal oxidation process, either through the interaction with point defects or as interstitial atoms in the SiO_2 network. Hydrogen can be considered as an important accidental impurity or an intentional additive in all forms of silicon dioxide which can passivate dangling bonds and produce high-quality interfaces [Rashkeev et al. 2001]. Besides, it can reduce electrical activity of point and extended defects by inducing grain boundaries and it saturates oxygen dangling bonds (NBOHC ≡Si−O•) or the threefold-coordinated SiO_2 defect (*E*′-center ≡Si•), through the chemical reactions:

$$\equiv Si{-}O\bullet \text{ (or } \equiv Si\bullet) + H^{o} \rightarrow \equiv Si{-}O{-}H \text{ (or Si−H)} \quad (3.1)$$

where " ≡ " denotes the three bonds and " • " represents the unpaired electron. Atomic hydrogen (H°) is unstable (mobile) above 130 K [Cannas et al. 2003b]. A variety of evidence strongly indicates that the dominant anneal mechanism for this atomic hydrogen is dimerization, ($H° + H° \rightarrow H_2$). Hydrogen can also enhance the diffusivity of impurities or other interstitial atoms such as oxygen by forming water molecules. Water molecules are known to form silanol ≡Si−O−H) groups even at room temperature:

$$\equiv\text{Si-O}\bullet + H_2 \rightarrow \equiv\text{Si-O-H} + H° \quad (3.2)$$

$$2\equiv\text{Si-O}\bullet) + H_2O \rightarrow 2(\equiv\text{Si-O-H}) + O° \quad (3.3)$$

Despite the wide interest in the behavior of H, paired H configurations (H_2) and H_2O in SiO_2, the understanding of the atomic scale processes remains limited and the microscopic identities of these electrically inactive H sites are the subject of intense debate. It is believed that the effectiveness of many defect generation and transformation processes depend critically upon sites where H can be trapped and released. We dedicate this section to presenting our results with hydrogen implanted SiO_2 layers.

3.1 CL of hydrogen implanted silica (SiO_2:H^+)

Besides the main luminescence peaks: red R, blue B, and UV an amplification of the yellow luminescence Y at the region between 560 nm (2.2 eV) and 580 nm (2.1 eV) has been recorded due to direct hydrogen implantation especially at RT, see Fig. 3.1. In both cases, LNT and RT, the hydrogen implantation diminishes the red luminescence. Other authors [Morimoto et al. 1996] have used nearly the same implantation parameters (dose and implantation energy) as used in this study, and they reported the PL emission band at around 2.2 eV without a detection of the 1.9 eV band. Similar results are also obtained with He^+ implantation [Morimoto et al. 1996]. As we present in hydrogen-implanted layers, Fig. 3.1, a yellow luminescence Y at $\lambda \approx 575$ nm (2.1 eV) is dominating the spectra and only a weak shoulder of the red luminescence appears. Here a high concentration of saturated bonds ≡Si−O−H or ≡Si−H) are expected, therefore the right hand side of eq. (3.1) is fulfilled where the NBOHC (≡Si−O•) and *E´*-center (≡Si•) are initially saturated by the excess hydrogen atoms. The ≡Si−O−H bond is a good candidate to form NBOHC at room temperature in hydrogen rich silica. The NBOHC is possibly produced by breaking the H bonds at high annealing temperatures (T_a>1000°C) or under electron irradiation [Kuzuu and Horikoshi 2005]. Direct hydrogen implantation or H_2O molecule formation on the surface or in the silica network are believed to be the main reasonable source of the Y luminescence [Fitting et al. 2005b]; that means there are two aspects for the origin of this band.

3.2 Hydride (≡Si−H) and hydroxyl (≡Si−H−O) in SiO_2:H^+

Hydrogen is a ubiquitous impurity in SiO_2, therefore some authors consider it an intrinsic defect. It is well known that hydrogen is present in all forms of silica. The wet oxide is proposed to contain around 10^{19} cm^{-3} OH groups (in the form of silanol or interstitial water molecules), while the typical OH concentration in dry oxides is only 10^{16} cm^{-3}.

Interstitial hydrogen does not form covalent bonds with the network, and the hydrogen molecule does not react with the defect-free silica lattice [Blöchl 2000]. It has no states in the band gap of silica. Thus it may be difficult to activate the hydrogen molecule with UV light in the absence of other defects. This result indicates that hydrogen molecules need to

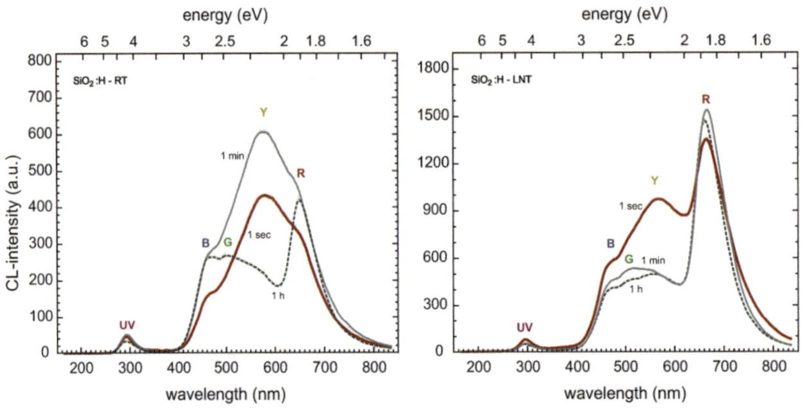

Fig. 3.1 Initial (1sec) and saturated (5h) and dose-dependent CL spectra of H^+ implanted SiO_2 layers recorded at room temperature (RT) and liquid nitrogen temperature (LNT).

interact with defects in silica before they can be activated. That means interstitial H_2 molecules could react at least with broken or strained silicon bonds, as

$$\equiv Si\cdots O-Si + H_2 \rightarrow \equiv Si-H + H-O-Si\equiv \quad (3.4)$$

or

$$D + H_2 \rightarrow \equiv Si-H + H-O-Si\equiv \quad (3.5)$$

where D is an unspecified defect site. As we see, the product of the majority of the chemical interactions proposed so far is saturated defects which can be a source (precursors) for radiation induced defects later. In addition, hydrogen processing of the glass has been found to greatly improve the radiation resistance because it is suspected to reduce the number of precursors of radiation-induced defects [Brichard 2003]. It has been believed that OH bonds make the silica system softer and better able to resist the creation of many kinds of defects [Kuzuua and Horikoshi 2005].

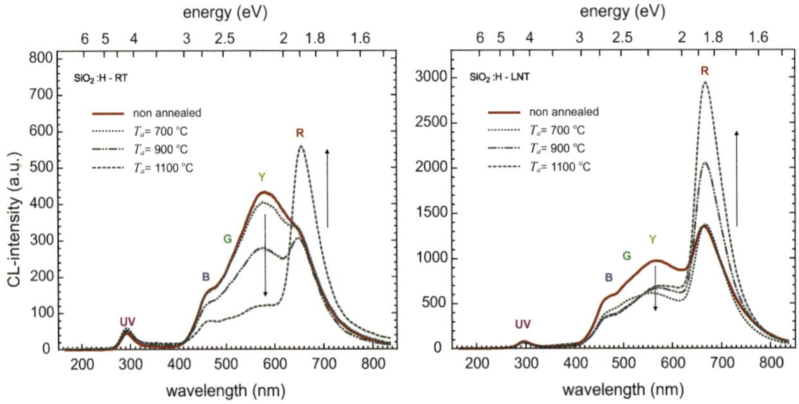

Fig. 3.2 Initial (1sec) CL spectra of H^+ implanted SiO_2 layer at different annealing temperatures, $700 \leq T_a \leq 1100$ °C, recorded at RT and LNT.

With additional hydrogen implantation we expect higher concentrations of both hydride (≡Si−H) and hydroxyl (≡Si−O−H) in the whole network which we consider as a first suspect for the dominant yellow luminescence in Fig. 3.1. If this hypothesis is correct, the yellow luminescence should possibly diminish by eliminating hydrogen from the system. Releasing hydrogen atoms even from amorphous material is previously reported by thermal treatment [Pan and Biswas 2004]. The samples have been thermally annealed up to relatively high temperature (T_a) so that we can state that we were able to break the hydrogen bonds and let an amount of hydrogen out. Fig. 3.2 shows a comparison between the non-annealed and those thermally annealed. We found a slight change in the intensity of the yellow luminescence at T_a=700 °C at both RT and LNT, which means that T_a=700 °C is not enough yet to make a significant change in ≡Si−H and ≡Si−O−H concentration. But by increasing the thermal annealing temperature to 900 and 1100 °C, we found a considerable change in the CL spectra. We see diminishing of the yellow luminescence and growing of the red luminescence R, leading us to the conclusion that T_a>900 °C can release hydrogen from both hydride and hydroxyl. The effective diffusion coefficient of hydrogen and the rate of ≡Si−O−H and ≡Si−H in hydrogen rich silica glass have been measured using Infrared spectroscopy [Lou et al. 2003]. It is found that the concentration of both ≡Si−O−H and ≡Si−H decreases due to sample thermal treatment, see Fig. 3.3. The decrease in hydroxyl quantity is very slow at 750 °C compared with other higher temperatures (1000, 1250 and 1500 °C). More and faster elimination of hydroxyl is achieved by increasing the temperature. A similar change in hydride quantity is also shown in Fig. 3.3. Our samples have been annealed for 3600 sec (the red vertical dashed line in Fig. 3.3) in vacuum, up to this period of time and T_a=1100 °C we can estimate that around 80% of hydride and hydroxyl have been eliminated from the SiO_2:H. In Fig. 3.4 (top), we signify the dose behavior of the yellow Y and the red R luminescence. The yellow band intensity shows higher initial level in the non annealed samples, it decreases by increasing T_a, but it passes a maximum at around 100 sec of electron beam irradiation. This means that other precursors for the yellow luminescence are produced. We consider short-term-living water molecule formation in the network to be one of these precursors. When H_2O molecules dissociate under the electron beam irradiation the yellow band starts to decrease.

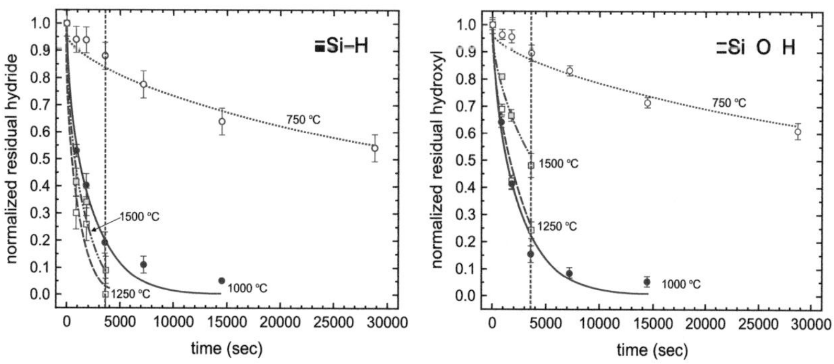

Fig. 3.3 Normalized residual quantities of hydride (≡Si−H) and hydroxyl (≡Si−O−H) as a function of heat treatment time in air. Open circle: 750 °C, filled circle: 1000 °C, open square: 1250 °C, filled square: 1500 °C, [Lou et al. 2003].

Contrary to the yellow luminescence, the red luminescence has much lower intensity in non-annealed samples and rises with increasing annealing temperature T_a until it shows the same dose behavior as the non-implanted wet a-SiO_2 layers as articulated in the previous section. We observe the same CL spectra and dose behavior of the red R luminescence in SiO_2:H as well as wet oxide SiO_2 samples at T_a=1100 °C, see Fig. 3.4 (bottom). Finally we can confirm the following production mode, eq. (3.6), of the non-bridging oxygen hole centers (NBOHC, ≡Si−O•), the source of the red R luminescence in wet oxide SiO_2, where hydrogen and hydroxyl are present.

$$\equiv\text{Si–O–H} \rightarrow \equiv\text{Si–O}\bullet + \text{H}^{o} \quad (3.6)$$

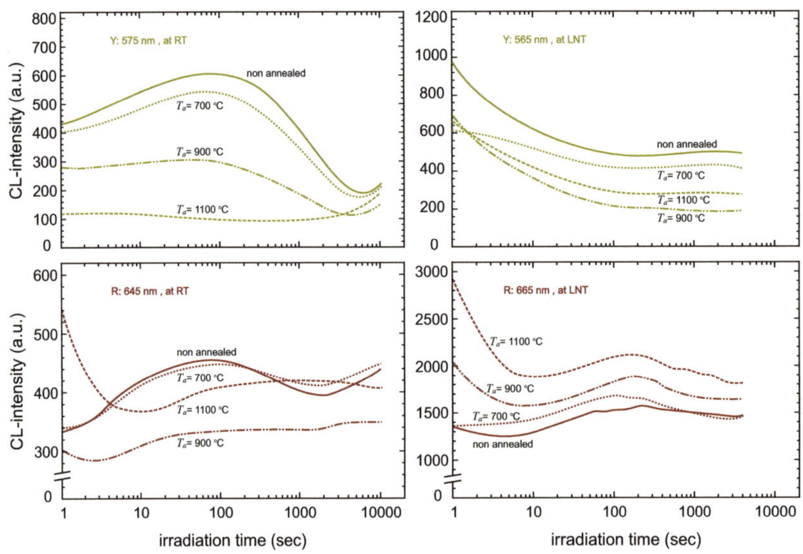

Fig. 3.4 The dose-dependent of the yellow band Y (top) and the red band R (bottom) in SiO_2:H at different annealing temperatures recorded at RT and LNT.

3.3 H_2O molecules and the yellow luminescence

The interaction of water molecules especially with the surfaces of amorphous silica is of great technological interest [Legrand 1998], and thus numerous studies have been devoted to this issue focusing especially on IR spectroscopy. It is suggested that the possible existence of small-membered (i.e. having a small number of members) Si−O rings on SiO_2 surfaces are expected to be the reactive centers for the interaction with water and other molecules [Mischler et al. 2005]. Additionally it is well known that water may dissociate on SiO_2 surfaces resulting in the formation of silanol (≡Si−O−H) groups. In particular it is frequently believed that the silanol groups are a result of the interaction of water molecules with small-membered rings [Mischler et al. 2005], see Fig. 3.5. Besides, some experimental results in the literature [Morimoto and Nozawa 1999] suggest that the photon irradiation of isolated ≡Si−O−H can lead to the formation of some hydrogen bonds between the hydroxyls and the H bonded ≡Si−O−H, which is decreased by heating to form once again isolated ≡Si−O−H and some H may be released.

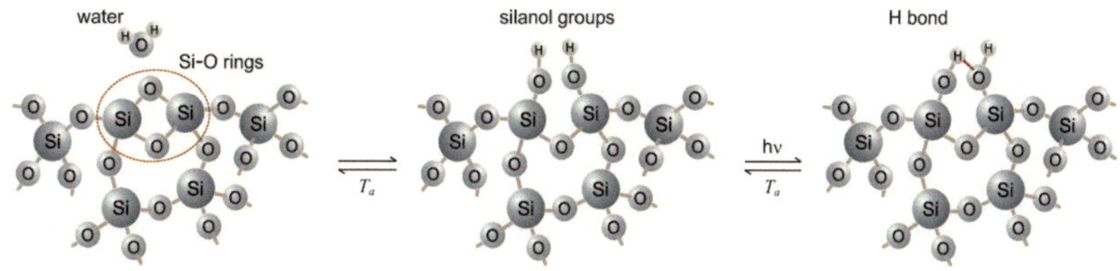

Fig. 3.5 The speculated equilibria showing the interaction of H_2O molecules with surface SiO_2 rings followed by a photochemical reaction of the ≡Si−O−H to the hydrogen bond. The dotted red line indicates the H bonding between H and O atoms, [modified after Mischler et al. 2005, Morimoto and Nozawa 1999].

Based on IR absorption spectra described by [Rinnert and Vergant 2003], the adsorption of water is favored by silicon dangling bonds (*E*´-center: ≡Si•) to form silanol groups not only on the surface but also in the silica network. The reaction between water molecules and the SiO_2 is supported too by the same authors, leading to the formation of two ≡Si−O−H.
With some complexities we were able to produce a thin layer of ice on the surface of pure wet SiO_2 layer, whose CL behavior have presented in Fig. 3.6. Here we could measure the CL spectra of ice together with the typical CL spectra of SiO_2, see Fig. 3.6. Very intense yellow Y luminescence has been detected, even higher than the red R luminescence of SiO_2. An additional sharper band in the UV range (λ≈370 nm) is also clearly seen. The width of this band is much smaller than the conventional *a*-SiO_2 band widths indicating a crystalline structured H_2O. The whole spectral shape presented in Fig. 3.6 is loses its outlined profile in quite short time. We see that it is no longer possible to detect a luminescence band after some thirty seconds, especially the sharp band at 370 nm is totally disappearing.
A photoluminescence band at 3.7 eV (≈340 nm) has been reported in water-treated sol-gel synthesized porous silica. The authors have correlated this PL emission band indirectly to isolated silanols especially in the surface region [Yao et al. 2001], but others favored more the interacting OH-related centers [Anedda et al. 2003b].

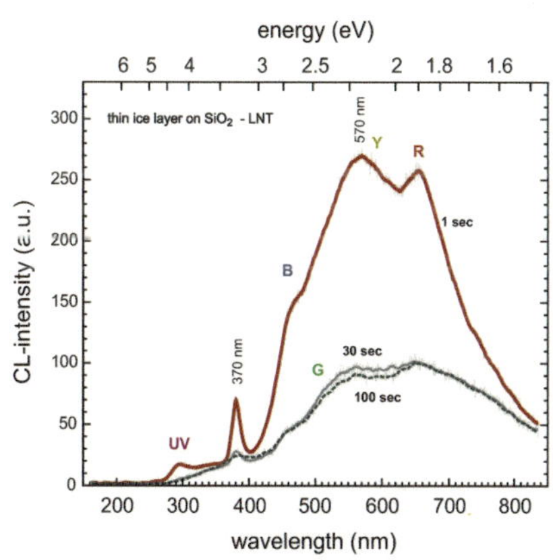

Fig. 3.6 CL spectra of a thin ice layer (H_2O) on SiO_2.

To determine whether the additional features presented in Fig. 3.6 belong to water molecules on the surface or not, we performed the same experiment where a thicker ice layer was produced on a metallic surface this time. To avoid any other influences coming from the substrate material, the metallic substrate was examined first; it gave absolutely no

CL signals in our sensitive detection region. The possibility of ice bilayers on metallic surfaces has been reported previously [Ogasawara et al. 2002]. It was found that half of the water molecules bind directly to the surface metal atoms and the other half are displaced toward the vacuum in the H-up configuration.
Ice layers on a metallic substrate show similar initial spectra with both 570 and 370 nm emitted CL bands; they start with very stable intensities but the intensities fall down rapidly due to the heat produced by the electron beam where the ice layer begins to melt then, see Fig. 3.7.

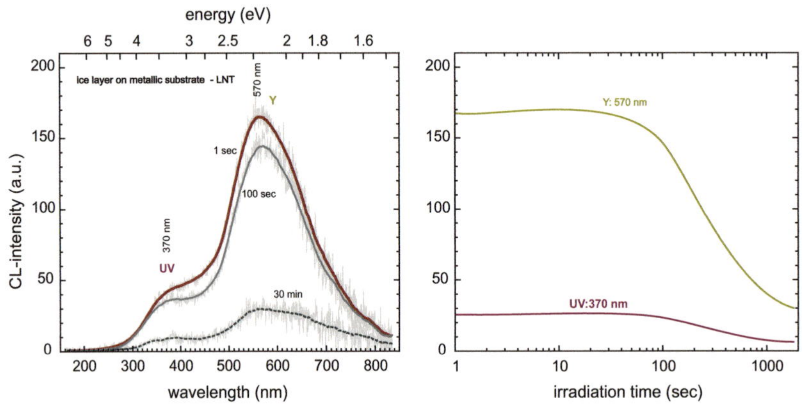

Fig. 3.7 CL spectra of thin ice (H_2O) layer on a metallic substrate (left), the dose behavior of the individual luminescence bands (right).

Thus we state that both the fast decreasing yellow Y band at 570 nm, 2.15 eV (formerly called green-yellow band G) as well as the long-term irradiation Y band is the same electronic state and all attributed to water. In the first case condensed water and ice sublimate at LNT from the surface whereas the longer irradiation Y band is due to water molecules formed in the SiO_2 network by radiolytic processes.

3.4 Hdrogen association in luminescence defects

Extrapolating from the facts presented up to now we can formalize a model for the different luminescence properties of the radiation induced defects in *a*-SiO_2, presented in Fig. 3.8. We assume that strained bonds ≡Si−O··Si≡ in dry oxide and the hydroxyl species (≡Si−O−H) in wet oxide are the prevailing main precursors of the red R luminescence associated with non-bridging oxygen hole center (NBOHC: ≡Si−O•).
During electron beam irradiation both precursors are transformed to NBOHC. We see that the NBOHC centers produced in dry oxide increase up to a certain concentration obtained by an equilibrium of center generation and electron beam induced dissociation to the *E´*-center (≡Si•) and mobile atomic oxygen O_{mob}. The production and the role of mobile oxygen have already been stressed by [Skuja et al. 2002 and Fitting et al 2002b]. There, a model and respective rate equations are given for the temperature and dose dependence of both the red R and the blue B bands. The re-association of mobile oxygen to the *E´*-centers and re-creation of the NBOHC will increase the role of mobile oxygen and hydrogen. Experiments had suggested that the ≡Si−O−H is resisting bond breakage effectively at relatively short irradiation time. Bond breakage might saturate only at sufficiently long irradiation time [Kuzuu and Horikoshi 2005]. Different properties are shown by the wet oxide in Fig.3.8.

Here the hydrogen is dissociated from the silanol group of the non-bridging oxygen bond, eq. (3.6). But then the red luminescence of the NBOHC is destroyed by further electron beam dissociation as in dry oxide too. The dissociated mobile hydrogen H_{mob} may react with the mobile oxygen O_{mob} to form molecules H_2, O_2, and H_2O on interstitial sites. These reactions have been recently described [Bakos et al. 2004a]. There the authors underlined that water and oxygen molecules are participating in various defect formation processes in thermally grown SiO_2 films as well as in synthetic silica glasses. Formation energies and energy barriers are obtained by first-principles calculations and compared for different reactions. A part of the H atoms on the right-hand side of eq. (3.6) must form H_2 molecules through the diffusion of H atoms in the silica network. In addition to H_2 molecules produced by this mechanism, interstitial H_2 molecules are expected to exist in the sample. These H_2 molecules and interstitial H_2 molecules could react with broken or strained bonds and form ≡Si−H and ≡Si−O−H pair as in eq. (3.4).

The ≡Si−H structure on the right hand side of eq. (3.4) can be a precursor of the *E´*-center through the process expressed in the reverse of eq. (3.1). The amount of H_2 molecules created by the irradiation must increase with increasing OH content. In addition to the creation of hydrogen molecules from the ≡Si−O−H species, interstitial H_2 molecules exist especially in the wet samples. Therefore, an excess amount of *E´*-centers, relative to that of NBOHC, is induced as shown in Fig. 3.8.

Water molecules may cluster in the bigger voids of the oxide, i.e., form hydrogen-bonded complexes with each other and the silica network's O atoms [Bakos et al. 2004a]. In such cases two H_2O molecules may react with each other forming once more OH bonds. Thus, the red luminescence is stabilized at some fraction of the number of OH bonds. This model of the hydrogen effect is consistent with our previous model of center transformation based on the mobile oxygen generation and re-association [Fitting et al. 2002b], and extends it by the reactions of H, OH, and H_2O with the radicals in the silica atomic network as shown in Fig. 3.8. This model is supported by investigations of the yellow Y luminescence, where the yellow luminescence at the beginning of irradiation at LNT is associated with sublimating ice from the sample surface rather more probably than due to a self-trapped exciton (STE) luminescence as often emphasized [Trukhin 1994]. Moreover, the yellow Y luminescence after longer irradiation (2 As/cm^2), especially in hydrogen implanted samples, could be associated with water molecules H_2O too, formed in radiolytic processes as demonstrated in Figs. 3.6 and 3.7.

4. Group IV elements implanted in SiO_2

Ion implantation into glasses results in network damage and in compositional changes, it modifies silica's physical properties such as density, refractive index, surface stress, hardness, and chemical durability. Compositional changes can also occur due, e.g., to radiation-enhanced diffusional losses of alkali ions, crystallization, phase separation, and H incursion. Many authors [Hosono et al. 1990, Morimoto et al 1996, Fitting et al. 2002b, Magruder et al. 2003] have implanted several kinds of ions in silica glass and found that ion implantation causes an increase in refractive index by 1%-6% owing to the compaction of surface region and to a chemical change in the structure of glass. It was deduced that this refractive index change is caused by the formation of Si\textendash Si homobonds, but not by the decrease in Si−O−Si bond angle which leads to compaction. In addition to the compaction, the chemical change in structure, and the formation of colloid particles, ion

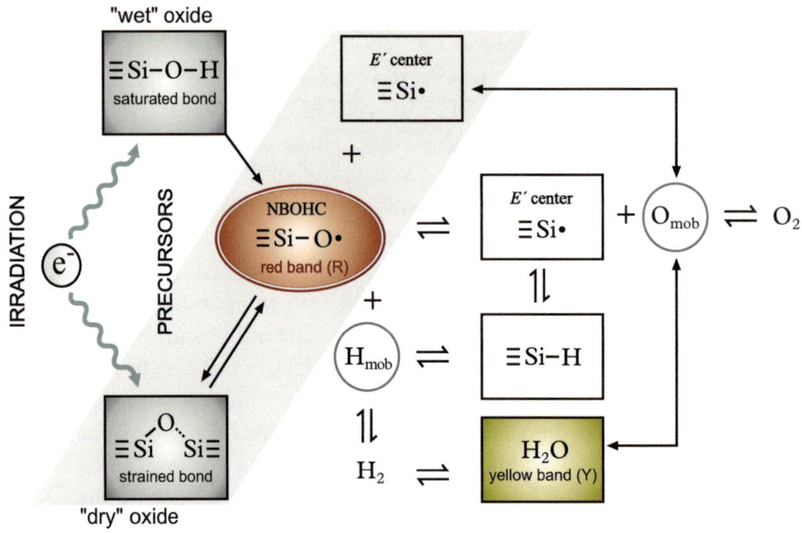

Fig. 3.8 Model of the red luminescent center (NBOHC) creation from different precursors in "wet" and "dry" oxide. The center destruction and recombination by radiolytic hydrogen and oxygen dissociation and re-association will lead to a dynamic equilibrium.

implantation in silica glass is always accompanied by the formation of defects, such as oxygen vacancy, *E´*-center, NBOHC, and peroxy radicals, resulting not only in changes to emission bands but also to the emission of new CL bands especially in the violet V or in the ultraviolet UV regions.

Before we start reviewing our results, it is appropriate to keep in mind that there are species which diffuse through the glass without modifying the structure of the matrix, and these are called non-interacting elements. There are both interstitial and substitutional non-interacting species. Species which modify the structure of the glass matrix are called interacting species [Minke and Jackson 2005]. Carbon (C), silicon (Si), Germanium (Ge), tin (Sn) and lead (Pb) are the dopants whose influence on silica's natural luminescence defects will be discussed in this section. They are examples of non-interacting substitutional species. Since these elements have similar bonding characteristics to silicon, they can replace silicon in the matrix of the glass, without significantly changing the network structure. Substitutional non-interacting elements diffuse much more slowly than interstitial elements. Ion implantation results allow deeper understanding of the relationship of the structure to dopand incorporations, which is important for the application of ion implantation wave guide formation in optoelectronic applications.

4.1 Silicon implantation SiO_2:Si^+

To get started with the investigation of the implanted samples, we prefer to recognize especially the surplus of atoms from the host material in this complex many body correlated system. We report in this section our observation of visible-light emission at room temperature from Si^+ implanted thermally grown SiO_2 layers on silicon substrates. Cathodoluminescence measurements were performed on silicon implanted samples using the same experimental parameters as used for the non implanted samples. As a result of comparison between the CL spectra of the pure and Si^+ implanted SiO_2, we see a significant blue B luminescence emission (460 nm ; 2.7 eV) and an intense broad luminescent band in the yellow Y region with a peak beyond 580 nm (2.1 eV) are observed especially after annealing at high temperature (T_a=900 °C), see Fig. 4.1. The ultra violet UV (290 nm ; 4.3 eV) and the red R luminescence (650 nm ; 1.9 eV) are also present but with less influence due to silicon implantation. Two additional luminescence bands can be anticipated, one in the green G region at 490 nm (2.5 eV) and another in the red region at around 750 nm (1.65 eV). Higher initial intensities in the thermally annealed samples were registered but all luminescence were saturated to the same level as of the non annealed samples. The green (490 nm ; 2.5 eV), yellow (580 nm ; 2.1 eV) and the additional red (750 nm ; 1.65) emission bands are associated with the presence of silicon nanoclusters in the silica matrix.

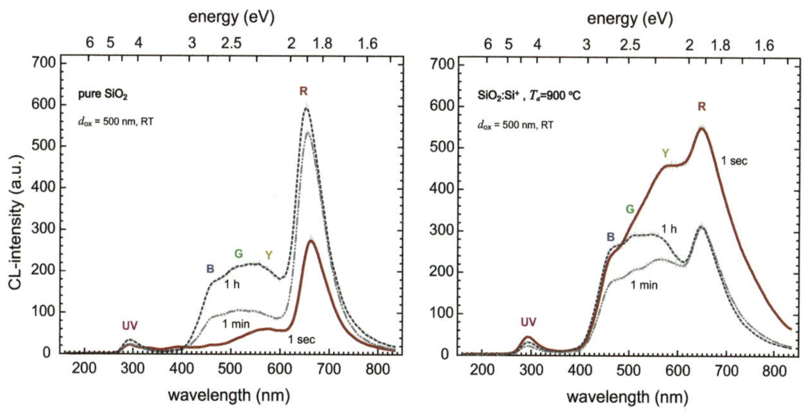

Fig. 4.1 CL spectra of pure and Si^+ implanted SiO_2 layers at room temperature (RT). The initial spectra (red colored) is labeled by (1 sec) and the saturated by (1 h).

The presence of silicon nanoclusters (crystalline and amorphous) is confirmed by transmission electron microscopy (TEM) and by means of EDX measurements.Recently, some authors presented room-temperature photoluminescence data from silica layers implanted with Si^+ ions of 160 keV energy excited using 292 nm excitation light from a 450 W xenon lamp [Mutti et al. 1995]. They showed the existence of a visible band peaked at 1.9 eV (620 nm) together with a broad band centered at lower energy 1.7 eV (730 nm) which was present only after annealing at 1100 °C. They ascribed the 1.9 eV band to *E´* defects created by ion implantation in the silica matrix, while they attributed the 1.7 eV band to the presence of silicon nanocrystals.

4.2 Germanium implantation $SiO_2:Ge^+$

Typical CL spectra of Ge^+-implanted silica layers at room temperature (RT) are shown in Fig. 4.2. The main ultraviolet (UV) and violet (V) luminescence bands at 295 nm (4.2 eV) and 410 nm (3.1 eV) respectively, and a green band around 535 nm (2.3 eV) are seen predominantly on non-annealed samples even at low temperature. The well-known red band appears also in our detection range but not as dominant band as in the standard SiO_2 spectra. Previously we have demonstrated that the spectra of Ge-doped amorphous SiO_2 layers are a mixture of SiO_2 and tetragonal GeO_2. Whereas the red luminescence at 1.9 eV from the NBOHC of the SiO_2 matrix is conserved, the larger amplitude of the violet band at 3.1 eV seems to be overtaken from tetragonal GeO_2 modification indicating a strong defect luminescence at the Ge dopant centers in the rutile-like tetragonal coordination [Barfels 2001].

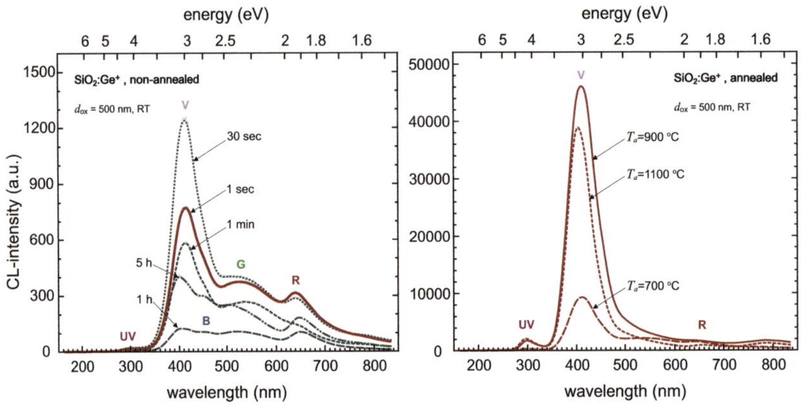

Fig. 4.2 CL-spectra of Ge^+-implanted (500nm) SiO_2 layers (implantation dose $D=5\times10^{16}$ cm^{-2} recorded at RT on the left hand side, demonstrating the huge violet band (V) at $\lambda\approx410$ nm: 3.1 eV. The thermal annealing of the samples was performed at three different annealing temperatures T_a=700, 900, 1100 °C, as shown on the right hand side.

The CL spectra of pure undoped *a*-SiO_2 and Ge^+-doped are similar to the local intrinsic point defect centers associated with the fundamental silicon dioxide defect structure. The energy positions and widths of the red R and the UV CL emissions are the same for both specimen types within the limits of experimental uncertainty, unless the violet band ($\lambda\approx410$ nm, 3.1 eV) is considered to be a well seen fingerprint of Ge related defects and covering the blue band ($\lambda\approx465$ nm, 2.7 eV) of pure SiO_2. According to an earlier model [Skuja 1998], the violet luminescence corresponds to the so-called twofold coordinated germanium luminescence center (=Ge••) which imperceptibly interacts with the host material atoms due to its poor correlation in the silica glass network. However, this band could be also associated with different phases of Ge, that is to Ge clusters as well nanocrystals located in the SiO_2 layer [Fitting et al. 2002b], which can remarkably grow in size with increasing post annealing temperature. In the absence of Ge impurities, the luminescent emission component observed between 3.1-3.3 eV in oxygen deficient silica has been attributed to the recombination of a hole trapped adjacent to a substitutional charge-compensated aluminum ion center [Stevens-Kalceff 1998].

Furthermore, Fig. 4.2 (right) shows the CL spectra of the Ge^+-implanted sample annealed at 700, 900, 1100 °C for 1 hour in dry nitrogen. The large emission band at 3.1 eV due to the germanium implantation is observed and the intensity of this peak increases up to a factor of 10-50 with increasing annealing temperature (T_a), but it decreases rapidly with increasing irradiation time. The concurrent changes in the various bands of the emission spectra due to the Ge implantation are shown in Fig. 4.3.

With increasing annealing temperature up to T_a=900 °C the CL intensity strongly increases. Exceeding the annealing temperature up to 1100 °C, i.e. to the original oxidation temperature, the CL intensity is reduced again and the green luminescence intensity at 535 nm is terminated (totally annealed), contrary to the violet (V) luminescence band which still shows an enormous presence in the CL detection range. Also we see that NBOHC fades away with increasing annealing temperature (Fig. 4.3). That could be somehow a reason of activation of various interstitial atoms at high temperatures, where electron spin resonance (ESR) experiments have shown that the thermally activated diffusion of mobile interstitial species can result in the annealing of defects involved in luminescent processes [Griscom 1990b]. As we stated, the violet luminescence is related to different states or phases of Ge, namely to GeO_2 dissolved in the near SiO_2 surface region and to Ge nanocrystals [Rebohle et al. 2002a] located in the SiO_2 layers, see Fig. 4.4, which may be partially oxidized at their interface to the surrounding amorphous SiO_2 matrix. The nanoclusters size are growing with annealing temperature from 2-4 nm at T_a=900 °C to 5-10 nm at T_a=1100 °C as shown in Figs. 4.5 and 4.6.

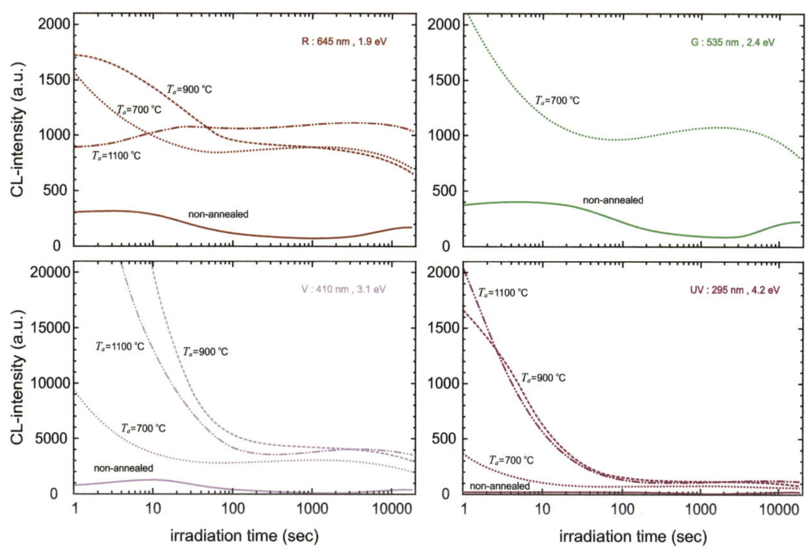

Fig. 4.3 CL bands red (R:1.9 eV), green (G:2.4 eV), violet (V:3.1 eV) and (UV:4.2 eV) from Ge^+-implanted SiO_2 layers after different annealing temperatures T_a as a function of irradiation time; CL measured at RT.

High resolution TEM micrographs shown in Fig. 4.5 reveal a spherical shape of Ge nanocrystals in silica, in contrast to the shape of nanocrystals in other crystalline host material. This is apparently the result of the anisotropy of the amorphous silica matrix. Further experimental analysis of the orientation relationships between the nanocrystals and the crystalline matrix shows that there is no fixed relationship of orientation between the nanocrystals and the host [Xu et al. 2005]. A closer look at the highly resolved area is obtained (marked by light colored circles) in Fig. 4.6 where higher magnification was applied. The white circles enclose some of the nanocrystals visible under this magnification. The crystalline structure (lattice) pattern of germanium nanoparticles is clearly distinguishable from the amorphous host, in some areas similar even smaller crystal lattices overlap each other. The host matrix remains in amorphous phase surviving the implantation and thermal annealing.
The size distribution of the Ge nanocrystals was obtained through a laborious TEM effort of a micrograph of very thin cross-sectional TEM specimen, and then followed by manually measuring the size of the nanocrystals. The result is shown in Fig. 4.7. The dark bar

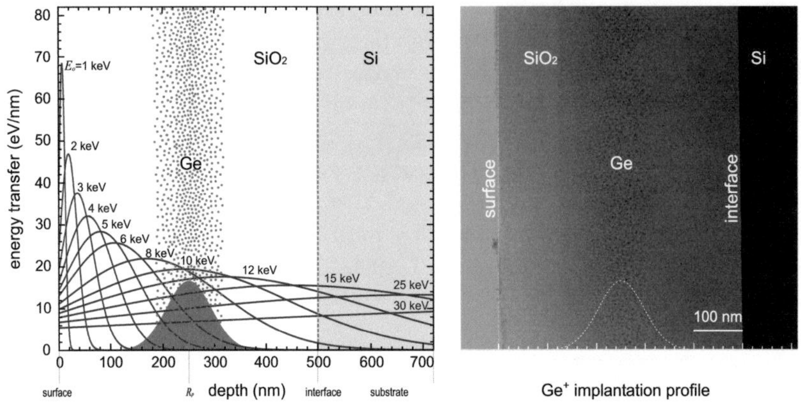

Fig. 4.4 Electron beam excitation densities in SiO_2 layers on Si substrate for different beam energies E_o allowing a CL depth profiling. Here we show the Ge^+ implanted SiO_2 in the mean projected range R_p=250 nm by an ion energy E_{Ge+}=350 keV shown by the shaded Gaussian shaped region. On the right hand side a scanning transmission electron microscope (STEM) image of the same sample showing the actual Ge cluster profile after thermal annealing.

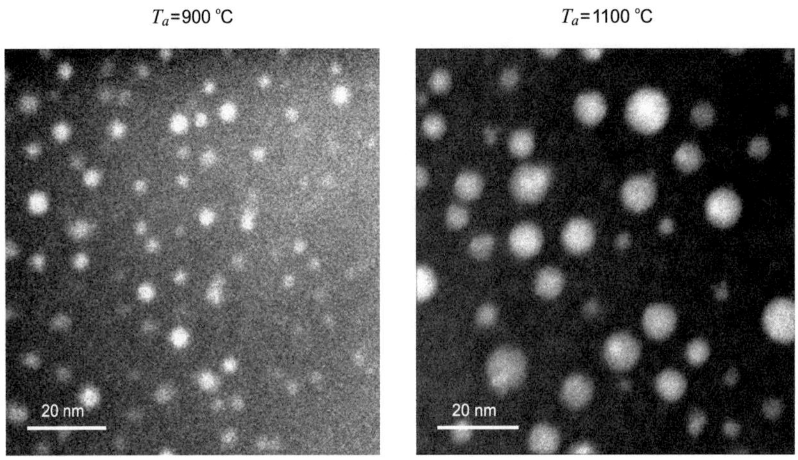

Fig. 4.5 Scanning transmission electron microscope (STEM) images of germanium implanted SiO_2 sample annealed at: T_a=900, 1100 °C, showing the actual size of the Ge clusters.

histogram shows the size distribution of Ge nanocrystals embedded in silica produced at T_a=1100 °C and the light bars are the size distribution of nanocrystals formed at T_a=900 °C. The Ge nanocrystals at higher temperatures are larger on average and have a wider size distribution than those formed at lower temperatures, as it was expected. The size distribution of the germanium particles in the silica system is near-Gaussian-shaped, corresponding to average diameters of 3 nm and 6 nm for T_a=900 and 1000 °C, respectively. The cluster density is also shown in Fig. 4.7, where the cluster concentrations are $N_c=4.6\times10^{17}$ and 2.6×10^{17} cm^{-3} for T_a=900 and 1100 °C, respectively. It is expected that thermally treating the samples is not the only reason for nanocluster formation but also

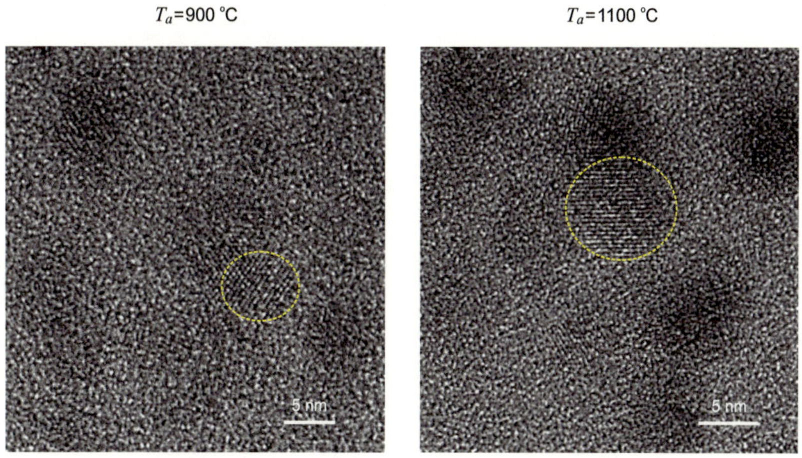

Fig. 4.6 HR-TEM micrograph of Ge-implanted SiO_2 layers after 1 h anneal at 900 , 1100 °C. Selective areas in the host matrix showing the growing of the crystalline Ge spots with increasing temperature.

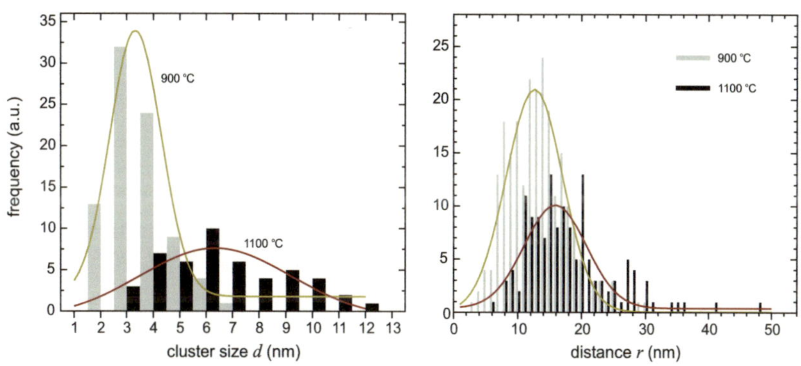

Fig. 4.7 Germanium cluster diameter and separation distance distributions: the correlated cluster concentrations are N_c=4.6×10^{17} and 2.6×10^{17} cm^{-3} for T_a=900 and 1100 °C, respectively.

heavy electron beam irradiation, where Ge atoms diffuse faster in the more damaged area caused by the irradiation. Large nanocrystals are thus formed, whereas in the area where Ge atoms are less mobile, smaller nanocrystals are formed.

Based on the assessments specified in this section , we are presenting a model in Fig. 4.8 of the annealing process forming Ge aggregates with an optimum size for a maximum luminescence at annealing temperatures near T_a=900 °C. For higher annealing temperatures T_a≥1100 °C the cluster or crystal growth has continued, thus their luminescent surface or surroundings in their sum has been reduced, i.e. the overall luminescence efficiency decreases. Simultaneously the specific depth of luminescent Ge clusters, however now in a lesser amount, has shifted towards the surface, whereas at greater depths we find bigger Ge crystallites but much less effective in luminescence than the smaller clusters.

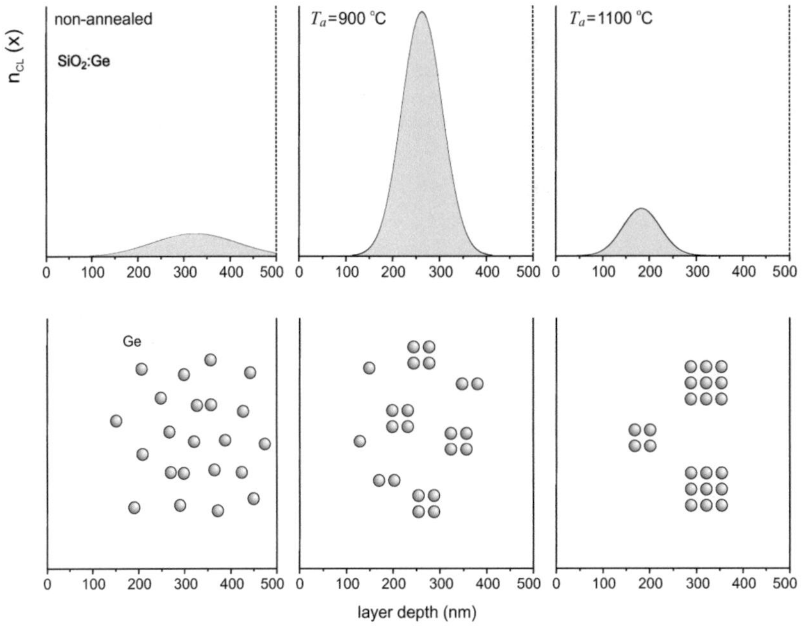

Fig. 4.8 Schematic presentation of the luminescent center depth profile as a function of post annealing temperature T_a (above) and the supposed depth distribution of Ge nanocrystals (below) with an optimum size for maximum luminescence (middle part).

4.3 Carbon implantation $SiO_2:C^+$

Carbon implantation, like other ion implantations, produces many different chemical reactions. It has been observed that the implanted carbon in silica can form carbon dioxide (CO_2) and carbon monoxide molecules (CO) [Perez-Rodriguez, et al. 2003]. It is intriguing that the CO and CO_2 formed in the silica matrix have a very different spectroscopic behavior from that of the gas phase molecules. These differences may reveal the unique physical statuses of the CO and CO_2 embedded in the matrix. C^+ implantation is usually combined with Si^+ implantations (Si rich SiO_2 layers), whereas in SiO_2 layers implanted only with C^+ ions, it has been evidenced early on that a severe drawback consists in the strong outdiffusion of the implanted carbon during the thermal process. This is caused by the formation of highly mobile CO species. Accordingly, these phenomena could be prevented by annealing under high-vacuum conditions. However, the most interesting approach to stabilize the implanted C profile is the previous existence of a high Si supersaturation, which enhances the interaction of both C and O atoms with the Si atoms, and prevents the formation of C−O bonds. In this respect, we discuss the CL spectra of C^+ implanted silica.

High dose sequential implantation of C^+ ions into the SiO_2 oxides followed by high-temperature annealing (T_a=900 °C) probably results in the mixture of violet-blue-green-yellow luminescence. The observed visible luminescence bands have been correlated with the implant and annealing conditions and with the microstructure of the processed films.

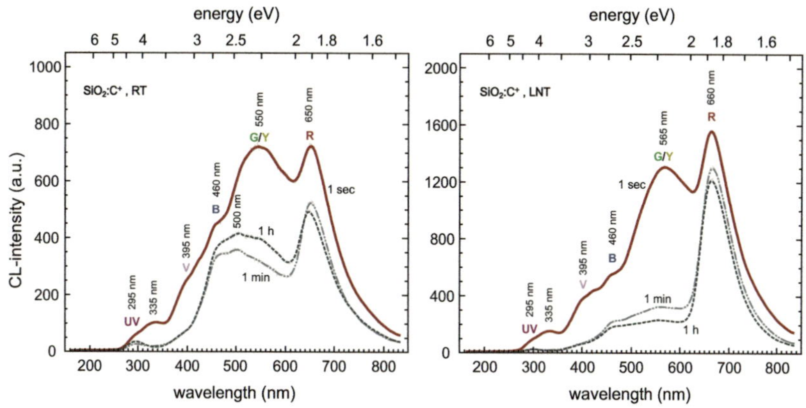

Fig. 4.9 CL-spectra of C^+-implanted (500nm) SiO_2 layers (implantation dose D=5×10^{16} cm^{-2} recorded at RT and LNT. The sample was thermally annealed at T_a=900 °C.

Fig. 4.9 demonstrate the CL spectra of C^+ implanted SiO_2 layers at room temperature (RT) and liquid nitrogen temperature (LNT). A significant difference due to the measured temperature change has not been registered. In general, it is more similar to the CL spectra of Si^+ implanted silica than Ge^+ implanted silica (Figs. 4.1 and 4.2). Here we see a significant blue B luminescence emission (460 nm ; 2.7 eV) and an intense broad luminescent band in the green-yellow G/Y region with a peak beyond 565 nm (2.2 eV). The ultra violet UV (290 nm ; 4.3 eV) and the red R luminescence (650 nm ; 1.9 eV) are also present but with lower intensity than in pure SiO_2. The expectation of another red luminescence at around 750 nm (1.65 eV) is also possible. The only difference between the CL spectra of Si^+ and C^+ implanted samples are that two additional luminescence bands in UV and V regions are excited. One is at 335 nm (3.7 eV) and the other at around 395 nm (3.1 eV). Luminescence at 335 nm is reported in AlGaN [Riemann et al. 2002], in $Lu_3Al_5O_{12}$ films [Zorenko et al. 2005] and even in crystalline SiO_2 (α-quartz) coated with $LiNbO_3$ [Siu et al. 1999], but never in normal or carbon implanted silica. The violet V luminescence comes into view at a lower wavelength, 394 nm, where this luminescence band was detected in the wavelength range 400-410 nm in other implanted silica layers, as we have already demonstrated in Ge^+ implanted SiO_2. We found this band in all ion implanted samples which means that the violet luminescence is not only created due to a specific ion kind implanted in silicas but part of it arises from the network damage caused by the ion beam bombardment. Here the intensity of the V band is lower compared to CL spectra of other ion implanted samples presented in this study. We propose that some carbon atoms have been diffused out of the network due to samples thermal treatments.

The intense room-temperature luminescent bands from the blue up to the yellow spectral region as a result of C^+ ion-implantation processes into SiO_2 layers have been reported by several authors [Zhao et al. 1998, Yu et al. 1998, Rebohle et al. 2001b]. There is a general consensus in assigning these bands to the formation of C-related nanoparticles. The green-yellow luminescence band (2.0-2.2 eV) was also observed in the C^+ implanted SiO_2 layers. In this case, the intensity of the luminescent band was well correlated with the contribution of carbon-related nanoclusters. A luminescence band at higher energies, in the range of 2.7 eV, has also been reported from carbon graphite-like nanoparticles embedded in SiO_2 layers synthesized either by ion implantation [Yu et al. 1998, Gonzalez-Verona et al. 2002] or by sputtering deposition of C-rich oxides [Zhang et al. 1996] followed by thermal annealing. The blue luminescent band is also characteristic of SiC-related crystalline nanostructures, as porous SiC [Ma et al. 2000]. Furthermore, some authors have analyzed the PL emission from C^+ implanted SiO_2 and they attributed the luminescent bands in the blue region to the formation of amorphous clusters of $Si_yC_{1-y}O_x$ complexes [Rebohle et al.2001b]. The microstructure of carbon implanted silica was investigated by Auger electron spectroscopy (AES) and transmission electron microscopy (TEM). Amorphous nanostructures with a size between 2 and 3.5 nm were found in a depth region between 80 and 150 nm below the oxide surface. Strong photoluminescence (PL) around 2.1 and 2.7 eV has also been observed after excitation at 4.77 eV as an indication of nanoclusters [Rebohle et al. 2001b].

4.4 Tin implantation $SiO_2:Sn^+$

The implantation of Sn ions into SiO_2 layers has been studied in connection with the formation of defects and nanostructures exhibiting intense visible and ultraviolet Cathodoluminescence (CL). The spectra of Sn^+ implanted SiO_2 is dominated by a strong violet V emission band with the intensity maximum at about 400 nm and a faint shoulder of the blue luminescence at its usual position in SiO_2 spectra (around 460 nm), see Fig. 4.10. It seems that another UV band is overlapped with the UV band detected previously at 290 nm, or it could be that the 290 nm luminescence band has shifted to a higher wavelength position at around 320 nm. An emission band at 320 nm is attributed to bulk tin dioxide [Lopes et al. 2005a, Calestani et al. 2005]. Tin dioxide (SnO_2) is an n-type semiconductor with a wide band gap (E_g=3.6 eV at 300 K) and is particularly important for many electronic applications. At low temperature (10 K) the intensity maximum of the UV band experiences a blue shift of about 10 nm.

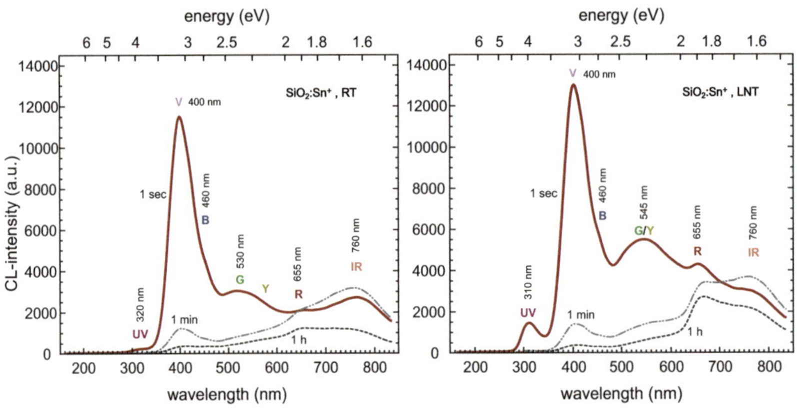

Fig. 4.10 CL-spectra of Sn^+-implanted (500nm) SiO_2 layers (implantation dose D=5×10^{16} cm^{-2}) recorded at RT and LNT. The sample was thermally annealed at T_a=900 °C.

A broad band G peaked at about 530 nm is revealed in room temperature (RT) CL spectra; this band was assigned to Sn nanobelts (or Sn rings) with a lateral dimension 50 nm to 1000 nm. Smaller nanobelts were assigned with longer wavelength position in the CL spectra. At LNT the G/Y shifts 15 nm forward to the red region. In addition, the formation of a rather dense array of Sn-rich nanoparticles presenting a narrow size dispersion and located within the oxide but very close to the SiO_2/Si interface has been observed [Lopes et al. 2005a]. More recently, it has been shown that the annealing atmosphere also influences the microstructure development of Sn^+ implanted silica layers, with significant effects in blue-violet PL response [Lopes et al. 2005a]. Sn^+ implanted samples were thermally annealed at T_a=900 °C. The CL bands in the V, B and G regions were associated with oxygen deficiency centers ODC created during the implantation and annealing processes [Rebohle et al. 2000], and probably assisted by the development of the nanoparticles system.

Data from literature [Hu et al. 2002, Hu et al. 2003] report on the evidence of broad PL optical bands from SnO_2 nanobelts in visible wavelength range from 400 nm to 600 nm. The nature of the transition is tentatively ascribed to nanocrystals inside the nanobelts or to Sn or O vacancies occurring during the growth which can induce trapped states in the band gap [Wu et al. 1997]. Other authors [Gu et al. 2003] present absorption and PL luminescence spectroscopy on SnO_2 nanoparticles showing an absorption edge at 300 nm. The same authors show two distinct PL emissions at 400 and 430 nm which are tentatively attributed to Sn interstitials or dangling bonds and to oxygen vacancies respectively.

The indication of the presence of doping-related oxygen deficient centers (ODC's) was obtained from the 5 eV absorption band. This band is supposed to arise from twofold coordinated silicon (=Si) cation sites in pure silica, and =Ge or =Sn sites in Ge^+ doped and Sn^+ implanted silica [Skuja 1992a, Anedda et al. 2001], as evidenced by polarized photoluminescence and lifetime data of the emission excited in this band [Skuja 1992a].

In particular, we showed that Sn^+ doping can give rise to strong and thermally stable luminescence bands. However, it is not clear whether the microscopic mechanisms involved are those proposed to be responsible for the photosensitivity of Sn^+ doped silica photoconversion of optically active defects induced by doping and structural compaction of the doped host network.

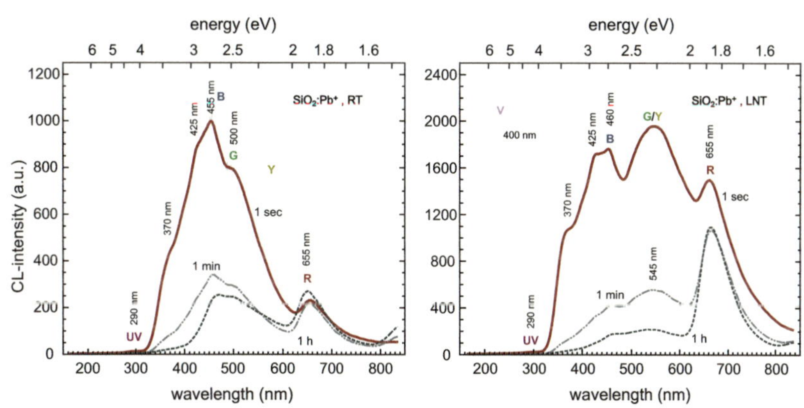

Fig. 4.11 CL-spectra of Pb^+-implanted (500 nm) SiO_2 layers (implantation dose D=5×10^{16} cm^{-2}) recorded at RT and LNT. The sample was thermally annealed at T_a=900 °C.

The red R luminescence (655 nm) is partially eliminated due to Sn^+ implantation. A probable IR CL band can be seen in Fig. 4.10. No evidence of CL or PL emission at 760 nm is, however, reported.

4.5 Lead implantation $SiO_2:Pb^+$

The CL spectrum of the Pb^+ implanted sample is shown in Fig. 4.11. Both Sn and Pb are classified as metallic substances in contrast to the other dopands presented in this section.

Pb^+ implantation creates defect centers providing more intense luminescence in the violet-blue region. Here, two UV bands are detected; one is the UV of the SiO_2 matrix at 290 nm with very low intensity and another, for sure due to Pb implantation at 370 nm. Contrary to the violet band detected in Ge and Sn implanted silicas, the violet band in Fig. 4.11 is shifted towards longer wavelengths (425 nm) and showing lower intensity than the blue luminescence at 455 nm. We were anticipating the existence of a luminescence band at exactly at 500 nm and even in pure SiO_2. Pb^+ implantation enhanced this band significantly at RT. The blue and the green bands suffered red shifts when the measurement's temperature was changed to LNT.

All bands labeled in Fig. 4.11 were going through destructive modes where their intensities dropped considerably in the first seconds of irradiation and they never recovered again as it is often happens in pure SiO_2 CL spectra. The bands in the shorter wavelength than the red are often ascribed to nucleation of the dopand atoms in silica, therefore we expect different forms of Pb aggregates which destroyed under the electron beam irradiation, can be the main precursor of these bands.

The famous red band (655 nm) has lost fractions of its intensity because of Pb^+ implantation but it goes via destructive and creation modes simultaneously, where the NBOHC is saturated by the dopands and followed by liberation from the dangling oxygen bonds under heavy electron bombardment.

5. Group VI elements implanted in SiO_2

The visible cathodoluminescence from Si and its substitutional atoms at room temperature and liquid nitrogen temperatures have been presented in the previous section. In this section we will report the luminescence emission characterizing oxygen ion-implantation in *a*-SiO_2 layers under electron beam excitation. Moreover some other elements (sulfur and selenium) from group IV which are supposed to replace some of the oxygen atoms in the silica matrix are important for better understanding of oxygen related radiation processes and the structure and electronic state of respective defects while oxygen diffusion and chemistry in SiO_2 are most important for silicon-based microelectronics.

5.1 Oxygen implantation SiO_2:O^+

The typical CL spectra of wet SiO_2 is dominated mainly by bands: red R (650 nm, 1.9 eV), blue B (460 nm, 2.7 eV), and UV (290 nm, 4.3 eV) besides we recognize a yellow band Y (570 nm, 2.2 eV) at LNT decaying very rapidly at the beginning of the electron beam irradiation and CL excitation, but appearing and increasing at RT after a longer time of irradiation. We could relate some of these bands to special luminescence defect centers. Defects rolled or influenced by addition of oxygen are shown in Fig. 5.1 where the CL spectra of an oxygen implanted SiO_2 layer are presented. Direct comparison between Fig. 5.1 and the CL on non-implanted (pure) SiO_2 shows that the changes are in the red luminescence region and no extra modification in the other regions of the spectra. This serves to confirm the origin and the structure of the NBOHC where more strained silicon-oxygen bonds can transfer to ≡Si−O• and more interstitial oxygen can be produced depending on the implantation doses. The red luminescence is not only associated with the NBOHC but also associated with interstitial oxygen [Skuja et al. 1994a].

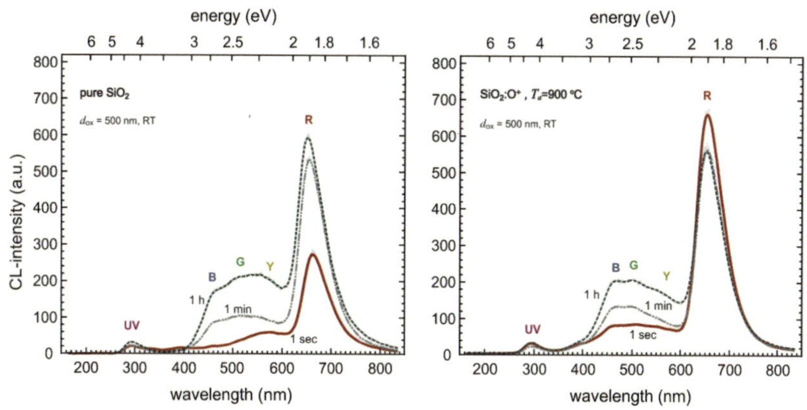

Fig. 5.1 CL spectra of non-annealed and annealed (T_a=900 °C) O^+ implanted SiO_2 layers at room temperature (RT). The initial spectra are labeled by (1 sec) and the saturated by (1 h).

5.2 Selenium implantation SiO_2:Se^+

Se^+ was isoelectrically implanted with regard to oxygen. Fig. 5.2 shows the CL spectra obtained from the SiO_2 layers implanted by Se^+ and annealed at 900 °C. Once again no change is found in the UV luminescence; it is appearing at the same position with low intensity, also the violet luminescence due to Se implantation this time at 410 nm. The blue luminescence is also located clearly at 460 nm. The red and yellow luminescences are enhanced by Se^+ implantation, and both bands seem to be from the same origin as in pure SiO_2 where both have the same tendency during electron beam irradiation at RT and LNT.

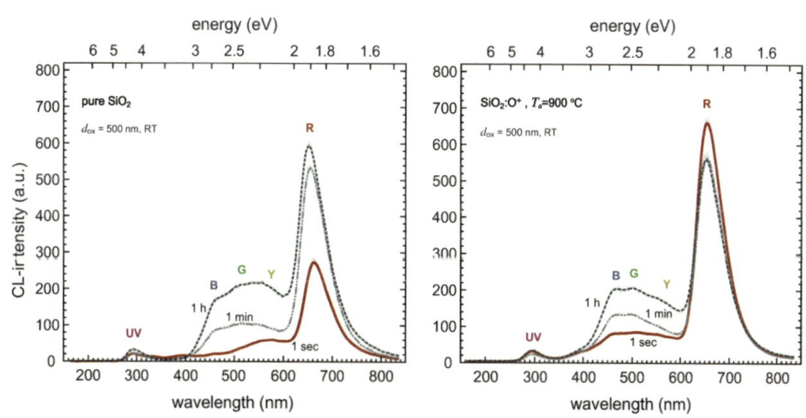

Fig. 5.2 CL spectra of non-annealed and annealed (T_a=900 °C) Se^+ implanted SiO_2 layers at room temperature (RT) and liquid nitrogen temperature (LNT). The initial spectra are labeled by (1 sec) and the saturated by (1 h).

5.3 Sulfur implantation SiO_2:S^+

Defect centers in sulfur-implanted silica layers differ considerably from those observed in other implanted samples. Fig. 5.3 shows the cathodoluminescence spectra of S^+ doped SiO_2 at room temperature (RT) and liquid nitrogen (LNT) as well as their time dependence.

Obviously, the high violet intensity V at ≈405 nm is assigned to sulfur S^+ implantation. Moreover, a sharp and intensive multi-step emission in the green-yellow-red-nearIR (500-820 nm) region is observed for these layers. The exact band positions in wavelengths and energies are given in Table 5.1.

The UV band (290 nm) has been observed at the same position in both samples as well as in pure SiO_2. Moreover, after longer irradiation of about 1 min, i.e. an electron beam dose of 0.3 As/cm^2, the multiplet structure disappears and the characteristic red band R (660 nm) of the NBOHC in SiO_2 becomes visible besides remaining components at the blue band B (460 nm) position and in the yellow region at 560 nm and 590 nm. On the other hand, the sulfur-associated violet band V (405 nm) still remains visible.

Analyzing the multiplet (MP) band structure according to the data listed in Table 5.1, we find that the energy differences between the sub-bands start from 0.14 eV in the green region and then decreases to 0.12 eV in the red region up to 0.11 eV in the near IR. Thus the mean step width amounts to about 120 meV. This energy difference may correspond to a series of almost equidistant vibration levels of non-saturated sulfur radicals ≡Si−S• or ≡Si−O−S• formed during implantation and thermal annealing analogously to the red R band center of the non-bridging oxygen (NBOHC) ≡Si−O• in pure SiO_2.

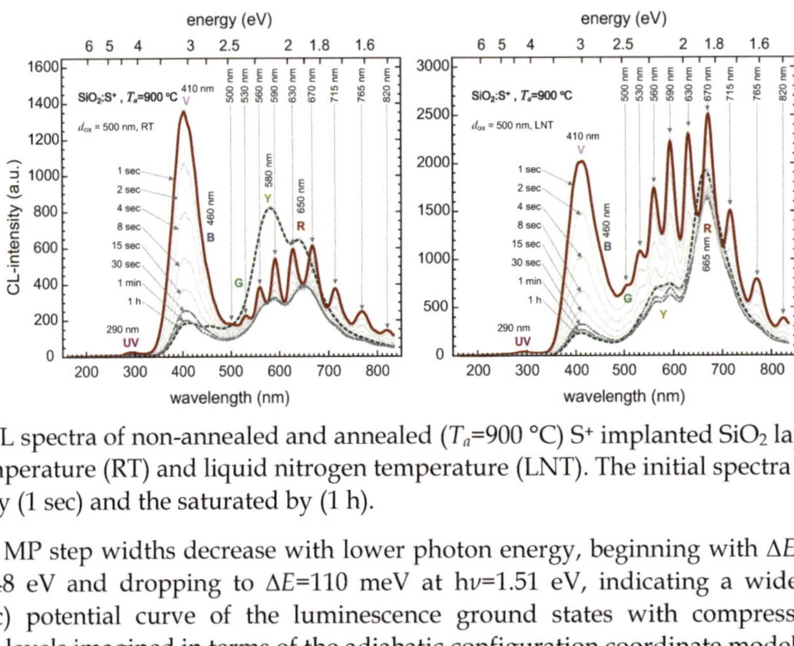

Fig. 5.3 CL spectra of non-annealed and annealed (T_a=900 °C) S^+ implanted SiO_2 layers at room temperature (RT) and liquid nitrogen temperature (LNT). The initial spectra are labeled by (1 sec) and the saturated by (1 h).

Even the MP step widths decrease with lower photon energy, beginning with ΔE=140 meV at hν=2.48 eV and dropping to ΔE=110 meV at hν=1.51 eV, indicating a widened (sub-quadratic) potential curve of the luminescence ground states with compressed higher vibration levels imagined in terms of the adiabatic configuration coordinate model.

5.4 Investigation of the multimodal luminescence

Before attributing this effect to sulfur implantation, we should prove whether the structured spectrum is indeed true or arises from some experimental artifacts. We considered the second argument in our first analysis, where contaminated layers especially at low temperature measurements, interference (Fabry-Perot type) in thin films, and monocrystals as quantum dots could also cause such effects temporarily. The measurements of the CL spectra of the S^+ implanted sample have been repeated many times and under different experimental conditions; the results were always the same. Samples annealed at higher temperatures still showed resolved multimodal bands, see Fig. 5.4, but the structured area was less sharp, indicating a destruction of the centers causing this effect. Then any experimental artifacts are definitely excluded.

Cluster formation is always expected especially when the samples are annealed at temperatures exceeding 900 °C. We found no traces of sulfur clusters in our samples. Fig. 5.5 shows a STEM micrograph of the S^+ implanted silica layer in the same magnification as used to detect Ge cluster (Fig. 5.4). S^+ implanted silica is rarely reported in the literature. Some authors reported their PL results of S^+ doped in 1 mm thick optical fiber disks; the sulfur content was 0.05 wt% which corresponds to $2{\times}10^{19}$ atoms/cm^3. They attributed a similar structured luminescence spectra but at a lower energy region (300-500 nm) to S_2 and S^+_2 interstitial molecules [Zavorotny et al. 2001, Gerasimova et al. 2002]. But afterwards they concluded that the irregular intensity distribution of vibrational components of the PL excitation band indicates that the color centers responsible for this band belong not to a diatomic molecule with one vibrational frequency but to a polyatomic molecule whose vibrational spectrum is formed by a combination of three vibrational frequencies.

SiO_2				SiO_2:S			
	λ/nm	$h\nu$/eV			λ/nm	$h\nu$/eV	ΔE/meV
UV	290	4.3		UV	290	4.30	
B	460	2.7		V	405	3.10	
Y	570	2.2			500	2.48	
R	660	1.9			530	2.34	140
				Y	560	2.21	130
					590	2.10	110
			MP		630	1.97	130
				R	670	1.85	120
					715	1.73	120
					765	1.62	110
				IR	820	1.51	110

Table 5.1 Luminescence bands and multiplet states (MP) in SiO_2 and sulfur implanted SiO_2:S.

It is for this reason that they later attempted the multimodal structured spectra with the SO_2 molecule [Gerasimova 2003]. Whether this hypothesis is correct or not, we see a direct connection between the multimodal structured luminescence and oxygen atoms where sulfur atoms are supposed to substitute oxygen in the matrix and then more oxygen in interstitial sites is expected because over-stoichiometric SiO_x with x>2 does not exist.

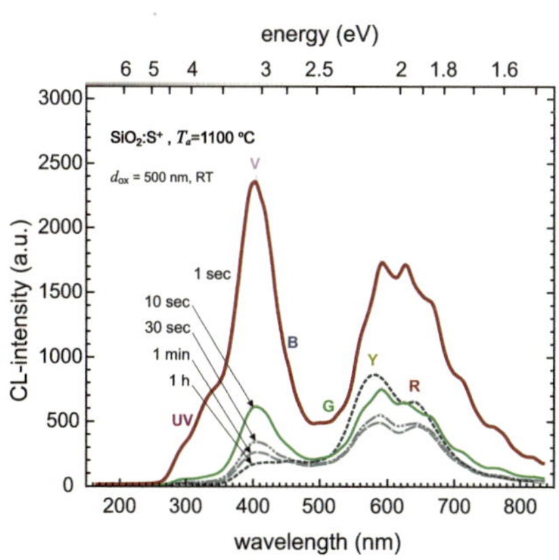

Fig. 5.4 Room temperature (RT) CL spectra of annealed S^+ implanted SiO_2 at higher temperature (T_a=1100 °C).

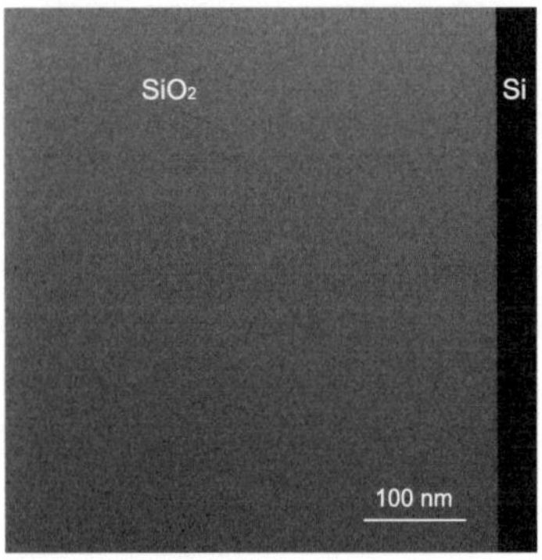

Fig. 5.5 STEM micrograph of the S^+ implanted silica layer annealed at 900 °C showing no evidence of sulfur clusters.

In order to avoid water formation and binding of oxygen we have chosen dry oxidized SiO_2 layers to run our test. Oxygen atoms were implanted in a thinner (d_{ox}=100 nm) dry SiO_2 layer with lower energy (20 keV) and lower doses (3×10^{16} cm^{-2}). Thus the overall CL intensity is about one order of magnitude lower than in 500 nm thick wet oxidized layers, Fig. 5.6. Surprisingly we found the same multiple regular-shaped structure from the green G over the yellow Y and red R regions into near infrared IR in the same peak positions, when the sample was annealed at 1000 °C. And no trace of the violet luminescence was to be seen, see Fig. 5.6. Such structural spectra are never recorded even by using different compositions of non-implanted SiO_2 layers.

This leads us to the conclusion that not sulfur but oxygen should be the source of these multimodal spectra. As already declared, oxygen is responsible for the red R luminescence in SiO_2.

Looking to the literature, we found excitation [Rolfe 1979] and emission spectra [Ewig and Tellinghuisen 1991] of the negatively charged oxygen molecule O^-_2 on interstitial sites in alkali halide crystals. The ground electronic state and several low-lying excited states of the superoxide ion O^-_2 have been studied by multi-configuration self-consistent fields (MCSCF), see Fig. 5.7. For comparison, the ground state of the neutral O_2 molecule was also considered. Parallel computations were carried out for the species in vacuo and in a simulated KCl crystal lattice (in lattio). Computed spectroscopic parameters are in good agreement with experiments for X and A states of O^-_2 in vacuo. In Fig. 5.7 there is also substantial agreement between the computed energy curves for both the ground X and the excited A states in a point-charge lattice and those measured in alkali halide lattices. Further, the spectroscopic parameters of the electron scattering resonance states in vacuo agree well with those of the analogous lattice-stabilized excited electronic states in the solid. There is a typical absorption from the ground state X to the excited state A of about $h\nu$=5.1 eV corresponding to the red R luminescence excitation associated with the NBOHC [Skuja 1994a]. Moreover, the related luminescent transition from $A\rightarrow X$ shows the red R luminescence at about $h\nu$=2 eV. Looking to vibronic levels within the potential configuration curves, we see levels of about 120 meV step-widened towards higher energies, i.e towards the green G region, Fig. 5.7. Thus we should prefer the explanation of the multimodal spectra by means of electronic-vibronic spectra transitions as given by the common configuration coordinate model of luminescence. All transitions appear in good agreement with the multimodal CL spectra of SiO_2:O and SiO_2:S layers shown here. Therefore we favor the interstitial O^-_2 model for the multimodal CL structure and may reject the concept of a photonic crystal structure as proposed in [Bailey et al. 2005] as well as effects of quantum confinement as reported recently for GaAs quantum dots in [Rodt et al. 2005]. Of course, in the latter case of GaAs or other nanostructure materials even other molecule formations should be taken into account as possibly responsible for different but similarly shaped multimodal structures of luminescence spectra.

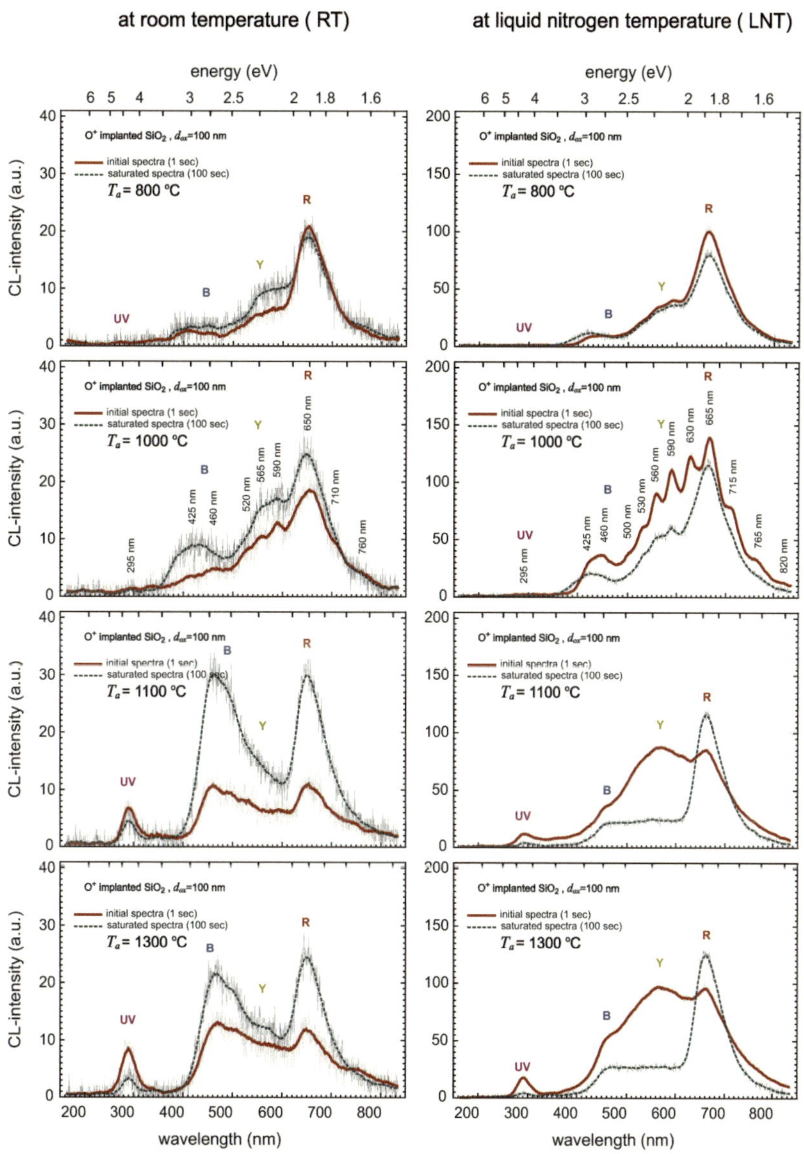

Fig. 5.6 CL spectra of non-annealed and annealed (T_a=700, 900, 1000, 1300 °C) O^+ implanted SiO_2 layers at room temperature (RT) and liquid nitrogen temperature (LNT). The initial spectra are labeled by (1 sec) and the saturated by (1 h).

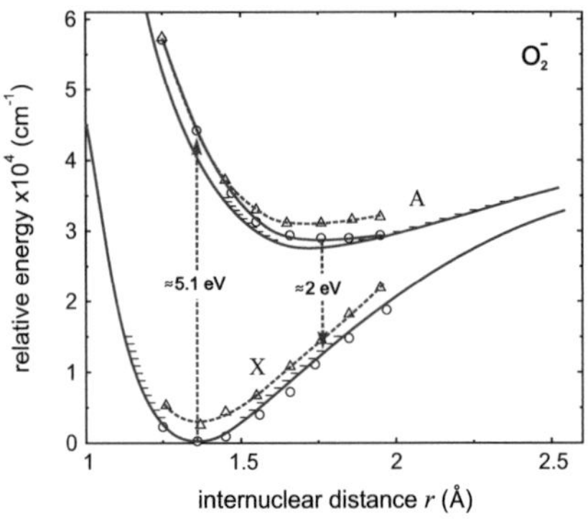

Fig. 5.7 Comparison of MCSCF in lattio (cycles) and in vacuo (triangles) energies with experimental Morse curves derived from analysis of absorption, excitation and luminescence spectra of O^-_2 in NaCl; vibronic levels are indicated by bars [Ewig and Tellinghuisen 1991]

6. Conclusions

The investigation of the present work are extended to various electronical and optical modifications of silica SiO_2 layers as they are applied in microelectronics, optoelectronics, as well as in forthcoming photonics. Electron irradiation of *a*-SiO_2 layers induces chemical defect reactions dependent on the sample oxidation procedure, thermal post-annealing and CL excitation temperature. The red luminescence R (650 nm ; 1.9 eV) is associated with the non-bridging oxygen hole center (NBOHC: ≡Si−O•) whereas the blue luminescence B (460 nm ; 2.7 eV) and the ultraviolet band UV (4.2 eV ; 295 nm) are attributed to Si-related oxygen deficient centers (Si-ODC). CL spectra of hydrogen-free (dry-oxidized) and hydrogen-rich (wet-oxidized) samples have shown significant differences especially in the red region. Hydrogen and hydroxyl groups during wet oxidization increase the intensity of the red luminescence center, where exposure of SiO_2 during the oxidation to water vapor results in the appearance of various OH species (Si−O−H, hydrogen bonded OH, H_2O and the Si−H bond). Thus the dose-dependence of the red R luminescence in wet and dry oxides differs significantly, decreasing in wet oxide from a high initial level down to saturation and increasing in dry oxide from almost zero to saturation. The initial decrease in the red luminescent NBOHC concentration is indicating a recombination of NBOHC with different types of hydrogenous species. In dry SiO_2 the strained bonds (≡Si−O··Si≡) and their electron beam-induced rupture are the precursors for the NBOHC. The mean lifetime of the red luminescence in wet oxide τ=(4.7±0.2)µs is almost the same as in dry oxide τ=(5.3±0.2)µs. Additional hydrogen ion implantation into SiO_2 diminishes the intensity of the red luminescence R because of hydrogen saturation of NBOHC, but magnifies the yellow luminescence Y to very high intensity, even higher than the intensity of the red luminescence. In hydrogen implanted SiO_2 the yellow Y luminescence appears after irradiation doses of about 2 As/cm^2 and we may propose that this band is due to water molecules in the SiO_2 network formed by mobile radiolytic hydrogen and oxygen. A comparison with the CL of water (ice) layers supports this hypothesis.

Understoichiometric SiO_x layers with $x \le 2$ of oxygen deficit or silicon surplus are prepared by means of thermal evaporation of silicon monoxide SiO_1 in different ambient oxygen

atmospheres. The stoichiometric degree varies between $1 \le x \le 2$ and was determined by FTIR calibration measurements by means of the respective shift of the Si−O−Si stretching mode. CL spectra of SiO_x begin to show first typical silica luminescence bands at a stoichiometric threshold of about $x>1.4$. Thermal annealing of these layers and electron bombardment lead to a partial phase separation of SiO_x into atomic Si fragments and clusters in a nearly stoichiometric SiO_2 matrix. This process is correlated with the different luminescent centers in amorphous SiO_2. Especially the yellow luminescence 2.1 eV seems to be associated with silicon hexamer rings in under-stoichiometric SiO_x. Probably they represent the first step of Si nanocrystal formation. Further on, Si clusters can grow to a size of several nanometers. The size distribution, shown by TEM micrographs, exhibits a most probable diameter size of 4 nm and a maximum diameter up to 10 nm. Our results provide evidence that at least two distinct defects are responsible for the yellow luminescence Y at 2.1 eV in silica. The first in wet and H-implanted SiO_2 is due to high H_2O molecule content which can be removed by thermal annealing until it behaves like normal wet SiO_2. The second defect for the yellow luminescence in dry SiO_2 (where almost no H or H_2O exist) could be due to oxygen deficient centers in the form of silicon fragments or small rings in the SiO_2 network.
To investigate whether the different luminescent centers are related to oxygen or to silicon, we have compared non-stoichiometric SiO_2 layers produced by direct oxygen or silicon ion implantation. Thermally oxidized SiO_2 layers of 100 and 500 nm thickness have been implanted with different kinds of ions with a dose 3×10^{16} and 5×10^{16} ions/cm^2, respectively, leading to an atomic dopant fraction of about 4 at.% in the middle, i.e. the half depth of the SiO_2 layers. Oxygen implantation as well as, on the other hand, direct silicon implantation led to an oxygen surplus as well as an oxygen deficit, respectively. As the main experimental results we could state: Oxygen surplus increases the red band R in SiO_2 but does not affect the blue band B. Silicon surplus increases the blue luminescence B, but reduces the red band R. So it is verified that the red luminescence R is an oxygen related center stated as NBOHC and the blue luminescence B is a silicon related oxygen deficient center Si-ODC.
In Ge^+-implanted SiO_2 a huge violet band V (3.1 eV ; 410 nm) appears and is associated with the Ge-related oxygen deficient center Ge-ODC. It dominates the luminescence spectra, covering the original blue band B (Si-ODC) of the SiO_2 matrix. Further on, the V band shows a large increase after thermal annealing. Thus thermal annealing around temperatures of T_a=900 °C increases the violet luminescence of the Ge-ODC by more than two orders of magnitude. Thus the Si-ODC's as well as the Ge-ODC's are formed by Si and Ge molecules clustering to certain low-dimension fragments: dimers, trimers and higher aggregates. With further annealing at higher temperatures, Si and Ge nanocrystals are formed embedded in the amorphous SiO_2 matrix, and the blue B and the violet V luminescence begin to decrease again. The nanocluster size grows with annealing temperature from 2-4 nm at T_a=900 °C to 5-10 nm at T_a=1100 °C.
Further on, heavy electron beam bombardment of SiO_2 with doses of 2.7 As/cm^2 and associated oxygen dissociation may also create Si nanoclusters in a size range of 2-10 nm with a most probable diameter of 4 nm. These Si nanoclusters with their quantum confinement are a source of luminescence in the near IR at 1.6 eV and probably at 1.3 eV too. Further ion implantations of group IV elements: C^+, Sn^+, Pb^+ which are thought to substitute Si in the silica matrix, show new bands and a general increase of the luminescence in the blue region. On the other hand, implantations of oxygen substitute elements of group VI

lead to an increase in the red spectral region. The ultraviolet UV and violet V luminescence are detected in mostly all ion implanted samples at around 290 nm and 410 nm, respectively, indicating not only extrinsic related ODC or extrinsic defects but also ion implantation induced defects in the SiO_2 matrix.
As a surprising peculiarity, the cathodoluminescence spectra of oxygen and sulfur implanted SiO_2 layers show, besides characteristic bands, a sharp and intensive multimodal structure beginning in the green region at 500 nm over the yellow-red region and extending to the near IR measured up to 820 nm. The energy step differences of the sublevels amount to an average of 120 meV and indicate vibronic-electronic transitions, probably, of O^-_2 interstitial molecules, as we could demonstrate by a respective configuration coordinate model. However, such "mysterious" multimodal luminescence spectra are observed occasionally in other material compounds too, and are many-fold in their interpretations by other authors ranging from photonic crystals and interference effects over discrete quantum dots and respective quantum confinement, even to our model of interstitial molecules and their electronic-vibronic luminescent transitions.

7. References

Anedda A. , Carbonaro C. M. , Serpi A. , Chiodini N. , Paleari A. , Scotti R. , Spinolo G. , Brambilla G. and Pruneri V. : *Vacuum ultraviolet absorption spectrum of photorefractive Sn-doped silica fiber preforms*, J. Non-Cryst. Solids 280 (2001) 287.

Anedda A. , Carbonaro C. M. , Clemente F. , Corpino R. , Grandi S. , Mustarelli P. and Magistris A. : *OH-dependence of ultraviolet emission in porous silica*, J. Non-Cryst. Solids 322 (2003b) 68.

Bailey R. C. , Parpia M. , Hupp J. T. : *Sensing via optical interference*, Materialstoday, Vol 8, Iss.4, April (2005), 46.

Bakaleinikov L. A. , Zamoryanskaya M. V. , Kolesnikova E. V. , Sokolov V. I. and Flegontova E. Yu. : *Silicon Dioxide Modification by an Electron Beam*, Physics of the Solid State 46 (2004) 1018.

Bakos T. , Rashkeev S. N. and Pantelides S. T. : *H_2O and O_2 molecules in amorphous SiO_2:Defect formation and annihilation mechanisum* , Phys. Rev. B 69 (2004a) 195206.

Bakos T. , Rashkeev S. N. and Pantelides S. T. : Optically active defects in SiO_2: *The nonbridging oxygen center and interstitial OH molecule*, Phys. Rev. B 70 (2004b) 75203.

Barfels T. : PhD dissertation , *Kathodolumineszenz amorpher und kristalliner Modifikationen von SiO_2 und GeO_2*, Rostock university (2001).

Blöchl P. E. : *First-principles calculations of defects in oxygen-deficient silica exposed to hydrogen*, Phys. Rev. B 62 (2000) 6158.

Brichard B. , Fernandez A. F. , Ooms H. , Borgermans P. and Berghmans F. : *Dependence of the POR and NBOHC Defects as Function of the Dose in Hydrogen-Treated and Untreated KU1 Glass Fibers*, IEEE Transactions on Nuclear Science 50 (2003) 2024.

Calestani D. , Lazzarini L. , Salviati G. and Zha M. : *Morphological, structural and optical study of quasi-1D SnO_2 nanowires and nanobelts*, Cryst. Res. Technol. 40 (2005) 937.

Cannas M. , Agnello S. , Boscaino R. , Costa S. , Gelardi F. M. , Messina F. : *Growth of H(II) centers in natural silica after UV laser exposure*, J.Non-Cryst. Solids, 322 (2003b) 90.

Cazaux J. : *Some considerations on the electric field induced in insulators by electron bombardment* , J. Appl. Phys. 59 (1986) 1418.

Ewig C. S. and Tellinghuisen J. : *Ab initio study of the electronic states of O^-_2 in vacuo and in simulated ionic solids*, J. Chem. Phys. 95 (1991) 1097.

Fitting H.-J. , Barfels T. , Trukhin A. N. , Schmidt B. , Gulans A. and von Czarnowski A. : *Cathodoluminescence of Ge^+, Si^+, and O^+ implanted SiO_2 layers and the role of mobile oxygen in defect transformation*, J. Non-Cryst. Solids 303 (2002b) 218.

Fitting H.-J. , Ziems T. , von Czarnowski A. and Schmidt B. : *Luminescence center transformation in wet and dry SiO_2*, Radiation Measurements 39 (2004) 649.

Fitting H.-J. , Ziems T. , Salh R. and von Czarnowski A. : *Luminescent defect dynamics in amorphous SiO_2:H*, phys. stat. sol. (c) 2 (2005a) 693.

Fitting H.-J. , Ziems T. , Salh R. , Zamoryanskaya M. V. , Kolesnikova K. V., Schmidt B. and von Czarnowski A. : *Cathodoluminescence of wet, dry and Hydrogen-implanted silica films*, J. Non-Cryst. Solids 351 (2005b) 2251.

Fitting H.-J. , Salh R. , Barfels T. and Schmidt B. : *Multimodal Luminescence Spectra of Ion-implanted Silica*, phys. stat. sol. (a) 202 (2005c) R142.

Gerasimova V. I. , Zavorotny Yu. S. , Rybaltovskii A. O. , Chernov P. V. , Sazhin O. D. , Khrapko R. R. and Frolov A. A. : *Color Centers in Sulfur-Doped Silica Glasses: Spectroscopic Manifestations of an S^+_2 Interstitial Molecular Ion*, Glass Physics and Chemistry 28 (2002) 5.

Gerasimova V. I. , Rybaltovskii A. O. , Chernov P. V. and Spasskii D. A. : *Color Centers in Sulfur-Doped Silica Glasses: Spectroscopic Manifestations of an SO_2 Interstitial Molecule*, Glass Physics and Chemistry 29 (2003) 232.

Goldberg M. : *Cathodoluminescence and cathode-electroluminescence of SiO_2-layers*, PhD thesis, University of Rostock, (1996).

Gonzalez-Verona O. , Garrido B. , Perez-Rodriguez A. , Morante J. R. , Bonafos C., Carrada M., Sanz L. F. , Gonzalez M. A. and Jimenez J. : *Analysis of the white emission from ion beam synthesised layers by in-depth resolved scanning photoluminescence microscopy*, Materials Science and Engineering B 91 (2002) 51.

Griscom D. L. : *Electron spin resonance investigations of defects and defect processes in amorphous silicon dioxide*, Rev. Solid State Sci. 4 (1990b) 565.

Gu F. , Wang S. F. , Song C. F. , Lu M. K. , Qi Y. X. , Zhou G. J. , Xu D. and Yuan D. R. ; *Synthesis and luminescence properties of SnO_2 nanoparticles*, Chem. Phys. Lett. 372 (2003) 451.

Helm C. R. and Deal B. E. (eds.) : *The Physics and Chemistry of SiO_2 and the Si-SiO_2 Interface*, Plenum, New York, (1993).

Hosono H. , Weeks R. A. , Imagawa H. and Zuhr R. : *Formation of oxygen-deficient type structural defects and state of ions in SiO_2 glasses implanted with transition metal ions*, J. Non-Cryst. Solids 120 (1990) 250.

Hu J. Q. , Ma J. X. L. , Shang N. G. , Xie Z. Y. , Wong N. B. , Lee C. S. and Lee S. T.: *Large-Scale Rapid Oxidation Synthesis of SnO_2 Nanoribbons*, J. Phys. Chem. B 106 (2002) 3823.

Hu J. Q. , Bando Y. and Golberg D. : *Self-catalyst growth and optical properties of novel SnO_2 fishbone-like nanoribbons*, Chem. Phys. Lett. 372 (2003) 758.

Iacona F., Franzo G. and Spinella C. : *Correlation between luminescence and structural properties of Si nanocrystals*, J. Appl. Phys. 87 (2000) 1295

Ivanda M. , Clasen R. , Hornfeck M. and Kiefer W. : *Raman spectroscopy on SiO_2 glasses sintered from nanosized particles*, J. Non-Cryst. Solids 322 (2003) 46.

Kajihara K. , Skuja L. , Hirano M. and H. Hosono : *Diffusion and Reactions of Hydrogen in F_2-Laser-Irradiated SiO_2 Glass*, Phys. Rev. Lett. 89 (2002) 135507.

Kolesnikova E. V. , Sitnikova A. A. , Sokolov V. I. and Zamoryanskaya M. : *Modification of Silicon Oxide by High Energy Electron Beam*, Solid State Phenomena 108-109 (2005) 729.

Kuzuu N. and Horikoshi H. : *X-ray induced absorption in fused silica containing various amounts of OH*, J. Appl. Phys. 97 (2005) 93508.

Ledoux G. , Gong J. , Huisken F. , Guillois O. and Reynaud C. : *Photoluminescence of size-separated silicon nanocrystals: Confirmation of quantum confinement*, Appl. Phys. Lett. 80 (2002) 4834.

Legrand A. P. : *The Surface Properties of Silica*, New York, Wiley, (1998).

Lehmann A. , Schumann L. and Hübner K. : *Optical Phonos in Amorphous Silicon Oxides*, phys. stat. sol. (b) 121 (1984) 505.

Lopes J. M. J. , Zawislak F. C. , Fichtner P. F. P. , Papaleo R. M. , Lovey F. C. , Condo A. M. and Tolley A. J. : *Formation of epitaxial β-Sn islands at the interface of SiO_2/Si layers implanted with Sn ions*, Appl. Phys. Lett. 86 (2005a) 191914.

Lopes J. M. J. , Zawislak F. C. , Fichtner P. F. P. , Lovey F. C. and Condo A. M. : *Effect of annealing atmosphere on the structure and luminescence of Sn-implanted SiO_2 layers*, Appl. Phys. Lett. 86 (2005b) 023101.

Lou V. , Sato R. and Tomozawa M. : *Hydrogen diffusion in fused silica at high temperatures*, J. Non-Cryst. Solids 315 (2003) 13.

Ma T., Xu J. , Du J. , Li W. , Huang X. and Chen K. : *Full color light emission from amorphous SiCx:H with organic-inorganic structures*, J. Appl. Phys. 88 (2000) 6408.

Magruder R. H. , Weeks R. A. and Weller R. A. : *Luminescence and absorption in type III silica implanted with multi-energy Si, O and Ar. ions*, J. Non-Cryst. Solids 322 (2003) 58.

Minke M. V. and Jackson K. A. : *Diffusion of germanium in silica glass* , J. Non-Cryst. Solids 351 (2005) 2310.

Mischler C. , Horbach J. , Kob W. and Binder K. : *Water adsorption on amorphous silica surfaces: a Car-Parrinello simulation study*, J. Phys.: Condens. Matter 17 (2005) 4005.

Morimoto Y. , Weeks R. A. , Barnes A. V. , Tolk N. H. and Zuhr R. A. : *The effect of ion implantation on luminescence of a silica*, J. Non-Cryst. Solids 196 (1996) 106.

Morimoto Y. and Nozawa S. : *Effect of Xe_2 light (7.2 eV) on the infrared and vacuum ultraviolet absorption properties of hydroxyl groups in silica glass*, Phys. Rev. B 59 (1999) 4066.

Mutti P., Ghislotti G. , Bertoni S. , Bonoldi L. , Cerofolini G. F. , Meda L. , Grilli E. and Guzzi M. : *Room-temperature visible luminescence from silicon nanocrystals in silicon implanted SiO_2 layers*, Appl. Phys. Lett. 66 (1995) 851.

Nicklaw C. J. , Pagey M. P. , Pantelides S. T. , Fleetwood D. M. , Schrimpf R. D. , Galloway K. F. , Wittig J. E. , Howard B. M. , Taw E., McNeil W. H. and Conley J. F. , Jr. : *Defects and Nanocrystals Generated by Si Implantation into a-SiO_2*, IEEE Transactions on Nuclear Science 47 (2000) 2269.

Ogasawara H. , Brena B. , Nordlund D. , Nyberg M. , Pelmenschikov A. , Pettersson L.G.M. and Nilsson A. : *Structure and Bonding of Water on Pt(111)*, Phys. Rev. Lett. 89 (2002) 276102.

Pacchioni G. and Mazzeo C. : *Paramagnetic centers in Ge-doped silica: A first-principles study*, Phys. Rev. B 62 (2000) 5452.

Pan B. C. and Biswas R. : *Simulation of hydrogen evolution from nano-crystalline silicon*, J. Non-Cryst. Solids 333 (2004) 44.

Perez-Rodriguez A. , Gonzalez-Varona O. , Garrido B. , Pellegrino P. , Morante J. R., Bonafos C. , Carrada M. and Claverie A. : *White luminescence from Si^+ and C^+ ion-implanted SiO_2 films*, J. Appl. Phys. 94 (2003) 254.

Prokes S. M. , Carlos W. E., Veprek S. and Ossadnik C. : *Defect studies in as-deposited and processed nanocrystalline Si/SiO_2 structures*, Phys. Rev. B 58 (1998) 15632.

Rashkeev S. N. , Fleetwood D. M. , Schrimpf R. D. and Pantelides S. T. : *Defect Generation by Hydrogen at the Si-SiO_2 Interface*, Phys. Rev. Lett. 87 (2001) 165506.

Rebohle L. , von Borany J. , Fröb H. and Skorupa W. : *Blue photo- and electroluminescence of silicon dioxide layers ion-implanted with group IV elements*, Appl. Phys. B 71 (2000) 131.

Rebohle L. , Gebel T. , Fröb H. , Reuther H. and Skorupa W. : *Ion beam processing for Si/C-rich thermally grown SiO_2 layers: photoluminescence and microstructure*, Appl. Surf. Sci. 184 (2001b) 156.

Rebohle L. , Gebel T. , von Borany J. , Skorupa W. , Helm M. , Pacifici D. , Franzo G. and Priolo F. : *Transient behavior of the strong violet electroluminescence of Ge-implanted SiO_2 layers*, Appl. Phys. B 74 (2002a) 53.

Riemann T. , Christen J. , Kaschner A. , Laades A. , Hoffmann A. , Thomsen C., Iwaya M., Kamiyama S. , Amano H. and Akasaki I. : *Direct observation of Ga-rich microdomains in crack-free AlGaN grown on patterned GaN/sapphire substrates*, Appl. Phys. Lett. 80 (2002) 3093.

Rinnert H. and Vergnat M. : *Structure and optical properties of amorphous silicon oxide films with different porosities*, J. Non-Cryst. Solids 320 (2003) 64.

Rodt S. , Seguin R. , Schliwa A. , Pötschke K. and Bimberg D. : International Conference on Luminescence ICL-05, Beijing, July 25-29, (2005), Book of Abstracts: WED-P-E-01.

Rolfe J. : *First excited state of the O_2^-* , J. Chem. Phys. 70 (1979) 2463.

Salh R. , von Czarnowski A. and Fitting H.-J. : *Multiplet luminescence of sulfur implanted silica-SiO_2:S*, phys. stat. sol.(a) 202 (2005) R53.

Schweigert I. V. , Lehtinen K. E. J. , Carrier M. J. and Zachariah M. R. : *Structure and properties of silica nanoclusters at high temperatures*, Phys. Rev. B 65 (2002) 235410.

Siu G. G. , Wu X. L. , Gu Y. and Bao X. M. : *Ultraviolet and blue emission from crystalline SiO_2 coated with $LiNbO_3$ and $LiTaO_3$*, Appl. Phys. Lett. 74 (1999) 1812.

Skuja L. N., Silin A. R. and Mares J. : *Decay time and properties of luminescence centers in vitreous silica*, phys. stat. sol (a) 50 (1978) 149.

Skuja L. : *Isoelectronic series of twofold coordinated Si, Ge, and Sn atoms in glassy SiO_2: a luminescence study*, J. Non-Cryst. Solids 149 (1992a) 77.

Skuja L. : *The origin of the 1.9eV luminescence band in glassy SiO_2*, J. Non-Cryst. Solids 179 (1994a) 51.

Skuja L. : *Direct singlet-to-triplet optical absorption and luminescence excitation band of the twofold-coordinated silicon center in oxygen-deficient glassy SiO_2*, J. Non-Cryst. Solids 167 (1994b) 229.

Skuja L. : *Optically active oxygen-deficiency-related centers in amorphous silicon dioxide*, J. Non-Cryst. Solids 239 (1998) 16.

Skuja L. , Hirano M. , Kajihara K. and Hosono H. : *Point defect creation by photochemical processes in glassy silica*, Phys. Chem. Glasses 43C (2002) 145.

Stevens-Kalceff M. A. : *Cathodoluminescence microcharacterization of the defect structure of irradiated hydratedand anhydrous fused silicon dioxide*, Phys. Rev. B 57 (1998) 5674.

Tandon P. : *Chemical annealing of oxygen hole centers in bulk glasses,* J. Non-Cryst. Solids 336 (2004) 212.

Torchynska T.V. : *Photoluminescence of Si nanocrystallites in different types of matrices,* J. Non-Cryst. Solids 352 (2006) 2484.

Trukhin A. N. : Self-*trapped exciton luminescence in α-quartz,* Nucl. Instr. and Meth. in Phys. Res. B 91 (1994) 334.

Trukhin A. N. , Goldberg M. , Jansons J. , Fitting H.-J. and Tale I. A. : *Silicon dioxide thin film luminescence in comparison with bulk silica,* J. Non-Cryst. Solids 223 (1998) 114.

Trukhin A. N. and Fitting H.-J. : *Investigation of optical and radiation properties of oxygen deficient silica glasses,* J. Non-Cryst. Solids 248 (1999) 49.

Trukhin A. N. , Fitting H.-J. , Barfels T. and von Czarnowski A. : *Cathodoluminescence and IR absorption of oxygen deficient silica-Influence of hydrogen treatment,* J. Non-Cryst. Solids 260 (1999) 132.

Trukhin A. N. , Jansons J. , Fitting H.-J. , Barfels T. and Schmidt B. : *Cathodoluminescence decay kinetics in Ge^+, Si^+, O^+ implanted SiO_2 layers,* J. Non-Cryst. Solids 331 (2003a) 91.

Tsu D. V. , Lucovsky G. and Davidson B. N. : *Effects of the nearest neighbors and the alloy matrix on SiH stretching vibrations in the amorphous SiO_r:H $(0<r<2)$ alloy system,* Phys. Rev. B 40 (1989) 1795.

Wilkinson A. R. and Elliman R. G. : *The effect of annealing environment on the luminescence of silicon nanocrystals in silica,* J. Appl. Phys. 96 (2004) 4018.

Wu X. , Zou B. , Xu J. , Yu B. , Tang G. , Zhang G. and Chen W., Nanostruct. Mat. 8 (1997) 179.

Xu Q. , Sharp I.D. , Liao C.Y. , Yi D. O. , Ager III J. W. , Beeman J. W. , Yu K. M. , Chrzan D. C. and Haller E. E. : *Germanium Nanocrystals Embedded in Sapphire,* Lawrence Berkeley National Laboratory, University of California, Paper LBNL-57463, (2005).

Yao B. , Shi H. , Zhang X. and Zhang L. : *Ultraviolet photoluminescence from nonbridging oxygen hole centers in porous silica,* Appl. Phys. Lett. 78 (2001) 174.

Yi L. X. , Heitmann J. , Scholz R. and Zacharias M. : *Phase separation of thin SiO layers in amorphous SiO/SiO_2 superlattice during annealing,* J. Phys. : Condens. Matter 15 (2003) S2887.

Yu Y. H. , Wong S. P. and Wilson I. H. : *Visible Photoluminescencein Carbon-Implanted Thermal SiO_2 Films,* phys. stat. sol. (a) 168 (1998) 531.

Zacharias M. , Heitmann J. , Scholz R. , Kahler U. , Schmidt M. and Bläsing J. : *Size-controlled highly luminescent silicon nanocrystals: A SiO/SiO_2 superlattice approach,* Appl. Phys. Lett. 80 (2002) 661.

Zacharias M. , Yi L. X. , Heitmann J. , Scholz R. , Reiche M. and Gösele U. : *Size-controlled Si nanocrystals for photonic and electronic applications,* Solis State Phenomena 94 (2003) 95.

Zavorotny Yu. S. , Lutsko E. V. , Rybaltovskii A. O. , Chernov P. V. , Sokolov V. O. and Khrapko R. R. : *Color Centers in Sulfur-Doped Silica Glass: Spectroscopic Manifestations of an Interstitial Molecule S_2,* Glass Physics and Chemistry 27 (2001) 331.

Zhang Q. , Bayliss S. C. and Frentrup W. : *The stable blue and unstable UV photoluminescence from carbon nanoclusters embedded in SiO_2 matrices,* Solid State Commun. 99 (1996) 883.

Zhao J. , Mao D. S. , Lin Z. X. , Jiang B. Y. , Yu Y. H. , Liu X. H. , Wang H. Z. and Yang G. Q. : *Intense short-wavelength photoluminescence from thermal SiO_2 films co-implanted with Si and C ions,* Appl. Phys. Lett. 73 (1998) 1838.

Zorenko Y. , Gorbenko V. , Voloshinovskii A. , Stryganyuk G. , Nedilko S. , Degoda V. and Chukova O. : *Luminescence of Sc-related centers in single crystalline films of $Lu_3Al_5O_{12}$ garnet,* phys. stat. sol. (c) 2 (2005) 105.

10

Nanocrystalline Porous Silicon

Sukumar Basu and Jayita Kanungo
IC Design & Fabrication Centre
Dept. of Electronics and Telecommunication Engineering, Jadavpur University
India

1. Introduction

Porous silicon can be considered as a silicon crystal having a network of voids in it. The nano sized voids in the bulk silicon result in a sponge-like structure of pores and channels surrounded by a skeleton of crystalline Si nano wires. Porous silicon (PS) is gaining scientific and technological attention as a potential platform mainly for its multifarious applications in sensing and photonic devices (Canham, 1997a; Pavesi & Dubos;1997; Dimitrov,1995; Tsamis et al., 2002; Archer & Fauchet, 2003; Barillaro et al.,2003). The extremely large surface to volume ratio ($\geq 500 m^2/cm^3$) of PS, the ease of its formation, control of the surface morphology through variation of the formation parameters and its compatibility to silicon IC technology leading to an amenability to the development of smart systems-on-chip sensors have made it a very attractive material. Due to these multi functional applications of PS, recently it has been proposed to be an educational vehicle for introducing nanotechnology and inter-disciplinary material science by eminent scientists working in this field. But in order to develop porous silicon based devices and their integration to electronic circuits the low resistance stable electrical contacts are necessary. However, unlike crystalline silicon the outstanding problem with PS is the instability of its native interface with a metastable $Si-H_x$ termination (Tsai et al.,1991). The metastable hydro-silicon can undergo spontaneous oxidation in ambient atmosphere and results in the degradation of surface structures. This also creates problems to get a stable Ohmic contact (Deresmes et al.,1995; Stievenard & Deresmes, 1995) which is again a very important factor regarding its commercial applications. Therefore passivation of surface is necessary to make stable porous silicon based devices. For that purpose substituting surface hydrogen by another chemical species has appeared desirable. Oxidations (Rossi et al, 2001; Bsiesy, et al 1991; Petrova-Koch et al., 1992) nitradation (Anderson et al., 1993) and halogenetion (Lauerhaas & Sailor, 1993) are found to be useful for PS surface passivation. Derivatisation by organic groups and polymer (Lees et al. 2003; Mandal et al. 2006), offers an alternative possibility to stabilize the material. Metals like Cu, Ag, In etc. were also used to modify the porous silicon surface to stabilize its photoluminescence properties (Andsager et al 1994; Steiner et al., 1994). Surface modification of PS using noble metals like Pd and Pt has also been studied recently (Kanungo et al. 2009a).
The details on PS are given in a comprehensive review published by Cullis et al. (Cullis et al., 1997) and in the handbook on Porous Silicon properties edited by Canham (Canham, 1997a). H. Foll et al. (Foll et al. 2002) and V. Parkhutik (Parkhutik, 1999) also elaborately reviewed the formation and applications of porous silicon.

2. Preparation of nanocrystalline porous silicon using different chemical methods

Several methods are developed to make the porous layer with wide variation of pore morphologies having the pore dimensions from micro to nanometers. Chemical etching of silicon using chemical solutions of HF, HNO_3 and water (Vasquez et al., 1992), $NaNO_2$ and HF or CrO_3 and HF(Beale et al., 1986; Zubko et al. 1999) are employed for PS formation. However, the most widely used method is the electrochemical etching of silicon crystal in an electrolyte solution of HF and ethanol or methanol (Saha et al., 1998; Kanungo et al., 2006) or HF and water or HF and N, N dimethyl formamide (DMF) (Archer et al., 2005) by passing current for a fixed duration of time. Hummel et al. (Hummel & Chang, 1992) utilized a new spark erosion technique for PS formation, which does not involve any aqueous solution or fluorine contaminants in air or in the other gases. Another interested development in this area is the magnetic field assisted anodization technique employed by T. Nakagawa et al. (Nakagawa et al., 1996). Recently Y. Y. Xu, et al. (Xu, et al. 2005) describe hydrothermal etching of crystalline silicon in HF containing ferric nitrate to obtain the large quantities of regular, uniformly distributed silicon nano pillars which are perpendicular to the surface and well separated from each other. In addition, perpendicular electric field assisted method, illumination assisted method, Hall effect assisted method, lateral electric field method, Buried P-layer assisted method and their combinations have been very recently reported (Samuel, BJ., 2010).

In most cases, the porous silicon structure is formed by electrochemical etching of Si wafers in electrolytes including hydrofluoric acid (HF) and ethanol. The cleaned, polished Si wafer surface is hydrophobic. Added absolute alcohol increases the wettability of the substrate and thus helps the electrolyte penetrating into the pores. So laterally homogenous current density can be maintained to result in the formation of uniform PS layers. The added ethanol also helps in removing the H_2 bubble from the sample surface formed during the anodization process. To get the uniform porous layers with high reproducibility, the applied anodic current density, etching time and the electrolyte concentration are controlled precisely during the process. The cathode of the anodization cell is generally made of platinum or other HF-resistant conductive material. Platinum is used as cathode and Si surface itself acts as the anode. In Fig. 1, a conventional single tank setup is shown both for vertical and horizontal field application.

To get the nanocrystalline porous silicon on n-type silicon, illumination method (Samuel, BJ., 2010). is the most popular way to generate holes required in the electrochemical etching process (Fig.2). However, the photo energy absorption by the atoms depends on the intensity of the illumination source, the distance from the source and the electrolyte environment. Therefore, only the surface layer under the illumination generates electron hole pairs. But the etching rate gradually decreases with time as it is very difficult to reach the illumination into the deep area of the pores.

An alternative Hall Effect assisted method (Fig.3a) can be used for n-type porous silicon formation (Samuel, BJ., 2010). In this arrangement the sample is exposed to the Hall Effect environment. By applying a very large bias voltage and a large magnetic field, the upper layer of the semiconductor is depleted of electrons and is then inverted from n-type to p-type (Fig. 3b). The accumulated hole on the upper layer can participate in the chemical reactions during the etching process. Two main advantages of this method are (i) the illumination sources are not required and (ii) no metal electrodes are needed for etching. Therefore, an etching container devoid of illumination and free of metal electrode can be well designed for the safety of handling the corrosive HF electrolyte so that the contamination from the metal electrode could be avoided.

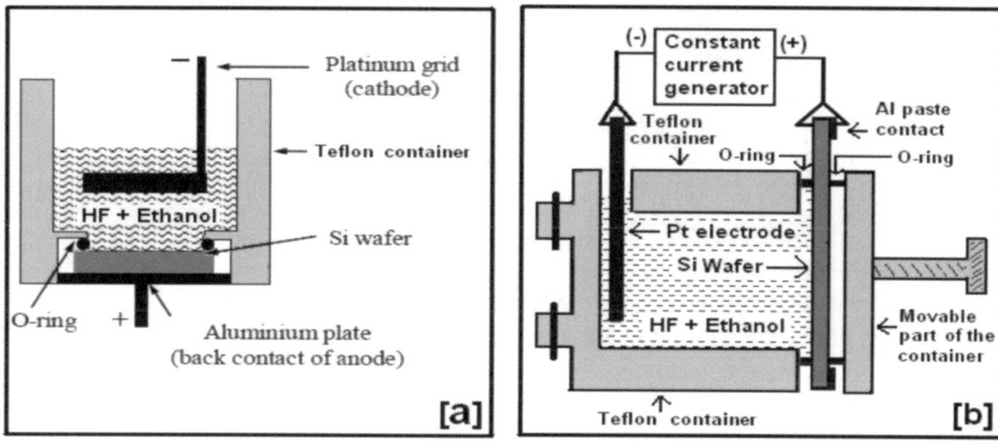

Fig. 1. Schematic of cross sectional view of the jig used for PS formation by electrochemical anodization by the application of electric field in (a) vertical and (b) horizontal mode.

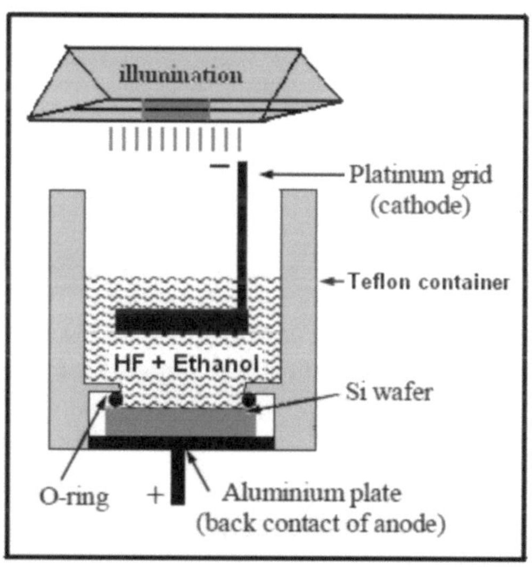

Fig. 2. Schematic of the experimental setup of the illumination assisted method.

A buried p-layer assisted method (Samuel, BJ., 2010) is also proposed for n-type porous silicon fabrication. The buried p-layer acts as the source of hole and is placed underneath the n-layer (Fig.4). The proposed buried p-layer assisted method can also be superimposed on the illumination assisted method during the electrochemical anodization.

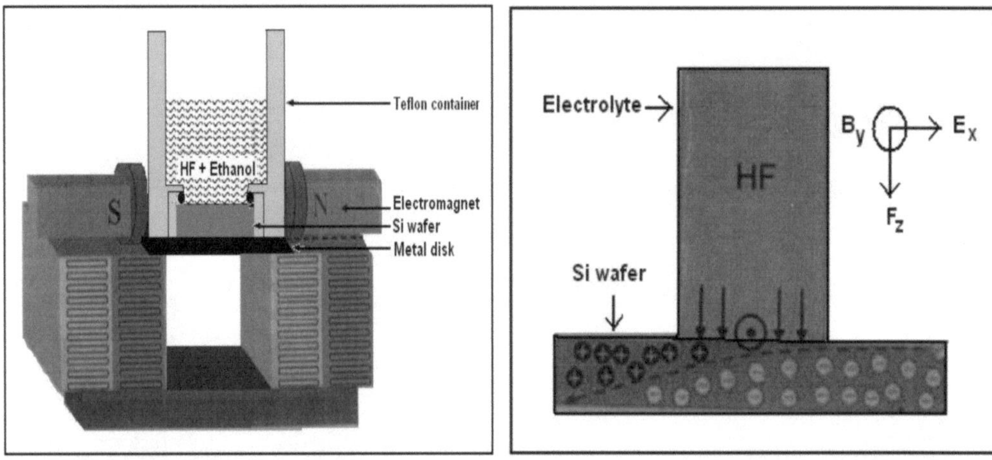

Fig. 3. Schematic of the (a) experimental setup of Hall effect assisted method and (b) the mechanism of pore formation [with the permission of Nova Science Publishers, USA].

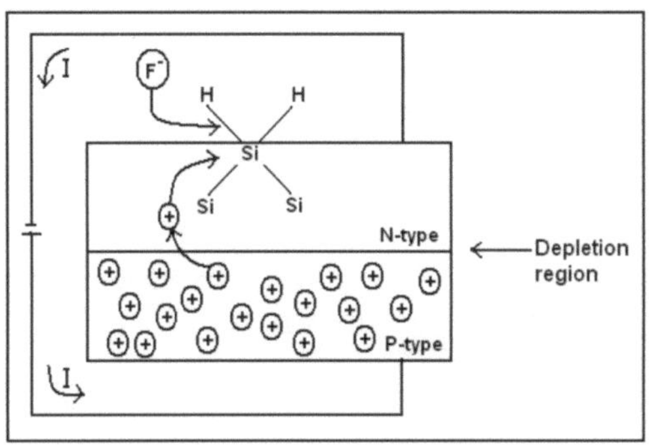

Fig. 4. The schematic mechanism of the biased pn structures for buried p-layer assisted method [with the permission of Nova Science Publishers, USA].

2.1 Mechanism of silicon dissolution and pore formation

Although the complete understanding of the Si dissolution mechanism is still under study, the mostly accepted theory describes that holes are required for pore formation. During the anodization process, the positively charged Si surface is oxidized by F^- ions followed by the formation of water-soluble H_2SiF_6 complex as shown in Fig.5. For the pore formation, the anodic reactions can be depicted as

$$Si + 6HF \rightarrow H_2SiF_6 + H_2 + 2H^+ + 2e^-. \quad (1)$$

However, the detail chemical steps during PS formation may be expressed as follows;

$$Si + 2HF + 2h^+ \rightarrow SiF_2 + 2H^+$$

$$SiF_2 + 4HF \rightarrow H_2 + H_2SiF_6 \quad \text{(for divalent anodic reaction)}$$

$$Si + 4HF + 4h^+ \rightarrow SiF_4 + 4H^+$$

$SiF_4 + 2HF \rightarrow H_2SiF_6$ (for tetravalent anodic reaction)

Different models are also proposed to explain the formation of porous silicon and the pore morphology. According to the model suggested by Kang and Jorne (Kang & Jorne, 1993) the distance between the pores varies as a square root of the applied voltage. The model by Beale et al (Beale et al. 1985) shown in Fig. 5a highlights a quantitative idea about the pore morphology. They pointed out that there is a space charge region between the PS layer and bulk silicon. This model further considers the pinning of the Fermi level at the Si/electrolyte interface. It is due to a large number of surface states that create Schottky barrier between the semiconductor and the electrolyte. The decrease of the Schottky barrier height occurs due to the dissolution of the pore tips because of the electric field generated in that region.

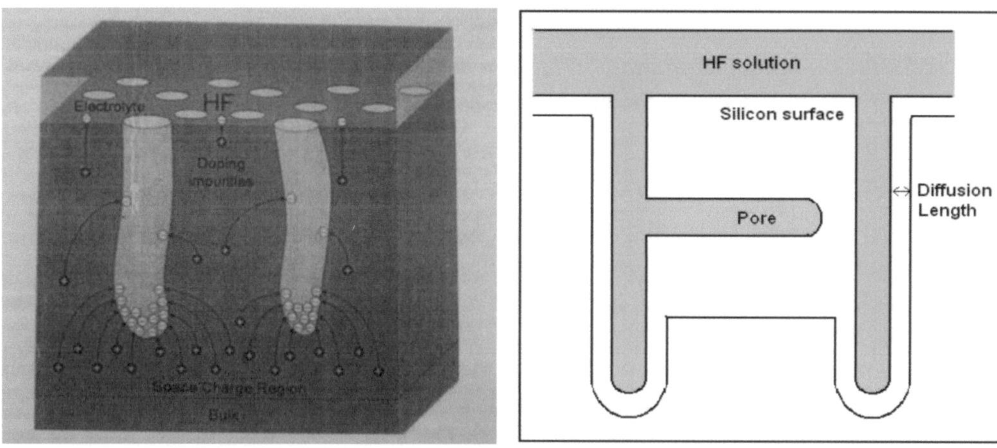

Fig. 5. The schematic of the (a) Beale model and (b) The diffusion – limited model (Samuel, BJ., 2010) [with the permission of Nova Science Publishers, USA].

The distribution of the electric field in the pore tips can be estimated by the model proposed by Zhang (Zhang, 1991) that also gives reasonable explanation of the localized dissolution of silicon. According to this model PS growth takes place through anodic oxidation and dissolution through direct etching of silicon in HF.

The diffusion-limited model in Fig. 5b proposed by Smith and Collins was the first computer simulated theoretical model. According to this theory the holes diffuse to the Si surface and react with the surface atoms (Samuel, BJ., 2010)..

Recently Lehamann proposed a model (Fig. 6) that considers the quantum confinement of charge carriers on the pore walls due to their small thickness and it may be responsible for the pore formation (Samuel, BJ., 2010).

The explanation of pore nucleation, pore structure and the importance of the formation parameters etc. during PS formation has been provided in the model proposed by Parkhutik and Shershulsky (Parkhutik & Shershulsky, 1992). According to this model the bottoms of the pores are covered by a virtual passive layer (VPL) that prevents a direct contact between the electrolyte and the substrate. When the electric field is applied the dissolution of VPL takes place and the pores are formed. The growth of VPL and its dissolution are exponentially related to the electric field strength and are dependent on the chemical reactivity of the material towards the electrolyte and other experimental parameters.

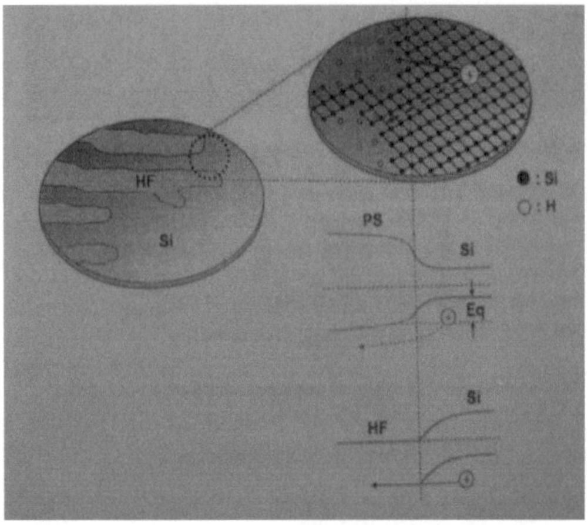

Fig. 6. The schematic of the proposed quantum model of Lehamann (Samuel, BJ., 2010) [with the permission of Nova Science Publishers, USA].

A new model on the mechanism of the pore growth suggests that the pore propagation into the crystal volume is a process that is controlled not only by the electrochemical factors, but also by mechanical stresses and hydrogen-related defects in Si. The PS growth is the alignment of the pore nucleation sites along the crystal defects of the Si surface. The defects serve as easy paths for the pore propagation. The dynamic stress during pore growth is another important issue. At the bottom of each pore, the dissolution reaction liberates the essential amount of hydrogen (Parkhutik, 1999). Outward movement of gas bubbles and products of Si dissolution together with inward propagation of fresh electrolyte are to produce essential hydrodynamic pressure inside the porous silicon layer. According to the theory of propulsion the essential tensile stresses are produced both in porous silicon and in Si substrates. Therefore, the micro cracks are formed in PS (Fig. 7) and that serve as easy path for further pore growth.

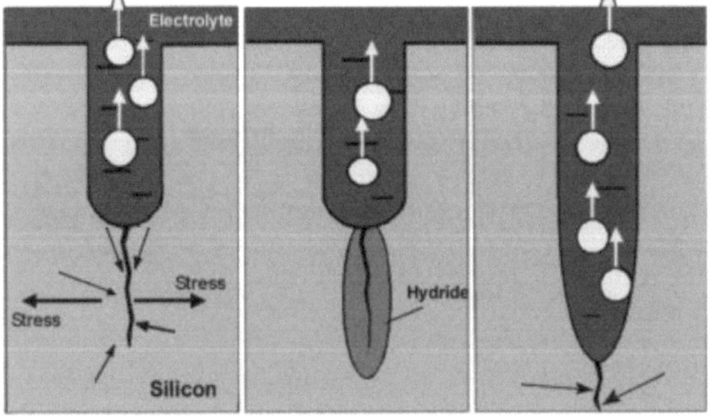

Fig. 7. Schematic of the pore growth according to the mechanical stress assisted mechanism of porous silicon (V. Parkhutik, 1999)

3. Properties of nanocrystalline porous silicon

The pore morphology of the PS layer characterized by the void fraction or porosity, the mean size of the pores, the pore size distribution, the interconnectivity of the pores, the passivation and mean size of the skeleton enclosing the pores depends on the formation parameters of PS, the additives in the electrolyte solution and the doping concentration of the silicon wafer. However, vertically grown uniform pores can also be developed in n-type c-Si using photolithography.

3.1 Structural properties of porous silicon

The main structural parameters of the porous silicon are the pore type, pore size, porosity and porous thickness. FESEM, TEM, AFM etc are commonly used to study the porous silicon structure. Fig. 8 shows the FESEM of nano and macro porous silicon as developed in the authors' laboratory.

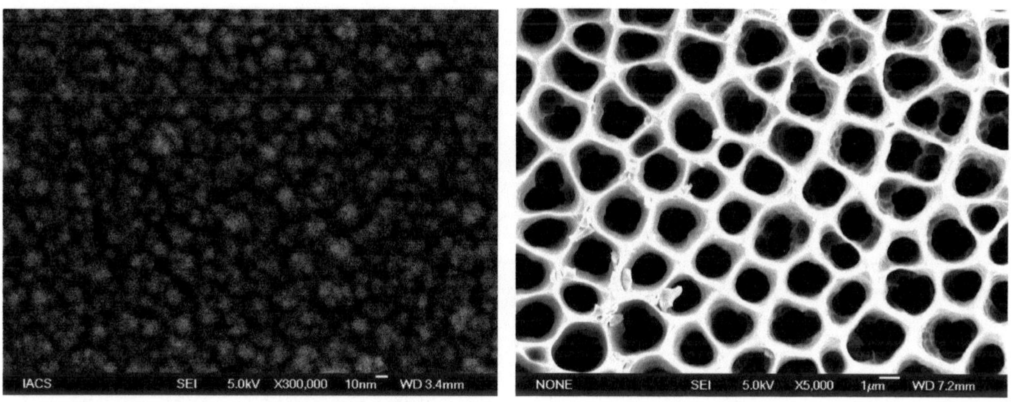

Fig. 8. FESEM of (a) nano porous and (b) macro porous silicon surfaces.

Most of the pores in porous silicon are closed at one end and they are inter connected to each other. Pores open at both ends i.e. the free standing porous silicon can also be realized in membrane structure. The cylindrical pore with 'branching' is very common in porous silicon. But for nano porous silicon a gradual decrease in pore size with depth in thick porous layer is observed. This arises due to the secondary dissolution since the top layer is exposed to the electrolyte longer than the bottom layer. Pore size is also a very important parameter for the PS based sensing devices. The nano, meso and macro porous silicon are defined in Table 1 according to the IUPAC guideline (Canham, 1997a),

Pore width	Pore type
< 2 nm	Micro porous/ nano porous
2 nm - 50 nm	Meso porous
> 50 nm	Macro porous

Table 1. IUPAC classification of pore size

The porosity of a sample is defined as the fraction of void in the porous structure. The porosity (P) and thickness (t) are determined by the relations:

$P = (m_1 - m_2) / (m_1 - m_3)$

$t = (m_1 - m_3) / (\rho \times A)$

Where m_1, the mass of the sample before porous silicon formation, m_2, the mass after formation and m_3, the mass of the sample after complete dissolution of the porous layer by 20% NaOH solution. ρ is the density of bulk silicon and A is the area of the porous silicon layer. The masses m_2 and m_3 are measured after drying the samples properly in the vacuum environment.

The porosity, thickness, pore diameter and microstructure of porous silicon depend on the formation parameters like composition of electrolyte, anodization current density, etching time, temperature, ambient humidity, wafer type and resistivity, illumination, and drying conditions (Table 2).

Increasing the values of	Porosity	Etching rate
HF concentration	Decreases	Decreases
Current density	Increases	Increases
Anodization time	Slightly increases	Slightly decreases
Wafer doping (p-type)	Decreases	Increases
Wafer doping (n-type)	Increases	Increases

Table 2. Effect of anodization conditions on the formation of porous silicon (Bisi et al. 2000).

3.2 Chemical properties of porous silicon

Porous silicon has a very high surface area. Therefore, it contains a high density of Si dangling bonds and impurities such as hydrogen and fluorine that are residuals from the electrolyte used during PS formation. After the formation, initially porous layer is covered by SiHx (x = 1, 2, 3) bond that remains on the surface even after rinsing and drying the samples. However, it disappears during annealing at 300-500°C (Vázsonyi et al. 1993).

The PS surface oxidizes spontaneously if it is kept in the ambient air for a few hours and the process continues until the whole surface is oxidized. Illumination and increased temperature enhance the oxidation reaction rate. A blue shift in the luminescence spectra occurs after oxidation of the surface (Hossain et al. 2000, Karacali et al. 2003). It also affects the electrical conductivity and the optical properties of the porous layer (Astrova et al. 1999).

The common impurity in freshly prepared porous silicon is fluorine. The concentration of fluorine depends on the type of electrolyte used. It is commonly found in the form of SiFx (x = 1, 2, 3). The fluoride concentration decreases with time (Petit et al. 1997), probably due to the replacement of SiF bonds by SiOH bonds through reactions with water vapor content of the ambient air.

3.3 Physical properties of porous silicon

3.3.1 Young's modulus

It is reported (Barla et al. in 1984) that PS should behave similar to the starting Si single crystal but the lattice mismatch occurs at the Si-PS interface due to its nanostructure and that may be the reason of different Young's modulus value of PS. The Young's modulus of porous silicon has been investigated by using X-ray diffraction, acoustic wave propagation and nanoindentation technique. The data measured from these different techniques are in good agreement and it is observed that for the PS layer the Young's modulus decreases with increasing porosity as shown in Table 3 (Bellet 1997). Also the micro hardness of the PS layer decreases with the increasing porosity (Duttagupta et al. 1997).

Sample	Method of investigation	Porosity (%)	Youn's modulus (GPa)
p^+	Acoustic technique	20	82.9
p^+	Acoustic technique	28	60
p^+	Acoustic technique	32	50.3
p^+	X-ray	34	41
p^+	X-ray	54	17
p^+	X-ray	74	11
p^+	Nanoindentation	80	5.5
p^+	Nanoindentation	90	0.87

Table 3. Young's modulus values of porous silicon deduced from different techniques. (Bellet, 1997)

3.3.2 Thermal conductivity

Nano porous silicon is a good thermal insulator. As reported in the literature the thermal conductivity of PS decreases with the porosity and in the case of nanoporous structures the thermal conductivity depends on the formation parameters. (Lang 1997).

Table 4 presents the experimental data for the thermal conductivity obtained from the literature. Electrically induced thermal wave were analyzed for thermal conductivity measurements by Lang et al. Photo thermal reflectance analysis and photoacoustic spectroscopy were also used by the other groups for the thermal conductivity measurements. However, the values given in the literature shows dramatic differences for different techniques.

Sample	Morphology	Porosity (%)	Thermal conductivity (as prepared) $Wm^{-1}K^{-1}$	Thermal conductivity (after 300^{0}C prepared) $Wm^{-1}K^{-1}$
p-type	Nano porous	40	1.2	1.3
p-type	Meso porous	45	80	2.7
n-type	Nano+ macro	53	1.75	1.85

Table 4. Experimental data of thermal conductivity of porous silicon (W Lang, 1997)

As reported in the literature, due to the small size of structures, the thermal conductivity of p-type nanoporous silicon is 1.2 $Wm^{-1}K^{-1}$, which is less than the thermal conductivity of the silicon oxide. This is because of the nanostructure that does not transport heat easily and also the pores prevent the heat transfer from one elemental crystallite to the next. But when the material is partially oxidized, the thermal conductivity increases slightly. It is because of the incorporation of the oxygen into the structure, due to which the porosity decreases and as a result heat transport increases.

But for meso porous silicon crystallites are connected by small bridges that are oxidized at low temperature and create heat barrier, which must be overcome during the heat transport process. As a result the thermal conductivity decreases.

3.3.3 Electrical resistivity

The electrical resistivity of the PS layer is very high due to its nano structure. Four probe method is very commonly used for the conventional determination of the resistivity of porous silicon. But PS is very sensitive to the ambient atmosphere. So it is difficult to determine the exact values for the PS resistivity (Ben-Chorin, 1997). The resistivity of porous silicon changes with the porosity of the sample as it depends on the quantum confinement, mobility and drift of the carriers, changes in the band structure, temperature and on the medium inside the pores (Parkhutik, 1999).

To a large extent the low porosity material behaves like intrinsic bulk Si. The barrier heights are similar to the bulk silicon which suggests that the Fermi level lies near the mid-gap position. But the drift mobility of the majority carriers is order of magnitude lower than the bulk silicon. (Ben-Chorin 1997). Thermally activated carriers move in extended states, but multiple trapping in the surface states disturbs their motion. Therefore, the drift mobility is very low. And as the traps are energetically distributed throughout the gap, hopping between these states at high frequencies gives rise to a temperature independent AC conductivity

A strong increase of resistivity (10^{10}-10^{12} Ωcm) of highly porous silicon was observed for high porosity samples due to the quantum confinement effect that results in a modulation of the effective bandgap. For these samples the AC conductivity is almost temperature independent at high frequencies (Ben-Chorin 1997). A non linear dependence of the conductivity with the applied voltage was observed for these samples. It is suggested that two parallel processes are involved in current conduction (Ben-Chorin 1997). The first one is the hopping transport of the carrier and the second one is the conduction due to the thermo-ionic emission .

Both the low and high porosity samples are very sensitive to the environment. The exposure of the porous layer to the vapour of the polar solvents affects the conductivity. The conductivity depends exponentially on the partial pressure of the solvent and on its dipole moment.

3.3.4 Optical and optoelectronic properties

3.3.4.1 Refractive index

The effective refractive index of PS depends on the porosity, the impurities on the surface of porous silicon and the refractive index of the materials filling the pores. The refractive index of an oxidized PS layer is lower than the non-oxidized layer having the same porosity.

3.3.4.2 Absorption co-efficients

The absorption coefficient of the PS layer indicates the correlation between the bandgap and the porosity that is mainly due to the quantum confinement effect. (Bisi et al. 2000, Behren & Fauchet 1997). The absorption coefficient of porous silicon can be measured by optical transmission, photoluminescence excitation and photo thermal deflection spectroscopy. For a free standing PS layer the optical transmission method is very sensitive. In this case Fabry-Perot interference fringes are used to determine the refractive index. The absorption coefficient is then calculated including both the multiple interference effect and the sample reflection. The absolute value of absorption coefficient can be obtained by the photoluminescence excitation method but not by photo thermal deflection spectroscopy (PDS). In fact, PDS is effective for estimating very low values of the absorption coefficient below the crystalline Si band gap.

As expected for an indirect band gap material, a linear behaviour of the transmission spectra of crystalline Si and p^+ PS were observed. But a non linear dependence of the absorption coefficient on photon energy was obtained for p-type porous silicon with an absorption blue shift with respect to crystalline Silicon as reported (Bisi et al. 2000). It is determined from the Raman experiment that the blue shift is correlated with the smaller average crystallite diameter and the quantum confinement is considered as the key factor for the absorption in porous silicon.

3.3.4.3 Photoluminescence

Porous silicon shows photoluminescence at wavelengths ranging from the ultraviolet (Lin et al., 1996) to the infrared due to radiative recombination of the carriers confined in nanoclusters of Si embedded into the walls of PS layer. However, the physical mechanisms of photoluminescence from porous silicon are still the subjects of controversy. The most available reports on PL bands of porous silicon are available in different spectral regions e.g. (i) ultraviolet band, (ii) Visible band and iii) Infrared PL band. The following are the characteristics of different PL bands of porous silicon.

i. Ultraviolet PL band (<500 nm):

a. UV PL band in room temperature occurs mainly due to the oxidized porous silicon
b. PL shows a red shift in the peak intensity with time.
c. PL get quenched when exposed to methanol.
d. PL intensity and peak intensity are stable under illumination.

ii. Visible PL band (500nm-800nm):

a. It requires silicon crystallites to get visible PL from porous silicon
b. Peak wavelength decreases as crystal size decreases.
c. PL intensity increases and a blue shift occurs with time.
d. Blue shift in peak wavelength occurs with surface passivation by hydrogen or by oxidation.

iii. Infrared PL band (800nm-2 μm):

a. The radiative electron capture at positively charged dangling bonds on the surface of the silicon branches is responsible for the infrared PL band.
b. Sometimes this band is referred as the tunable IR band as the peak energy of this IR band is shifted as a function of porosity.
c. The relative intensity of the IR PL band increases with time.
d. The PL intensity is totally quenched at higher temperatures of around 450°C.

The following models of the photoluminescence from PS have been proposed following the classification by Canham (Cullis & Canham, 1991),

a. Quantum confinement:

Quantum confinement within the nanometer size silicon branches of the porous silicon is responsible for direct band-to-band recombination and thus the origin of PL (Streetman, 1990). With decreasing crystal size a blue shift is observed (Canham, 1990a) because the reducing crystal size causes a further widening of the band gap. The band-gap energy shift of silicon with temperature followed by the thermal broadening can explain the observed temperature dependence of the PL (Lockwood et al. 1992).

b. Nano-crystal surface states:

Due to quantum confinement effect (Koch, 1993) absorption of the carriers occurs in the silicon crystallites and as a result the recombination centers are formed due to an adjustment of the bond lengths and bond angles by silicon atoms at the surface of the crystallite. This adjustment leads to the formation of a number of traps or so called surface states within the band gap. Probably due to the confinement of the carriers in these states radiative recombine occurs by a tunneling mechanism and gives rise to PL.

c. Specific defects and spurious molecules:

The surface hydrides on the branches of the porous silicon (Tsai, 1991) are possibly responsible for the 'red' band PL. The tunable visible PL in PS may be attributed to SiHx groups (Wolford et al., 1983). This may be further confirmed from the fact that on heat treatment (Prokes et al , 1992) hydrogen is desorbed from the porous silicon and thus the quenching of the PL with temperature (Cullis et al, 1997) due to presence of non-radiative dangling bonds. Further there is a report that surface molecules, specifically siloxene, a by-product of the anodization process and remaining on the surface of the porous silicon (Hamilton , 1995), may be responsible for 'red' PL band (Fuchs et al., 1992). After annealing at 400°C siloxene has an attractive luminescent property similar to porous silicon (Stathis & Kastner , 1987) .

d. Structurally disordered phases:

A significant amount of amorphous silicon was reported to contain within porous silicon (Perez et al., 1992). It is known that hydrogenated amorphous silicon (a-Si:H) can produce visible PL (Vasquez, 1985) and so it was suggested that a-Si:H might be responsible for 'red' PL band (Vasquez et al. 1992) The observed tunability of the PL was further suggested as due to introduction of hydrogen into the amorphous silicon in varying amounts (Cullis & Canham,1991).

3.3.4.4 Electro luminescence

The third form of the luminescence is the electro luminescence, which is tunable by an electric field applied on the sample (Halimaoui et al. 1991). PS produces different luminescent properties. 'The origin of luminescent phenomena is the quantum confinement in the nanostructure Si wires of porous silicon.' With increasing porosity crystallites dimension decreases and the band gap increases thereby shifting the peaks in the luminescence spectra (Read et al. 1992). As reported in the literature, the EL spectra show a blue shift with an increasing bias. However, the mechanisms of the electro luminescence in PS are not studied in detail. In fact, the absence of reliable electrical contact to PS due to the rough surface of silicon was so far the major difficulty in doing this study.

3.3.4.5 Cathodoluminescence

During the structural SEM or TEM study on PS Cathodoluminescence could be observed. The peak of Cathodoluminescence (CL) is extremely weak and unstable. But when SiO_2 is present stable and brighter CL could be achieved (Williams 1997). Piillai et al. (Piillai et al., 1992) suggested that during the SEM or TEM study the KeV electron beam may excite several high energy defect states of an unspecified molecules attached to the PS surface and their optical de-excitation gives the CL emission.

4. Surface stabilization of NPS by different methods

In spite of all other advantages, the major barrier preventing commercial applications of PS is the instability of its native interface with metastable $Si-H_x$ termination. The metastable hydro-silicon undergoes spontaneous oxidation in ambient atmosphere and results in the degradation of surface structures. This creates problem for taking good electrical contacts on PS, important for its commercial applications (Fig. 9). Therefore, the passivation of PS surface is necessary to make it possible to fabricate the stable devices based on porous silicon.

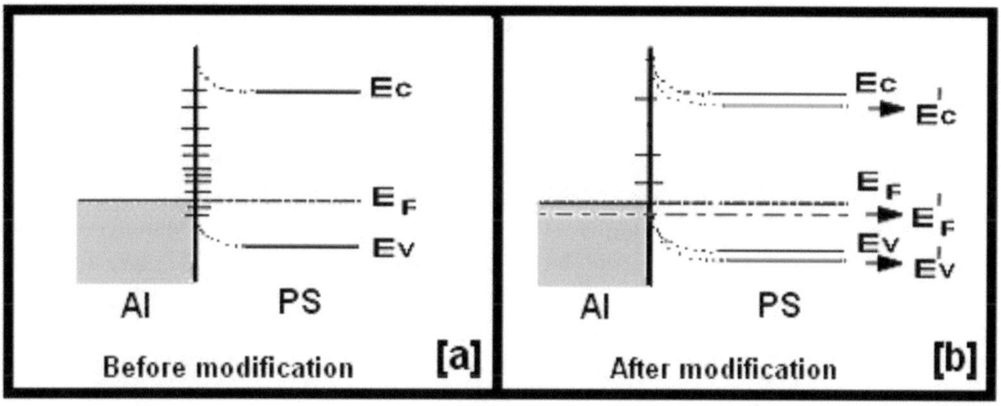

Fig. 9. Band diagram of Al/PS junction (for Ohmic contact to p type semiconductor Φm>Φs) (a) before surface modification (Fermi level is pinned by the PS surface states) and (b) after surface modification (Fermi level become unpinned and it can move under the applied bias).

In 1965, Beckmann (Beckmann , 1965) observed that during a long storage in ambient air the PS film gets oxidized. This chemical conversion is slow and basically similar to the ageing of Si wafers i.e. a native oxide layer forms on the surface of pores. Due to the ageing effect, the structural (Astrova et al. 2002), electrical and optical properties (Karacali et al., 2003) of PS showed continuous change with the time. The growth of the native oxide was completed approximately after one year (Petrova, et al., 2000). In order to avoid the transient period of ageing, substitution of surface hydrogen by another chemical species appeared desirable. Chemical, anodic, dry and wet thermal oxidation processes were investigated and presented by Rossi et al.in 2001. Anodic oxidation showed some improvement against ageing in air but rapid thermal oxidation appeared to give more optimum results, in providing PS samples with good electronic surface passivation and improved stability against ageing for optical measurements. For stabilization of the porous silicon surface, derivatisation by organic groups and polymer offers an alternative possibility to oxidation. Nitradation and halogenetion were also studied to stabilize the PS surface. Metals like Cu, Ag, In etc. were also used to modify the porous silicon surface to stabilize its photoluminescence properties. The metal electrode contact to the unmodified PS surface for the electrical measurements may exhibit an unstable and rectifying behaviour, because PS surface states that arise due to its nano structure can act as the recombination centers. Therefore, the passivation of defect states in porous silicon is necessary to get a stable and reliable electrical contact to PS. There is an existing challenge for the fabrication of reproducible low resistance Ohmic contacts on porous silicon and it is indispensable for porous silicon based electronic devices. The PS layer contains huge number of volume traps and interface states, resulting in the Fermi-level pinning at the PS–Si interface. This may block the electrical response of the PS–Si structure. In a recent work the problem of contact on porous silicon was bypassed by taking contacts from the back of the silicon substrate for capacitive sensors. A few reports are available on the formation of electrical contact to porous silicon. Zimin et al. (Zimin et al., 1995) reported a lateral Al ohmic contact to n type PS having a contact resistivity of the order of 10^{-3} to 10^{-2} Ω cm^2 for low porosity and low resistivity samples but they observed a rectifying behaviour for high resistive PS, both n and p type. Martin Palma and co-workers (Martin-Palma et al., 1999) reported the same for the Al-PS-Si-Al sandwich structure, which showed rectifying behavior even after prolonged exposure to the atmosphere. There are also reports on metallic contacts to PS using Au, In, Au-In, In-Sn, Al etc and all the contacts showed Schottky behavior (Angelescu & Kleps, 1998). Electroless Ni deposition was studied for getting metal contact to PS and an Ohmic behaviour was observed only for low bias voltages. H. A. Andersson et al. (Andersson et al., 2008) performed an experiment to examine the morphology and the properties of Ni obtained by electro less deposition on p-type PS and subsequent formation of Ni silicide by heat treatment. They showed that both rectifying and Ohmic contacts could be formed between electro less deposited Ni and PS depending upon the heat treatment conditions. Reports (Jeske et al., 1995) are also available on the use of electro less deposited Au, Cu and Ni contact to porous silicon. Although Au and Cu showed some positive results Ni was proved to be ineffective for Ohmic contact.

The stabilization of PS surface by noble metals for reliable metal contact has been recently reported by us (Kanungo et al. 2009a). We used the noble metal (Pd, Pt, and Ru) ions to stabilize the surface of porous silicon. Porous silicon surfaces were modified at room temperature by electroless method using acidic aqueous solutions of $PdCl_2$, $RuCl_3$ and K_2PtCl_6 to passivate the PS surface by the noble metals like Palladium (Pd), Ruthenium (Ru) and Platinum (Pt) respectively. To remove the native oxide layer from the PS surface, prior to modification the samples were dipped into 10% HF solution for 10 sec and were then immediately dipped into the different chloride solutions. Subsequently the samples were rinsed gently by DI water and dried followed by annealing in air at 110^0C inside an electric oven for 10 min to evaporate the residual solvent present in the samples. The flow chart of the surface modification steps is given in Fig. 10.

To optimize the solution strength and modification time a number of experiments was carried out with different concentrations (0.1 M, 0.05 M, 0.01 M, 0.005 M, 0.001 M) of the solution and different modification time (5 sec – 180 sec). The stability and reproducibility were checked by taking I-V measurements.

The chemical modification steps are cited as follows:

$$[M]Cl_Z \longrightarrow [M]^{Z+} + ZCl^-$$

Where, [M] = Pd, Ru and Pt.

$$[M]^{Z+} \longrightarrow [M]\ (\text{Island}) + Zh^+$$

$$Si + 2H_2O + 2h^+ \longrightarrow Si(OH)_2 + 2H^+$$

$$Si(OH)_2 \longrightarrow SiO_2 + 2H^+ + 2e^-$$

$$2H^+ + 2e^- \longrightarrow H_2$$

$$2Cl^- + 2H^+ \longrightarrow 2HCl$$

$[M]^{Z+}$ formed by decomposition of $[M]Cl_Z$ in an aqueous acidic solution is reduced to [M] metal islands by a chemical reduction process ((Jeske et al., 1995; Porter et al. 2002) and h^+ are released. Subsequently PS surface gets oxidized to SiO_2 by h^+. The Pd modified samples showed the better consistency & stability compared to Ru and Pt modified samples as determined by I-V measurements.

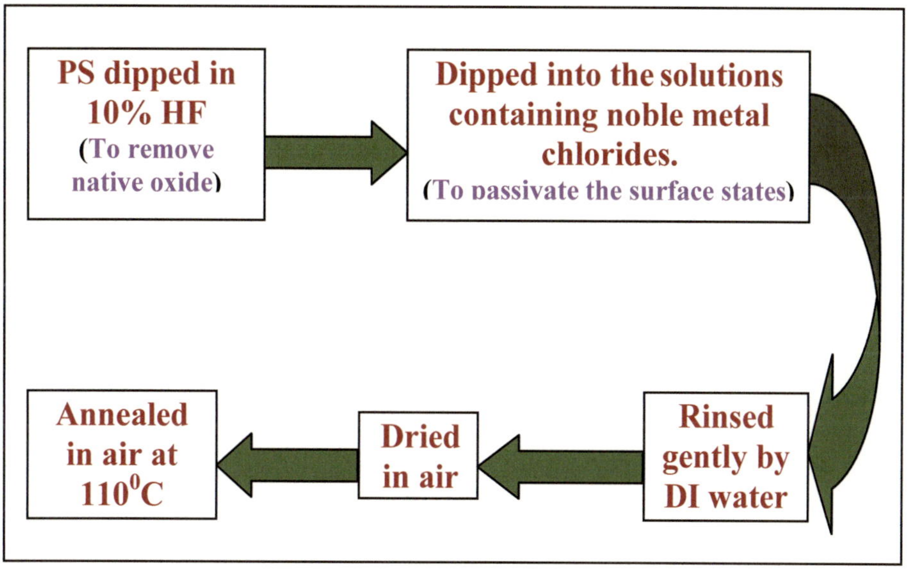

Fig. 10. Flow chart for porous silicon surface modification

This study indicates that the noble metal dispersed on the PS surface helps in two ways. Firstly, by the metal dispersion Si surface gets oxidized and a thin layer of SiO_2 is formed (Steiner et al., 1994) and it was verified from EDAX and XPS results (Kanungo et al. 2010a). Secondly, the oxidized PS surface contains much less defect states due to the passivation effect by the oxide layer. Therefore, it helps in favourable current conduction and also improves the stability of PS to a large extent. From the consideration of work function Al should give Ohmic contact to PS. But the large density of surface states on unmodified PS creates barrier against the current flow and thus a rectifying behaviour is displayed by it as shown in Fig. 9. But after modifications the PS surface states are reduced to a large extent by surface oxidation and Al contact becomes Ohmic in nature for the lateral structure (Fig. 11). On the otherhand, the sandwich structure (Fig. 12) exhibits a rectifying behaviour due to the presence of PS/Si heterojunction but the rectification is much improved after modification due to the effective passivation of the recombination states. However, the possible apprehension that Pd, Ru and Pt may create a continuous metal layer on the PS surface is ruled out by digital X-ray mapping and EDAX line scan analysis (Kanungo et al. 2010a) that show distinctly a discontinuous dispersion of noble metals.

Specific contact resistance of Al to Pd modified PS was measured by the transmission line model (TLM) method and the value obtained was of the order ~ 10^{-1} ohm-cm^2 (Kanungo et al., 2009a).

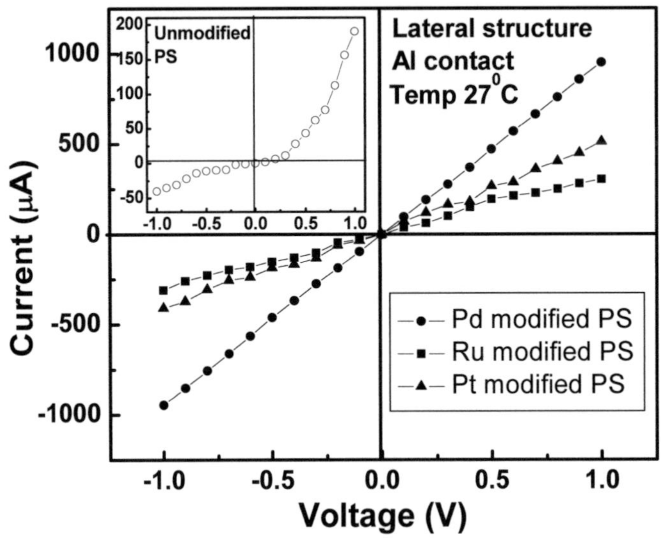

Fig. 11. I-V characteristics of unmodified (inset) and Pd, Ru and Pt modified PS for lateral structure using Al contact. [Unit is same for both the graphs].

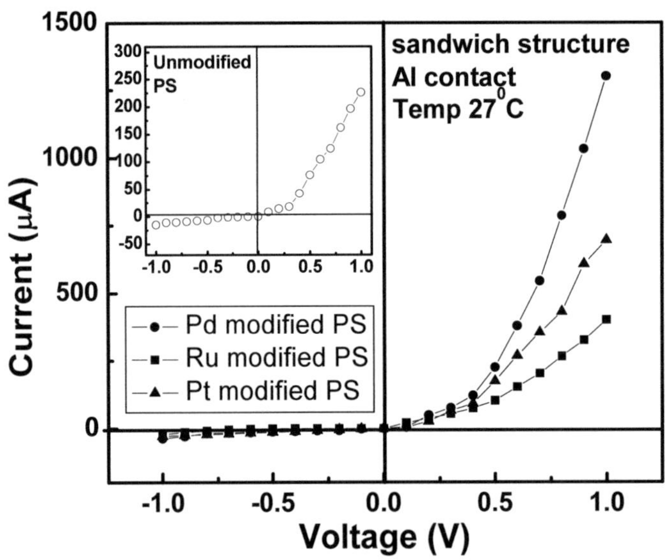

Fig. 12. I-V characteristics of unmodified (inset) and Pd, Ru and Pt modified PS for sandwich structure using Al contact. [Unit is same for both the graphs].

5. Applications of porous silicon

The PS structure was discovered by Arthur Uhlir at Bell Lab, USA in 1956 (Ulhir, 1956) followed by Turner (Turner,1958) and Memming and Schwandt (Memming & Schwandt, 1966). In 1971, Watanabe and Sakai (Watanabe & Sakai, 1971) demonstrated the first application of PS for device isolation in integrated circuits. In 1986 Takai & Itoh (Takai & Itoh, 1986) introduced the silicon-on-insulator (SOI) in integrated circuits technology. The silicon on sapphire technology (SOS), and silicidation of porous silicon were reported by Ito et al. in 1989 (Ito et al., 1989). In 1984 Pickering et al. (Pickering et al., 1984) have observed the low temperature photoluminescence in PS. But the discovery of room-temperature photo- and electroluminescence by L. Canham in 1990 boosted the research on porous silicon because of the huge potential in silicon-based integrated sensing and photonic devices(Canham, 1990, 1997b, 1997c,1997d; Canham et al.,1991).
Also there is a number of publications with the physical and chemical properties, and different applications of PS such as solar cells (Menna, 1997; Aroutiounian , 2004) integrated light-emitting devices on silicon chips (Bondarenko et al., 1997a, 1997b; Nakajima et al. 2004), selective chemical, gas and humidity sensors (Sailor, 1997), pressure sensors (Pramanik et al., 2005), bio-sensors based on functionalized PS substrates (Dzhafarov et al. 2004; DeLouise, 2004), electrolyte-insulator-semiconductor capacitor (EISCAP) based on silicon and PS for detecting different organic materials.
Another important characteristic of PS is the ability to tailor the material morphology for the desired applications. Fine control can be achieved by specifying the etching parameters so that micropores with different pore sizes can be tailored according to the condition required. Sometimes, it is also important to produce pores of graded size on the same substrate to achieve multiple targets sensing capability on the same chip. Graded morphology is the key to multiple gas/vapor sensing capability within the same array and the possibility of an 'e-nose'. A compact and sensitive PS-based electronic nose could bring greater reliability and repeatability into the professional sensing of food and wine (Marsh, 2002).

5.1 Optoelectronic applications

5.1.1 LEDs

The first publication on the formation of the electro-luminescent devices based on porous silicon was made by Richter et al. (Richter et al., 1991) and later by Koyama and Koshida (Koshida & Koyama, 1992). It is no doubt that silicon has a great importance in micro electronics integration technology. But Si is an indirect band gap semiconductor with very poor emission property. Although the radiative recombination exists in Si the quantum efficiency at room temperature is very low. In contrast, the discovery of the visible photoluminescence in porous silicon made it an interesting material for optoelectronic application. Steiner et al. (Steiner et al. 1993) introduced LEDs with porous p-n contact with improved efficiency. However these samples are suffering from poor stability problem while working in the open environment. But the environmental stability of LEDs is much improved when the pulsed mode of operation is used. Lazarouk et al. (Lazarouk et al., 1994) reported the formation of very stable Al/PS LEDs. The diodes emit white light under the reverse bias, possibly due to the excitation of surface plasmons in oxide near the edges of Al metallization stripes where the maximum local electric field is concentrated. In spite of the fact that there is an essential progress in the fabrication of porous silicon LEDs further work is necessary to improve their stability, quantum efficiency and speed.

5.1.2 Photodetectors

Porous silicon photoconductors are commonly fabricated by depositing aluminium film on top of oxidized porous silicon structure. The passivation of the surface by oxidation could improve the external quantum efficiency of a porous silicon photodiode (Tsai, 1993). Porous silicon also can be used as anti-reflecting coating for p-n photocells because of its reduced reflectivity of incident light from its surface. As suggested by Krueger et al. (Krueger et al. 1997) multilayer PS structures could be used as selectively absorptive filters to modify the spectral sensitivity of photo detectors.

5.1.3 Porous silicon solar cell

Solar cell converts solar radiation into electricity by photovoltaic effect. High optical absorption and low carrier recombination are two necessary conditions for efficient power conversion in a solar cell.

For the last two decades there is a steady growth of 15-30 % in the photovoltaic market and it is expected to increase more. However, the cost of photovoltaic energy source is still very high than the conventional energy sources. This cost can be reduced by high volume of production, reducing the material cost and improving the efficiencies.

The 90% of the solar cells are fabricated from crystalline silicon. But silicon is an indirect band gap semiconductor and its optical adsorption coefficient is also relatively low. So in thin film silicon light absorption will be much less. Therefore, the cost of silicon wafer becomes the 40% of its final module. Recently there is a huge shortage of silicon wafer in the photovoltaic market.

The nanocrystalline porous silicon thin film, on the other hand, has relatively high optical absorption coefficient. Therefore, using a thin film of porous silicon, the efficiency of a solar cell can be improved to a large extant as well as the shortage of silicon can also be avoided. Actually, a porous silicon coating reduces the reflectance and improves the light absorbance in the useful photovoltaic spectral range. Indeed, the effectiveness of the porous layer as the anti reflecting coating of the solar cell is already confirmed (Strehle et al. 1997). In tandem solar cell structure the use of micro porous silicon as an active element on a non degenerate p-type crystalline silicon has been reported. Porous silicon photovoltaic device was also fabricated by anodization of crystalline silicon as reported in the literature.

5.2 Biological applications of porous silicon

The biological incompatibility of Si material with living organisms is well known. But unlike bulk Si porous silicon becomes biologically active and restorable as demonstrated by Canham. Further research has shown that the porous Si may become an important biomaterial and reliable interface between the Si integrated circuit and the biological ambient. Bayliss et al. (Bayliss et al., 1997) reported that the nanocrystalline Si could be used as growth supports for cell cultures. Application of porous silicon in biological sensing is based on the immobilization of specific bio molecules at its surface that register changes in electrical and optical properties of PS as a result of their interaction with the biological complexes to be detected. Significant quenching of photoluminescence from PS as a result of antigen-antibody interaction is reported by Starodub (Starodub et al. 1996). Lin and co-workers (Lin et al., 1996, 1997) worked on the displacement of interference fringes in optical absorption or in luminescence spectra of porous silicon thin films for the absorption of chemical or biological molecules at the pore walls.

Silicon nanoparticles and porous silicon have several valuable biological properties e.g. biocompatibility, biodegradation capability, and high penetration power. Since the silicon nanoparticles can increase the permeability of cellular membranes, high doses of the pharmaceutical substances can be bound with the particles. Silicon nanoparticles can softly remove bacterial dental plaque without affecting the enamel and therefore can be used as a toothpaste additive. The silicon nanoparticles/porous silicon is considered as a fine preservative in food products because they can inhibit the development of pathogenic microorganisms. So, the introduction of silicon nanoparticles into the organism can be a novel idea. In fact, it is a subject of active research in many laboratories of the world. The properties of polycrystalline or porous silicon nanoparticles are also being explored in the regime of cancer research. The silicon nanoparticles bound with medicinal substances are added into the nutrient medium where the cells of human larynx tumour or mouse fibroblast cells are being cultivated. Russian researchers found that the nanoparticles in concentrations over three mg per ml hinder the growth of both normal and tumor cells. Porous silicon particles are more effective than polycrystalline silicon for this purpose.

5.3 Porous silicon pressure sensor

Pressure sensors are widely used in automobiles, robotics, space, and process industries, in addition to VLSI and biomedical applications. However, the commonly used metal strain gauges and capacitive manometers suffer from low sensitivity and bulky size and so they are not very suitable for sophisticated pressure sensing applications.

To improve the pressure sensor technology silicon can be used as a high precision, high strength, and high reliability mechanical material (Peterson, 1982). It can be micromachined to form various mechanical microstructures like diaphragm, cantilevers, nozzles and grooves and can also be readily integrated with the signal-processing unit. However such sensors suffer from low sensitivity, which makes them incompetent for low-pressure applications. The use of nanocrsytalline silicon as an active material improves the performance of pressure sensors without considerably increasing the cost and complexity of fabrication. It has been observed that nanocrystalline porous silicon has the potential for improving the performance of the pressure sensors due to its increased piezoresistive coefficient of about 54.8% (Toriyama et al., 2002). Tailoring the dimensions of the nanocrystallites, which depends on the formation parameters, may modify the performance of the nanocrystalline porous silicon pressure sensor.

Reports are available (Pramanik & Saha, 2006a, 2006b) on the response of the porous silicon membrane under pressure for possible improvement of piezoresistive coefficient due to quantum confinement in the nanocrystalline structure of porous silicon. In fact, there is a change in the band structure of porous silicon induced by pressure and it affects the piezoresistive coefficient.

5.4 Porous silicon gas sensor

Porous silicon is one of the most promising materials in the field of gas sensor technology. It can be used in gas sensors either as an active or a passive layer for gas sensing.

For passive application, the PS layer acts as the heat isolator or sacrificial layer in the hotplate technology (Fürjes et al., 2004). The low power consumption and the short thermal transient times are the remarkable advantages of using porous silicon for thermally isolated micro hotplates. Additionally, due to its very high surface to volume ratio the sacrificial PS layer can be quickly and perfectly etched away.

For the active application of PS, it acts as the gas sensor layer with electrical, optical or thermal readout possibility. A simple cheap gas detectors that are super-sensitive, could be based on resistive PS elements (Pancheri et al., 2003). Researchers at the University of Brescia, have patented a technique using a porous silicon membrane on alumina substrate which could sense NO_2 as low as 100 parts per billion (ppb) at room temperature using low power and with a minimum interference from the contaminant organic vapors (Baratto et al., 2000).

Recently, a conductometric porous silicon gas sensor was used for rapid and reversible detection of gas analyte like CO (<5 ppm), NO_x (<1 ppm), SO_2 (<1 ppm) and NH_3 (500 ppb) (Lewis et al. 2005). A hybrid nano porous layer on macro porous silicon was used for the sensor study. To enhance the sensor performance a thin layer of gold or tin oxide nanostructured framework was developed on the PS surface by the electroless deposition of Au or Sn.

There are few reports on PS–Si structures as resistor type ethanol (Han et al., 2001), NO_2, organic vapour (Salgado et al., 2006) and H_2 (Rahimi & Iraji zad, 2006) sensors. Seals and Gole (Seals et al., 2002) reported a PS based resistor type gas sensor for the detection of HCl, NH_3, and NO at 10 ppm level. But the full explanations of the sensor operations and theoretical interpretations were not discussed. However, Green et al. (Green & Kathirgamanathan, 2002) tried to explain the gas sensitivity of the PS-Si layered structure with parallel electrodes. Luongo and co-workers (Luongo et al., 2005) reported on a Pd doped PS, as a resistor type sensor for H_2 detection at room temperature. The basis of their sensor operation is the volume change of the Pd particles dispersed on porous silicon layer leading to a reduction in the layer impedance due to closer contact with each other. Further, a study on the same structure was carried out by (Sekhar et al., 2007) to find out the influence of anodization current density and etching time during PS formation on the response time and stability of Pd doped sensors. They also investigated the role of catalyst thickness on the sensor response.

Barillaro et al (Barillaro et al., 2005) reported a field-effect transistor like structure for gas sensing. As the floating gate controls the resistor value by modulating the interface potential barrier the PS resistance is not involved in the measurement, thus avoiding the contact problems.

There are some contact free methods (vibrating capacitor and surface photo voltage (SPV) measurements) to measure the surface and interface potentials and potential barriers. The vibrating capacitor has long been used for investigating semiconductor gas sensor surfaces (Mizsei , 2005; Souteyrand et al. 1995; Nicolas et al. 1997). This simple and exact method is sensitive to the contact potential difference (CPD) between a vibrating reference electrode and the surface to be investigated. An attempt was made to use PS–Si interface as a sensor structure together with vibrating capacitor readout (Polishchuk et al., 1998). An amorphous silicon on silicon structure, similar to the PS–Si system was reported for alcohol sensor (Sueva et al., 2002). Zhang et al. (Zhang et al., 2000) reported NO_X sensor with a Pd-Pt/PS/Al (MIS) structure with higher response than Pd-Pt/Si/Al configuration but with longer response time. V.G. Litovchenko and co workers (Litovchenko et al., 2004) studied the mechanism of hydrogen, oxygen and humidity sensing using Cu-Pd/PS/Si structures. They coated polymer film on the PS surface and studied the chemisorptions of the gas molecules. An attempt was also made to theoretically interpret the mechanism using an energy band diagram of Schottky like structure. There is a report (Kwok et al., 1999) on PS based Schottky diode sensor with Al electrode to sense the volatile organic compounds. L pedrero et al. (Pedrero et al., 2004) reported on PdO/PS structure for sensing ammonia and other reducing gases.

As already mentioned instability is the most important problem related to porous silicon. As a chemically active high surface/volume ratio material it can be oxidized easily. Because of this chemical instability the physical properties of PS may change with time. However, the oxidized PS may be more stable and may contain less interface states responsible for the Fermi-level pinning. The oxygen adsorption may also result in an increase of conductance (Khoshnevis et al., 2006).

The authors reported porous silicon based room temperature hydrogen sensor (Kanungo et al., 2009b). The Metal-Insulator-Semiconductor (MIS) sensors were fabricated using both unmodified and surface modified porous silicon. Pd-Ag (26%) was chosen as the top noble metal electrode to fabricate the Pd-Ag/PS/Si/Al sensor structure. The junctions were characterized by I-V studies and were confirmed to behave as Schottky devices. They were subsequently used for hydrogen sensing at room temperature. At higher temperatures the junction deteriorated most probably due to the damage of PS surface. The modified sensors showed improvements over the unmodified samples in terms of response, time of response, time of recovery and stability as shown in the Table 5. The superior performance was observed for Pd modified sensors showing 84% response, 8 sec response time and 207 sec recovery time on exposure to 1% hydrogen in nitrogen carrier gas.

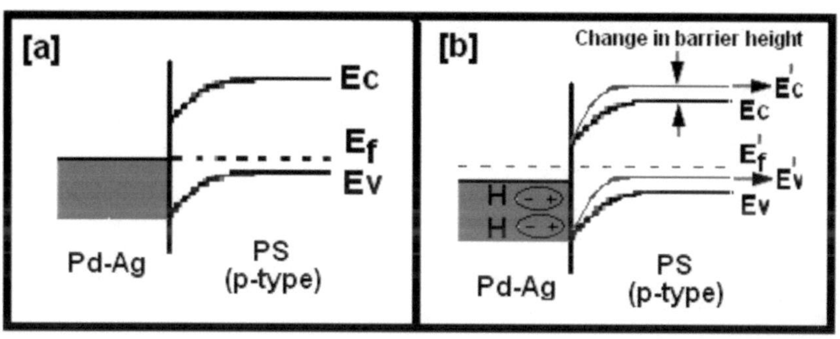

Fig. 13. Band diagram (not to scale) of Pd-Ag/PS junction (a) in absence and (b) in presence of hydrogen. A decrease in metal work function due to the formation of a dipole layer at the interface by the diffused hydrogen increases the barrier height at the metal / PS (p-type) junction.

The mechanism of hydrogen sensing was attributed to the dipole formation at the metal-PS interface in presence of the reducing gas that decreases the work function of the metal. As a result, the Schottky barrier height increases for the metal/p-type semiconductor junction and reduces the current through the noble metal/PS junctions (Fig. 13).

The modified samples improved the gas response behaviour after the formation of dispersed metal islands that passivate the PS surface and catalyze the dissociative adsorption of more hydrogen molecules. The defect free interface further helps in producing strong dipole during hydrogen sensing. Our study confirms Pd metal as the most effective modifier of PS surface for the superior performance of the Pd-Ag/PS/p-Si/Al hydrogen sensors.

The porosity of the PS was varied from 40% to 65% to study the effect of porosity on hydrogen sensor performance (Kanungo et al., 2010b). Both unmodified and Pd modified porous silicon sensors of different porosity were characterized for gas sensing. For unmodified sensors gas response increased with increasing porosity and finally got saturated. On the other hand, the Pd modified sensors showed improvement in gas response with increasing porosity up to 55% and then deteriorated (Table 6). The structural characteristics of the Pd modified sensors by EDAX line scan analysis revealed that the incorporation of metal islands increased with the increasing porosity. The gas response depends on the effective surface area of the dispersed Pd, which increases with increasing porosity. But further enhancement in metal deposition (above 55% porosity in this case), may reduce the effective surface area of the dispersed Pd (Yamazoe, 1991). As a result the decomposition of the hydrogen molecule to atomic hydrogen on the surface of the catalytic Pd islands during gas sensing also decreases. Possibly for this reason the gas response behaviour decreases with higher porosity PS, higher than 55% in the present investigation.

Metal used	Biasing voltage (V)	Max response (%)	Response time (Sec)	Recovery time (Sec)
Unmodified	0.6	58	120	1055
Pd	0.2	84	8	207
	0.6	76	11	219
Ru	0.5	64	35	689
Pt	0.6	61	106	1020

Table 5. Response, response time and recovery time for unmodified and modified porous silicon hydrogen sensors in 1% hydrogen in nitrogen as carrier gas at room temperature.

Porosity of the sample (%)	Unmodified PS Biasing voltage=0.6V Temperature 27^{0}C			Pd Modified PS Biasing voltage=0.2V Temperature 27^{0}C		
	Response (%)	Response Time (Sec)	Recovery Time (Sec)	Response (%)	Response Time (Sec)	Recovery Time (Sec)
40	42	169	1176	66	54	318
45	47	147	1132	72	23	276
50	53	132	1084	79	14	231
55	58	120	1055	84	8	207
60	59	114	1047	77	17	249
65	60	108	1033	68	29	286

Table 6. Response, response time and recovery time for unmodified and modified porous silicon hydrogen sensors with different porosities, in presence of 1% hydrogen in nitrogen and at 27^{0}C.

6. Factors related to the improved performance of porous silicon devices

The performance of photonic and gas sensor devices is related to the grain size, porosity and thickness of the porous thin film. The use of catalytic metal electrode and the modification of the sensor surface are other two important factors for an improved chemical sensor.

6.1 Grain size

For a nano crystalline material the fraction of atoms at the grain boundary increases due to decrease in crystal size. As a result the grain boundaries contain a high density of defects like vacancies and dangling bonds that can play an important role in the transport properties of electrons of nano materials. As the grain boundaries are metastable states they want to reduce their energy either by exchange of electrons or by sharing the electrons with other atoms. Hence the surface reactivity (or chemical reactivity) increases. Different models are proposed to explain the dependence of crystal size and the high sensitivity of nanocrystalline sensors and it was found that the sensitivity is proportional to 1/D, where D is the average grain size. Further, nanocrystalline structure can reduce the operating temperature of the sensors (Rothschild & Komen, 2004).

6.2 Porosity and thickness of the porous thin films

Dependence of PS solar cell parameters on its microstructure i.e. the initial porosities and thickness are reported in the literature (Menna & Tsuo, 1997).

For a porous silicon pressure sensor, the change in resistance on application of pressure has been reported to depend on the variation of porosities and thickness of the porous silicon layer (Pramanik & Saha, 2006a, 2006b).

In a compact sensing layer, gases cannot penetrate into the layers and the gas sensing reaction is confined to the surface. In case of the porous films, gases can access the entire volume of the sensing layer and therefore the gas sensing reactions can take place at the surface of the individual grain, at the grain boundaries and at the interface between grains and electrodes. Therefore, porous layer is more suitable for gas sensing as compared to compact ones (Basu et al., 2005; Xu et al., 1991; Tiemann, 2007; Sakai et al., 2001; Basu et al., 2008). The thickness of a thin film plays a great role in the sensor response. F. H. Babaei et al. (Babaei et al., 2003) proposed a model to establish a general mathematical relationship between the steady state sensitivity of the sensor and the thickness of the sensitive film used. It was shown that the sensitivity drops exponentially as the thickness of the sensitive film increases. On the other hand, some groups reported that for certain combinations of the structural parameters like porosity; cracks etc, the gas sensitivity of the sensors could increase with thickness. For porous silicon all these parameters can be controlled during its preparation.

6.3 Effect of noble metal catalyst as contact electrode

The performance of chemical sensor can be improved to a large extent by incorporation of noble catalytic metals on the porous layer. It increases the rate of chemical reactions. In fact, it does not change the free energy of the reactions but lowers the activation energy.

There are quite a few reports of the applications of nanoporous noble metal thin films as the electrode contact for gas sensing (Ding et al., 2006; Lundstrom et al., 2007). The nanoporous noble metal thin films have significant role on hydrogen sensing. The nano holes can provide much more surface area, which in turn helps in rapid adsorption/desorption processes and diffusion into the porous thin film interface.

Pure Pd is a good catalyst for sensing hydrogen, methane and other reducing gases (Armgarth & Nylander, 1982). But there are some drawbacks associated with the use of pure Pd metal due to blister formation because of the irreversible transition from the α phase of palladium to the β phase hydride at low H_2 and at 300 K (Wang & Feng, 2007; Hughes et al.,

1987). In addition, the response time more than 10 min for Pd-sensors is too slow to allow real-time monitoring of flowing gas streams. To overcome these problems Pd is alloyed to a second metal (Ag at 26 %) and is used for H_2 or hydrocarbon sensing. Pd-Ag alloy thin film has some special properties for use in gas sensors and they are reported as follows:
The rate of hydride formation is very low for Pd-Ag alloy compared to pure Pd.
Since the solubility of hydrogen is favorable up to 30 % of Ag in Pd Ag atom does not hinder the diffusion of hydrogen.
The OH formation barrier energy is higher in presence of Pd-Ag alloy.
The mechanical properties of polycrystalline Pd-Ag alloy are better than Pd.

6.4 Effect of noble metal dispersion on the surface of porous silicon

The sensitivity of a semiconductor sensor could be improved by surface modification through highly dispersed catalytic platinum group metals like Pd, Ru and Pt (Vaishampayan et al., 2008; Cabot et al., 2002). These additives act as activators for the surface reactions (Zhua et al., 2005; Rumyantseva et al., 2008).
F. Volkenstein (Volkenstein, 1960) provides an idea of how the adsorbate affects the overall band structure of the modified matrix. Additionally, the chemical nature of the modifier and its reactivity in acid-base or redox reactions may play an important role (Korotcenkov, 2005). The dispersed catalyst actually activates the spillover process. Therefore, the functional parameters such as sensitivity, response time, recovery time and selectivity improve significantly through surface modification by noble metals.
It was found from the literature that Pd modification is very effective to reduce the operating temperature and to achieve a high response of a gas sensor. Depending on the factors like grain size, porosity and the thickness of the thin film, porous silicon can appreciably be used as a gas sensor operating at low temperature (Mizsei, 2007).
Improved gas response behaviour of a palladium doped porous silicon based hydrogen sensor was reported by Polishchuk et al. (Polishchuk et al. 1998). C. Tsamis and co workers (Tsamis et al., 2002) reported on the catalytic oxidation of hydrogen to water on Pd doped porous silicon. K. Luongo and coworkers. (Luongo et al., 2005) reported an impedance based room temperature H_2 sensor using Pd doped nano porous silicon surface. P K Sekhar and co workers (Sekhar et al, 2007) tried to find out the influence of anodization current density and etching time during PS formation on the response time and stability of Pd doped sensors. They also investigated the role of catalyst thickness on the sensor response. Rahimi et al. (Rahimi et al., 2006) studied Pd growth on PS by electroless plating and the response to hydrogen for both lateral and sandwich structures using gold contact.
As already mentioned in section 5.4 the authors also worked on noble metal modification of porous silicon and development of room temperature hydrogen sensors. The modified sensors showed significant improvements over the unmodified ones in terms of sensor response, time of response, time of recovery and long term stability.

7. Summary and conclusion

In this chapter the preparation of nanocrystalline porous silicon (PS) by electrochemical anodization has been described and different models have been mentioned to improve the quality of PS film. The mechanism of porous silicon formation has also been cited. Structural, chemical, optical & electrical properties of porous silicon have been mentioned. Optical, optoelectronic, biological and chemical gas sensor applications of PS have been discussed. The merits and demerits of reported work on porous silicon have been critically discussed. The factors that make porous silicon a special material for some specific applications are illustrated. The common problem of instability of nanocrystalline porous silicon surface, related to the large density of surface states and recent approaches to stabilize the PS surface have been highlighted using the example of gas sensor applications. The mechanism of surface modification using noble metal ions has been clarified. The effect of porosity on the sensor parameters has also been explained with the specific example of hydrogen gas sensor studies.

In conclusion, the basic concepts of the importance of nanocrystalline silicon over crystalline silicon for recent applications in different areas of science & technology have been high lighted. The simple method of chemical surface modification using noble metal ions to stabilize porous silicon as demonstrated using hydrogen gas sensor devices needs special mention.

8. References

Anderson, R.C., Muller, R.S. & Tobias, C.W. (1993) Chemical Surface Modification of Porous Silicon, *J.Electrochem.Soc.* (USA) Vol. 140, pp. 1393.

Andersson, H.A., Thungstrom, G. & Nilsson, H. (2008) Electroless deposition and Silicidation of Ni contacts into p-type Porous Silicon. *J Porous Mater,* Vol. 15, pp. 335.

Andsager, D., Hilliard, J. & Nayfeh, M.H. (1994) Behavior of porous silicon emission spectra during quenching by immersion in metal ion solutions, *Appl. Phys. Lett.*Vol. 64, pp. 1141.

Angelescu, A. & Kleps, I. (1998) Metallic contacts on porous silicon layers. *IEEE Conf.* pp 447.

Archer, M. Christophersen, M. & Fauchet, P.M. (2005) Electrical porous silicon chemical sensor for detection of organic solvents, Sens. Actuators B Vol.106, pp. 347.

Archer, M. & Fauchet, P.M. (2003) Electrical Sensing of DNA hybridization in porous silicon layers. *Phys Stat Solidi A,* Vol.198, pp. 503.

Armgarth, M. & Nylander, C. (1982) Blister formation in Pd gate MIS hydrogen sensors, IEEE *Electron Devices. Lett.EDL* Vol.3, pp. 384.

Aroutiounian, V.M., Martirosyan, K. & Soukissian, P. (2004) Low reflectance of diamond-like carbon/porous silicon double layer antireflection coating for silicon solar cells. *J. Phys. D* 37, No.19, pp. L25.

Astrova, E.V., Ratnikov, V.V., Remenyuk, A.D. & Shul'pina I.L. (2002)Starins and crystal lattoce defects arising in macroporous silicon under oxidation. *Semiconductors* Vol.36, pp. 1033.

Astrova, EV., Voronkov, VB., Remenyuk, AD. & Shuman, VB. (1999) Variation of the parameters and composition of thin films of porous silicon as a result of oxidation: ellipsometric studies. *Semiconductors* Vol. 33, No.10 , pp. 1149-1155.

Babaei, FH. & Orvatinia, M., (2003) Analysis of thickness dependence of the sensitivity in thin film resistive gas sensors, *Sensors and Actuators, B,* Vol. 89, pp. 256.

Baratto, C., Sberveglieri, G., Comini, E., Faglia, G., Benussi, G., La Ferrara, V., Quercia, L., Di Francia, G., Guidi, V., Vincenzi, D., Boscarino, D. & Rigato, V. (2000) Gold-catalysed porous silicon for NOx sensing, Sens. Actuators B, Vol. 68, pp. 74.

Barillaro, G., Diligenti, A. , Marola, G. & Strambini, L.M. (2005) A silicon crystalline resistor with an adsorbing porous layer as gas sensor, Sens. Actuators, B, Vol. 105, pp. 278.

Barillaro, G., Nannini, A.& Pieri, F.. (2003) APSFET : a new, porous silicon based gas sensing devices, Sensors. Actuators, B, Vol.93, pp. 263.

Barla, K., Herino, R., Bomchil, G., Pfiser, JC. & Freund, A. (1984) Determination of lattice parameter and elastic porprties of porous silicon by X-ray diffraction. *J. Cryst. Growth* Vol. 68, No. 3, pp. 727-732.

Basu, P. K., Bhattachayya, P., Saha, N., Saha H. & Basu, S. (2008) The superior performance of the electrochemically grown ZnO thin films as methane sensor, *Sensors and Actuators, B,* Vol. 133, pp. 357.

Basu, S. & Hazra, S.K. ZnO (2005) p-n homojunctions for hydrogen gas sensors at elevated temperature, *Asian Journal of Physics*, Vol.14, pp. 65.

Bayliss, SC., Harris, PJ., Buckberry, LD.& Rousseau, C. (1997) *Mater Sci Lett.* Vol. 16, pp.737

Beale, M. I. J., Benjamin, J. D., Uren, M. J., Chew, N. G. & Cullis, A. G. (1986) The formation of porous silicon by chemical stain etches, J. Crys. Growth Vol. 75, pp. 408.

Beale, M.I.J., Chew, NG., Uren, MJ., Cullis, AG.& Benjamin, JD. (1985) *Appl Phys Lett*, Vol.46, pp. 86.

Beckmann, K.H. (1965) Investigation of the chemical properties of stain films on silicon by means of infrared spectroscopy. *Surface Science*, Vol.3, pp. 314.

Behren JV & Fauchet PM (1997) Absorption coefficient of porous silicon. In: *Properties of porous silicon.* Canham L (ed), pp. 229-233, INSPEC, London.

Bellet, D. (1997) Drying of porous silicon. In: *Properties of porous silicon.* Canham L (ed): pp. 38-43 and 127-131. INSPEC, London

Ben-Chroin, M. (1997) Resistivity of porous silicon. In*: Properties of porous silicon.* Canham L (ed) pp. 38-43 and 165-175, INSPEC, London.

Bisi O, Ossicini S & Pavesi L (2000) Porous silicon: a quantum sponge structure for silicon based optoelectronics. *Surface Science Rep.* Vol.38, pp.1-126.

Bondarenko, V.P. & Yakovtseva, V.A. (1997) Microelectronic applications of porous silicon. In: *Properties of porous silicon.* L. Canham (ed), pp. 343, INSPEC, London,

Bondarenko, V.P. & Yakovtseva V.A. (1997) Optoelectronic applications of porous silicon. In: *Properties of porous silicon.* L. Canham (ed), pp. 356, INSPEC, London,.

Bsiesy, A., Vial, J.C., Gaspard, F., Herino, R., Ligeon, M,. Muller, F., Romestain, R.,Wasiela, A., Halimaoui, A. & Bomchil, G., (1991) Photoluminescence of high porosity and of electrochemically oxidized porous silicon layers, Surf.Sci. (Netherlands) Vol. 254, pp. 195

Cabot, A., Vila,` A. & Morante, J.R. (2002) Analysis of the catalytic activity and electrical characteristics of different modified SnO2 layers for gas sensors, *Sensors and Actuators, B*, Vol.84, pp. 12.

Canham, L, Editor (1997a.) *Properties of porous silicon*, INSPEC - The Institution of Electrical Engineers ; United Kingdom.

Canham, L.T. (1997b) Biomedical applications of porous silicon. In: *Properties of porous silicon.* L. Canham (ed), pp. 371, INSPEC, London,.

Canham, L.T. (1997c) Storage of porous silicon. In: *Properties of porous silicon.* L. Canham (ed), pp. 44, INSPEC, London.

Canham, L.T. (1997d) Visible photoluminescence from porous silicon. In: *Properties of porous silicon.* L. Canham (ed), pp. 249, INSPEC, London,.

Canham, L.T., Houlton, M.R., Leong, W.Y., Pickering, C. & Keen, J.M. (1991) Atmospheric impregnation of porous silicon at room temperature. *J. Appl. Phys.* Vol.70, pp. 422.

Canham, LT. (1990) Silicon quantum wire array fabricated by electrochemical and chemical dissolution of wafers, *Appl Phys Lett,* Vol 57, No 10, pp 1046-1048

Cullis, AG. & Canham, LT. (1991) Visible light emission due to quantum size effects in highly porous crystalline silicon, *Nature,* Vol 353, pp 335-338

Cullis, AG., Canham, LT. & Calcott, PDJ. *(1997)* The structural and luminescence properties of porous silicon, *J App Phys,* Vol 82, No 3, pp 909-965

DeLouise, L.A. & Miller, B.L. (2004) Quantitative assessment of enzyme immobilization capacity in porous silicon. *Anal. Chem.* Vol.76, No.23, pp. 6915.

Deresmes, D., Marissael, V., Stievenard, D. & Ortega, C., (1995) Electrical behaviour of aluminium-porous silicon junctions, *Thin Solid Films.*Vol. 255, pp. 258.

Dimitrov, D.B. (1995) Current-voltage characteristics of porous silicon layer. *Physical Review B,* Vol. 51, pp. 1562.

Ding, D. Chen, Z; Lu, C., (2006) Hydrogen sensing of nanoporous palladium films supported by anodic aluminum oxides, *Sensors and Actuators, B* , Vol.120, pp. 182.

Duttagupta, SP., Chen, XL., Jenekhe, SA. & Fauchet, PM. (1997) Microhardness of porous silicon films and composites. *Solid State Comm.* Vol.101, No. 1, pp. 33-37.

Dzhafarov, T.D., Oruc, C. & Aydin, S. (2004) Humidity-voltaic characteristics of Au - porous silicon interfaces. *J. Phys. D* Vol.37, pp. 404.

Fo¨ll, H., Christophersen, M., Carstensen, J. & Hasse, G. (2002) Formation and application of porous silicon, *Materials Science and Engineering R* Vol. 39, pp. 93.

Fuchs, HD., Stutzmann, M., Brandt, MS., Rosenbauer, M., Weber, J. & Cardona, M. *(1992)* Visible luminescence from porous silicon and siloxene, *Physica Scripta,* Vol T45, pp 309-313

Fürjes, P., Dücső, Cs., Ádám, M., Zettner, J. & Bársony, I. (2004) Thermal characterisation of micro-hotplates used in sensor structures, Superlattices Microstruct. Vol. 35, pp. 455.

Green, S. Kathirgamanathan, P. (2002) Effect of oxygen on the surface conductance of porous silicon: towards room temperature sensor applications Mater. Lett. Vol. 52, pp. 106.

Halimaoui, A., Oules, C., Bomchill, G., Bsiesy, A., Gaspard, F., Herino, R., Ligeon, M. & Muller, F. (1991) Electroluminescence in the visible range during anodic oxidation of porous silicon films. *Appl.Phys. Lett.* Vol.59, No.3, pp. 304-306.

Hamilton, B. *(1995)* Porous silicon, *Semicond Sci Technol,* Vol 10, pp 1187-1207

Hossain, SM., Chakraborty, S., Dutta, SK., Das, J. & Saha, H. (2000) Stability in photoluminescence of porous silicon. *J. Lumin.,* Vol.91, pp. 195-202.

Hughes, R.C., Schubert, W.K., Zipperian T.E., Rodriguez J.L. & Plut, T.A. (1987) Thin film palladium and shiver alloys and layers for metalinsulator-semiconductor sensors, *J. Appl. Phys.,* Vol.62, pp. 1074.

Hummel R.E. & Chang S S., (1992) Novel technique for preparing porous silicon, *Appl. Phys. Lett.*Vol. 61, pp. 1965.

Ito, T., Yamama, A., Hiraki, A. & Satou, M. (1989) Silicidation of porous silicon and its application for the fabrication of a buried metal layer. *Appl. Surf. Sci.* Vol. 41-42, pp. 301.

Jeske, M., Schultze, J.W. Thonissen, M. Munder,v (1995) Electrodeposition of metals Into porous silicon. *Thin solid films,* Vol.255, pp. 63-66.

Kang Y, Jorne J. (1993) *J Electrochem Soc* Vol.140:2258.

Kanungo, J. Pramanik, C. Bandopadhyay, S. Gangopadhyay, U. Das, L. Saha, H. & Gettens, R.T.T. (2006) Improved Contacts On Porous Silicon Layer By Electroless Nickel Plating and Copper Thickening, *Semicond Sci and Technol.,* Vol. 21, pp. 964.

Kanungo, J., Maji, S., Saha, H. & Basu, S. (2009a) Stable Aluminium Ohmic Contact to Surface Modified Porous Silicon. *Solid-State Electronics* Vol. 53,pp. 663–668.

Kanungo, J., Saha H. & Basu, S. (2009b) Room temperature Metal-Insulator -Semiconductor (MIS) Hydrogen Sensors Based On Chemically Surface Modified Porous Silicon. *Sensors and Actuators B,* Vol.140, pp. 65–72.

Kanungo, J., Maji, S., Mandal, A.K., Sen , S., Bontempi , E., Balamurugan, A.K., Tyagi, A.K., Uvdal, K., Sinha, S., Saha H. and Basu. S. (2010a) Investigations on the surface morphology of the modified porous silicon using noble metal ions. *J. of Applied Surface Science,* Vol.256, pp. 4231- 4240.

Kanungo, J., Saha, H. & Basu, S. (2010b) Effect of porosity on the performance of surface modified porous silicon hydrogen sensors. *Sensors and Actuators B,* Vol.147, pp. 145–151.

Karacali, T., Cakmak, B. & Efeoglu, H. (2003) Aging of porous silicon and the origin of blue shift. *Optics Express* , Vol.11, pp. 1237-1242.

Khoshnevis, S., Dariani, R.S., Azim-Araghi, M.E., Bayindir, Z. & Robbie, K. (2006) Observation of oxygen gas effect on porous silicon-based sensors, *Thin Solid Films,* Vol. 515, pp. 2650.

Koch, F. *(1993)* Models and mechanisms for the luminescence of porous Si, *Mat Res Soc Symp Proc,*Vol 298, pp 319-329

Korotcenkov, G. (2005) Gas response control through structural and chemical modification of metal oxide films: state of the art and approaches, *Sensors and Actuators, B,* Vol.107, pp. 209.

Koshida, N. & Koyama, H. (1992) *Appl Phys Lett.,* Vol.60, pp. 347.

Krueger, M., Marso, M., Berger, MG., Thonissen, M., Billat,S., Loo, R., Reetz, W., Luth, H., Hilbrich, S., Arensfischer, R. & Grosse, P. (1997) *Thin Solid Films,* Vol. 297, pp. 241.

Kwok, W.M., Bow, Y.C., Chan, W.Y., Poon, M.C., Han, P.G. & Wong, H. 1999 Study of porous silicon gas sensor, Electron devices meeting, 1999. Proceedings., *IEEE Hong Kong,* June Vol.26,1999.

Lang, W. (1997) Thermal conductivity of porous silicon. In: *Properties of porous silicon*. Canham L (ed) pp. 128-141, INSPEC, London

Lauerhaas, J.M. & Sailor, M.J. (1993) The Effects of Halogen Exposure on the Photoluminescence of Porous Silicon. *Mat. Res. Soc. Symp. Proc. (USA),* Vol. 298, pp. 259.

Lazarouk, S., Baranov, I., Maiello, G., Proverbio, E., deCesare, G. & Ferrari, A. (1994) *J Electrochem Soc.,* Vol.141, pp. 2556.

Lees, I.N., Lin, H., Canaria, C.A., Gurtner, C., Sailor, M.J. & Miskelly, G.M. (2003) Chemical stability of porous silico surfaces electrochemically modified with functional alkyl species, *Langmuir,* Vol. 19, pp. 9812.

Lewis, S.E., DeBoer, J.R., Gole, J.L. & Hesketh, P.J. (2005) Sensitive selective and analytical improvements to a porous silicon gas sensor, *Sens. Actuators, B,* Vol. 110, pp. 54.

Lin, J., Yao, GQ., Duan, JQ. & Qin, GG. *(1996)* Ultraviolet light emission from oxidized porous silicon, *Solid State Communications,* Vol 97, No 3, pp 221-224

Lin, VS., Motesharei, K., Dancil, KS., Sailor, MJ.& Ghadiri, MR. (1997) *Science,* Vol. 278, pp.840.

Litovchenko, V.G., Gorbanyuk, T.I., Solntsev, V.S. & Evtukh, A.A. (2004) Mechanism of hydrogen, oxygen and humidity sensing by Cu/Pd-porous silicon-silicon structure, *Appl.Surf.Sci.* Vol. 234, pp. 262.

Lockwood, DJ. Aers, GC., Allard, LB., Bryskieicz, B., Charbonneau, S., Houghton, DC., McCaffrey, JP. & Wang, A. *(1992)* Optical properties of porous silicon, *Can J Phys,* Vol 70, pp 1184-1193

Lundstrom, I., Sundgren, H., Winquist, F., Eriksson, M., Rulcker, C.K. & Spetz, A.L (2007) Twenty-five years of field effect gas sensor research Linkoping, *Sensors and Actuators, B,* Vol.121, pp. 247

Luongo, K., Sine, A. & Bhansali, S. (2005) Development of a highly sensitive porous Si-based hydrogen sensor using Pd nano-structures, *Sens. Actuators, B,* Vol. 111–112, pp. 125.

Mandal, N.P., Sharma, A. & Agarwal, S.C. (2006) Improved stability of nanocrystalline porous silicon after coating with a polymer, *J.Appl.Phys.* Vol. 100, pp. 024308.

Marsh, G. (2002) Porous silicon a useful imperfection, Materials Today, Vol.5, pp. 36.

Martin-Palma, R.J., Perez-Rigueiro, J., Guerrero-Lemus, R., Moreno, J.d. & Martinez-Duart, J. (1999) Ageing of aluminium electrical contacts to porous silicon. *J Appl Phys.* Vol. 85, pp. 583.

Memming, R. & Schwandt, G. (1966) Anodic Dissolution of Silicon in Hydrofluoric Acid Solutions, *Surf Sci.,* Vol.4, pp. 109.

Menna, P Tsuo, S (1997) Solar Cells using porous silicon. In: *Properties of porous silicon.* Canham L (ed), pp.384-389, INSPEC, London:.

Mizsei, J. (2007) Gas sensor applications of porous Si layers, *Thin solid films,* Vol.515, pp. 8310.

Mizsei,J. (2005) Vibrating capacitor method in the development of semiconductor gas sensors, Thin Solid Films, Vol. 490 17.

Nakagawa, T., Koyama, H. & Koshida, N. (1996) Control of structure and optical anisotropy in porous Si by magnetic-field assisted anodization, *Appl. Phys. Lett.* Vol.69, pp. 3206.

Nakajima, Y., Uchida, T., Toyama, H., Kojima, A., Gelloz, B. & Koshida, N. (2004) A solid-state multicolor light-emitting device based on ballistic electron excitation. *Jap. J. Appl. Phys.* Part 1 Vol.43, No.4B, pp. 2076.

Nicolas, D. Souteyrand, E. Martin, J.R. (1997) Gas sensor characterization through both contact potential difference and photopotential measurements, Sens. Actuators, B, Vol. 44, pp. 507.

Pancheri, L., Oton, CJ., Gaburro, Z., Soncini, G. & Pavesi, L. (2003) Very sensitive porous silicon NO2 sensor, Sens. Actuators B, Vol. 89, pp. 237.

Parkhutik V (1999) Porous silicon - mechanism of growth and applications. *Solid-state Electron.* Vol.43, pp. 1121-1141.

Parkhutik VP. & Shershulsky VI. (1992) *J Phys D, Appl Phys,* Vol.25, pp. 1258.

Pavesi, L. & Dubos P, (1997) Random porous silicon multilayers application to distributed Bragg reflectors and interferential Fabry Perot filters. *Semicond Sci and Technol.*, Vol. 12, pp. 570.

Pedrero, O.L., Pena-Sierra, R. & Romero Paredes, RG. 2004 Gas sensor based on porous silicon and palladium oxide clusters. *1st International conference on Electrical and Electronics Engineerin: (ICEEE)*: Acapulco, Mexico, September 8-10, 2004.

Perez, JM., Villalobos, J., McNeill, P., Prasad, J., Cheek, J., Kelber, J., Estrera, JP., Stevens, PD. & Glosser, R. *(1992)* Direct evidence for the amorphous-silicon phase in visible photoluminescent porous silicon, *Appl Phys Lett*, Vol 61, No 5, pp 563-565

Peterson, K. (1982) Silicon as a mechanical material, *Proc. Of IEEE*, Vol.70, No.5, pp.420.

Petit, D., Chazalviel, JN., Ozanam, F. & Devreux, F. (1997) Porous silicon structure studied by nuclear magnetic resonance. *Appl. Phys. Lett.* Vol.70, No. 2, pp. 191-193.

Petrova, E.A., Bogoslovskaya, K.N., Balagurov, L.A. & Kochoradze, G.I. (2000) Room temperature oxidation of porous silicon in air. *Mater. Sci. Eng. B* Vol 69-70, pp. 152.

Petrova-Koch V., Muschik, T., Kux, A., Meyer, B.K., Koch, F. & Lehmann, V., (1992) Rapid-thermal-oxidized porous Si–The superior photoluminescent Si, *Appl. Phys. Lett.*Vol. 61, pp. 943.

Pickering, C., Beale, M.I.J., Robbins, D.J., Pearson, P.J. & Greef, R. (1984) Optical studies of the structure of porous silicon films formed in p-type degenerate and non-degenerate silicon. *J. Phys. C* Vol.17, pp. 6535.

Pillai, SM., Xu, Z Y., Gal, M., Glaisher, R., Phillips, M. & Cockayoe, D. (1992) *Jpn. J. Appl. Phys.* Vol.31, , pp. L1702.

Polishchuk, V., Souteyrand, E., Martin, J.R., Strikha, V.I. & Skryshevskya, V.A. (1998) A study of hydrogen detection with palladium modified porous silicon, *Anal. Chim. Acta*, Vol. 375, pp. 205.

Pramanik, C. Islam, H. Saha, H. (2005) Design, fabrication, testing and simulation of porous silicon based smart MEMS pressure sensor. *IEEE Computer Society, 18th Internation Conference on VLSID'05* , pp. 235.

Pramanik, C. & .Saha H. (2006) Low pressure piezoresistive pressure sensors for biomedical applications", *Materials and Manufacturing Processes*, Vol.21, No. 3, pp.233-238.

Pramanik, C. & Saha, H. (2006) Piezoresistive pressure sensing by porous Silicon Membrane,*IEEE Sensors*, Vol.6, pp.301-309.

Prokes, SM., Glembocki, OJ., Bermudez, VM., Kaplan, P., Friedersdorf, LE. & Searson, PC. *(1992)* SiHx excitation - an alternative mechanism for porous Si photoluminescence, *Physical Review B*,Vol 45, No 23, pp 13788-13791

Rahimi, F. & Irajizad, A. (2006) Effective factors on Pd growth on porous silicon by electroless-plating: response to hydrogen, *Sens. Actuator, B*, Vol. 115, pp. 164.

Read, AJ., Needs, RJ., Nash, KJ., Canham, LT., Calcott, PDJ. & Qteish, A. (1992) First-principles calculation of the electronic properties of silicon quantum wires. *Phys. Rev. Lett.* Vol.69, No. 8, pp. 1232-1235.

Richter, A., Steiner, P., Kozlowski, F. & Lang, W. (1991) *IEEE Electron Device Lett*, Vol.12, pp. 691.

Rossi, A.M., Amato, G., Camarchia, V., Boarnio, L. & Borini, S., (2001) High-quality porous-silicon buried waveguides. *Appl. Phys. Lett.* Vol.78, No.20, pp. 3003.

Rothschild, A. & Komen, Y. (2004) The effect of grain size on the sensitivity of nanocrystalline metal-oxide gas sensors, *J. Appl. Phys.*, Vol.95, pp. 6374.

Rumyantseva, MN., Kovalenko, VV., Gas'kov, AM. & Pagnier, T. (2008) Metal-Oxide Based Nanocomposites as Materials for Gas Sensors, ISSN 1070-3632, *Russian Journal of General Chemistry*, Vol.78, pp. 1081.

Sailor, M.J. (1997) Sensor applications of porous silicon. In: *Properties of porous silicon*. L. Canham (ed), pp. 364, INSPEC, London,.

Sakai, G., Baik, N.S., Miura, N. & Yamazoe, N. (2001) Gas sensing properties of tin oxide thin films fabricated from hydrothermally treated nanoparticles—dependence of CO and H_2 response on film thickness, *Sensors and Actuators, B*, Vol. 77, pp. 116.

Samuel, BJ., Editor (2010) *Nanoporous Materials - Types, Properties and Uses*, Nova Scientific Publishers, Inc.,New York.

Seals, L., Tse, L.A., Hesketh, PJ. & Gole, JL. (2002) Rapid, reversible, sensitive porous silicon gas sensor, *J. Appl. Phys.* Vol. 91, pp. 2519.

Sekhar, P.K., Sine, A. & Bhansali, S. (2007) Effect of varying the nanostructured porous-Si process parameters on the performance of Pd-doped hydrogen sensor, *Sens. Actuators, B*, Vol. 127, pp. 74.

Souteyrand, E., Nicolas, D., Queae, E. & Martin, J.R. (1995) Influence of surface modifications on semiconductor gas sensor behaviour, Sens. Actuators, B, Vol. 26, pp. 174.

Starodub, NF., Fedorenko, LL., Starodub, VM., Dikij, SP. & Svectinikov, SV. (1996) *Sensors and Actuators*, Vol. 35-36, pp.44.

Stathis, JH. & Kastner, MA. *(1987)* Time-resolved photoluminescence in amorphous-silicon dioxide, *Phys Rev B*, Vol 35, pp 2972-2979

Steiner, P., Kozlowski, F.& Lang, W. (1993) *Appl Phys Lett.*, Vol.62, pp. 2700.

Steiner, P., Kozlowski, F., Wielunski, M. & Lang, W. (1994) Enhanced Blue-Light Emission from an Indium-Treated Porous Silicon Device *Jpn.J.Appl.Phys.* Vol.33, pp. 6075.

Stievenard, D. & Deresmes, D., (1995) Are electrical properties of an aluminium-orous silicon junction governed by dangling bonds, *Appl. Phys. Lett.*, Vol. 67, pp. 1570.

Strehle, S., Sarti, D., Krotkus, A., Grigiras, K. & Levy-Clement, C. (1997) *Thin Solid Films*, Vol. 297, pp. 296-298.

Takai, H. & Itoh, T. (1986) Porous silicon layers and its oxide for the silicon-on-insulator structure, *J. Appl. Phys.* Vol.60 pp. 222.

Tiemann, M. (2007) Porous Metal Oxides as Gas Sensors, *Chemistry – A European Journal*, Vol.13, pp. 8376.

Toriyama, T., Tanomoto, Y. & Sugiyama, S. (2002) Single crystal silicon nanowires piezoresistors for mechanical sensors, *IEEE Journal of MEMS*, Vol.11, No.5,pp. 605-611.

Tsai, C., Li, KH., Campbell, JC. & Tasch, A. (1993) *Appl Phys Lett.*, Vol.62, pp. 2818.

Tsai, C., Li, KH., Sarathy, J., Shih, S., Campbell, JC., Hance, BK., & White, JM. *(1991)* Thermal treatment studies of the photoluminescence intensity of porous silicon, *Appl Phys Lett*, Vol 59, No 22, pp2814-2816

Tsamis, C., Tsoura, L., Nassiopoulou, A.G., Travlos, A., Salmas, C.E., Hatzilyberis, K.A. & Ndroutsopoulos, G.P. (2002) Hydrogen Catalytic Oxidation Reaction on Pd-Doped Porous Silicon, *IEEE Sensors Journal*, Vol.2, pp. 89.

Turner, D.R. (1958) Electro polishing of Silicon in Hydrofluoric Acid Solutions, J Electrochem Soc., Vol.105 , pp.402-408.

Ulhir, A. (1956) Electrolytic shaping of germanium and silicon, *Bell Syst. Tech. J.* Vol.35, pp. 333.

Vaishampayan, MV., Deshmukh, RG. & Mulla, I.S. (2008) Influence of Pd doping on morphology and LPG response of SnO_2, *Sensors and Actuators, B*, Vol.131, pp. 665.

Vasquez, RP., Fathauer, RW., George, T., Ksendzov, A. & Lin, TL. (1992) Electronic-structure of lightemitting porous Si, *Appl Phy Lett,* Vol. 60, No. 8, pp. 1004-1006

Vasquez, RP., Madhukar, A. & Tanguay, JAR. *(1985)* Spectroscopic ellipsometry and X-ray photoelectron-spectroscopy studies of the annealing behaviour of amorphous Si produced by Si ion-implantation, *J Appl Phys,* Vol 58, No 6, pp 2337-2342

Vázsonyi ÉB., Koós M., Jalsovzsky, G. & Pócsik, I. (1993) The role if hydrogen in luminescence of electrochemically oxidized porous Si layer. *J. Lumin.*, Vol. 57, pp. 121-124.

Volkenshtein, F.F., Electronic Theory of Catalysis on Semiconductors, Moscow: Fizmatgiz, 1960.

Wang, M. & Feng, Y. (2007) Palladium–silver thin film for hydrogen sensing, *Sensors and Actuators, B*, Vol.123, pp. 101.

Watanabe, Y. & Sakai, T. (1971) Application of a thick avoided film to semiconductor devices, *Rev.Electron. Commun. Labs.* Vol.19, pp. 899.

Williams, GM. (1997) Cathodoluminescence properties of porous silicon. In: *Properties of porous silicon*. Canham L (ed), pp. 270-275, INSPEC, London.

Wolford, DJ., Scott, BA., Reimer, JA. & Bradley, JA. *(1983)* Efficient visible luminescence from hydrogenated porous silicon, *Physica B,* Vol 117, pp 920-922

Xu, C., Tamaki, J., Miura N. & Yamazoe, N. (1991) Grain size effects on gas sensitivity of porous SnO_2-based elements, *Sensors and Actuators, B*, Vol.3, pp. 147.

Xu, Y.Y., Li, X.J., He, J.T., Hu, X. & Wang, H.Y. (2005) Capacitive humidity sensing properties of hydrothermally-etched silicon nano-porous pillar array Sens. Actuators B Vol.105, pp. 219.

Yamazoe, N. (1991) New approaches for improving semiconductor gas sensors, *Sens. Actuators B*, Vol.5, pp. 7-19.

Zhang, W., Vasconcelos, E.A., Uchida, H., Katsube, T., Nakatsubo, T. & Nishioka, Y. (2000) A study of silicon Schottky diode structures for NO_X gas detection, *Sens. Actuator, B,* Vol.65, pp. 154.

Zhang, XG. (1991) *J Electrochem Soc* , Vol.140, pp. 2258.

Zhua, B.L., Xie, C.S., Zeng, D.W., Song, W.L. & Wang, A.H. (2005) Investigation of gas sensitivity of Sb-doped ZnO nanoparticles, *Materials Chemistry and Physics* Vol.89, pp. 148.

Zimin, S.P., Kuznetsov, V.S. & Prokaznikov, A.V. (1995)Electrical characteristics of aluminium contacts to porous silicon. *Appl Surf Sci.* Vol. 91, pp. 355.

Zubko, V.G., Smith, T.L. & Witt, A.N. (1999) Silicon Nanoparticles and Interstellar Extinction, *The Astrophysical Journal Letters*, Vol. 511, pp. L57.

11

Nanocrystalline Porous Silicon: Structural, Optical, Electrical and Photovoltaic Properties

Ma.Concepción Arenas[1], Marina Vega[2], Omar Martínez[3] and Oscar H. Salinas[4]

[1]*Departamento de Ingeniería Molecular de Materiales, Centro de Física Aplicada y Tecnología Avanzada, Universidad Nacional Autónoma de Mexico, Santiago de Querétaro, Querétaro*

[2]*Centro de Geociencias, Universidad Nacional Autónoma de México, Querétaro*

[3]*Universidad Politécnica de Guanajuato, Av. Universidad Norte S/N, Juan Alonso Cortazar, Guanajuato*

[4]*Universidad Tecnológica Emiliano Zapata del Estado de Morelos, Morelos. México*

1. Introduction

In the electronics industry, crystalline silicon is the most widely used semiconductor. A great variety of electronic devices can be built based on silicon, from discrete to low and very high scale integrated circuits. One of the most common commercial applications is the photovoltaics; in fact, silicon is found in 90% of solar cells for terrestrial applications. Silicon has an indirect band gap of 1.12 eV and emits light weakly near the infrared region by phonon assistance. This fact limits its use in optoelectronic applications, where semiconductors such as GaAs and InP dominate this field. However, the photoluminescent and electroluminescent properties of porous silicon (Bisi et al., 2000; Canham, 1990) have generated great interest in optoelectronic devices. Porous silicon was accidentally discovered by Uhlir at Bell laboratories in the middle of the 1950 s. He found that, under the appropriate conditions of applied current and solution composition, silicon did not dissolve uniformly but instead fine pores (holes) were produced. Porous silicon (PS) is a semiconductor in nanocrystalline form (NPS), which could improve some properties of silicon, given that some properties, such as the electricals, are determined by the short range order of atoms, rather than by the long range order. In recent years, the great interest in NPS has been due to the high surface area of porous silicon, which has been useful as a model of crystalline silicon surfaces in spectroscopic studies (Anderson et al., 1990; Bisi et al., 2000; Dillon et al., 1992; Gupta et al., 1991), as a precursor to generate thick oxide layers on silicon and as a dielectric layer in capacitance chemical sensors. In addition, NPS has been applied in optoelectronics, micro-optics, energy conversion, environmental monitoring, wafer technology, micromachining and biotechnology. In the photovoltaic field, the application of NPS has not been widely explored. The first NPS cell with a crystalline silicon substrate was reported in 1992 by Smestad (Smestad et al., 1992). Nevertheless, until now, its main application in solar cells has been as a non-reflecting layer (Vitanov et al., 1997). The use of semiconducting polymers to form hybrid heterojunctions with NPS represents another potential application of porous silicon; it has been found that the conductor polymer increases the effective conductivity of NPS as a function of the polymer penetration into the silicon pores (Jung et al., 1995; Moreno et al., 1999). This kind of heterojunction can improve the stability of light emitting devices (LEDs) and has formed structures with interesting photoluminescence (Nguyen et al., 2003; Parkhutik et al., 1994) and electroluminescence properties (Bsiesy et al., 1995). In this chapter, several topics are discussed, ranging from the study of NPS structure formation from monocrystalline silicon by electrochemical etching in acid electrolyte solution to the construction and evaluation of optoelectronic devices based on these structures, including the formation and electrical evaluation of hybrid heterojunctions based on nanocrystalline silicon structures and semiconducting polymers.

2. General aspects of nanocrystalline porous silicon

Nanocrystalline porous silicon (NPS) is composed of silicon wires and pores, and thus, it is a material with a large surface area (Bisi et al., 2000). It is obtained by the electrochemical dissolution of silicon (Si) wafers in aqueous hydrofluoric acid (HF) solution or by electroless etching in an HF solution containing an oxidizing agent (Arenas et al., 2006; 2008; Smith & Collins, 1992). The anodic dissolution in HF solutions is the main process used for this effect, where darkness or illumination is necessary for p-type or n-type silicon substrates, respectively, to achieve the etching process. In this section, a short description of NPS preparation and its structural and morphological characterization is presented.

2.1 NPS formation process

One of the most important advantages of porous silicon is its simple and easy preparation (Yu & Wie, 1992). Since the first studies in the middle of the 1950 s, NPS has been mainly obtained by the electrochemical dissolution of silicon wafers based on HF solution. This process consists of applying an anodic current to the silicon wafer in an HF electrolyte contained in a Teflon cell, which is composed of a silicon working electrode and a Pt counter electrode. The dissolution of silicon requires holes supplied from its valence band at its surface; thus, for n-type silicon, light is necessary to photogenerate holes in contrast to p-type silicon. A constant current density can adjust the porosity, thickness, uniformity and reproducibility of samples (Eddowes, 1990; Galun, 1995; Smith & Collins, 1992). Ethanol is added to the HF electrolyte to counteract the hydrophobic character of silicon and facilitate the infiltration of HF inside the pores. In addition, ethanol helps to remove the hydrogen bubbles generated during pore formation on the silicon surface. In the past, many studies of porous silicon formation have been limited exclusively to an electrochemical characterization of its current-voltage (I-V), relationships with the Schottky diode model of the semiconductor/electrolyte interface playing a predominant role. Although additional analysis techniques have been used recently to study porous silicon, its formation still arises from the I-V relationships, and a basic knowledge of silicon electrochemistry is essential to understand this process (Bard, 1986; Conway et al., 1983; Holmes, 1962; Vijh, 1973). Several reaction mechanisms for pore formation have been proposed (Bomchil, 1983; Eddowes, 1990; Lehmann & Föll, 1990; Smith & Collins, 1992; Turner, 1958). However, the most widely accepted was proposed by Lehmann and Gösele, as shown in Fig. 1. This mechanism is based on the oxidation of bonds on the silicon surface with the capture of holes and electron injection in four steps (Lehmann & Gösele, 1991):

1. A hole reaches the surface for a nucleophilic attack on Si-H bonds by fluoride ions from the HF electrolyte (i.e., the hole can then migrate on to a Si-H bond), thus releasing a proton.

2. A second attack is accomplished by another fluoride ion, causing the evolution of molecular hydrogen and electron injection into the substrate. The attack of the Si radical by fluoride causes electron injection into the silicon conduction band and the formation of a Si-F bond.
3. The next few steps involve the sequential removal of $-SiF_2$ by the replacement of protons in the highly oxidized silicon with the concurrent injection of an electron into the conduction band. A chemical reaction occurs in which HF is added across the one remaining Si-Si bond to release SiF_4 into solution, and in this step, a tetrafluoride (SiF_4) molecule is produced.
4. The tetrafluoride molecule reacts with two HF molecules and H_2SiF_6 is the final product in solution.

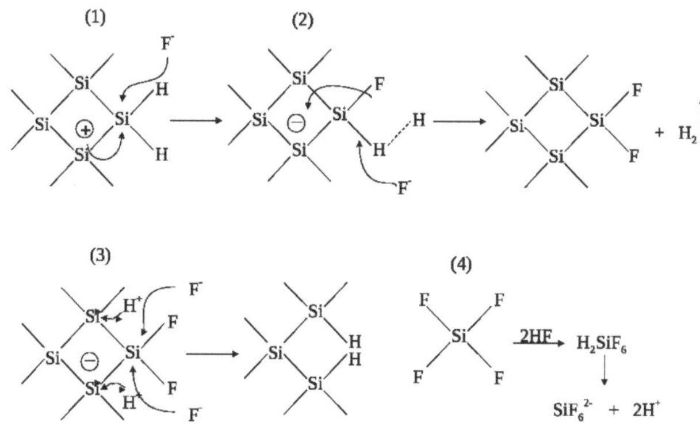

Fig. 1. Schematic of dissolution mechanism of silicon proposed by Lehmann and Gosele (Lehmann & Gösele, 1991).

The dissolution mechanism is complex (Eddowes, 1990). In this reaction, only two holes are required to remove each silicon atom while one molecule of hydrogen is formed. Hydrogen is assumed to result from the reaction of protons or HF with an intermediate of the silicon oxidation reaction:

$$\mathrm{Si}(II) + 2H^{+} + \mathrm{Si}(IV) + H_2$$

During pore formation, the following anodic semireactions occur:

$$\mathrm{Si} + 6\mathrm{HF} \rightarrow H_2\mathrm{Si}F_6 + H_2 + 2H^{+} + 2e^{-}$$

However, an electropolishing process can also exist if the current density reaches the critical value, and this process can be described with the reaction:
Si + 6HF $\rightarrow H_2SiF_6 + 4H^{+} + 4e^{-}$
The final properties, such as porosity and thickness of the nanocrystalline porous silicon layer, are very sensitive to many experimental variables, such as the doping density of the silicon substrate, the current density and the etching time, among others (Lehmann & Gösele, 1991; Searson et al., 1992). The porosity is an important parameter that determines the optoelectronic properties of NPS. Thus, it is necessary to determine posority.
The porosity (P) is defined as the fraction of voids (air) within the nanocrystalline porous layer. This parameter could be estimated by gravimetry as described in Fig. 2, which shows the porosity produced by removed or dissolved silicon mass (m_r) during the etching divided by the total mass of etched silicon (m_T). In the porosity equation, m_1 is the silicon mass before etching, m_2 is the mass of the silicon substrate after etching (including porous silicon) and

m_3 is the substrate mass after dissolving the nanocrystalline porous silicon with a KOH (3 M) solution. In the same way, the film thickness of the nanocrystalline porous silicon layer can be estimated by considering the silicon density (ρ=2328 mg/cm^3) and etched surface (S), S=1 cm^2 as shown in Fig. 2.

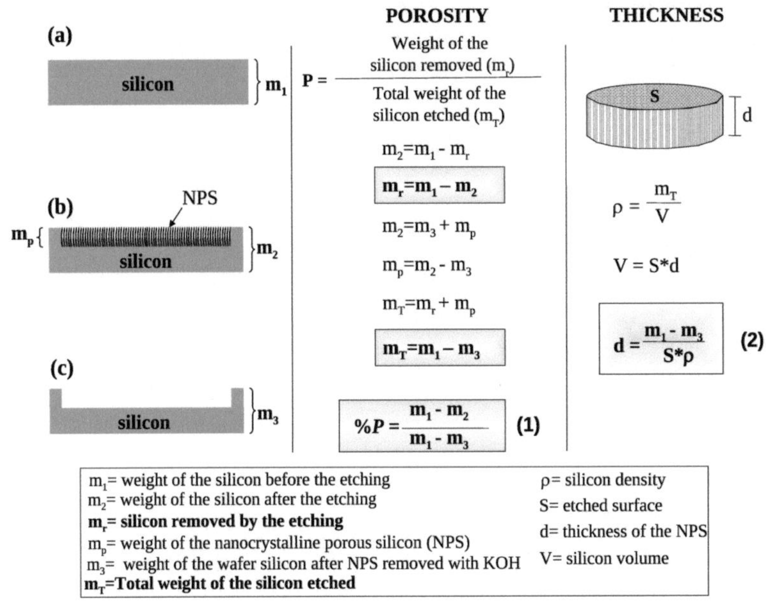

Fig. 2. Schematic representation of the method to calculate the porosity and thickness of nanocrystalline porous silicon by gravimetry (Bisi et al., 2000).

The porosity depends on the electrochemical etching conditions, such as the HF concentration and the type and resistivity of the silicon substrate. It has been observed that porous silicon porosity increases with current density at a given HF concentration. However, at a given current density, porosity decreases with HF concentration because HF is hydrophobic with silicon, which increases the difficulty of penetration/infiltration in the pores at high HF concentrations (Bisi et al., 2000). Fig. 3 shows the influence of the silicon substrate on the porosity and thickness of NPS layers. The porosity of the NPS layer obtained from a p-type substrate increases directly with the applied density current. In addition, it changes according to the silicon resistivity. More porous NPS layers are obtained from a silicon substrate of 10 Ω-cm of electric resistivity, whereas less porous NPS layers are obtained from lower resistivity silicon (0.002 Ω-cm). Starting from n-type silicon, the porosity increases as a function of current density, but these NPS layers are less porous than those obtained from p-type silicon. However, the film thickness (Fig. 3b) shows a linear behavior as a function of the current density, where values of greater thickness can be obtained with n-type silicon compared to p-type. In other words, with 10 Ω-cm n-type substrates, thick NPS layer with low porosity can be obtained, which, is related to an accelerated electrochemical etching. The discussion in the following sections correspond to NPS films from p-Si and n-Si obtained at the same conditions (20 mA, 1:1 etanol: HF), unless otherwise noted.

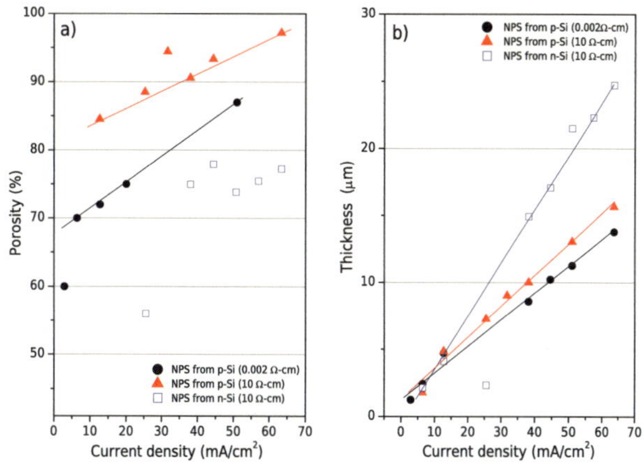

Fig. 3. Experimental a) porosity and b) thickness vs. current density of NPS films determined by the gravimetric method.

2.2 Structural characterization

2.2.1 Grazing-Incidence X-ray Diffraction

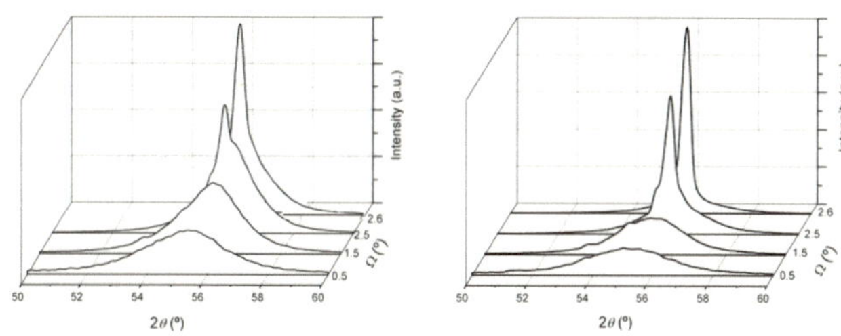

Fig. 4. Grazing-Incidence X-ray Diffraction of the nanocrystalline porous silicon from a) p-Si and b) from n-Si of approximately 10 μm (Arenas et al., 2008).

With Grazing-Incidence X-ray Diffraction (GID), it is possible show that, as a result of the electrochemical treatment of crystalline silicon, a nanocrystalline thin film is obtained, as shown in Fig. 4. At small incidence beam angles, $\Omega < 2.5^o$ a broad reflection corresponding to the film is observed, whereas, at $\Omega = 2.6^o$ only the crystalline silicon diffraction peak is shown as an acute signal. Both kinds of signals are detected at $\Omega = 2.5^o$. A broad reflection indicates that its average crystal size is smaller than that of in the crystalline silicon according to the Scherrer equation (Patterson, 1939):

$$L = \frac{0.9\lambda}{Bcos\theta} \quad (1)$$

where λ is the wavelength of the incident X-rays (0.15406 nm), B is the half value breadth of the diffracted beam and θ is the Bragg angle. Thus, it was possible to estimate an average crystal size, L, for the films. The results indicate that the average crystal sizes are in the nanometer range: L=3.5 nm for the film obtained from p-Si and L=3.1 nm from n-Si.

2.2.2 Surface molecular structure

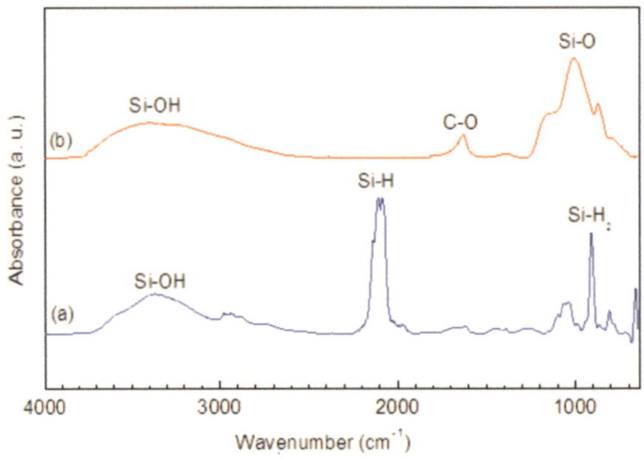

Fig. 5. Infrared spectra of NPS a) as prepared and b) oxidized at 300 ^oC.

The nanocrystalline silicon surface is saturated by hydrides (H^-) due to the nature of the electrochemical etching. The functional groups $Si - H_x$ (x=1, 2, 3) present in NPS can be identified by IR, as shown in Fig 5. The characteristic bands correspond to the molecular vibration of the bonds $Si - H_2$ at 910 cm^{-1} and Si-H at 2087 cm^{-1} (Beckmann, 1965; Bisi et al., 2000). It is possible to break these bonds easily by heating, light or exposure to gases and liquids, even after long periods of time under storage. Chemical stabilization of the surface is reached by controlled thermal oxidation; at 300 ^oC (Fig. 5b), the Si-H band disappears and a new band at 1000 cm^{-1} assigned to the Si-O bond vibration is observed (Fauchet et al., 1997). Other bands due to surface reactivity with the environment are summarized in Table 1.

2.2.3 Morphology by SEM and AFM

Nanocrystalline silicon film is a highly porous structure. It is composed of wires formed perpendicular to the surface, as can be observed using scanning electronic microscopy (SEM), as shown in Fig. 6a. Their length and diameter depend on the formation conditions and, therefore on the crystalline silicon type. From p-Si, wires with a diameter estimated by atomic force microscopy (AFM) between 130 and 160 nm are obtained, as shown in Fig. 6b, whereas, from n-Si, wire diameters are about three times smaller, between 30 and 60 nm, as shown in Fig. 6c.

3. Optical properties of NPS

Optical properties such energy band gap (Eg) and refractive index (n) of the NPS depend on the etching conditions, current density, electrolyte composition and type of silicon substrate (p-Si or n-Si). At high porosities, a blue shift of the optical absorption of NPS can be observed (Sagnes et al., 1993), i.e., the Eg and n can be modified.

Functional group	Wavenumber (cm^{-1})	Assignment
OH	3610, 3452	$Si-OH$ y H_2O stretching
CH	2958, 2927, 2856	CH_3, CH_2 y CH stretching
$SiHF_3$	2314, 2245, 2206	—
SiH	2248, 2197, 2136, 2116, 2087, 2089, 2115, 2070	O_3-SiH,SiO_2-SiH, $Si_2Si-SiH$, Si_2H-SiH, and Si_3-SiH stretching
SiH_2	2110	Stretching
CO	1630,1720	—
CH_3	1463	Asymmetric stretching
$SiCH_3$	1230	Bending
SiO	1050,1056-1160	$O-Si-O$ stretching
$Si-O-Si$	1070	Asymmetric stretching
SiH	979/948	$Si_2-H-SiH$ bending
SiH_2	880-906, 908, 910	Scissoring
SiH_2	856, 845	Wagging
SiO	827-832	$O-Si-O$ bending
SiH	661	Wagging
SiH	624-650	Si_3-SiH bending
$Si-Si$	616, 620	Stretching

Table 1. Vibrational frequencies of NPS species by IR spectroscopy (Bisi et al., 2000).

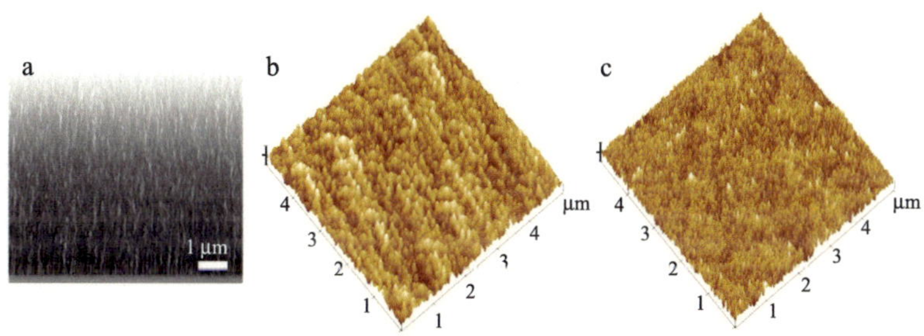

Fig. 6. Morphology of the nanocrystalline porous silicon film: a) SEM image of the transversal section from n-Si; b) 3D AFM image from n-Si, and c) 3D AFM image from n-Si. Scale in z= 30 nm/div.

3.1 Energy band gap of free standing NPS

The relationship of the optical transmittance (T) and absorption coefficient (α) of a semiconductor with thickness d is (Schroder, 1998):

$$T = \frac{(1-R)^2 e^{-\alpha d}}{1-R^2 e^{-2\alpha d}}. \tag{2}$$

If the reflectance of the semiconductor is almost null, $R \approx 0$ and equation 2 is simplified:

$$T = e^{-\alpha d}, \quad \alpha = -\frac{1}{d} \ln T. \tag{3}$$

The absorption coefficient is related to the semiconductor optical band gap (Eg) (Sze, 1990):

$$(\alpha h v)^{\gamma} = C(Eg - hv), \tag{4}$$

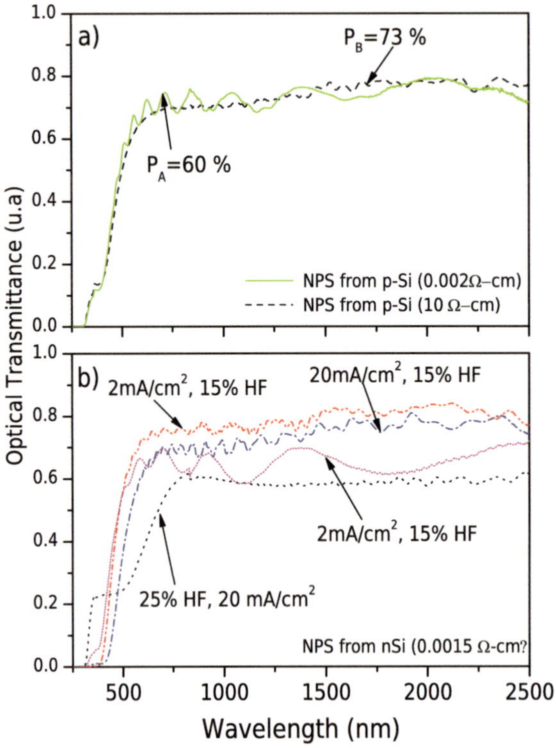

Fig. 7. Optical transmittance of free standing NPS films prepared using different anodization conditions.

where h υ is the photonic energy, C is a constant and γ takes values of 2 for direct transitions and 1/2 for indirect transitions.

The optical transmittance of free standing NPS films changes according to the porosity value and the anodization parameters, as shown in Fig. 7. Free NPS films with porosities of 60% and 70% show similar transmission spectra, but interference fringes that are well defined are only observed in less porous NPS films. The interference fringes are produced due to the light interference in the film. At low HF concentrations, the films show about 60% to 70% transmission, unlike the films prepared at high HF concentrations (Fig. 7b). The direct and indirect band gap of these free standing films are shown in Fig. 8 and Table 2. For direct transitions, the experimental Eg of the films is about 2.64 - 2.7 eV closer to the 3.3 eV value reported for powdered NPS supported between two transparent substrates by Feng (Feng & Tsu, 1994). The direct transition gives an Eg approximately 1.88 - 2 eV closer to those reported for NPS used in optoelectronic applications (Lugo et al., 1998; Peng et al., 1996; Sagnes et al., 1993). High concentrations (25%) of HF produce free standing NPS with larger Eg, which is distinctive of highly porous films, whereas, in low concentrations (15%), the less porous films show lower Eg values. Sagnes found that the Eg value can also be modified by the nanocrystalline size (Sagnes et al., 1993).

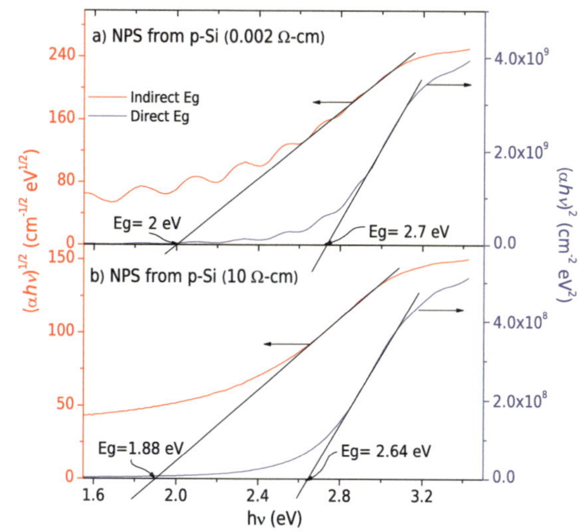

Fig. 8. a) Direct and b) indirect band gap of free standing NPS films obtained from p-Si.

NPS films from (ρ = Ω-cm)	HF (%)	J (mA/cm^2), % porosity, [Time, s]	Direct Eg (eV)	Indirect Eg (eV)
p-Si (0.002)	~14 %	**2**, 60%	2.7	2.0
p-Si (10)	25 %	**20**, 73%	2.64	1.88
n-Si (0.0015)	25 %	**20**	1.79	1.64
	15 %	**20**	2.85	2.64
	15 %	**2**, [1200]	2.96	2.9
	15 %	**2**, [480]	3.08	2.66

Table 2. Direct and indirect band gap of free standing NPS films prepared using different conditions.

3.2 Complex refractive index of NPS on silicon substrate

The optical reflectance of NPS on a silicon substrate is about 20 to 40% depending on the anodization conditions (Fig. 9). The main difference is in the amount of interference fringes, which are sensitive to the current density and thickness of NPS films (Arenas et al., 2010). This optical property determines important film parameters such as its refractive index and its extinction coefficient.

The NPS film is composed of air and silicon. Therefore, its refractive index (n) should range between 1.0 and 3.4. If the NPS is highly porous, n is closer to the refractive index of the air, and its value increases when the NPS films present low porosity (Arenas et al., 2010; Kordás et al., 2004). An algorithm is proposed to calculate the complex refractive index of NPS on crystalline silicon by means of reflectance spectra only (Arenas et al., 2010). Additionally, the NPS film thickness is determined by the normal reflectance spectra without information from its transmission spectrum (i.e., the separation of NPS films from the silicon substrate is not necessary). The algorithm is based on the analytical relations established by Heavens (Heavens, 1965) to obtain both the complex refractive index and thickness of an absorbing

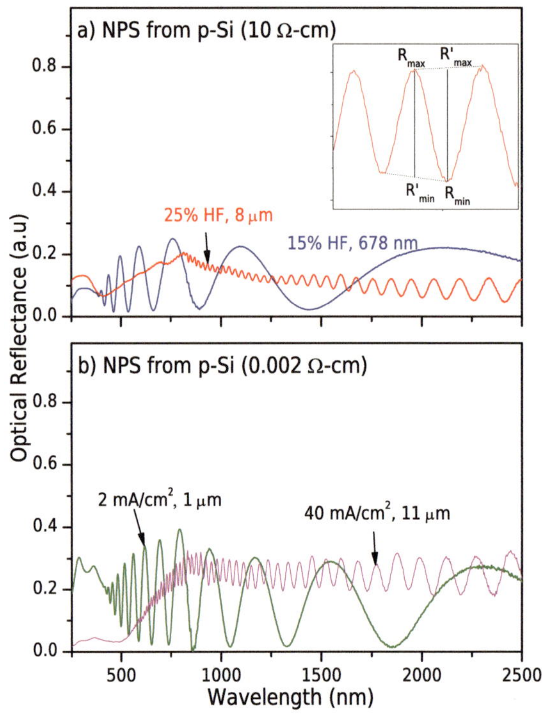

Fig. 9. Optical reflectance of NPS supported on a silicon substrate.

thin film on an absorbing substrate. Based on this model, some simplifications are introduced at different wavelengths. The model considers the maxima and minima of the interference fringes of reflectance spectrum (see inset Fig. 9a), and R_{max} and R_{min} are the assumed values for the continuous maxima (minima) envelopes using the following equation (Arenas et al., 2010):

$$\frac{1}{R_r} = \frac{4g_1g_2(1 - g_1^2 - g_2^2x^2 + g_1^2g_2^2x^2)x}{g_1^4 - 2g_1^2g_2^2x^2 + g_2^4x^4}. \tag{5}$$

where R_r is defined as (Manifacier et al., 1976; Swanepoel, 1983)

$$\frac{1}{R_r} = \frac{R_{max} - R_{min}}{R_{max}R_{min}}. \tag{6}$$

The model is only valid in the range of 1000 to 2500 nm, where the absorption is almost constant. For the UV-Vis region, other considerations must be taken into account (Swanepoel, 1983; Torres et al., 2003). It was applied to NPS films obtained from p-Si substrates, as shown in Fig. 10. It was found that the refractive index of NPS films with 60, 70, 72 and 87% porosity is a function of this parameter, where n decreases at high porosities of NPS, which is consistent to the trend reported by Bisi (Bisi et al., 2000).

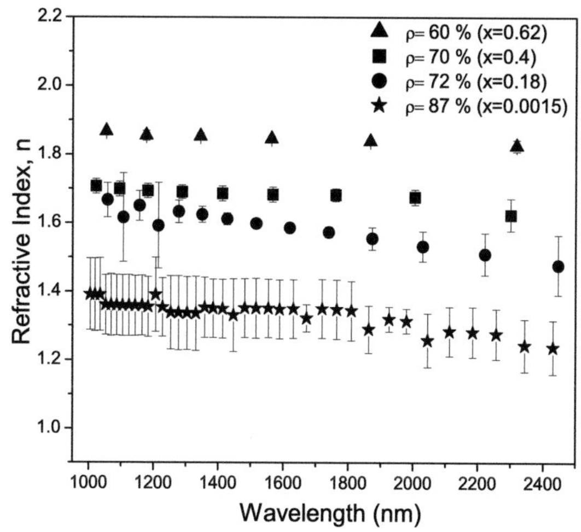

Fig. 10. Experimental refractive index of NPS supported on a silicon substrate (Arenas et al., 2010).

4. Electrical properties

4.1 Photoconductivity

The electrical current is the amount of electrons flowing in a unit of time. An electron is the elemental electrical charge in matter and its absence is called a hole. They are referred to as free charge carriers because they are free to move from one point to another inside matter. It is necessary to be aware that the particles that actually move in matter are electrons, and hole movement is only a consequence. This movement leads to electron flow; a continuous electron flow is one of the definitions of electrical current. When charge is flowing, there is an electrical current. Under most circumstance, charges in movement are confined within a limited region, in the interior of a metallic wire, the cylindrical volume of a neon tube, or the transversal section of an electron beam into a kinescope, for instance. Quantifying the current is the total charge crossing an area perpendicular to the flow per unit of time (s):

$$Q = \int_{t1}^{t2} I dt \tag{7}$$

Current originates from charge flow, and, inside of a conductor, these charges are electrons. Therefore, electron movement is the phenomenon responsible for electrical current. The coulomb is a magnitude that is defined by means of the elemental charge, which is the charge that the electron has, -1.602 $\times 10^{-19}$ C: 1 Coulomb = 6.242 $\times 10^{18}$ electrons

For an idea of the size of a coulomb, approximately 6242197253433208489 electrons are needed to form a charge of 1 C. Therefore, the amount of electrons passing a specific point dat a given current level can be calculated. For example, if a current of eight amperes is crossing a cross sectional area over five seconds, approximately 2.497 $\times 10^{20}$ electrons flowed across the area.

Electrons movement inside matter has no velocity or acceleration constant due to interatomic interactions and the material characteristics at the atomic level.Thus, this movement is based on a phenomenon called diffusion or drift. The average velocity of electrons is approximately 4 m/h.

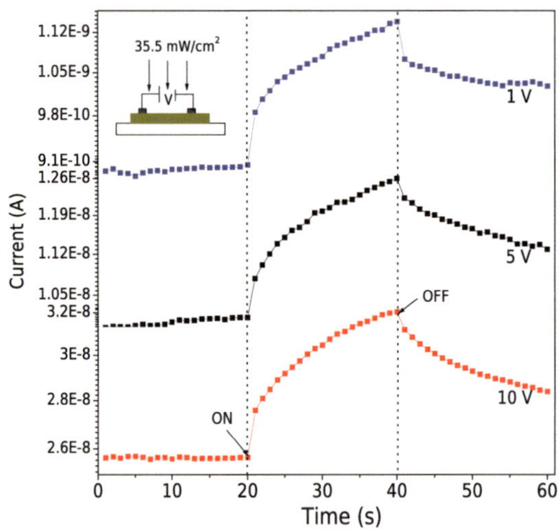

Fig. 11. Current as a function of time of the NPS film at three applied electrical voltages.

An electrical field, internal or external, is necessary to move electrons. Therefore, applying an electrical potential to the bulk or surface can create an electrical current. An electromotive source is used to apply a constant or variable external electrical field. It is a device that converts chemical, mechanical or solar energy into electrical energy, which is necessary to maintain a constant electrical charge flow. An electromotive source of one volt will perform one joule of work over every charge of one coulomb passing through it. Photoresponse characterization was performed for nanocrystalline porous silicon thin films to understand its sensitivity to light exposure, as shown in Fig. 11. The technique consisted of applying a fixed electrical potential on the material surface and recording the current under dark conditions after illumination for a determined period of time and continuing after another period of time without illumination. The current intensity of the NPS thin film is almost the same under dark and illuminated conditions, as shown in Fig. 11. Based on the plot of I vs. t, it is possible to estimate the total generated charge under an applied electrical potential under both dark and illuminated conditions using Eq. 7. The area under the curve is the result of the defined integral. The difference between the generated charge under light and under illumination is charge generated by light exposure, where

$$generatedcharge = chargeunderlight + chargeunderdark. \qquad (8)$$

When a material is highly conductive, there are fower electrons due to illumination compared to electrons in the conduction band. If the material is almost intrinsic, the generation under light must be higher, but it also depends on the material band gap and the wavelength of light. The nanocrystalline porous silicon layer used in the characterization was almost intrinsic.

However, the charge generation was low, and thus, the material was not sensitive to this wavelength. The charge generated under illumination corresponded to $2.14450152 \times 10^{10}$ electrons (Fig. 11).

4.2 Metal - NPS junction

Metal - semiconductor (M - S) contacts are an essential part of the electrical performance of any semiconductor device because they are the communication channel between the world inside the device and the world outside of it, which is a simple way to describe the role of M - S contacts. The metal layer enables current to flow out of the device, allowing the external circuit to work with the purpose for which it was designed. M - S contacts are fabricated by depositing a metal layer over the semiconductor surface. Thus, both materials are in intimate contact (Table 3). Metal deposition can be done by vapor deposition (Carroll et al., 1966), metal sputtering (Murarka, 1983) and electroless techniques (Wolfgang, 1991), for example. Vapor is used in manufacturing to deposit thicker layers. If the device structure requires a thinner layer, the sputtering technique is used. Electroless techniques are also used for thinner layers. However, they are not used often in manufacturing due to cycle time and process issues. A technique is selected depending on the semiconductor device structure and application. Depositing a metal on the semiconductor implies forming an interface between two materials. The characteristics of this interface present many different issues, such as mechanical, physical, chemical, and mainly electrical issues. Why is necessary to deposit a metal layer over the semiconductor surface? The purpose is to protect to the semiconductor surface from the environmental conditions. Most semiconductors are easily oxidized; metals are less susceptible to oxidation. Therefore, the purpose of the metal layer is the protection of the semiconductor surface and mechanical support for the external terminal connections. Otherwise, the device performance would be unstable and easily degraded. Mechanical, electrochemical and physical issues at the interface impact the main behavior of interest: the electrical behavior.

Ohmic behavior	Rectifying behavior
Height barrier equals to zero(Rhoderick & Williams, 1998; Salinas et al., 2006; Sze, 1990)	Barrier height positive and not equal to zero.
Minimum resistance stated at the M - S interface.	Maximum or high resistance state.
Electrons and holes are able to flow to or from the semiconductor	Electrons can flow in only one direction.

Table 3. Electrical behavior of M - S contacts based on barrier height level.

Therefore, whether the current can flow to the external circuit depends on the electrical characteristics. The crystalline silicon industry is mature and manufactures electronic devices, and thus, the M - S interface issues have already been overcome or at least minimized. When a metal and a semiconductor are in intimate contact because the work functions of both materials are different, a built-in barrier is created at their interface (Salinas et al., 2006), which is called a Schottky barrier. The barrier height is directly related to the difference in the Fermi levels between the metal and semiconductor material. The larger the difference is, the higher will be the barrier height is. The barrier opposes the flow of the free charge carriers from one side to another. The electrical behavior of M - S contacts is identified depending on the barrier height (Salinas et al., 2006; Sze, 1990).

When a new semiconductor material is proposed to build electronic devices, research on the M - S interface must be done. For nanocrystalline porous silicon the panorama is not as clear as that for crystalline silicon. The electrical characterization of M - S with different metal layers must be done. If the Schottky barrier is equal or close to zero, an ohmic contact is expected. The current can flow inside or outside the device with minimum opposition, and the relationship between electrical potential (V) and current (I) is governed by *Ohm's* law (Salinas et al., 2006; Sze, 1990), the contact is considered ohmic. If the barrier height is not close to zero, a rectifying contact can be expected. An ohmic contact affects the electrical performance of the device with a minimum or insignificant impact. There is a condition of minimum resistance across the contact, and therefore, free charge carriers can flow in or out of the device. However, rectifying contacts play an important role in different applications. In addition to these two types of contacts, a third type of contact could be formed if the semiconductor is heavily doped. In this special case, the Schottky barrier is sufficiently thin to let carries tunnel across it instead of jumping to overcome the barrier. There are many considerations to keep in mind during the analysis of M-S behavior. One consideration, for example, is the interfacial states, which are present at the mechanical junction of the contact, such as unbonding atoms, a rough surface, and mechanical damage during the metal deposition. For an ideal M - S contact, interfacial states are not taken into account. If this assumption works, no deep analysis is needed. Otherwise, a different characterization technique must be used to find the electrical behavior of the interfacial states (Rhoderick & Williams, 1998). For ideal conditions, Schottky theory explains the interface behavior and establishes the method to estimate the barrier height value.

This theory is called Schottky in honor of the German physicist Walter H. Schottky, who developed it. According to Schottky theory:

If $\Phi_{metal} < \Phi_{p-semiconductor}$, a rectifying barrier must be formed at the interface.
If $\Phi_{metal} > \Phi_{p-semiconductor}$, an ohmic contact exists rather than rectifying behavior.
If $\Phi_{metal} > \Phi_{n-semiconductor}$, a rectifying barrier must be formed at the interface.
If $\Phi_{metal} < \Phi_{n-semiconductor}$, an ohmic contact exists rather than rectifying behavior.

Characteristics of the I vs. V curve of a Schottky junction can be described by the following equation (Rhoderick & Williams, 1998):

$$I = I_O \left[\exp\left(\frac{qV}{nkT} \right) - 1 \right], \tag{9}$$

where I_o is the reverse saturation current, which can be experimentally determined. If the transport mechanism for electrical current is given by thermoionic emission theory, the barrier height (ϕ_b) of the junction can be defined by the following equations:

$$\begin{aligned} I_O &= aA^{**}T^2 \exp\left(-\frac{q\phi_b}{kT} \right), \\ \phi_b &= -\frac{kT}{q} \ln\left(\frac{I_o}{aA^{**}T^2} \right). \end{aligned} \tag{10}$$

where A^{**} is the modified Richardson constant, which depends on the effective mass of electrons in the semiconductor (Rhoderick & Williams, 1998), T is the absolute temperature, a is the contact area, and k is the Boltzmann constant. In practice, this junction hardly meets the equation and can be described with the modified equation:

$$I = I_O \exp\left(\frac{qV}{nkT}\right)\left[1 - \exp\left(-\frac{qV}{kT}\right)\right], \tag{11}$$

where the ideality factor of the diode, n, is almost independent of the electrical potential (V) and is greater than 1. The equation can be simplified as

$$I = I_O \exp\left(\frac{qV}{nkT}\right) \quad when \quad V > 3kT/q, \tag{12}$$

From this last equation, the parameters I_o and n can be obtained from the intersection and slope of the straight line of the plot of ln I vs. V. However, it is recommended to obtain them from the plot of $\ln I/[1 - \exp(-\frac{qV}{kT})]$ $vs.$ V of Eq. 11 because the straight line involves all values of V and not only the zone of V greater than 3kT/q, which can determine the value I_o with accuracy. The deviation of linearity due to other transport mechanisms is better seen when plotting $\ln I/[1 - \exp(-\frac{qV}{kT})]$ $vs.$ V. Therefore, these recommendations are taken into account in this study to handle the experimental data of developed junctions.
Aluminum (Al), copper (Cu) and gold (Au) have work functions of Φ_{Al}=4.3 eV (Brabec et al., 2001), Φ_{Cu}= 4.6-4.7 eV (Rhoderick & Williams, 1998) and Φ_{Au}=5.1 eV (Brabec et al., 2001), respectively. To generate contacts of crystalline silicon,p-Si (10 Ω-cm) with an acceptor density of 10^{15} cm^{-3} = 10^{21} m^{-3} and nSi(10 Ω-cm) with a donor concentration of 10^{14} cm^{-3} = 10^{20} m^{-3} were used.
According to the $I/[1 - \exp(-qV/kT)]$ $vs.$ V curves of the metal contacts of p-type and n-type silicon with aluminum and copper (not shown here), the exponential behavior of the current in the potential range of -1 to 1 V is similar to a rectifier, and the rectifier ratio (F_R) at a given potential can be estimated with the following equation:

$$F_R = \frac{I_{max}}{I_{min}} \tag{13}$$

In the Cu:p-Si:Al contact, the rectifier behavior is governed by the Cu:p-Si contact because the Al:p-Si showed ohmic behavior:

- F_R is about 1.0×10^2 at $\pm$ 1 V.
- In the potential range of -1 V to 0.04 V, the reverse saturation current I_o is 1.57×10^{-6} A and n= 1.04
- Between 0.06 to 0.18 V, I_o is 2.1×10^{-6} A and n= 2.4
- The deviation of the ideal n value (n=1) could be due to the presence of the interfacial layer or recombination in the depletion region.
- Above 0.18 V a serial resistance 1239 ohms was determined by the procedure described in (Pierret & Neudeck, 1989).
- The high serial resistance could be due to the physical contact between copper and silicon.

The parameters of the Cu:n-Si:Al are the following:

- F_R is about 18 at $\pm$ 1 V.
- Under reverse bias, the linear behavior of the current indicates a decrement of the barrier height potential due to the interfacial layer.

- Between -1 to 0.05V, I_o is 1.34× 10^{-6} A and n=1.09. Therefore, the current is given by I= $1.34 \times 10^{-6} \exp(qV/(1.09kT))$.
- At 0.04 to 0.14 V, I_o is 8× 10^{-7} A and n= 2.0. The current is given by I= 8× 10^{-7} A $\exp(qV/(2.0kT))$.
- The high value of n indicates that the current is limited by the recombination in the depletion zone, which can be described by;

$$I_r = I_{ro}exp\left(\frac{qV}{2kT}\right)\left[1 - exp\left(-\frac{qV}{kT}\right)\right], \tag{14}$$

  where I_{ro} depends directly of the depletion weight.
- At high injection potential, the serial resistance is approximately 1799 ohms.

Fig. 12 displays the barrier height (ϕ_b) distribution of the silicon contacts with aluminum and gold metals. For the determination of the ϕ_b, it was assumed that the electrical current is governed by the thermoionic emission mechanism. Therefore, Eq. 10 was used. The Richardson constants (A^{**}) taken into account were 32 Acm^{-2} K^{-2} for p-Si and 112 Acm^{-2} K^{-2} for n-Si (Rhoderick & Williams, 1998).

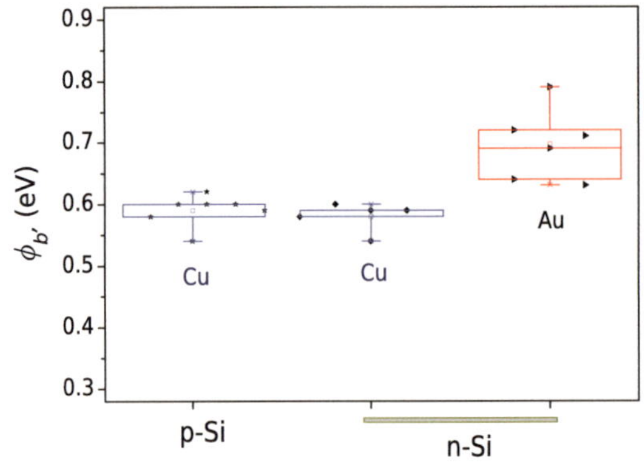

Fig. 12. Barrier height of metal contacts based on silicon.

5. NPS photovoltaic devices

5.1 Fundamental equations of a solar cell

A solar cell produces electrical energy by the absorption of solar irradiation without a secondary process. The electrical parameters of a photovoltaic device under dark conditions are given by (Sze, 1990)

$$I = I_O\left[\exp\left(\frac{qV}{nkT}\right) - 1\right], \tag{15}$$

where I is the current flow through the device under the influence of an electrical potential in direct bias V, I_O is the reverse saturation current, n is the diode ideality factor, k is the

Boltzmann constant, q is the electron charge, and T is the temperature.
If $qV/nkT > 3$, the exponential term of the diode equation is predominant. Therefore, direct bias of the I vs. V curve is governed by

$$I = I_o \exp\left(\frac{qV}{nkT}\right) \Longrightarrow \ln I = \ln I_o + \frac{qV}{nkT}. \quad (16)$$

where I_o and n can be estimated with the extrapolation at V = 0 and the slope of the plot ln I vs. V, respectively.
Considering the resistances of the device, the equation diode is modified as

$$I = I_O \left[\exp\left(\frac{q(V - IR_s)}{nkT}\right) - 1\right] + \frac{q(V - IR_s)}{R_{shunt}}. \quad (17)$$

where R_s and R_{shunt} are the serial and shunt resistances.

Under illumination, the current is given by the following equation:

$$I = I_O \left[exp\left(\frac{q(V - IR_s)}{nkT}\right) - 1\right] + \frac{q(V - IR_s)}{R_{shunt}} - I_L. \quad (18)$$

where I_L is the electrical current under illumination conditions.
The current under illumination for an arbitrary photovoltage is

$$I = I_O \left[\exp\left(\frac{qV}{nkT}\right) - 1\right] - I_L. \quad (19)$$

where I_{sc} is the short circuit current at $V = 0$.
If $I = 0$, Equation 19 is simplified to obtain (V_{oc}):

$$V_{OC} = \frac{kT}{q} \ln\left[\frac{I_L}{I_O} + 1\right]. \quad (20)$$

The conversion efficiency, η, is given by

$$\eta = \frac{P_{max}}{A * P_{in}} * 100 = \frac{I_{sc} * V_{oc} * FF}{A * P_{in}} * 100, \quad (21)$$

where V_{oc}, is the open circuit voltage, I_{sc} the short circuit current, V_{max}, I_{max} and P_{max} are the voltage, current and power maxima, respectively, FF is the fill factor, A is the effective area (m^2) and P_{in} is the incident irradiation (W/m^2).

5.2 Photovoltaic NPS based devices

NPS is widely used in optoelectronic applications (e.g., photonic and electroluminiscent devices). This nanocrystalline porous material has been used as a reflector layer in solar cell devices due its large light-trapping. Few works on the photovoltaic effect of NPS (Arenas et al., 2005; 2006;a; Smestad et al., 1992) indicate the need for continued research in this field to understand the mechanism charge carrier transport in NPS according to the type of silicon substrate, which is part for its fabrication.
NPS devices from p-Si and n-Si were fabricated using aluminum as the back contact and copper as the front contact. Both devices depicted the exponential behavior of the current under dark conditions, as shown in Fig. 13. The graphic adjusted to a diode rectifier with a

high confidence level. Experimentally, linear current behavior was found in the metal contacts of the Cu:NPS film and Al:p-Si substrate. Therefore, the rectifier behavior in the p-Si device is only attributed to the NPS:p-Si interface. In the NPS:n-Si device, the rectification contribution was mainly due to the Cu:NPS, shown in Table 4. The rectification ratio at $\pm$ 1 V was on the order of 10^3 for both devices. In fact, the NPS layer modified the electrical parameters of the silicon devices, J_o decreased by four orders of magnitude and the resistance increased one order of magnitude. In all devices, the n values was far from that of an ideal diode, suggesting that the current transport was limited by the depletion zone (Pierret & Neudeck, 1989; Rhoderick & Williams, 1998).

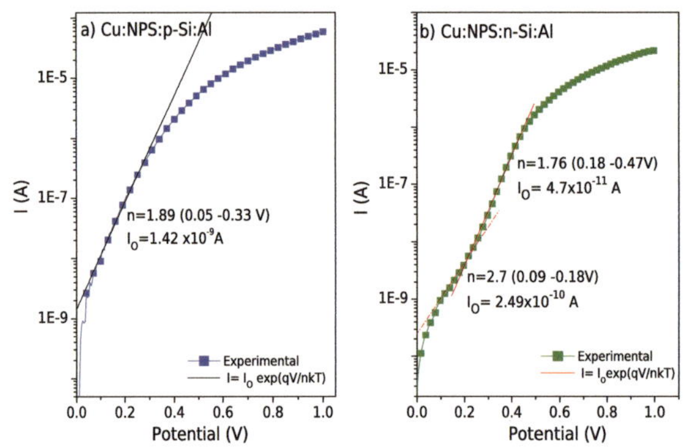

Fig. 13. Current - voltage curves under dark conditions of NPS devices based on p-Si and n-Si.

Under illumination, the photovoltaic effect is evident in the NPS devices, as shown in Fig. 14. The current density is about 0.13 to 0.32 mA/cm^2, and the open circuit voltage average is 235 mV for NPS:p-Si devices and 330 mV for NPS:n-Si devices. The photovoltaic effect was also observed in silicon devices without an NPS layer, suggesting that it is caused by the Schottky diode of the copper with the semiconductors. A thicker NPS film under the silicon substrate shows a similar behavior, indicating that the photovoltage is based on Cu:NPS and the Cu:n-Si junctions (Arenas et al., 2008).

Devices	J_O (mA/cm^2)	F_R at $\pm$1V	n (at V)	R_s (ohms)	J_{sc} (mA/cm^2)	V_{oc} (mV)	FF	η (%)
p-Si	0.17	1×10^2	2.03 (0.04-0.17)	1239	—	—		
NPS:p-Si	1.59×10^{-4}	2×10^3	1.89 (0.05-0.33)	8557	0.13	235	0.33	0.016
n-Si	0.15	18	3.04 (0.07-0.29)	1799	0.32	355	Lower	
NPS:n-Si	2.8×10^{-5}	1×10^3	2.7 (0.09-0.18)	18441	0.2	330	Lower	0.96

Table 4. Electrical parameters of NPS devices based on p-Si and n-Si.

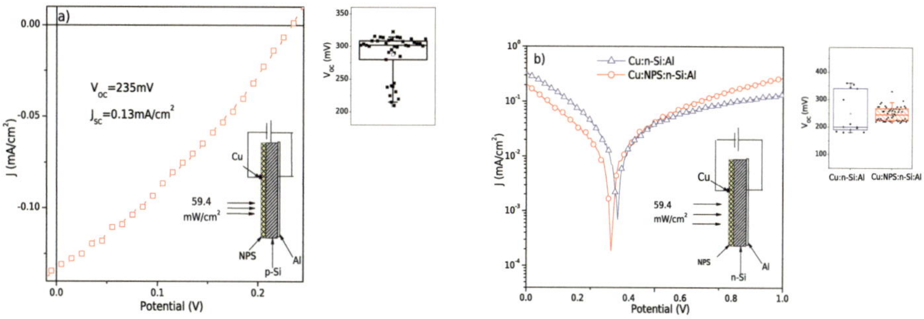

(a) NPS from p-Si (b) NPS from n-Si

Fig. 14. J vs. V curves under illumination conditions of NPS devices from p-Si and n-Si (Arenas et al., 2008).

The contribution of the photocurrent and photovoltage in the heterojunction was monitored by the spectral response, as shown in Fig. 15. The relevant points for NPS:p-Si devices are described below:

- The photocurrent and photovoltage spectra are similar in the range of 1 to 3.5 eV of photon energy.
- Two zones are well defined, the first in the infrared region (1-1.47 eV) and the second in visible region (1.47 eV -3.25 eV).
- In the infrared region, the contributions are due to the absorption of bulk p-silicon, where the maximum peak consists of the energy band gap of bulk silicon.
- The contribution of NPS is evident in the visible zone, where the NPS presents high absorption (Eg 1.8 eV).
- Four smaller interferences (steps) are shown in the range of 2.11 to 2.63 eV. The average between these steps is about 0.17 eV $\pm$ 0.02.
- Similar steps were observed in the photovoltage response of the NPS device based on aluminum, which were related with the distribution sizes of the nanocrystalline silicon in the NPS layer (Yan et al., 2002).
- Two minima are seen at 1.47 eV and 1.85 eV. The first decrement of energy is due to the end of the contributions of bulk silicon and the start of the contributions of the NPS. The second decrement is due to the radiative recombination of charge carriers caused by the photoluminescence process (Wang et al., 1993; Zhang et al., 1993).

For NPS:n-Si devices, the photovoltage and photocurrent spectral response were very different than that of NPS devices fabricated from the p-Si substrate:

- The Cu:n-Si and Cu:NPS:n-Si devices showed similar behavior in terms of spectral response.
- Only the sharp peak at 1.2 eV is displayed in both spectra. It suggests that the energy band gap of NPS is similar to the energy band gap of silicon substrate or well, the contribution of the NPS to the photovoltaic effect is negligible. The absence of photocurrent from the NPS layer is attributed to the recombination of charge carriers due to the dangling bonds (Hwang et al., 2011).

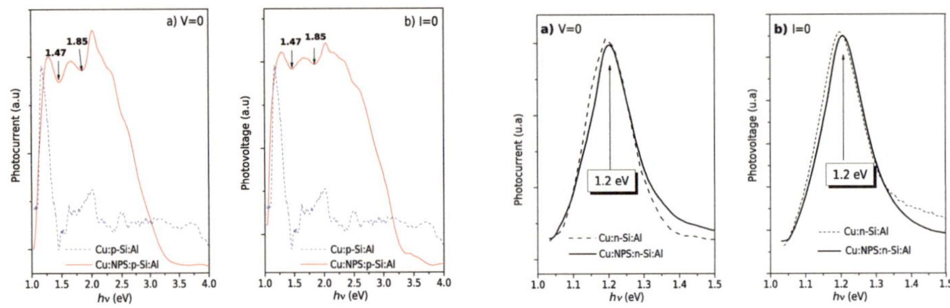

(a) NPS from p-Si (b) NPS from n-Si

Fig. 15. Photocurrent and photovoltage of NPS devices from p-Si and n-Si (Arenas et al., 2008).

An energy diagram for NPS from p-Si (Fig. 16) is shown with the experimental data of Eg ($\approx$ 1.88 eV) and the electronic afinity of NPS ($\chi \approx$ 3.6 eV (Peng et al., 1996)). The data for crystalline silicon were also taken into account (Eg=1.12 eV): $E_F = \approx 4.99$ eV for p-Si of 10 Ω-cm (Sze, 1990). The internal electrical field originated at the interface of the NPS:p-Si junction causes the opposite charge carriers to reach their respective metal contacts: electrons to Cu through NPS and holes to Al through p-Si. The photovoltage or photocurrent responses of the device were produced by the photogeneration of both electrons and holes in p-Si for photon energies greater than 1.12 eV and in NPS for energies greater than 1.8 eV.

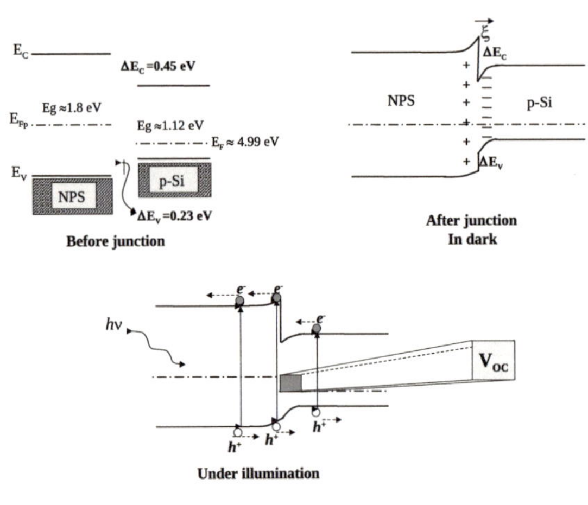

Fig. 16. Flat energy band diagrams of NPS devices based on p-Si before and after intimate contact and under illumination conditions.

5.3 Hybrid photovoltaic NPS:polypyrrole devices

A novel hybrid heterojunction based on NPS and polypyrrole (PPy) was proposed as a promising heterojunction for solar cell applications (Arenas et al., 2005; 2006;a; 2008). The conducting polymer improved the electroluminescent and photoluminescent properties of NPS (Antipán & Kathirgamanathan, 2000; Bsiesy et al., 1995; Halliday et al., 1996; Parkhutik et al., 1994). The nanocrystallinity and the pore sizes are important parameters of the NPS layer because of their influence on the topography of the PPy:NPS devices and consequently the final performance of the PPy:NPS:n-Si devices (Arenas et al., 2006a):

- First, the photovoltaic response is present in PPy:n-Si devices without any NPS layer (V_{oc}=135 mV, J_{sc}=8.58 mA/cm^2).
- The linear I - V curve trace under light is due to the high serial resistance (10^4 ohms), and the efficiency conversion reached was 0.96%, as shown in Fig. 17a.
- The rough topography of the tip-like morphology of PPy:NPS devices leads to lower values of V_{oc}=60 mV and J_{sc}= 9.73×10^{-3} mA/cm^2 compared to PPy:n-Si. The efficiency conversion was approximately 2×10^{-4}%.
- A smooth and agglomerated morphology led to the following electrical parameters of the devices: V_{oc}=95 mV and J_{sc}=0.13×10^{-3} mA/cm^2.

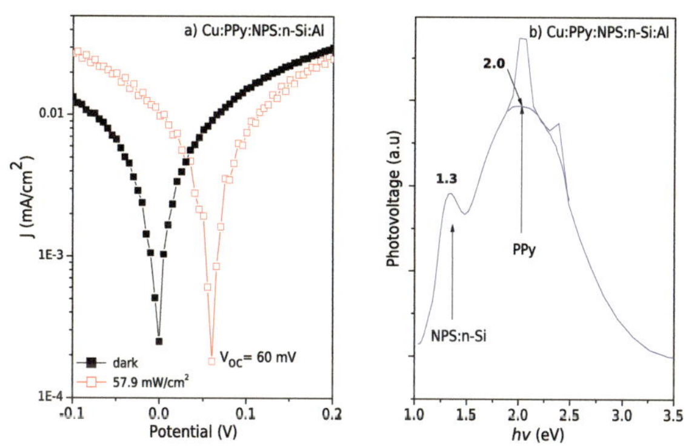

Fig. 17. a) J vs. V in dark and illumination conditions and b) photovoltage spectra of an NPS device based on polypyrrole.

The photovoltaic spectra displayed two peaks between 1 and 3 eV, as shown in Fig. 17b. The first acute peak is in the energies of 1 to 1.47 eV, and the second broad peak is at 1.47 to 3 eV, related to the contributions of the n-Si and PPy layers, respectively. The maximum peak at 1.9 eV corresponds to the energy band gap of PPy and is indicative of both components of the photogeneration of the charge carriers. The internal electrical field in the PPy:n-Si slightly aids the photogeneration of charge carriers.

6. Conclusion

This chapter focused on the preparation, characterization and systematic electrical evaluation of NPS based photovoltaic devices. The large surface area of NPS makes it a promising material for optoelectronic devices. The main structure of NPS is based on silicon crystals of nanometric size, which depend on which silicon type is used. Its experimental energy band gap of 1.8 eV leads to an absorption range in the visible spectra, which is an advantage if it is required as an active absorbing material in solar cells. The results shown in this chapter demonstrate that NPS could represent a good alternative to develop solar cells based on hybrid heterojunctions. However, it is necessary to continue researching strategies to dope NPS to increase its electrical conductivity and therefore improve the conversion efficiency of hybrid devices.

7. Acknowledgments

Antonio del Rio and Hailin Hu from CIE-UNAM by their advice, and IACOD-DGAPA (I1102611 project) for the financial support.

8. References

Anderson, R.C., Muller, R.S. & Tobias, C.W. (1990). Investigations of porous silicon for vapor sensing, *Sensors and Actuators A: Physical* 23: 835-839.

Antipán J. & Kathirgamanathan, P. (2000). White light electroluminescence from P Si devices capped with poly(thiophene)(s) as top contact, *Synthetic Metals* 110 (3): 233-240.

Arenas, C., Hu, H., del Río, J. A., Salinas, O.H. & Sánchez, A. (2005). Porous silicon with conducting polymers for photovoltaic applications, *Proceedings of 20th European Photovoltaic Solar Energy Conference and Exhibition*, Spain, 220-222.

Arenas, M. C., Hu, H., Del Río, J. A., Sánchez, A. & Nicho-Díaz, M. E. (2006). Electrical properties of porous silicon/polypyrrole heterojunctions, *Solar energy materials & solar cell* 90: 2413-2420.

Arenas, M. C., Hu, H., del Río, J. A. & Nicho, M. E. (2006). Morphology Study of a hybrid structure base on porous silicon and polypyrrole, *Proceedings of Materials Research Society Symposium* 939: 0939-O03-22.

Arenas, M.C., Hu, H., Del Río, J. A. & Salinas O. H. (2008). Photovoltage and J-V features of porous silicon, *Revista Mexicana de Física* 54(5): 391-396.

Arenas, M. C., Hu, H., Nava, R., & del Río, J. A. (2010). Determination of the complex refractive index of porous silicon layers on crystalline cilicon substrates, *International Journal of Modern Physics B* 24(24): 4835-4850.

Bard, A. J. (1986). *Encyclopedia of Electrochemistry of the Elements*, Dekker, New York.

Beckmann, K.H., (1965). Investigation of the chemical properties of stain films on silicon by means of infrared spectroscopy, *Surface Science* 3(4): 314-332.

Bisi, O., Ossicini, S. & Pavesi, L. (2000). Porous silicon: a quantum sponge structure for silicon based optoelectronics, *Surface science reports* 38: 1-126.

Brabec, C.J., Shaheen, S.E., Fromherz, T., Padinger, F., Hummelen, J.C., Dhanabalan, A., Jassen, R.A.J. & Sariciftci, N.S. (2001). Organic photovoltaic devices produced from conjugated polymer / methafullerene bulk heterojunctions, *Synthetic Metals* 121: 1517-1520.

Bsiesy, A., Nicolau, Y.F., Ermolieft, A., Muller, F. & Gaspard, F. (1995). Electroluminescence from n^+-type porous silicon contacted with layer-by-layer deposited polyaniline, *Thin Solid Films* 255:43-48.

Bomchil, G., Herino, R., Barla, K. & Pfister J. C. (1983). Pore size distribution in porous silicon studied by adsorption isotherms, *Journal of Electrochemical Society* 130: 1611-1614.

Canham, L.T. (1990). Silicon quantum wire array fabrication by electrochemical and chemical dissolution of wafers, *Applied Physics Letter* 57: 1046-1048.

Carroll F. Powell, Joseph H. Oxley, Johj M. & Blocher Jr., (1966), *Vapor Deposition*, John Wiley & Sons.

Conway, B.E., Bockris, J.O M., Yeager, E., Khan, S.U.M. & White, R.E. (1983). *Comprehensive Treatise of Electrochemistry* Plenum, New York.
Dillon, A. C., Robinson, M. B., Han, M. Y. & George S. M. (1992). Diethylsilane Decomposition on silicon surfaces studied using Transmission FTIR spectroscopy, *Electrochemical Society* 139: 537-543.
Eddowes, M. J. (1990). Anodic dissolution of p- and n-type silicon: Kinetic study of the chemical mechanism, *Journal of Electroanalytical Chemistry* 280: 297-311.
Fauchet, P.M., Tsybescov, L., Duttagupta, S.P. & Hirshmann, K. D. (1997). Stable photoluminescence and electroluminescence from porous silicon, *Thin Solid Films* 297 (1-2): 254-260.
Feng, Z. C. & Tsu, R. (1994). *Porous silicon*, World Scientific.
Galun, E., Reuben, C., Matlis, S., Tenne, R. & Levy-Clement, C. (1995). Morphology of n-type macroporous silicon: doping density dependence, *Journal of Physical Chemistry* 99: 4132-4140.
Gupta, P., Dillon, A.C., Bracker, A.S. & George S.M. (1991). FTIR studies of H_2O and D_2O decomposition on porous silicon surfaces, *Surface Science* 245: 360-372.
Halliday, D. P., Holland, E. R., Eggleston, J. M., Adams, P.N., Cox, S.E. & Monkman, A.P. (1996). Electroluminescence from porous silicon using a conducting polyaniline contact, *Thin Solid Films* 276(1-2): 299-302.
Heavens, O. S. (1965). *Optical properties of thin solid films*, Dover Publications.
Holmes, P. J., (1962). *The Electrochemistry of Semiconductors*, Academic, London.
Hwang, J.D., Hwang, S.B., Chou, C.H. & Chen, Y.H. (2011). Investigation of opto-electronic properties on gradient-porosity porous silicon layer, *Thin Solid Films* 519 (7): 2313-2316.
Jung, K. G., Schultze, J. W., Thönissen, M. & Münder, H. (1995). Deposition of electrically conducting polybithiophene into porous silicon, *Thin Solid Films* 255: 317-320.
Kordás, K., Pap. A. E., Beke, S. & Leppävuori, S., (2004). Optical properties of porous silicon (Part I) - Fabrication and investigation of single layers, *Optical Materials* 25(3): 251-255.
Lehmann V., Föll, H. (1990). Formation Mechanism and properties of electrochemically etched trenches in n-type silicon, *Journal of Electrochemical Society* 137: 653-659.
Lehmann, V. & Gösele, U. (1991). Porous silicon formation: a quantum wire effect, *Applied Physics Letters* 58: 856-858.
Lugo, J. E., del Río, J. A. & Tagüe na-Martínez J. (1998) Surface contribution to the effective optical properties of porous silicon, *Solar Energy Materials and Solar Cells* 52(3-4): 239-249.
Manifacier, J.C., Gasiot, J. & Fillard, J.P. (1976). A simple method for the determination of the optical constants n, k, and the thickness of a weakly absorbing thin film, *Journal of Physics E: Scientific Instruments* 9(11): 1002.
Moreno, J. D., Marcos, M.L. & Agulló-Rueda F. (1999). A galvanostatic study of the electrodeposition of polypyrrole into porous silicon, *Thin Solid Films* 348: 152-156.
Murarka, S.P. (1983). *Silicides for VLSI Applications*, Academic Press, Inc.
Nguyen, T.P., Le Rendu, P. & Cheah K. W. (2003). Optical properties of porous silicon/poly(p phenylene vinylene) devices, *Physica E* 17: 664-665.
Parkhutik, V.P., Diaz Calleja, R., Matveeva, E.S. & Martinez-Duart, J.M. (1994). Luminescent structures of porous silicon capped by conductive polymers, *Synthetic Metals* 67: 111-114.

Patterson, A.L. (1939).The Scherrer Formula for X-Ray Particle Size Determination, *Physical Review 56* 15: 978-982.

Peng, C., Hirschman, K.D, Fauchet, P.M. (1996). Carrier transport in porous silicon light-emitting devices, *Journal of Applied Physics* 80(1): 295-300.

Pierret, R.F. & Neudeck, G. W. (1989). *The PN Junction Diode*, Addison-Wesley Publishing Company.

Rhoderick, E.H. & Williams, R.H. (1998). *Metal âĂŞ Semiconductor contacts*, Claredon Press Oxford.

Sagnes, I., Halimaoui, A., Vincent, G. & Badoz, P.A., (1993). Optical absorption evidence of quantum size effect in porous silicon, *Applied Physics Letters* 62(10): 1155-1157.

Salinas, O. H., López-Mata, C., Hu, H., Nicho, M. E. & Sánchez, A. (2006). Metal contact properties of poly3-octylthiophene thin films, *Solar Energy Material & Solar Cells* 90: 760-769.

Schroder, D. K. (1998). *Semiconductor material and device characterization*, John Willey & Sons.

Searson, P. C., Macaulay, J. M. & Ross, F. M. (1992). Pore morphology and the mechanism of pore formation in n-type silicon, *Journal of Applied Physics* 72: 253-258.

Smestad, G. & Kunst, M. (1992). Photovoltaic response in electrochemically prepared photoluminescent porous silicon, *Solar Energy Materials and Solar Cells* 26: 277-283.

Smith, R. L. & Collins, S. D. (1992). Porous silicon formation mechanisms, *Journal of Applied Physics* 71: R1–R22.

Swanepoel, R. (1983). Determination of the thickness and optical constants of amorphous silicon, *Journal of Physics E: Scientific Instruments* 6: 1214.

Sze, S.M. (1990), *Physics of semiconductor devices*, John Wiley & Sons.

Torres-Costa, V., Gago, R., Martín-Palma, R.J., Vinnichenko, M., Grötzschel, R., & Martínez-Duart, J.M. (2003). Development of interference dilters based on multilayer porous silicon structures, *Materials Science and Engineering: C* 23(6-8): 1043-1046.

Turner, D. R. (1958). Electropolishing silicon in hydrofluoric, *Journal of Electrochemical Society* 105: 402-408.

Vijh, A.K. (1973). *Electrochemistry of Metals and Semiconductors*, Dekker, New York.

Vitanov, P., Kamenova, M., Tyutyundzhiev, N., Delibasheva, M., Goranova, E. & Peneva M. (1997). High-efficiency solar cell using a thin porous silicon layer, *Thin solid Films* 297: 299-303.

Wang, X., Huang, D., Ye, L., Yang, M., Hao, P., Fu, H., Hou, X. & Xie, X. (1993). Pinning of photoluminescence peak positions for light-emitting porous silicon: An evidence of quantum size effect, *Physical Review Letters* 71(8): 1265-1267.

Wolfgang R. (1991). *Electroless Nickel Plating*, ASM International.

Yan, F., Bao, X.-M. & Gao, T. (2002). Photovoltage spectra of silicon/porous silicon heterojunction, *Solid State Communications* 91(5): 341-343.

Yu, Z. & Wie, C.R. (1992). Fabrication of MSM photoconductor on porous Si using micromachined silicon mask, *Electronics Letter* 28: 911-913.

Zhang, S.-L., Ho, K.-s., Hou, Y., Qian, B., Diao, P. & Cai, S. (1993). Steplike behavior of photoluminescence peak energy and formation of p-type porous silicon, *Applied Physics Letters* 62(6): 642-644.

12

Porous Silicon Integrated Photonic Devices for Biochemical Optical Sensing

Ilaria Rea[1], Emanuele Orabona[1,2], Ivo Rendina[1] and Luca De Stefano[1]
[1]*Institute for Microelectronics and Microsystems - Unit of Naples Research National Council, Naples*
[2]*Dept. of Physics, University of Naples "Federico II", Naples Italy*

1. Introduction

In the last few years, great efforts have been spent in the development of integrated microsystems, devices of few square centimeters in size including microsensors, microfluidic components, reaction chambers, detectors, and so on. More than a simple ensemble of devices, this is a new research field that combines the properties and characteristics of different materials to find innovative and affordable solutions in applications such as sensing, biotechnology, analytical chemistry. The device miniaturization not only means lower costs through mass production, but also improvement in terms of analysis time, simplicity of use and decrease in consumption of materials (reagents and analytes) (Chandrasekaran et al., 2007). The integrated devices are largely made of silicon but can also include a microfluidic systems; for this reason, their technology is based both on the techniques used in integrated circuit manufacturing and on "soft" fabrication methods (Xia & Whitesides, 1998).
In this chapter, we describe the fabrication and the characterization of integrated photonic devices based on nanostructured silicon for biochemical optical sensing. The porous silicon (PSi) is fabricated by electrochemical etching of doped crystalline silicon in an aqueous solution of hydrofluoridric acid. It can be simply described as a network of air holes in a silicon matrix: its dielectric properties, and in particular the refractive index, depend on the content of void, which can be accurately controlled by tuning the process parameters, so that different structures (Fabry-Perot interferometer, Bragg mirror, optical microcavity, aperiodic multilayered sequences) showing good quality optical responses can be obtained. Like other porous materials, PSi is an ideal platform for biosensing due to its high specific surface area (~100 $m^2 \cdot cm^{-3}$) which assures an efficient interaction with the species to detect. However, the integration of PSi sensing structures in a microsystem is not straightforward: its surface instability and the low compatibility with alkaline treatments, frequent in devices fabrication, are severe limitations in this application field. In this chapter, we analyze these technological limits and propose solutions that have led to the realization of innovative and high-performant integrated devices using porous silicon as functional platform in bio-analysis experiments.

2. Properties of porous silicon

PSi is a very versatile material due to its peculiar morphological, physical, and chemical properties: evidence of this is the huge number of papers about PSi features and devices based on this nanostructured material that appear in the literature every year. One reason for this clear success is the easy fabrication of sophisticated optical multilayers, such as one-dimensional photonic crystals, by a simple, but not trivial, computer-controlled electrochemical etching process.

2.1 Porous silicon fabrication by electrochemical etching

Porous silicon was discovered in 1956 by the Uhlirs at Bell Labs, USA, during a study on the electropolishing of crystalline silicon in an HF-based solution. They observed the formation of a deposit "tentatively supposed to be a Si suboxide" (Uhlir, 1956). The scientific community was not much interested in porous silicon until to 1990 when Leigh Canham, working at the Defence Research Energy in England, demonstrated an efficient tunable room temperature light output from the material (Canham, 1990). In the years later, thousands of papers were published on porous silicon and its potential applications in microelectronics, optoelectronic devices, chemical and biological sensing.

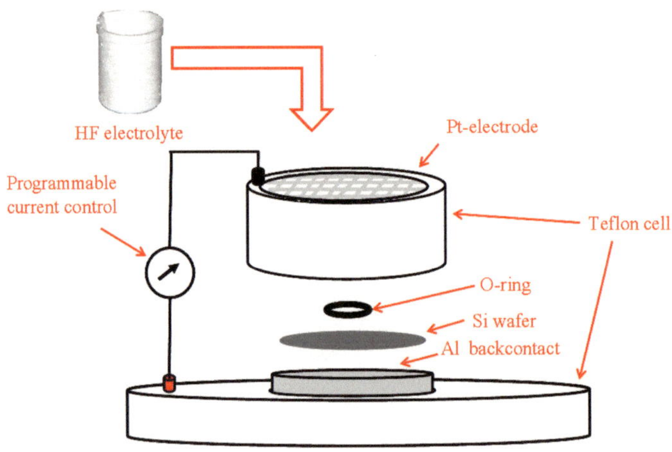

Fig. 1. Electrochemical etching setup.

In most cases, the porous silicon structure is formed by electrochemical dissolution of doped crystalline silicon wafers in hydrofluoric acid (HF) based solution. The cell used for the electrochemical etching is schematized in Figure 1. The silicon wafer is the anode; it is placed in back-side contact on an aluminum plate while the front side is sealed with an O-ring and exposed to the anodising electrolyte. The cathode is made of platinum. The anodisation cell is made of a highly-acid resistant polymer such as polyvinylidene fluoride (PVDF) or polytetrafluoroethylene (PTFE). For highly resistive silicon substrate (> few mΩ/cm) an evaporated metal backside contact is necessary to ensure the Schottky contact between the aluminum plate and the semiconductor.

Figure 2 shows the anodic I-V curve of silicon in HF based solution; the different dissolution regions of the curve are labeled A-C. Pore formation takes place in region A. At anodic overpotentials in excess of the current "peak," region C, silicon electropolishes. At intermediate overpotentials, region B, a "transition" zone exists where pore formation and electropolishing compete for control over the surface morphology. The resulting structure within this region is generally porous in nature but the pore diameters increase rapidly as the electropolishing potential is approached.

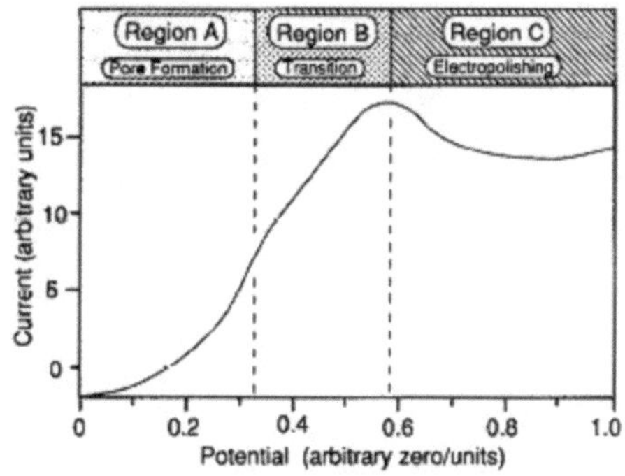

Fig. 2. Anodic I-V curve for silicon in HF. In the region A pore formation occurs. In region C there is the silicon electropolishing. The region B is a transition zone where pore formation and electropolishing compete (Smith & Collins, 1992).

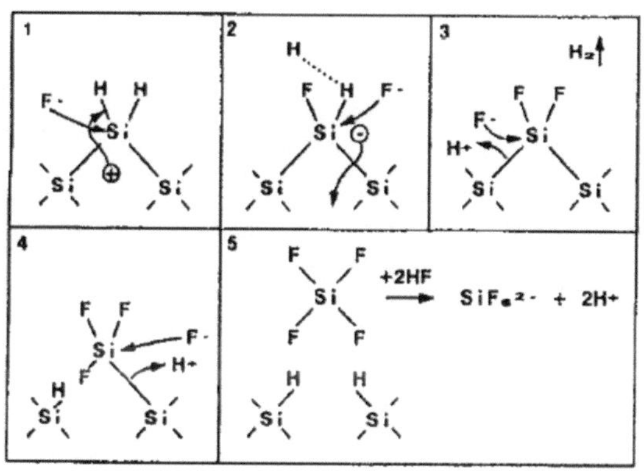

Fig. 3. Dissolution mechanism of silicon in hydrofluoridric acid (HF) (Lehmann & Gösele, 1991).

The exact dissolution chemistries of silicon are still in question, although it is generally accepted that holes are required in the initial steps for both electropolishing and pore formation. The Figure 3 illustrates the chemical dissolution mechanism suggested by Lehmann and Gösele ((Lehmann & Gösele, 1991), that has received great attention. If a hole of the silicon substrate has the sufficient energy to reach the surface, a nucleophilic attack on Si-H bond by fluoride ion can occur and a Si-F bond is formed (step 1 in Figure 3). Due to the polarizing influence of the Si-F bond, another F– ion can attack and bond under generation of an H_2 molecule and injection of one electron into the substrate (steps 2 and 3). The polarization induced by the Si-F groups reduces the electron density of the Si-Si backbonds; these weakened bonds will now be attacked by HF or H_2O (step 4) in a way that the silicon surface atoms remain bonded to hydrogen atoms (step 5). The silicon tetrafluoride molecule reacts with the HF to form the highly stable H_2SiF_6.

In the case of n-type silicon, where holes are minority carriers, the electrochemical dissolution of the material strongly depends on the hole/electron pair generation by illumination. In fact, when lightly doped (majority carrier concentration below ~10^{18} cm^{-3}) n-type silicon is anodised in the dark, the formation of PSi is observed only at high voltages (>5 V). If the anodisation is perform under illumination, PSi is formed at lower potentials (<1 V).

2.2 Porous silicon morphology

The PSi shows a great variety of morphologies dependent on the doping type and level of the silicon substrate and the electrochemical etching parameters. Usually for a given substrate and electrolyte, only one type of pore structure can be obtained.

The IUPAC (International Union of Pure and Applied Chemistry) guidelines define ranges of pore sizes that exhibit characteristic absorption properties (Sing et al., 1985): pores characterized by a diameter ≤ 2nm define microporous silicon; for sizes in the range 2-50 nm the PSi is mesoporous; pores diameters > 50 nm are typical of macroporous silicon.

Highly doped (10^{19} cm^{-3}) p- or n- silicon anodized in aqueous HF solution usually forms mesopores with pore sizes from 20 nm to 50 nm (Figure 4 (a)). Several groups have demonstrated that complex optical devices can be designed with mesoporous silicon (Mulloni & Pavesi, 2000). In the case of lightly doped (below ~10^{18} cm^{-3}) p-type silicon, the porous size distribution is normally found in the range 1 - 5 nm, in the lower limit for mesopores (Figure 4 (b)). The influence of the HF concentration has been studied (Ruike et al., 1996): it has been found that the pore size distribution broadens and the mean pore radius increases when the HF concentration in the etching solution is decreased from 55 to 37 %, but no significant changes are observed for lower HF concentrations. The electrochemical etching of lightly doped n-type substrates in the dark results in the formation of a macroporous material with radii in the micrometer range. Under sample illumination mesopores and/or micropores are formed, whereas macropores are still found (Figure 4 (c)).

2.3 Porosity and refractive index determination

The most important parameter of the PSi is the porosity, defined as the fraction of void within the porous layer. The easiest way to determine the porosity is by weight measurements. The wafer is weighed before the anodic reaction (m_1), after the anodic reaction (m_2) and finally after dissolution of the porous material in a molar NaOH aqueous solution (m_3). The porosity is given by the equation:

$$P = \frac{m_1 - m_2}{m_1 - m_3} \quad (1)$$

After the removal of the porous layer, it is also possible to determine its thickness by profilometric analysis. Another interesting technique, that allows to determine simultaneously both the thickness and the porosity of a PSi layer without destroying the material is the spectroscopic ellipsometry (SE) (Tompkins & McGaham, 1999). The method is based on the measurement of the change in the polarization state of the light over the spectral range after the reflection from the sample surface. Ellipsometry measures the complex reflectance ratio (ρ) defined by:

$$\rho = \frac{R_p}{R_s} = \tan\psi e^{i\Delta} \quad (2)$$

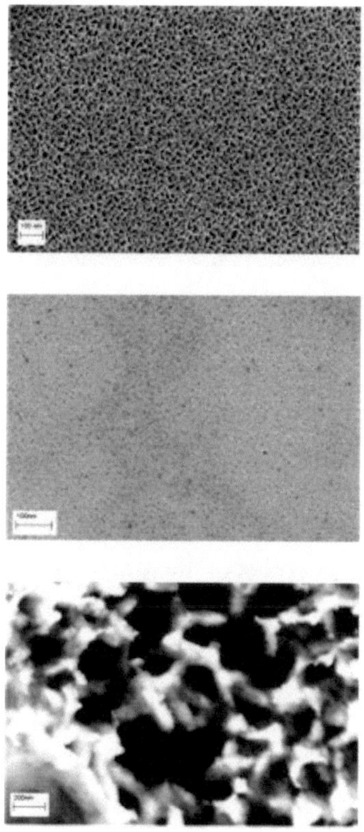

(a)

(b)

(c)

Fig. 4. SEM images illustrating different PSi morphologies. (a) p+ <100> silicon; the pores have an average width of 30 nm. (b) p <100> silicon; the pores size is about 2 nm. (c) n <100> silicon; the material consists of two parts, micro and macropores are found.

where R_p and R_s are the complex reflection coefficients of the light polarized parallel and perpendicular to the plane of incidence. Thus, ψ and Δ are, respectively, the amplitude ratio and the phase shift between s and p components of polarized light. From experimentally determined ρ, it is possible to obtain information about the properties of the material by performing proper model calculations. The use of SE in the near IR-UV spectral range for studying PSi has been largely reported in literature (Zangooie et al., 1999; Pickering et al., 1993). The PSi can be modeled by using a sequence of sublayers, each one constituted by air and crystalline silicon, in order to take into account the inhomogeneity along the layer normal (Zangooie et al., 1998) and adopting the Bruggeman effective medium approximation (EMA) (Aspnes et al., 1979) given in the Eq. 3:

$$(1-P)\left(\frac{\varepsilon_{Si}-\varepsilon_{PSi}}{\varepsilon_{Si}+2\varepsilon_{PSi}}\right)+P\left(\frac{\varepsilon_{void}-\varepsilon_{PSi}}{\varepsilon_{void}+2\varepsilon_{PSi}}\right)=0 \tag{3}$$

where P, ε_{si}, ε_{PSi}, and ε_{void} are the layer porosity, the dielectric constants of silicon, porous silicon and void, respectively. This approximation is acceptable because the size of the PSi pores is much smaller than the wavelengths of incidence light in the near IR-UV regions; in this range, the electromagnetic radiation does not distinguish between silicon and void, and it is possible to treat the PSi as a homogeneous medium.

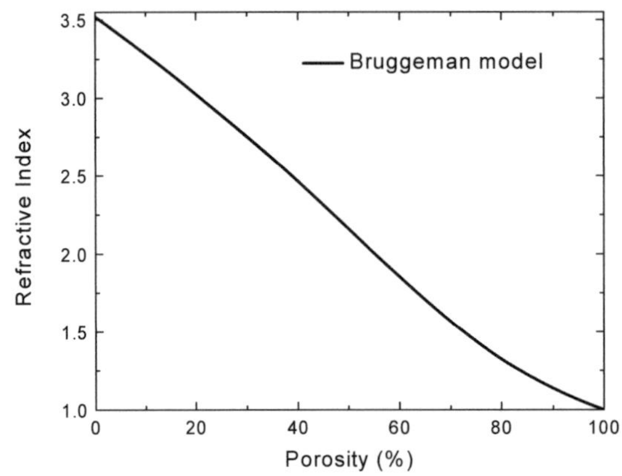

Fig. 5. Dependence of the PSi refractive index on the porosity given by the Bruggeman effective medium approximation.

The root square of the dielectric constant is the complex refractive index, $\tilde{n}=n+ik$ where n is the refractive index and the imaginary part k is the extinction coefficient. k is related to a, the linear absorption coefficient, by the relation $k=a\lambda/4\pi$ where λ is the optical wavelength. The Figure 5 shows the dependence of the PSi refractive index on porosity given by the Bruggeman model.

2.4 Thermal oxidation of porous silicon

The PSi is a material characterized by a high chemical reactivity; if stored in ambient air, the texture becomes partially oxidized and both the refractive index and the extinction coefficient change. To stabilize the PSi and eliminate the problem of aging, the thermal oxidation of the structure is used. The oxidation reduces or completely removes the Si from the skeleton substituting it by SiO_2, that isotropically grows also into the pores. The Figure 6 shows the SEM images of PSi layer (a) and oxidized PSi layer (b).

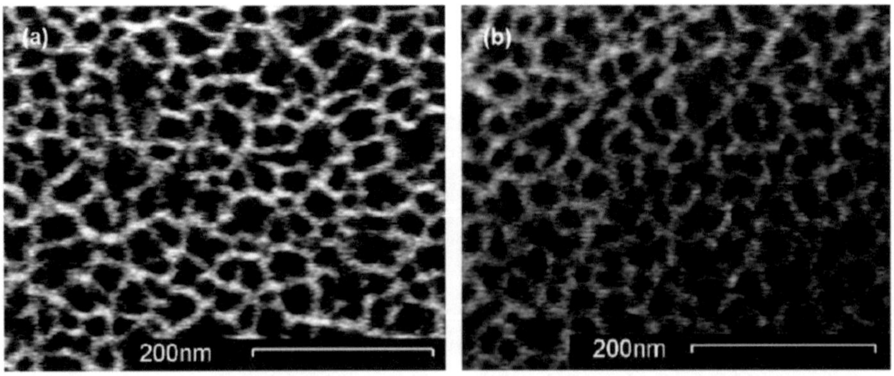

Fig. 6. SEM images of p^+ PSi layer (a) and p+ PSi layer pre-oxidized at 300°C for one hour following by an oxidation step at 900 °C in wet O_2 for 1 hour (b) (Pirasteh et al., 2006).

It is possible to observe that after the oxidation the pore shape is conserved but the size is reduced. In particular, the mean value of the pores width from 18 nm becomes 13 nm. Since the pores density is the same and the pore size decreased, the porosity after oxidation is lower than the porosity before oxidation (Pirasteh et al., 2006). The relationship between the porosities of a PSi layer before and after the thermal oxidation is expressed by the equation:

$$(1-P_{ox})=(1+1.27x)(1-P) \tag{4}$$

where P is the porosity before oxidation, P_{ox} the porosity after the oxidation and x is the oxidation extent which is a function of the temperature T and the duration t of the process:

$$x(T,t)=0.44\left(\frac{SiO_{2\,final}}{Si_{initial}}\right) \tag{5}$$

PSi is generally oxidized in a pure O_2 atmosphere by a two step thermal treatment. The first oxidation step, defined as pre-oxidation, at low temperature (300 - 400 °C) is required in order to selectively oxidise the silicon backbonds thus assuring an easy propagation of the oxidant into the silicon structure (Yon et al., 1987). Moreover, the pre-oxidation reduces the damage of the material texture in the second oxidation step at higher temperatures between 800 and 900 °C (Pap et al., 2005).

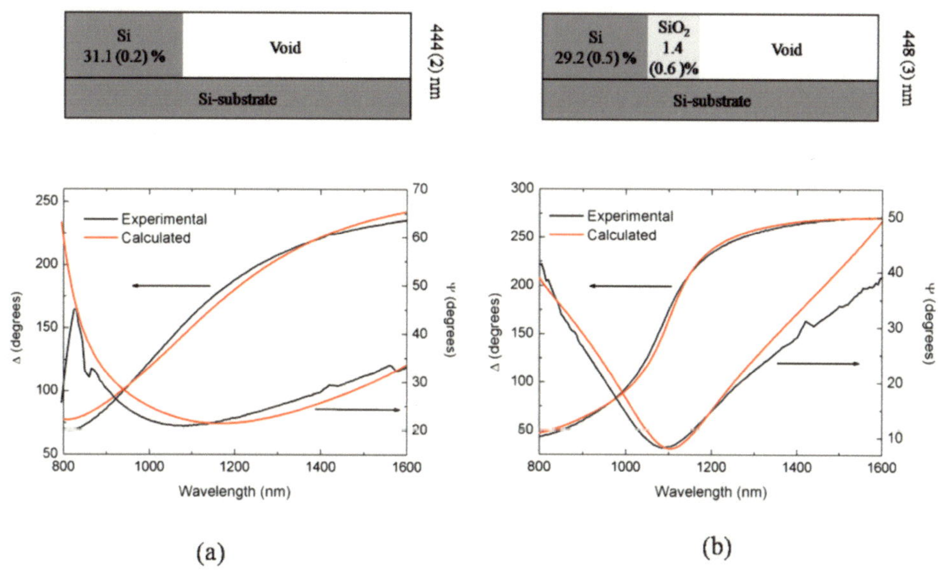

Fig. 7. Schematic diagram of the model layer used in the ellipsometric analysis and calculated ψ, Δ spectra compared with the experimental ones before (a) and after (b) the pre-oxidation at 400° C for 30 min of the p^+ PSi layer.

A p^+ PSi layer pre-oxidised at 400°C for 30 min has been characterized by spectroscopic ellipsometry before and after the thermal process. In Figure 7 are reported the schematic diagrams of the models used in the analysis and the calculated ψ and Δ spectra compared with the experimental ones before (a) and after (b) the pre-oxidation of the material. The value of the variable x has been estimated to be about 0.02; the oxidation interests only the surface of the material. A short (3-5 min) thermal treatment at 900°C completely oxidized the structure; in this case x=1 and the Eq. 5 becomes $(1-P_{ox}) = 2.27(1-P)$.

3. Porous silicon photonic devices

In the last few years, PSi resonant photonic structures as Fabry-Perot interferometers (Dancil et al., 1999), Bragg reflectors (Snow et al., 1999), optical microcavities (De Stefano et al., 2003), and Thue-Morse sequences (Moretti et al., 2006) have been intensively studied by several research groups in particular for their photonic properties as interference filters.
The refractive index profile of a PSi multilayered structure can be realized by choosing the proper current density profile during the electrochemical etching of crystalline silicon. This is possible because the PSi fabrication process is self-stopping; an as etched PSi layer is depleted of holes and any further etching only occurs at the pores tips (Lehman, 2002). In this section the PSi optical structures, fabricated for experimental purposes, are described.

3.1 Fabry perot interferometer

A single layer of PSi optically acts as a Fabry-Perot interferometer. In Figure 8 (a) the reflectivity spectrum of a PSi layer under white light illumination is reported. The maxima in the reflectivity spectrum appear at wavelengths λ_m which satisfy:

$$m = 2nd / \lambda_m \tag{6}$$

where m is an integer, d is the film thickness and n is the average refractive index of the layer (Lin et al., 1997; Anderson et al., 2003). Assuming that the refractive index is independent on the wavelength over the considered range, the maxima are equally spaced in the wavenumber $(1/\lambda_m)$. When m maxima are plotted as a function of the wavenumber, each point lies on a straight line whose slope is two times the optical path of the interferometer, as it is shown in the Figure 8 (b).

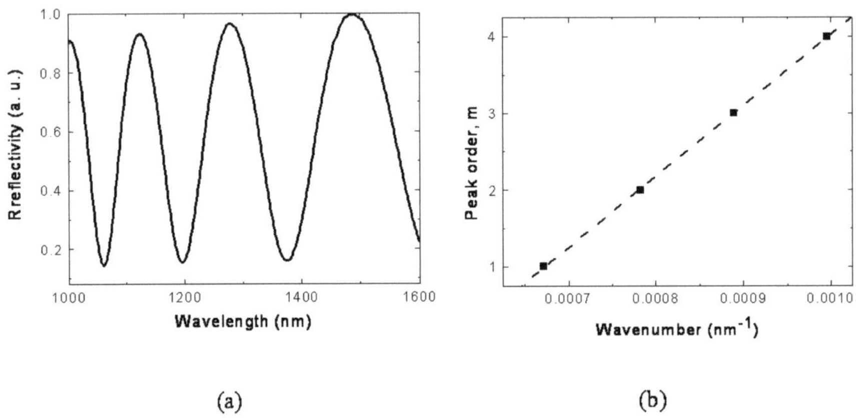

Fig. 8. (a) Reflectivity spectrum of a PSi layer realized by the electrochemical etching of p^+ crystalline silicon in a solution of 15 % hydrofluoric acid applying a current density of 115 mA/cm^2 for 11 s (P=69 %; n=1.593 @1.2 μm; d=2.9 μm); (b) m-order peaks are plotted as function of the wavenumber. The optical path of the interferometer has been estimated to be (4620±40) nm.

3.2 Bragg mirror and optical microcavity

The Bragg mirror is a periodic structure made alternating layers of high (n_H) and low (n_L) refractive index, whose thicknesses satisfy the relation $2(n_H d_H + n_L d_L) = m\lambda_B$, where m is the order of the Bragg condition (Figure 9 (a)). The layer stack is usually denoted as [HL]N, where N is the number of periods. The periodicity gives to the structure a photonic band gap (PBG) behavior characterized by the property to forbid the transmission of the light at fixed wavelengths. The reflectivity spectrum of a Bragg mirror is thus characterized by the presence of a stop band centered around the Bragg wavelength λ_B (Figure 9 (b)). For a given number of periods, the height and width of the reflectivity stop band increases by increasing the index ratio H/L. A low index contrast can be compensated by a higher number of periods.

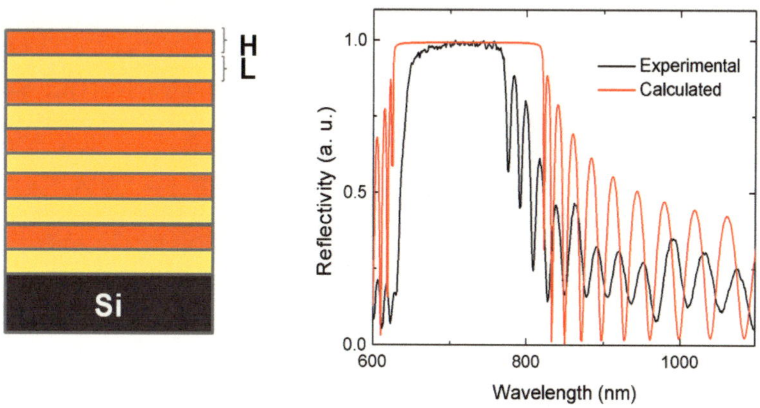

(a) (b)

Fig. 9. (a) Schematic of a Bragg mirror. (b) Experimental normal incidence reflectivity spectrum from a Bragg mirror (black line) compared with the calculated one (red line).

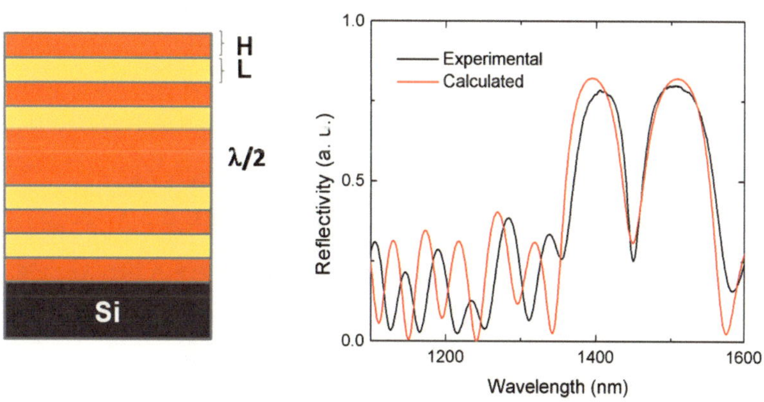

(a) (b)

Fig. 10. (a) Schematic of an optical microcavity. (b) Experimental normal incidence reflectivity spectrum of a microcavity (black line) compared with the calculated one (red line).

An optical microcavity is a $\lambda/2$ layer sandwiched between two distributed Bragg mirrors (Figure 10 (a)). The reflectivity spectrum of a microcavity is characterized by a transmittance peak in the photonic stop band (Figure 10 (b)). The Q factor of the microcavity is defined as $Q=\lambda/\Delta\lambda$, where λ is the wavelength of the resonance peak and $\Delta\lambda$ is the full width half maximum (FWHM) of the resonance. This parameter is used to evaluate how the light is confined in the PBG structure. The calculated reflectivity spectra of the structures reported in Figure 9 e 10 have been reproduced by a transfer matrix method (Muriel & Carballar, 1997), also taking into account the wavelength dispersion of the silicon refractive index.

3.3 Thue-Morse sequences

A quasi-crystal (QC) does not have a geometrical periodicity but is still deterministically generated. Even if these structures do not have a translational symmetry, they show several interesting physical properties such as the photonic band gaps, some resonance frequencies, and some high localized states (Soukoulis & Economou, 1982). Thue-Morse (T-M) (Liu, 1997) sequence is one of the most common examples of one dimensional QC. The T-M one dimensional structure is constituted by the sequence of two layers A and B with refractive index n_A (n_B) and thickness d_A (d_B). Applying the substitution rules A→AB and B→BA [26] all subsequent orders can be deduced, as follow: S_0=A, S_1=AB, S_2=ABBA, S_3=ABBABAAB, S_4=ABBABAABBAABABBA, and so on.

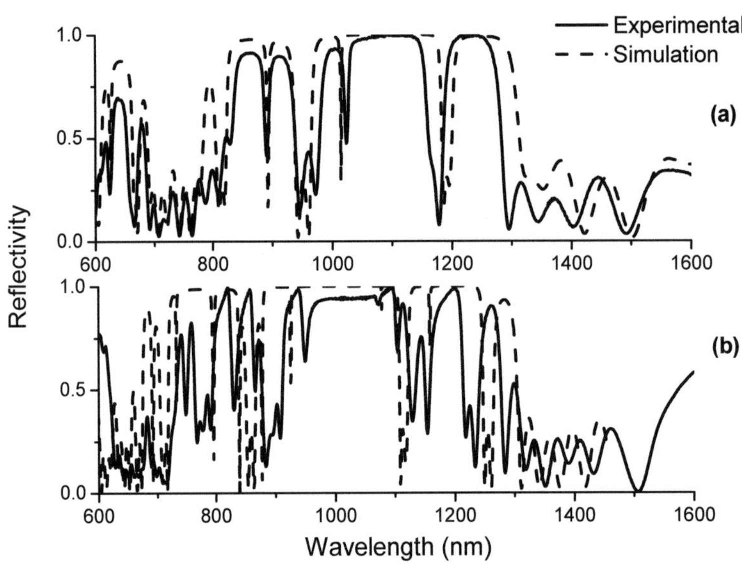

Fig. 11. Experimental (solid line) and calculate (dash line) reflectivity for S_6 T-M structure (a) and S_7 T-M structure (b). The measurements have been taken at normal incidence.

The layers number of S_N is 2N, where N is the T-M order. Dielectric T-M structures up to 128 layers have been fabricated by using PSi technology (Moretti et al., 2006). The high porosity layers are characterized by a porosity p_A=81%, with an average refractive index $n_A \cong 1.3$ and a thickness $d_A \cong 135$ nm.

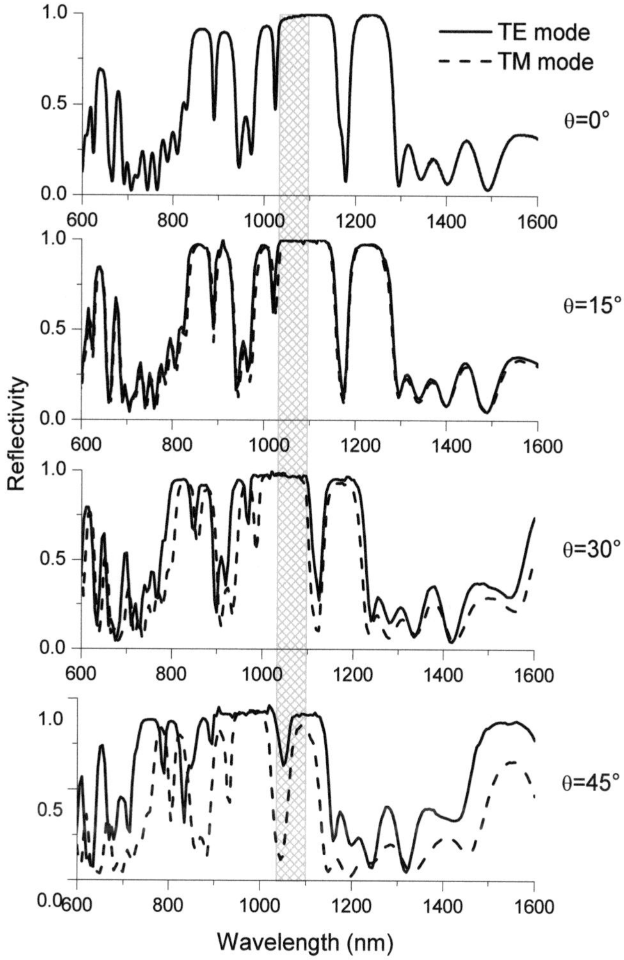

Fig. 12. Experimental reflectivity spectra of S_6 T-M structure for the transverse electric (TE) mode (solid line) and the transverse magnetic (TM) mode (dash line) for different incident angles.

The low porosity layers are characterized by a porosity p_B=56%, with a effective refractive index n_B≅1.96 and a thickness d_B≅90 nm. The thickness d_i of each layer was designed to satisfy the Bragg condition $n_i d_i = \lambda_0/4$ where n_i is the average refractive index and λ_0=700 nm. In Figure 11 the experimental (solid line) and calculated (dash line) reflectivity spectra are shown in case of S_6 (11-a), and S_7 (11-b) T-M structures. The good control in the fabrication process of the devices is demonstrated by the agreement between the measured and calculated spectra. The not perfect matching can be ascribed to non-uniformities of thickness and porosities of layers along the etching direction. The Figure 12 reports the experimental reflectivity spectra of the S_6 T-M structure for both the TE (solid line) and the TM (dash line) polarization for different incident angles up to 30°. The grey area highlights a PBG region of 70 nm, centred at 1100 nm which exists in the incident angle range between -30° and 30°.

4. Biochemical sensing based on porous silicon photonic devices

Biosensors are devices able to detect chemical and biological species or microorganisms. They can be used in many applications such as clinical diagnostics, environmental monitoring, and food quality control. A biosensor is constituted by a biomolecular probe, which can selectively recognize a biochemical target, immobilized on a transducer that converts the biological interaction into a signal. Different sensor platforms with remarkable detection properties, based on potentiometric, amperometric, magnetic and optical transducers have been developed. Among others, photonic sensing devices offer specific features which make them very attractive when not unique. Optical measurements are not invasive and can be used in harsh environments: they do not require electrical contacts that can cause explosions or fire. Opto-instrumentation is immune to electromagnetic interferences so that they are highly requested for applications where electrical currents could be harmful such as *in vivo* monitoring inside a patient body. Moreover, optical sensors are generally characterized by very short analysis time and high sensitivity. The optical detection principles are based either on fluorescence-labelled systems or on direct optical monitoring. Fluorescence detection has been the main approach in medical diagnostic, biotechnology and drug discovery for a long time. Recently, a new class of optical label-free sensors has been proposed, which uses direct optical methods (photoluminescence (Chan et al., 2001), surface plasmon resonance (SPR) (Homola et al., 1997; Jung et al., 1998), reflectivity (Mace et al., 2006), interference (Brandenburg & Henninger, 1994)) and label-free natural probes such as DNA sequences and proteins.

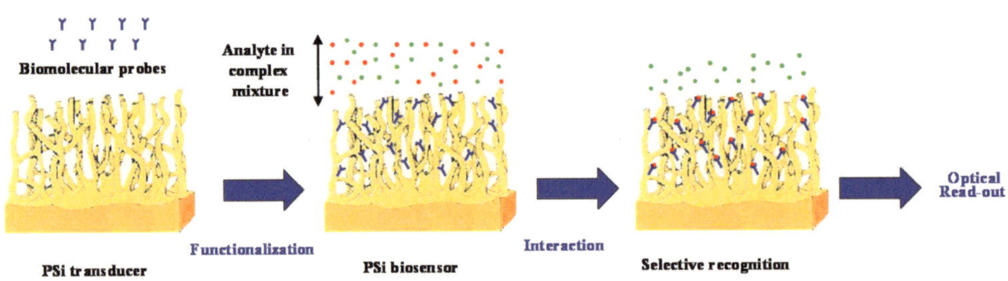

Fig. 13. A PSi biosensor is obtained immobilizing a biomolecular probe on the surface of a PSi transducer by using a proper functionalization process. The device is able to selectively recognize a target analyte in complex mixture.

The PSi is an ideal material as optical biosensor transducer since its large surface area can be easily modified by proper chemical passivation process. The label-free sensing mechanism is based on the change of the PSi refractive index on exposure to the substances to be detected, due to their infiltration into the pores; the consequence of the refractive index variation is a change in the reflectivity/transmittivity spectrum of the devices. The first measurement so far has been reported by Lin et al. who measured the shift of PSi Fabry-Perot fringes to detect the DNA hybridization (Lin et al., 1997). Since then, there has been a rapid growth of research activities in this area.

A key step in the fabrication of the PSi biosensor is the functionalization (*i.e.* covalent immobilization of the bioprobe) of its surface with the biological probe in order to make the device selective for the species of interest (Figure 13). The two main approaches used for the surface functionalization are the *ex situ* immobilization and the *in situ* synthesis.

4.1 *Ex situ* immobilization on PSi surface

The major drawback of the "as etched" PSi is its chemical instability: it is well known that the hydrogen-terminated PSi surface is slowly oxidized at room temperature by atmospheric oxygen, resulting in a blue shift of the optical spectrum (Canham, 1997). Moreover, it has been shown that a PSi device can be dissolved on exposure to alkaline solutions very often used in experiments of functionalization and biomolecular recognition (Anderson et al., 2003). The simplest method used to stabilize the PSi surface is the thermal oxidation of the structure. As already reported in section 2.4, the thermal oxidation at around 900°C completely oxidizes the PSi device inducing a reduction in pore diameter and porosity. Alternatively, a chemical oxidation in pyridine solution can be used: the process does not affect the pore size but it produces an oxide worse than the thermal one in terms of density and surface roughness. In both cases a change of the surface wettability, from hydrophobic to hydrophilic, is observed.

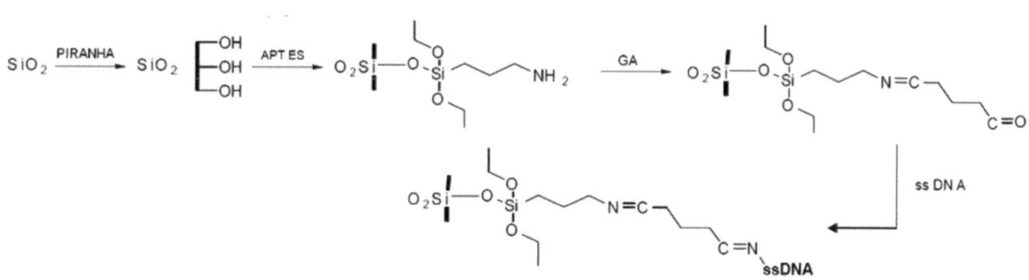

Fig. 14. Scheme of the functionalization process of a PSi Bragg mirror surface with DNA probe single strand (25 mer).

An example of surface functionalization with a bioprobe synthesized *ex situ* is reported in Figure 14. A PSi Bragg mirror, stabilized by means of a thermal oxidation at 900°C, was immersed in Piranha solution (4:1 sulfuric acid to hydrogen peroxide) for 40 min in order to assure the formation of Si-OH bonds. The chip was rinsed with deionized water and dried in a nitrogen stream. Then, the PSi surface was silanized in a 5 % solution of APTES (3-aminopropyltriethoxysilane) and anhydrous toluene for 30 min at room temperature. After the reaction time, the chip was washed twice in toluene to remove loosely physisorbed APTES, and baked at 100 °C for 10 min. The chip was thus immersed in a 2.5% glutaraldehyde (GA) solution in 20 mM HEPES buffer (pH 7.4) for 30 min. The GA reacts with the amino groups on the silanized surface and coats the internal surface of the pores with another thin layer of molecules. The PSi surface was incubated overnight at 4°C with a 200 µM amino terminated DNA single strand solution (30 µL).

From the optical point of view, the thermal oxidation causes blue-shifts of the reflectivity spectrum of the device of about 69 nm (Figure 15-b) due to the lower value of the SiO_2 refractive index (n_{ox}=1.46) with respect to the Si refractive index (n_{Si}=3.9). On the contrary, the functionalization steps (APTES+GA, and DNA probe) produce red shifts of the reflectivity spectra of 16 and 10 nm (Figures 15-c and 15-d), respectively, due to an increasing of the average refractive index of the layers.

4.2 *In situ* synthesis of oligonucleotides on PSi surface

The *in situ* synthesis of oligonucleotide (ON) on PSi surface is an alternative method to the traditional immobilization of DNA fragments. *In situ* technique allows to obtain high ON surface coverage and flexibility in the choice of the probe sequence.

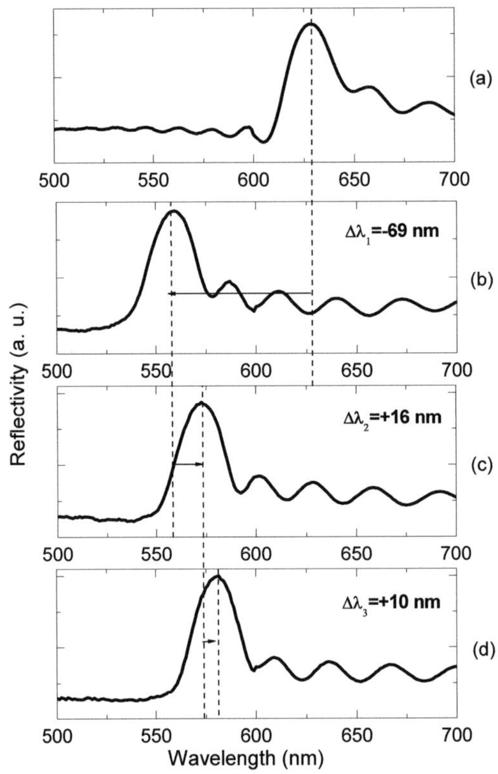

Fig. 15. Reflectivity spectra of the Bragg mirror before (a) and after (b) the thermal oxidation process, after the APTES and GA functionalization (c), and after the DNA probe immobilization (d).

Here, we report the ON synthesis on a PSi Bragg mirror obtained by alternating 20 high (H) refractive index layers (low porosity) and low (L) refractive index layers (high porosity) (Rea et al., 2010). The Bragg mirror was produced by electrochemical etching of highly doped p^+-silicon, <100> oriented, 0.003 Ω cm resistivity, 400 μm thick, using a HF (50% in weight):ethanol = 1:1 solution in the dark and at room temperature. A current density of 200 mA/cm^2 for 1.2 s was applied to obtain the low refractive index layers (n_L=1.542; d_L=125 nm) with a porosity of 72%, while a current density of 100 mA/cm^2 was applied for 1.4 s for the high index layers (n_H=1.784; d_H=108 nm) with a porosity of 64%. The structure has been thermally oxidized against uncontrolled environmental aging and corrosion in alkaline biological solutions. After the oxidation process, the PSi device was immersed in freshly prepared Piranha solution for 40 min at room temperature, rinsed with deionized water, and dried in a stream of nitrogen gas. This treatment creates Si-OH groups on the PSi surface. The structure was introduced in a suitable column reactor before the synthesis. DNA oligomer, constituted by thymine (T), was assembled on the chip following standard phosphoramidite chemistry by ten growing cycles thus obtaining polymer bound T10 (Figure 16).

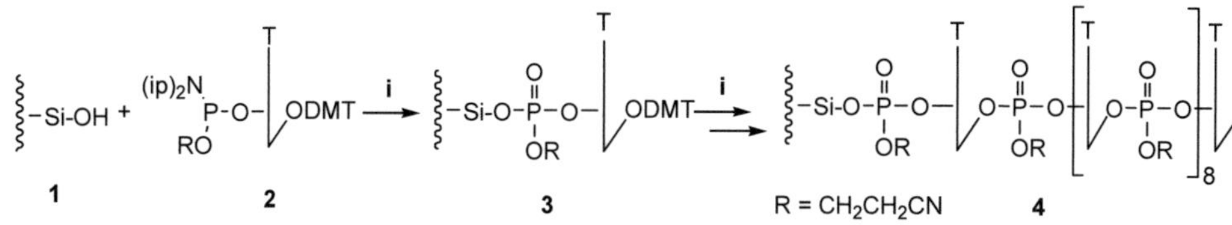

Fig. 16. Scheme of the solid phase synthesis of the 10 bases oligonucleotide **4** on the PSi-OH surface **1** using 5'-dimethoxytrityl-thymidine-phosphoramidite **2**; i: standard automatic synthetic cycle (Rea et al., 2010).

In order to quantify the surface functionalization, we have removed the 5'-dimethoxytrityl (DMT) protecting group from the support-bound 5′-terminal nucleotide by using the deblocking solution of trichloroacetic acid in dichloromethane (3% w/w). The release of the protecting group generates a bright red-orange colour solution in which the quantity of the DMT cation could be measured on-line by UV-VIS spectroscopy at 503 nm (ε = 71700 $M^{-1}cm^{-1}$). The Figure 17 shows the DMT analysis performed on the PSi device after each synthesis cycle: the amount of DMT indicated reaction yields over 98%. These values resulted almost steady during the ON growing process, confirming the stability of the chip surface and the high accessibility of ON 5'-OH end groups By averaging over these values, we have estimated a functionalization degree of 3.25 nmol/cm^2. The presence of ON chains bonded on the chip has been also verified by spectroscopic reflectometry. The biological molecules, attached to the PSi pore walls, induce an increase in the average refractive indexes of the layers, causing a red-shift in the reflectivity spectrum of the Bragg mirror. The magnitude of the shift increases with the increase of the pores surface coverage with the organic matter. The reflectivity spectra of the PSi multilayered structure before and after the ON synthesis are reported in Figure 18. A red-shift of 11 nm has been measured.

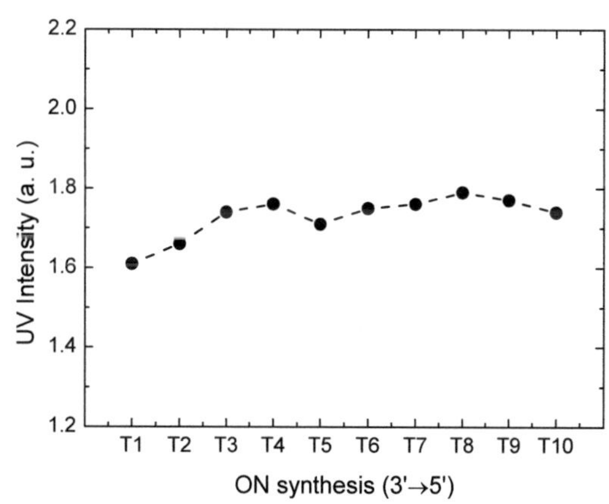

Fig. 17. DMT measurements performed on the sample after each synthesis cycle.

5. Integrated microfluidic porous silicon array

The microarray technology has demonstrated a great potential in drug discovery, genomics, proteomics research, and medical diagnostics (Pregibon et al., 2007; Poetz et al., 2005;

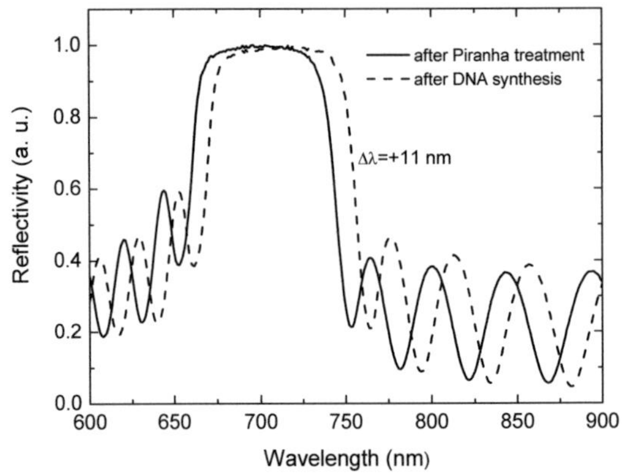

Fig. 18. Reflectivity spectra of the Bragg mirror before (solid line) and after (dash line) the oligonucleotide synthesis.

Nishizuka et al., 2003). The key issue is the very high throughput of these devices due to the large number of samples that can be simultaneously analyzed in a single parallel experiment. Further advantages are fast time analysis and the consumption of very small amount of reagents. The microarray technology is based on the immobilization of a large number of highly specific recognition elements on a solid platform. Different types of platform surfaces have already been explored; the most common examples are derivatized glass and gold/aluminium substrates (MacBeath & Schreiber, 2000; O'Connor & Pickard, 2003). Silicon, and silicon related materials, is by far the most important and diffuse material for lab-on-chip applications due to the high development of the integrated circuits technology. Recently, porous silicon substrates have been proposed for reverse phase protein and DNA microarray (Ressine et al., 2007; Chen et al., 2009; Yamaguchi et al., 2007): small sensing area with high detection efficiency is the key feature in both applications, in which quantitative signals are generated by fluorescence and infrared spectroscopy, respectively. Alternatively, we have studied the fabrication process and the optical characterization by reflectometry of a microarray of PSi photonic devices as functional platform for label-free detection of biomolecular interactions (Rea at al., 2010b). The array support has been integrated with a microfluidic circuit made of polydimethylsiloxane (PDMS) which strongly reduces the functionalization time, chemical and biological products consumption, while it preserves all the features of the PSi label-free optical detection.

5.1 Fabrication and optical characterization of the PSi Bragg mirror microarray

The integration of the PSi elements in a microarray is not straightforward. To this aim a proper technological process has been designed. The process flow chart of the PSi μ-array fabrication is schematized in Figure 1. The silicon substrate was a highly doped p^+-type wafer with a resistivity of 0.01 Ω cm, <100> oriented and 400 μm tick. Silicon nitride has been used as masking material during the electrochemical etching since it shows a better resistance against the HF solution with respect to photoresist, which effectively protects the silicon only for 2-3 min (Tao & Esashi, 2004). The silicon nitride film, 1.6 μm thick, was deposited by PECVD on the substrate (Figure 19 (a)). A standard photolithographic process was used to pattern the silicon nitride film (Figure 19 (b)), which has been subsequently etched by RIE process in CHF_3/O_2 atmosphere (Figure 19 (c)). Finally, the silicon wafer was electrochemically anodized in a HF-based solution (50 wt. % HF : ethanol = 1:1) in dark and at room temperature (Figure 19 (d)). We have realized the Bragg reflectors by alternating high (H) refractive index layers (low porosity) and low (L) refractive index layers (high porosity); a current density of 80 mA/cm^2 was applied to obtain low refractive index layers (n_L=1.6) with a porosity of 71 %, while one of 60 mA/cm^2 was applied for high index layers (n_H=1.69) with a porosity of 68 %. The device was then fully oxidized in pure O_2.

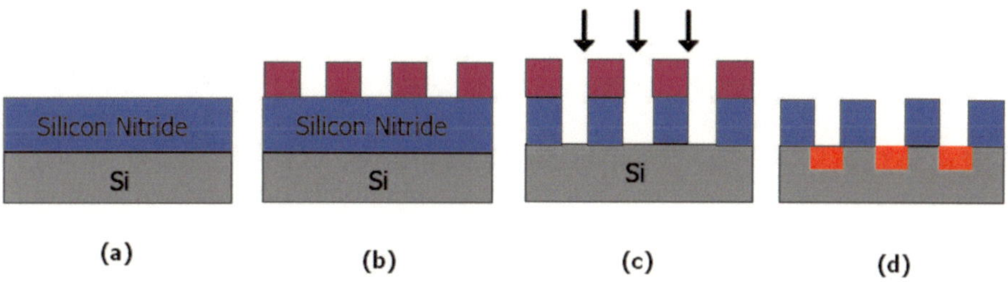

Fig. 19. Technological steps of the PSi μ-array fabrication process.

The optical microscope image of the microarray and the reflectivity spectra of some Bragg mirror elements are reported in Figure 20. The diameter of each element is of 200 μm, but it can be reduced to about 1 μm, by changing properly the photolithographic mask. The reflectivity spectra at normal incidence of the Bragg devices are characterized by a resonance peak at 627 nm and a FWHM of about 25 nm. The spectra demonstrate also the uniformity of the electrochemical etching on the whole microarray surface.

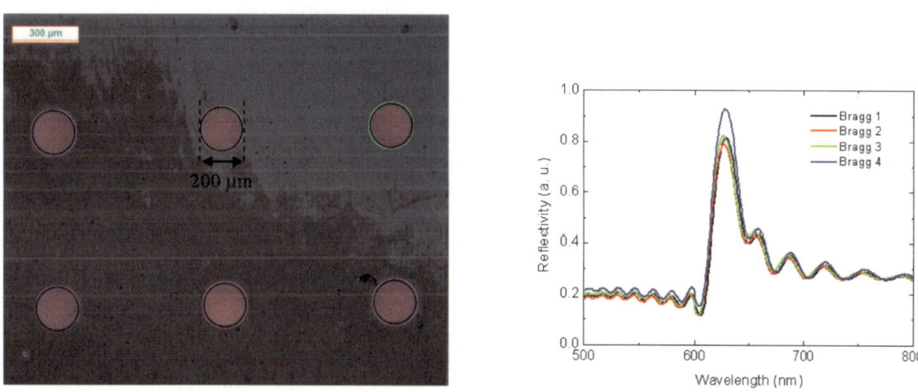

Fig. 20. Optical microscope image of the microarray and reflectivity spectra of the PSi Bragg mirrors.

5.2 Integration of the PSi array with a microfluidic system

The microfluidic system was designed by a computer aided design software. The pattern was printed 10 times bigger than its real size on a A4 paper by a laser printer (resolution 1200 dpi) and then transferred on a photographic film (Maco Genius Print Film) by a photographic enlarger (Durst C35) reversely used. The designed fluidic system was replicated by photolithographic process on a 10-µm thick negative photoresist (SU-8 2007, MicroChem Corp.) spin-coated for 30 s at 1800 rpm on a silicon substrate. After the photoresist development (SU-8 developer, MicroChem Corp.), the silicon wafer was silanized on exposure to chlorotrimethylsilane (Sigma-Aldrich Co.) vapour for 10 min as anti-sticking treatment. A 10:1 mixture of PDMS prepolymer and curing agent (Sylgard 184, Dow Corning) was prepared and degassed under vacuum for 1 hour. The mixture was poured on the patterned wafer and cured on a hot plate at 75°C for 3h to facilitate the polymerization and the cross-linking process. After the PDMS layer peeling, inlet and outlet holes were drilled through it in order to allow the access of liquid substances to the system. Finally, the PDMS layer was rinsed in ethanol in a sonic bath for 10 min. The surfaces of PDMS layer and microarray, whose PSi elements were thermally oxidized, were activated by exposing to oxygen plasma for 10 sec to create silanol groups (Si-OH) as shown in the schematic reported in Figure 21, aligned under a microscope using an x-y-z theta stage, and sealed together. After the sealing with the PDMS system, the PSi elements of the array have been functionalized with DNA single strand, as described in section 4.1. The microfluidic circuit allows to use only few microlitres (~5 µl) of biologicals with respect to the tens of microlitres used in the case of not integrated devices. Moreover, the incubation time has been also reduced from eight to three hours. After the bio-functionalization with DNA probe, we have studied the DNA-DNA hybridization by injecting into the microchannel 200 µM of complementary sequence. Figure 22 shows the reflectivity spectra of a PSi Bragg mirror after the DNA functionalization and after the complementary DNA interaction. A red-shift of 5.0 nm can been detected after the specific DNA-DNA interaction. A negligible shift, less than 0.2 nm (data not reported in the figure), is the result of a control measurement which has been done exposing another functionalized microchannel to non-complementary DNA, demonstrating that the integrated PSi array is able to discriminate between complementary and non-complementary interactions.

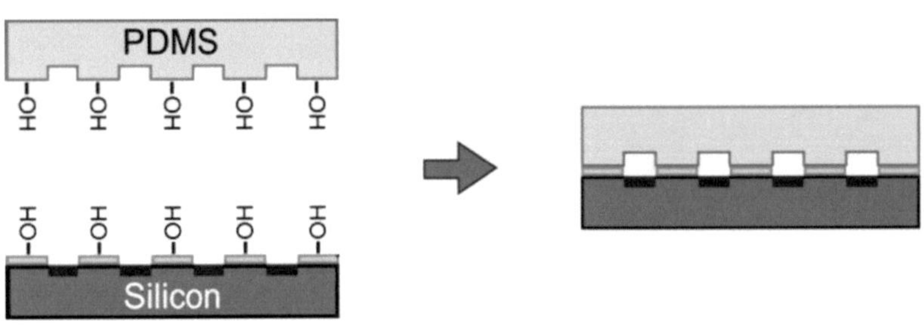

Fig. 21. Scheme of the fabrication process used to integrate the PSi array with a PDMS microfluidic system.

6. Conclusion

The PSi technology allows the fabrication of different multilayered devices with complex photonic features such as optical resonances and band gaps. These photonic structures, functionalized with a biomolecular probe able to selectively recognize a biochemical target, have been successfully used as label-free optical biosensors. The sensing mechanism is based on the increase of the PSi refractive index due to the infiltration of the biological

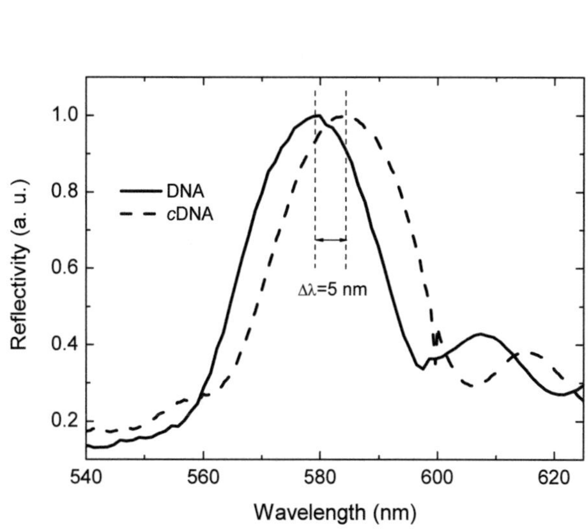

Fig. 22. Reflectivity spectra of a PSi Bragg mirror after the DNA probe attachment (solid line), and after the hybridization with the complementary DNA (dash line).

substances into the nanometric pores of the material; the consequence of the refractive index change is the shift of the reflectivity spectrum of the photonic devices. Since PSi technology is compatible with the microelectronic processes, it can be easily used as functional platform in the fabrication on integrated microsystems. As example, we have reported the realization of a PSi microarray for the detection of multiple DNA-DNA interactions. The array, characterized by a density of 170 elements/cm^2, has been integrated with a microfluidic system made of PDMS which allows to reduce the consumption of the chemical and biological substances.

7. References

Anderson, S.H.C.; Elliot, H.; Wallis, D.J.; Canham, L.T. & Powell, J.J. (2003). Dissolution of different forms of partially porous silicon wafers under simulated physiological conditions. *Phys. Status Solidi A*, Vol. 197, pp. 331-335.

Anderson, M.A.; Tinsley-Brown, A.; Allcock, P.; Perkins, E.A.; Snow, P.; Hollings, M.; Smith, R.G.; Reeves, C.; Squirrell, D.J.; Nicklin, S. & Cox, T.I. (2003). Sensitivity of the optical properties of porous silicon layers to the refractive index of liquid in the pores. *Phys. Stat. Sol. A*, Vol. 197, pp. 528-533.

Aspnes, D.E. & Theeten, J. B. (1979). Investigation of effective-medium models of microscopic surface-roughness by spectroscopic ellipsometry. *Physical Review B*, Vol. 20, pp. 3292-3302.

Brandenburg, A. & Henninger, R. (1994). Integrated optical Young interferometer. *Applied Optics*, Vol. 33, pp. 5941-5947.

Canham, L.T. (1990). Silicon quantum wire array fabrication by electrochemical and chemical dissolution of wafers. *Appl. Phys. Lett.*, Vol. 57, pp. 1046-1048.

Chan, S.; Horner, S.R.; Fauchet, P.M. & Miller, B.L. (2001). Identification of Crom Negative Bacteria using nanoscale silicon microcavities. *J. Am. Chem. Soc.*, Vol. 123, pp. 11797-11798.

Chandrasekaran, A.; Acharya, A.; You, J.L.; Soo, K. Y.; Packirisamy, M.; Stiharu, I. & Darveau, A. (2007). Hybrid integrated silicon microfluidic platform for fluorescence based biodetection. *Sensors*, Vol. 7, pp. 1901-1915.

Chen, L.; Chen , Z.T.; Wang , J.; Xiao , S.J.; Lu, Z.H.; Gu, Z.Z.; Kang, L.; Chen, J.; Wu, P.H.; Tang, Y.C. & Liu, J.N. (2009). Gel-pad microarrays templated by patterned porous silicon for dual-mode detection of proteins. *Lab on a Chip*, Vol. 9, pp. 756-760.

Dancil, K.P.S.; Greiner, D.P. & Sailor, M.J. (1999). A porous silicon optical biosensor: Detection of reversible binding of IgG to a protein A-modified surface. *J. Am. Chem. Soc.*, Vol. 121, pp. 7925-7930.

De Stefano, L.; Rendina, I.; Moretti, L. & Rossi, A.M. (2003). Optical sensing of flammable substances using porous silicon microcavities. *Mater. Sci. Eng. B*, Vol. 100, pp. 271-274.

Grosman, A. & Ortega, C. (1997). Chemical composition of 'fresh' porous silicon, In: *Properties of Porous Silicon*, Edited by L.T. Canham, pp. 145-153, INSPEC, London.

Homola, J.; Ctyroky, J.; Skalsky, M.; Hradilova, J. & Kolarova, P. (1997). A surface plasmon resonance based integrated optical sensor. *Sensors and Actuators B*, Vol. 39, pp. 286-290.

Jung, L.S.; Campbell, C.T.; Chinowsky, T.M.; Mar, M.N. & Yee, S.S. (1998). Quantitative interpretation of the response of surface plasmon resonance sensors to adsorbed films. *Langmuir*, Vol. 14, pp. 5636-5648.

Lehmann, V. & Gösele, U. (1991). Porous silicon formation: a quantum wire effect. *Appl. Phys. Lett.*, Vol. 58, pp. 856.

Lehmann, V. (2002). *Electrochemistry of Silicon*, Wiley-VCH Verlag GmbH & Co, pp. 17–20.

Lin, V.S.Y.; Motesharei, K.; Dancil, K.P.S.; Sailor, M.J. & Ghadiri, M.R. (1997). A porous silicon based optical interferometric biosensor. *Science*, Vol. 270, pp. 840.

Liu, N.H. (1997). Propagation of light waves in Thue-Morse dielectric multilayers. *Phys. Rev. B*, Vol. 55, pp. 3543-3547.

MacBeath, G. & Schreiber, S.L. (2000). Printing proteins as microarrays for high-throughput function determination. *Science*, Vol. 289, pp. 1760-1763.

Mace, C.R.; Striemer, C.C. & Miller, B.L. (2006). Theoretical and experimental analysis of arrayed imaging reflectometry as a sensitive proteomics technique. *Anal. Chem.*, Vol. 78, pp. 5578-5583.

Moretti, L.; Rea, I.; Rotiroti, L.; Rendina, I.; Abbate, G.; Marino, A. & De Stefano L. (2006). Photonic band gaps analysis of Thue-Morse multilayers made of porous silicon. *Optics Express*, Vol. 14, pp. 6264-6272.

Mulloni, V. & Pavesi, L. (2000). Porous silicon microcavities as optical chemical sensors. *Appl. Phys. Lett.*, Vol. 76, pp. 2523.

Muriel, M. A. & Carballar, A. (1997). Internal field distributions in fiber Bragg gratings. *IEEE Photonics Technol. Lett.*, Vol. 9, pp. 955-957.

Nishizuka, S.; Chen, S.T.; Gwadry, F.G.; Alexander, J.; Major, S.M.; Scherf, U.; Reinhold, W.C.; Waltham, M.; Charboneau, L.; Young, L.; Bussey, K.J.; Kim, S.Y.; Lababidi, S.; Lee, J.K.; Pittaluga, S.; Scudiero, D.A.; Sausville, E.A.; Munson, P.J.; Petricoin, E.F.; Liotta, L.A.; Hewitt, S.M.; Raffeld, M. & Weinstein, J.N. (2003). Diagnostic markers that distinguish colon and ovarian adenocarcinomas: identification by genomic, proteomic, and tissue array profiling. *Cancer Research*, Vol. 63, pp. 5243-5250.

O'Connor, D.C. & Pickard, K. (2003). *Microarrays and Microplates: Applications in Biomedical Science*, Edited by S. Ye & I.N.M. Day, pp. 65-72, BIOS Scientific Publishers, Oxford.

Pap, E.; Kordás, K.; Tóth, G.; Levoska, J.; Uusimäki, A.; Vähäkangas, J.; Leppävuori, S. & George, T.F. (2005). Thermal oxidation of porous silicon: study on structure. *Appl. Phys. Lett.*, Vol. 86, pp. 041501.

Pickering, C.; Canham, L.T. & Brumhead, D. (1993). Spectroscopic ellipsometry characterization of light-emitting porous silicon structures. *Appl. Sur. Sci.*, Vol. 63, pp. 22-26.

Pirasteh, P.; Charrier, J.; Soltani, A.; Haesaert, S.; Haji, L.; Godon, C. & Errien, N. (2006). The effect of oxidation on physical properties of porous silicon layers for optical applications. *Applied Surface Science*, Vol. 253, pp. 1999-2002.

Poetz, O.; Ostendorp, R.; Brocks, B.; Schwenk, J.M.; Stoll, D.; Joos, T.O.; Templin, M.F. (2005). Protein microarrays for antibody profiling: Specificity and affinity determination on a chip. Proteomics, Vol. 5, pp. 2402-2411.

Pregibon, D.C. ; Toner, M. & Doyle, P.S. (2007). Multifunctional encoded particles for high-throughput biomolecule analysis. *Science*, Vol. 315, pp. 1393-1396.

Rea, I.; Oliviero, G.; Amato, J.; Borbone, N.; Piccialli, G.; Rendina, I. & De Stefano L. (2010). Direct synthesis of oligonucleotides on nanostructured silica multilayers. *The Journal of Physical Chemistry C*, Vol. 114, pp. 2617-2621.

Rea, I.; Lamberti, A.; Rendina, I.; Coppola, G.; De Tommasi, E.; Gioffrè, M.; Iodice, M.; Casalino, M. & De Stefano L. (2010b). Fabrication and characterization of a porous silicon based microarray for label-free optical monitoring of biomolecular interactions. *Journal of Applied Physics*, Vol. 107, pp. 014513.

Ressine, A.; Corin, I.; Järås, K.; Guanti, G.; Simone, C.; Marko-Varga, G. & Laurell, T. (2007). Porous silicon surfaces - A candidate substrate for reverse protein arrays in cancer biomarker detection. *Electrophoresis*, Vol. 28, pp. 4407-4415.

Ruike, M.; Houzouji, M.; Motohashi, A.; Murase, N.; Kinoshita, A. & Kaneko, K. (1996). Pore structure of porous silicon formed on a lightly doped crystal silicon. *Langmuir*, Vol. 12, pp. 4828-4831.

Sing, K.S.W.; Everett, D.H.; Haul, R.A.W.; Moscou, L.; Pierotti, R.A.; Rouquerol, J. & Siemieniewska, T. (1985). Reporting physisorption data for gas solid systems with special reference to the determination of surface-area and porosity. *Pure Appl. Chem.*, Vol. 57, pp. 603-619.

Smith, R. L. & Collins, S. D. (1992). Porous silicon formation mechanisms. *J. Appl. Phys.*, Vol. 71, R1.

Snow, P.A.; Squire, E.K.; Russell, P.S.J. & Canham, L.T. (1999). Vapor sensing using the optical properties of porous silicon Bragg mirrors. *J. Appl. Phys.*, Vol. 86, pp. 1781.

Soukoulis, C.M. & Economou, E.N. (1982). Localization in one-dimensional lattices in the presence of incommensurate potentials. *Phys. Rev. Lett.*, Vol. 48, pp. 1043-1046.

Tao, Y. & Esashi, M. (2004). Local formation of macroporous silicon through a mask. *J. Micromech. Microeng.*, Vol. 14, pp.1411-1415.

Tompkins, H. G. & McGaham, W. A (1999). *Spectroscopic Ellipsometry and Reflectometry*. Ed. John Wiley & Sons.

Uhlir, A. (1956). Electrolytic Shaping of Germanium and Silicon. *The Bell System Technical Journal*, Vol. 35, pp. 333-347.

Xia, Y. & Whitesides, G.M. (1998). Soft lithography. *Annu. Rev. Mater. Sci.*, Vol. 28, pp.153-184.

Yamaguchi, R.; Miyamoto, K.; Ishibashi, K.; Hirano, A.; Said, S.M.; Kimura, Y. & Niwano, M. (2007). DNA hybridization detection by porous silicon-based DNA microarray in conjugation with infrared microspectroscopy. *J. Appl. Phys.*, Vol. 102, Article num. 014303.

Yon, J.J.; Barla, K.; Herino, R. & Bomchil, G. (1987). The kinetics and mechanism of oxide layer formation from porous silicon formed on p-Si substrates. *J. Appl. Phys.*, Vol. 62, pp. 1042-1048.

Zangooie, S.; Jansson, R. & Arwin, H. (1999). Ellipsometric characterization of anisotropic porous silicon Fabry-Perot filters and investigation of temperature effects on capillary condensation efficiency. *J. Appl. Phys.*, Vol. 86, pp. 850-858.

Zangooie, S.; Bjorklung, R. & Arwin, H. (1998). Protein adsorption in thermally oxidized porous silicon layers. *Thin Solid Films*, Vol. 313, pp. 825-830.

13

Life Cycle Assessment of PV Systems

Masakazu Ito
Tokyo Institute of Technology
Japan

1. Introduction

According to reporting by the Intergovernmental Panel on Climate Change (IPCC), global warming brings a variety of adverse effects including record-high temperatures, flooding due to increased rainfall, expansion of arid areas and a higher risk of drought, and stronger typhoons. Accordingly, it is necessary to mitigate emissions of greenhouse gases (GHGs; CO_2, CH_4, N_2O and others), which cause global warming. However, as GHGs are invisible, the amounts in which they are released are generally unclear.

Life cycle assessment (LCA) - the main topic of this chapter - is useful in calculating emissions. Although it is not ideally suited for evaluation on a macro scale (investigation from a global viewpoint, for example), it is highly appropriate for micro-scale analysis (e.g., consideration of products and generation systems). The results of LCA can clarify major emissions, thereby enabling consideration of measures for their reduction.

This chapter discusses LCA in relation to photovoltaic (PV) systems. First, an overview is given and the scheme of LCA is described, and evaluation indices, LCA limitations, inventory analysis, impact assessment and interpretation are outlined. Then, guidelines for LCA in regard to PV systems are discussed with a focus on important matters for related evaluation. Next, the collection of LCA data is outlined, and finally, calculations from example papers are introduced in relation to LCA for PV modules, PV systems and balance of system (BOS) technologies.

2. What is LCA?

Life cycle assessment (LCA) is an approach to environmental management system implementation involving the quantitative evaluation of a product's overall environmental impact. Energy requirements and CO_2 emissions throughout the whole life cycle of the product (including its manufacture, transport, use, disposal, etc.) are estimated in order to enable such evaluation, and the results can be used for related environmental assessment. However, since life cycle is related to a broad range of variables and is complicated, it is difficult to comprehend the exact significance of the results. Accordingly, it is very important to set a purpose for the evaluation. An LCA operator should implement research that matches the purpose and interpret the outcomes appropriately.

The research and analysis scheme for LCA consists of the four stages shown in Fig. 1 as follows: 1. goal and scope definition; 2. inventory analysis; 3. impact assessment; and 4. interpretation. The results of inventory analysis are referred to as life cycle inventory (LCI) data. LCA is applicable to any product or service, but its results are affected by objects, assumptions, data availability and accuracy. Hence, it is impossible to generalize the method in a very clear way. As a result, LCA operators and users must properly understand the limitations of LCA and the assumptions that can be drawn from its results. The essentials of LCA are standardized in ISO 14040 and ISO 14044, which stipulate the details and basic points of the approach.

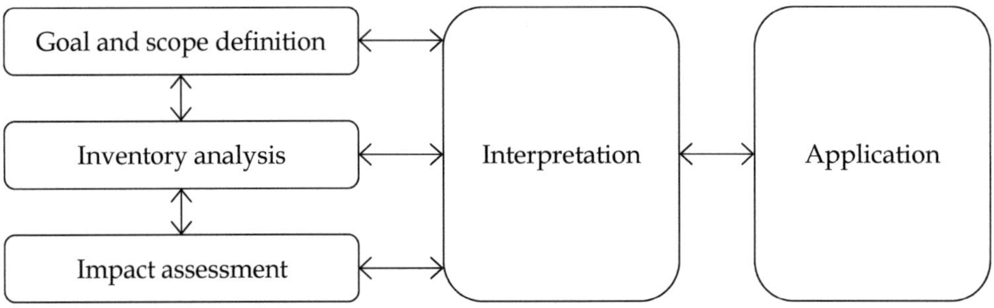

Fig. 1. Scheme of LCA

3. LCA for photovoltaic systems

In any LCA study, the purpose depends on the operator. However, when the operator evaluates a photovoltaic (PV) system, the main research point or characteristic relates to energy generation. This is a significant difference between PV systems and other products. When a building developer discusses new energy supply systems (e.g., in relation to buildings with low carbon emissions and high energy efficiency), LCA can highlight the potential of PV systems and useful materials. This is expected to provide two advantages, the first of which is PV system optimization. When a developer studies the installation of a PV system, the environment of the installation site must be considered. To ensure optimization, a variety of variables (e.g., cost and CO_2 emissions) are discussed. If LCA is used, the system can be optimized from an environmental viewpoint.

The second advantage is comparability. When comparing energy generation technologies (e.g., when researching the possible installation of a PV system as a supply of alternative energy as opposed to other generation systems, or when installing energy supply systems based on multiple generation technologies), the evaluation methods and rules applied must be uniform. In such cases, LCA can provide quantitative results, thereby enabling comparison of each technology on an equal footing.

3.1 Evaluation indices

In LCA study, evaluation indices are decided based on the purpose at hand. As PV systems generate electricity, the new index of energy payback time (EPT or EPBT) can be evaluated. EPT expresses the number years the system takes to recover the initial energy consumption involved in its creation throughout its life cycle via its own energy production. An equation for estimating EPT is shown below. The total initial energy for PV systems in Equation (1) is calculated using LCA, and the annual power generation aspect is described in Sections 4 and 5.

$$\text{EPT [years]} = \frac{\text{Total primary energy use of the PV throughout its life cycle [kWh]}}{\text{Annual power generation [kWh/year]}} \tag{1}$$

The CO_2 emission rate is a useful index for determining how effective a PV system is in terms of global warming. Generally, this index is used for comparison between generation technologies. As a PV system does not operate in the same way as a tree, there is no payback of CO_2 emissions as such. However, some research on comparisons between PV systems and other fossil fuel generation technologies have used CO_2 payback time as a metric. In these studies, PV systems were viewed as an alternative to fossil fuels and as offering a corresponding reduction in CO_2 emissions, which allowed calculation of the CO_2 payback time. However, this paper does not deal with the concept of CO_2 payback time.

$$\text{CO}_2 \text{ emission rate } [\text{g-CO}_2/\text{kWh}] = \frac{\text{Total CO}_2 \text{ emission during life-cycle } [\text{g CO}_2]}{\text{Annual power generation } [\text{kWh/year}] \times \text{Lifetime } [\text{year}]} \quad (2)$$

3.2 Boundaries of LCA

As using different boundaries obviously creates different results, defining and making boundaries known is important. Figure 2 shows typical boundaries for LCA of a PV system from the mining of its raw materials to its final disposal. The next consideration is the boundary for each stage. Boundaries involve products and services related to the item's life cycle. As the details vary in each case, it is important to fit the definition to the purpose of the product. For example, factors including the type of PV module used, efficiency, array, foundation, installation method and operation method should be identified to build a suitable system. Indirect factors should also be considered as much as possible.

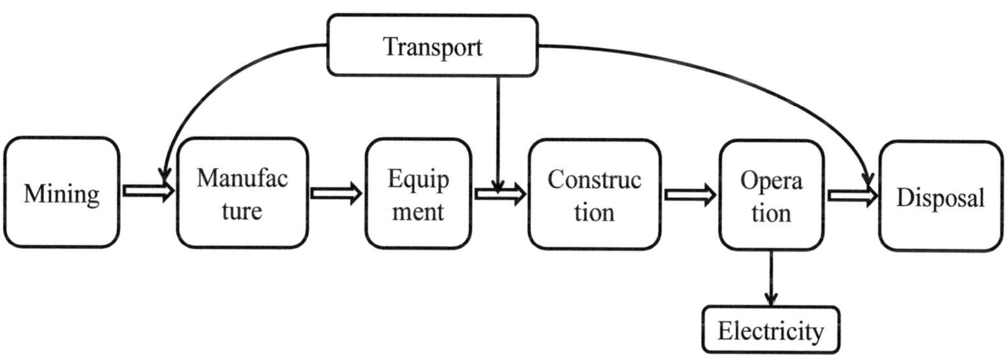

Fig. 2. Boundaries of LCA for a PV system

3.3 Inventory analysis

Inventory analysis is performed to evaluate the amounts of environment-influencing materials consumed or produced during the object's life cycle. It involves pinpointing the processes involved in the life cycle and evaluating them quantitatively, then identifying all related environment-influencing materials. The object's data are subsequently evaluated as a whole. However, as it is difficult to collect all information on related processes, the results may have simplified or missing data. Accordingly, it is important to understand the applicable boundaries, the quality of data and the assumptions involved in calculation when performing LCA study.

3.4 Impact assessment

Impact assessment consists of three processes; classification, characterization and weighting. In classification, environment-influencing materials are categorized in terms of related influence events. For example, CO_2 will be categorized as producing global warming, sulfur oxide (SOx) will be categorized as producing acid rain, affecting public health and so on. Impact potential is calculated based on inventory analysis. In research on energy payback time, the amount of energy consumed is calculated and classified. In research on CO_2 emission rates, emissions are calculated and classified into a suitable category.

In characterization, amounts of output materials are calculated with characterization factors to produce impact category indicators. In particular, input energy is calculated in terms of electricity or calorific value. Greenhouse gas emissions are calculated in terms of CO_2 equivalents (CO_2eq) using global warming potential (GWP) figures as defined by the IPCC. For example, in the case of a power conditioning system (PCS), the weight of each material would be determined as relevant data, and the energy requirements/CO_2 emissions of the production process would be ascertained. Then, input and output data would be calculated using inventory analysis, and the results indicating the energy requirement and CO_2eq values would be calculated to provide the impact category indicator.

Weighting is not stipulated in international standardization because it is considered difficult to form a single indicator for the different areas of global warming potential and ozone depletion potential. However, a simple comparison method is still needed. The two possible methods for this are damage evaluation and environmental category weighting by estimation. Whichever is used, the weighting must be transparent.

3.5 Interpretation

The results of LCA may depend on research boundaries and approaches to inventory analysis. Accordingly, in related interpretation, the effects of operation methods should be discussed. Usually, the data used in LCA include estimates and referred information. For this reason, if the data affect the results significantly, sensitivity analysis should be included.

4. LCA guidelines for PV systems

Recently, a set of LCA guidelines for PV systems titled "Methodology Guidelines on Life Cycle Assessment of Photovoltaic Electricity" was published by the International Energy Agency Photovoltaic Power System Programme (IEA PVPS), Task 12, Subtask 20. This is an informative and useful resource for LCA operators of PV systems that helps with the evaluation difficulties outlined in Section 3. This section describes a number of important considerations covered in the guidelines for evaluating PV systems.

4.1 Lifetime

Lifetime is difficult to quantify because most PV systems introduced are still in operation or were produced in the early stages of the technology's development. However, many researchers have studied the life expectancy of PV systems. The guidelines follow the results of papers outlining such research, and set the lifetimes shown in Table 1.

PV modules	30 years for mature module technologies
Inverters	15 years for small plants or residential PV systems; 30 years with 10% part replacement every 10 years for large plants
Structure	30 years for rooftop- and facade-mounted units, and between 30 to 60 years for ground-mounted installations on metal supports. Sensitivity analysis should be performed.
Cabling	30 years

Table 1. List of lifetimes (data from IEA/PVPS Task 12)

4.2 Irradiation data

Irradiation data depend on the location and tilt angle of PV modules. Accordingly, the two main recommendations given are analysis of industry averages/best-case systems and analysis of average systems installed on the grid network.

4.3 Performance ratio

The performance ratio (PR) depends on the type of installation. In general, the value rises with lower temperatures and monitoring of PV systems for early detection of defects. Task 12's recommendation is 75% for rooftop-mounted and 80% for ground-mounted latitude-optimal installations. Alternatively, actual performance data can be used where available.

4.4 Degradation

Most PV modules degrade year by year to an extent that is still an active topic of research, especially for thin-film PV systems. However, 0.5% per year seems to be a typical number for crystalline silicon PV modules. Accordingly, the guidelines set the degradation rate for flat-plate PV modules. Mature module technologies are considered to maintain 80% of their initial efficiency at the end of the 30-year lifetime under the assumption of linear degradation during this time.

5. Collection of LCA data

LCA data are usually categorized into foreground and background types. Foreground data relates to the materials from which products are made, such as arrays, foundations and cable. Background data relate to materials that are indirectly involved, such as array steel, foundation cement and cable copper. Foreground data are usually provided by producers, while database values are used for background data due to the difficulty of collecting such information. Such databases summarize the input and output data for various materials. For example, LCA data for galvanized steel in an LCA database would show that the unit is 1 kg; the input data are the weight of coal, limestone, iron ore, natural gas, crude oil and so on used in production; and the output data are the weight of related emissions of CO_2, nitrogen oxide (NOx), SOx, biochemical oxygen demand (BOD) and so on.

These data can be obtained from an LCA database or by using LCA software. Ecoinvent (Switzerland) and the Life Cycle Assessment Society of Japan (JLCA) have well-known LCA databases. The Ecoinvent resource is an inventory database with more than 4,000 entries developed from research for the company's environment reports, summaries of references and questionnaire surveys. The JLCA database includes inventory data, impact category indicators and reference data, which are based on a five-year project implemented by the

New Energy and Industrial Technology Development Organization (NEDO). Although inventory data are limited to about 280 entries, these are typical data obtained in collaboration with industry associations, thus making them highly reliable. There are also approximately 300 reference data entries made by industry associations themselves.
Calculation for small systems or products can be performed manually, but this is difficult for large systems. Accordingly, LCA software is produced to support such operations. As this type of software generally already includes LCI data, the operator does not need to input individual values. SimaPro developed by PRé Consultants, GaBi Software by PE International and MiLCA by the Japan Environmental Management Association for Industry are examples of such programs.
Irradiation data are also required for LCA calculation in regard to PV systems. If it is possible to use actual long-term generation data for such systems, there is no need for irradiation data. However, environmental reporting is needed before a PV system is installed. If irradiation data are available, PV system generation can be estimated and pre-LCA can be evaluated. Meteonorm developed by Meteotest (Switzerland) is a well-known irradiation database. It also provides a function to calculate irradiation in relation to tilted planes, thereby eliminating the need to use complex metrological models. A further resource is the System Advisor Model (SAM) energy analysis software developed by the National Renewable Energy Laboratory (NREL, USA), which also includes a function for calculating PV system generation. METPV and MONSOLA developed by NEDO are other irradiation databases with data related exclusively to Japan.

LCA databases	Ecoinvent, JLCA
LCA software	SimaPro, GaBi, MiLCA
Irradiation databases	Meteonorm, System Advisor Model, METPV

Table 2. List of databases

6. LCA calculations from example papers

This section introduces four interesting papers on PV system LCA and their results. The studies in question addressed PV modules, rooftop systems, balance of system (BOS) technology and large PV systems.

6.1 LCA study on PV modules

This paper describes PV module LCA with a focus on emissions, including not only greenhouse gases (GHGs) but also NOx, SOx, cadmium (Cd) and heavy metals. The results for GHGs are summarized, and heavy metals form the main topic of the paper.
The use of cadmium telluride (CdTe) PV modules is growing rapidly because of their high efficiency and low price. However, Cd can have adverse health effects, and there is now a tide of concern regarding the safety of CdTe PV modules. However, this paper indicates that emissions from such modules are much lower than those of oil power plants on a like-for-like basis. LCA is a good method for highlighting this type of finding.
Data on GHGs, NOx and SOx are summarized in the paper assuming three cases: Case 1: current electricity mixture for silicon (Si) production from the CrystalClear project and the Ecoinvent database; Case 2: combination of the Co-ordination of Transmission of Electricity (UCTE) grid mixture and the Ecoinvent database; and Case 3: the U.S. grid mixture and the

Franklin database. In Case 1, GHG emissions of Si modules for the year 2004 are 30 - 45 g CO_2eq/kWh, and the EPT is 1.7 - 2.7 years. These figures are for rooftop installation. The GHG emissions and EPT of a CdTe frame without PV modules are 24 g CO_2eq/kWh and 1.1 years for ground-mounted installations. CdTe has about half the GHG emissions of crystalline Si. A summary is shown in Table 4.

Paper title	Emissions from photovoltaic life cycles[1]
Author(s)	Fthenakis, V.M. Kim, H.C. and Alsema, E.
Journal	Environmental Science & Technology 2008; 42 (6): 2,168 – 2,174
Irradiation	1,700 kWh/m^2/year, 1,800 kWh/m^2/year
PV type	ribbon-Si, multi-Si, mono-Si, CdTe
System configuration	0.75 – 0.8 performance ratio, rooftop- and ground-mounted
Lifetime	30 years
Results	20 - 55 g CO_2eq/kWh, 40 - 190 mg NOx/kWh, 60 - 380 mg SOx/kWh (reading from figure)
Year	2006

Table 3. Summary of the paper

PV type	Assumption	GHG emissions	EPT
Si modules	Rooftop-mounted, 0.75 PR, 1,700 kWh/m^2/yr	30 – 45 g CO_2eq/kWh	1.7 - 2.7 years
CdTe	Ground-mounted 0.8 PR, 1,800 kWh/m^2/yr 30-year lifetime	24 g CO_2eq/kWh	1.1 years

Table 4. GHG emissions and EPT

PV type and fuel type	Atmospheric Cd emissions
Ribbon-Si	0.8 g/GWh
mc-Si	0.9 g/GWh
Mono-Si	0.9 g/GWh
CdTe	0.3 g/GWh
Hard coal	3.1 g/GWh
Lignite	6.2 g/GWh
Natural gas	0.2 g/GWh
Oil	43.3 g/GWh
Nuclear	0.5 g/GWh
Hydro	0.03 g/GWh
UCTE average	4.1 g/GWh

Table 5. Atmospheric Cd emissions

[1]Fthenakis VM, Kim HC, Alsema E. (2008). Emissions from photovoltaic life cycles. Environmental Science & Technology; 42 (6): 2,168 – 2,174

Life-cycle atmospheric Cd emissions for PV systems from electricity and fuel consumption are also evaluated for ribbon-Si, mc-Si, mono-Si, CdTe, hard coal, lignite, natural gas, oil, nuclear, hydro, and UCTE average, and the results are given as 0.8, 0.9, 0.9, 0.3, 3.1, 6.2, 0.2, 43.3, 0.5, 0.03 and 4.1 g/GWh (10^9 Wh), respectively, shown in Table 5. Compared to the emissions from oil at 43.3 g/GWh, PV system emissions are much lower.

Atmospheric emissions of arsenic (As), cadmium (Cd), chromium (Cr), lead (Pb), mercury (Hg) and nickel (Ni) are also evaluated. The CdTe PV module shows the highest level of performance, and replacing the regular grid mix with it affords significant potential to reduce these atmospheric heavy-metal emissions.

6.2 LCA study on BOS in a 3.5 MW PV system (USA)

This paper was published in 2006. At the time, there were not many large PV systems such as those operating at over a megawatt. Accordingly, this study provided worthwhile LCA results. Even now, it is difficult to find such a detailed LCA study focusing on BOS. The investigation did not include PV modules.

Paper title	Energy payback and life-cycle CO_2 emissions of the BOS in an optimized 3.5 MW PV installation[2]
Author(s)	Mason, J. E. Fthenakis, V. M. Hansen, T. and Kim, H.C.
Journal	Progress in Photovoltaics, 2006. 14 (2): 179 - 190
Location/country	Springerville, AZ/USA
Irradiation	1,725 kWh/kW (actual performance data used for LCA), approx. 2,100 kWh/m^2/yr (average)
PV capacity/PV type	3.5 MW/mc-Si
System configuration	Ground-mounted fixed flat-plate system
Lifetime	PV metal support structure: 60 years; inverters and transformers: 30 years (parts: 10 years)
Results	BOS: 542 MJ/m^2, 29 kg CO_2eq/m^2, 0.21 years of EPT, $940 US/kW
Year	2006

Table 6. Summary of the paper

The 3.5 MW Tucson Electric Power (TEP) Springerville PV plant is located in eastern Arizona, USA. The high elevation of this site and its low-temperature environment enables higher efficiency for its PV modules, which are the crystalline silicon type. Electricity from the plant is used to power a water pump at a coal-fired plant. PV support structures are anchored to the ground with 30-cm nails, thereby eliminating the need for concrete foundations. The structures' design wind speed is 160 km/h. The annual average AC electricity output in 2004 was 1,730 kWh/kW. The arrays each weigh 46.6 kg (including 5.44 kg of Al frame), cover an area of 2.456 m^2 and have a rated efficiency of 12.2%. The modules are the frameless PV type.

The total installed cost of the BOS components is $940/kW. This does not include financing or end-of-life dismantling and disposal expenses. However, the salvage value is assumed to equal the costs of dismantling and disposal. The corresponding cost for inverters and related

[2]Mason JE, Fthenakis VM, Hansen T, Kim HC. (2006). Energy payback and life-cycle CO_2 emissions of the BOS in an optimized 3.5 MW PV installation. Progress in Photovoltaics; 14 (2): 179 - 190

support software is \$400/kW, that for the wiring system is \$300/kW, and that for the PV support structures is \$150/kW.

The life expectancy of the PV metal support structures is assumed to be 60 years. Inverters and transformers are considered to have a life of 30 years, but parts amounting to 10% of the total mass must be replaced every 10 years.

The total primary energy in the BOS life cycle is 542 MJ/m^2. Using the average US energy conversion efficiency of 33% produces an EPT of 0.21. Under the average irradiation of the US (1,800 kWh/m^2/year), the EPT becomes 0.37 years.

Fig. 3. PV system power plant of Tucson Arizona Public Service (photo by author)

6.3 LCA study on the 2 MW Hokuto mega-solar plant (Japan)

This paper describes a comparative study on LCA for 20 different types of PV systems. Usually, comparative PV studies use different types of PV modules, but each module has only one or two pieces. However, in this PV project, each PV module is about 10 kW, making it necessary to evaluate the array size rather than the module size. On the other hand, the LCI data used were not for the PV modules themselves; they were from the NEDO PV project[3,4], which researched LCA for six types of PV modules including mono-crystalline silicon (mono-Si), amorphous silicon (a-Si)/mono-Si, multi-crystalline silicon (mc-Si), a-Si, micro-crystalline silicon (μc-Si)/a-Si and copper indium selenium (CIS). The data are listed in Table 8.

The installed PV modules are shown in Table 9. Six crystalline Si, one a-Si/mono-Si, seven mc-si, one a-Si, two μc-Si/a-Si, and two CIS PV modules were installed and evaluated. The

[3]NEDO, *Research and development of fabrication technologies for Life-Cycle Assessment of PV systems* (2009)

[4]Komoto, K. Uchida, H. Ito, M. Kurokawa, K. Inaba, (2008). A. Estimation of energy payback time and CO2 emissions of various kind of PV systems. Proceedings of 23rd EUPVSEC; 3,833 - 3,835

table shows that mc-Si PV modules have average or higher efficiency, while sc-Si PV modules are lower than average. This should be noted and understood, as pointed out in the paper.

The results showed an energy requirement ranging from 19 to 48 GJ/kW and an energy payback time of between 1.4 and 3.8 years. CO_2 emissions were between 1.3 and 2.7 t CO_2/kW, and CO_2 emission rates ranged from 31 to 67 g CO_2/kWh. The multi-crystalline (mc-Si) and CIS types showed good results. In particular, the CIS module generated more electricity than expected with catalogue efficiency. The single-crystalline silicon PV module did not produce good results because, considering the energy requirement, installed sc-Si PV modules do not have high efficiency.

Paper title	A comparative study on life cycle analysis of 20 different PV modules installed at the Hokuto mega-solar plant[5]
Author(s)	Ito, M. Kudo, M. Nagura, M. and Kurokawa, K.
Journal	Progress in Photovoltaics: Research and Applications, Volume 19, Issue 3
Location/country	Hokuto City, Japan
Irradiation	1,725 kWh/m^2/year at a 30-degree tilt angle
PV capacity/PV type	600 kW/mc-Si, sc-Si, a-Si/sc-Si, thin-film Si, CIS, μc-Si/a-Si
System configuration	Ground-mounted fixed flat-plate system
Lifetime	30 years; inverters: 15 years
Year	2011

Table 7. Summary of the paper

	Module efficiency in reference	Energy requirement	CO_2 emissions
PV module			
mono-Si	14.3%	3,986 MJ/m^2	193.5 kg CO_2/m^2
a-Si/mono-Si	16.6%	3,679 MJ/m^2	178.0 kg CO_2/m^2
mc-Si	13.9%	2,737 MJ/m^2	135.2 kg CO_2/m^2
a-Si (in 2000)	-	1,202 MJ/m^2	54.3 kg CO_2/m^2
a-Si/μc-Si	8.6%	1,210 MJ/m^2	67.8 kg CO_2/m^2
CIS	10.1%	1,105 MJ/m^2	67.5 kg CO_2/m^2
10 kW inverter		0.57 GJ/kW	43 kg CO_2/kW
Cable, conduit		1,068 GJ/600 kW	62.0 t CO_2/600 kW
Array (galvanized steel)		22.5 GJ/t	1.91 t CO_2/t

Table 8. LCI data from NEDO, Japan, on PV modules

[5]Ito, M. Kudo, M. Nagura, M. and Kurokawa, K. (May 2011). A comparative study on life cycle analysis of 20 different PV modules installed at the Hokuto mega-solar plant, *Progress in Photovoltaics: Research and Applications*, Volume 19, Issue 3

Type	Nominal power [W]	Module efficiency [%]	Capacity [kW]
mono-Si	84	13.2	30
a-Si/mono-Si	186	15.9	30
mono-Si	160	12.6	10
mono-Si	160	12.6	10
mono-Si	150	11.8	10
mono-Si	200	12.0	30
mono-Si	173	12.0	30
mc-Si	167	12.6	30
mc-Si	179	14.0	100
mc-Si	167	13.2	30
mc-Si	180	12.3	10
mc-Si	190	13.0	10
mc-Si	240	12.4	30
mc-Si	170	13.5	10
a-Si	60	6.1	30
μc-Si/a-Si	110	8.8	10
μc-Si/a-Si	130	8.3	10
CIS	70	8.8	30
CIS	125	11.2	3

Table 9. List of installed PV modules

Fig. 4. The 2 MW Hokuto mega-solar plant

6.4 LCA study on a VLS-PV (very large-scale PV) power plant in the desert (IEA PVPS)

This research differs from the other papers because it involved a simulation study rather than an actual system. However, the concept is interesting. It focused on a huge PV system that can generate the same amount of power as an existing power plant.

The concept of the VLS-PV was developed under IEA/PVPS Task 8. The objectives of Task 8 are to examine and evaluate the potential of very large-scale photovoltaic power generation (VLS-PV) systems. It was started in 1998, and the approaches of related evaluation are from technological, financial, environmental and local people's viewpoints. LCA is also performed as part of Task 8 to evaluate the potential of VLS-PV plants. It is assumed that very large-scale PV systems are installed in desert areas. This section discusses LCA studies on the VLS-PV system.

Paper title	Energy from the Desert: Very Large Scale Photovoltaic Systems: Socio-economic, Financial, Technical and Environmental Aspects[6]
Author(s)	Komoto, K. Ito, M. Van Der Vleuten, P. Faiman, D. Kurokawa, K.
Publisher	Earthscan
Location/country	Gobi Desert/China
Irradiation	2,017 kWh/year at a 30-degree tilt angle
PV capacity/PV type	1,000 MW/mc-Si, sc-Si, a-Si/sc-Si, thin-film Si, CIS, CdTe
Lifetime	30 years; inverters: 15 years
System configuration	Ground-mounted fixed flat-plate system
Year	2009

Table 10. Summary of the paper

The VLS-PV systems evaluated would have a capacity of 1 GW, and six kinds of PV modules were supposed: mono-crystalline silicon (mono-Si), multi-crystalline silicon (mc-Si), amorphous silicon/single-crystalline silicon hetero junction (a-Si/sc-Si), amorphous silicon/micro-crystalline thin-film silicon (thin-film Si), copper indium diselenide (CIS) and cadmium telluride (CdTe). The array structures were assumed to be conventional ones with concrete foundations. For comparison, an earth-screw approach is also discussed.

The installation site was assumed to be in Hohhot in the Gobi Desert in Inner Mongolia, China. Annual irradiation there was assumed to be 1,702 kWh/m^2/year, and the in-plane irradiation at a 30-degree tilt angle was 2,017 kWh/m^2/year. The annual average ambient temperature was 5.8°C. Most of the equipment for the VLS-PV system was assumed to have been manufactured in Japan and transported by cargo ship. However, the foundation and steel for the array structure were assumed to have been produced in China. For these materials, land transport was assumed over a distance of 600 km, and marine transport was assumed to have covered 1,000 km. The lifetime of the VLS-PV system was assumed to be 30 years, while that of the inverter was 15 years. It was assumed that after the end of the equipment's lifespan, all of it would be transported to a wrecking yard and used as landfill.

[6]Komoto, K. Ito, M. Van Der Vleuten, P. Faiman, D. Kurokawa, K. (September 2009). *Energy from the Desert -Very Large Scale Photovoltaic Systems: Socio-economic, Financial, Technical and Environmental Aspects, Earthscan,* ISBN-13: 978-1844077946

From comparison, the value for the concrete foundation was 2,458 kt CO_2/system, and that for the earth-screw approach was 2,597 kt CO_2/system. Usually, the earth-screw option would be favorable, but not in this study. The paper pinpoints the reason as the low efficiency of steel production in China. If this study had also included the recycling stage, the results would have been different, as steel can be recycled easily.

Energy consumption was from 35 to 46 TJ/MW, and CO_2 emissions were from 2,300 to 3,200 t CO_2/MW. The energy consumption of CIS was the smallest among the six types of PV modules, and the CO_2 emission of mc-Si was the smallest. Figures 5 and 6 show the EPT and CO_2 emission rate of the VLS-PV system. It was calculated that the EPT would be 2.1 - 2.8 years, and the CO_2 emission rate would be 52 - 71 g CO_2/kWh. This means that the VLS-PV would be able to recover its energy consumption in the lifecycle within three years and provide clean energy for a long time. Furthermore, the CO_2 emission rate of the VLS-PV system would be much smaller than that of a fossil fuel-fired plant. In particular, a PV system generates during the day when fossil fuel-fired plants are also operational.

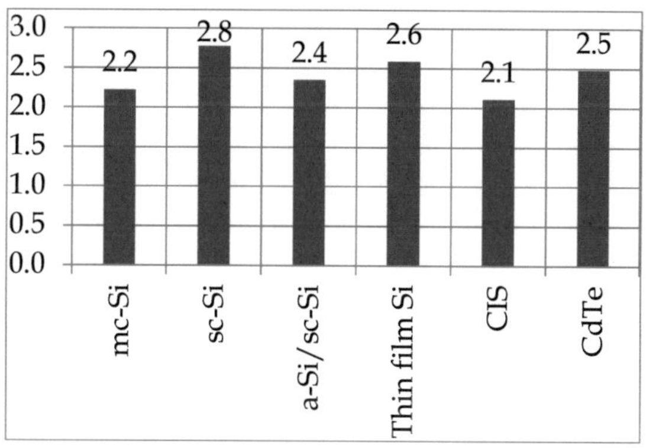

Fig. 5. Energy payback time [years] of the VLS-PV system with six types of PV module

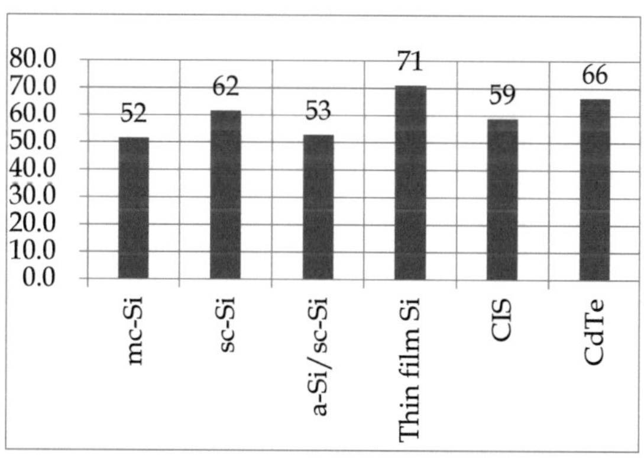

Fig. 6. CO2 emission rate [g CO2/kWh] of the VLS-PV system with six types of PV module

7. Summary

This paper describes life cycle assessment (LCA) of PV systems. An overview of LCA is given in Section 2 outlining the method for quantitatively determining the environmental effects of products. Section 3 describes LCA for PV systems, outlining evaluation indices, boundaries of LCA, inventory analysis, impact assessment and interpretation. Section 4 details LCA guidelines for PV systems, outlining important considerations for related evaluation. Section 5 deals with the collection of LCA data, and outlines ways to obtain the large amounts of information required for such analysis.

Section 6 presents LCA calculation by introducing related example papers. LCA of PV modules, PV systems and BOS is described. According to VM. Fthenakis et al., greenhouse gas (GHG) emissions from Si modules are 30 - 45 g CO_2eq/kWh, and the EPT of such modules is 1.7 - 2.7 years. The corresponding figures for CdTe frames without PV modules are 24 g CO_2eq/kWh and 1.1 years for ground-mounted installations. According to JE. Mason et al., the total primary energy in the BOS life cycle is 542 MJ/m^2. Taking the average US energy conversion efficiency of 33%, the EPT is 0.21 years. From the author's paper, the results showed an energy requirement ranging from 19 to 48 GJ/kW and an energy payback time of between 1.4 and 3.8 years. CO_2 emissions were from 1.3 to 2.7 t CO_2/kW, and CO_2 emission rates ranged from 31 to 67 g CO_2/kWh. According to Komoto et al., the EPT of a VLS-PV system would be 2.1 - 2.8 years, and the CO_2 emission rate would be 52 - 71 g CO_2/kWh.

Since the lifetime of PV systems exceeds 20 years, a low ETP means that a system can recover the energy required to pay for itself more quickly. These figures for CO_2 emission rates are also much lower than those for fossil fuel plants. It can therefore be concluded that PV systems have significant potential to mitigate global warming.

8. Abbreviations

a-Si	Amorphous silicon
BOD	Biochemical oxygen demand
BOS	Balance of system
Cd	Cadmium
CdTe	Cadmium telluride
CH_4	Methane
CIS	Copper indium diselenide
CO_2	Carbon dioxide
EPBT	Energy payback time
EPT	Energy payback time
g CO_2eq	Grams of carbon dioxide equivalents
GHG	Greenhouse gas
GJ	Gigajoule = 1,000,000,000 J
GW	Gigawatt = 1,000,000,000 W
GWh	Gigawatt hour = 1,000,000,000 Wh
GWP	Global warming potential
IEA	International Energy Agency
IPCC	Intergovernmental Panel on Climate Change
JLCA	Life Cycle Assessment Society of Japan
kWh	Kilowatt hour = 1,000 Wh
LCA	Life cycle assessment
LCI	Life cycle inventory
mc-Si	Multi-crystalline silicon

MJ	Megajoule = 1,000,000 J
MW	Megawatt = 1,000,000 W
N_2O	Nitrous oxide
NEDO	New Energy and Industrial Technology Development Organization
NOx	Nitrogen oxide
NREL	National Renewable Energy Laboratory
PCS	Power conditioning system
PR	Performance ratio
PV	Photovoltaic, Photovoltaic System
PVPS	Photovoltaic power system programme
ribbon-Si	Ribbon silicon
sc-Si	Single-crystalline silicon
Si	Silicon
SOx	Sulfur oxide
TEP	Tucson Electric Power
TJ	Terajoule
UCTE	Union of the Co-ordination of Transmission of Electricity
VLS-PV	Very large-scale photovoltaic power generation system
μc-Si	Micro-crystalline silicon

9. References

Benoit, C. UQAM/CIRAIG & Mazijn, B. (2009). *Guidelines for Social Life Cycle Assessment of Products*, United Nations Environment Programme, ISBN: 978-92-807-3021-0, France

Ito, M. Kudo, M. Nagura, M. and Kurokawa K. (May 2011). A comparative study on life cycle analysis of 20 different PV modules installed at the Hokuto mega-solar plant, *Progress in Photovoltaics: Research and Applications*, Volume 19, Issue 3

Mason, J.E. Fthenakis, V.M. Hansen, T. Kim, H.C. (2006). Energy payback and life-cycle CO_2 emissions of the BOS in an optimized 3.5 MW PV installation. *Progress in Photovoltaics;* 14 (2): 179 - 190

Fthenakis, VM. Kim, HC. Alsema, E. (2008). Emissions from photovoltaic life cycles. *Environmental Science & Technology;* 42 (6): 2,168 - 2,174

Jungbluth N. (2005). Life cycle assessment of crystalline photovoltaics in the Swiss Ecoinvent database. *Progress in Photovoltaics;* 13 (5): 429 - 446

de Wild-Scholten, MJ. Alsema, EA. ter Horst, EW. Bächler, M. Fthenakis, VM. (2006). A cost and environmental impact comparison of grid-connected rooftop and ground-based PV systems. *Proceedings of EUPVSEC-21;* 3,167 - 3,173; Dresden

Kato, K. Murata, A. Sakuta, K. (1998). Energy pay-back time and life-cycle CO_2 emission of residential PV power system with silicon PV module. *Progress in Photovoltaics;* 6 (2): 105 - 115

Fthenakis, VM. and Kim, HC. (2007). CdTe photovoltaics: Life cycle environmental profile and comparisons, *Thin Solid Films;* 515 (15): 5,961 - 5,963

Jungbluth, N. (2005). Life cycle assessment of crystalline photovoltaics in the Swiss Ecoinvent database, *Progress in Photovoltaics;* 13 (5): 429 - 446

Ito, M. Kato, K. Sugihara, H. Kichimi, T. Song, J. Kurokawa, K. (2003). A preliminary study on potential for very large-scale photovoltaic power generation (VLS-PV) system in the Gobi Desert from economic and environmental viewpoints. *Solar Energy Materials & Solar Cells 75*, pp. 507 - 517

JEMAI LCA Pro, Japan Environmental Management Association for Industry

Ito, M. Kudo, M. Kurokawa, K. (2007). A Preliminary Life-Cycle Analysis of a Mega-solar System in Japan, *Proceedings of PVSEC-17*, p. 508

Development of Technology Commercializing Photovoltaic Power Generation Systems, Research and Development of Photovoltaic Power Generation Application Systems and Peripheral Technologies, Survey and Research on the Evaluation of Photovoltaic Power Generation (2001). NEDO. p. 45

Alsema, E. Fraile, D. Frischknecht, R. Fthenakis, VM. Held, M. Kim, HC. Pölz, W. Raugei, M. de Wild Scholten, M. (2009). Methodology Guidelines on Life Cycle Assessment of Photovoltaic Electricity, Subtask 20 "LCA," IEA PVPS Task 12

Komoto, K. Uchida, H. Ito, M. Kurokawa, K. Inaba, A. (2008). Estimation of Energy Payback Time and CO_2 Emissions of Various Kinds of PV Systems, *Proceedings of 23rd EUPVSEC* pp. 3,833 - 3,835

Osterwald, CR. Adelstein, J. del Cueto, JA. Kroposki, B. Trudell, D. and Moriarty, T. (2006). Comparison of Degradation Rates of Individual Modules Held at Maximum Power, *Proceedings of 4th WCPEC* pp. 2,085 - 2,088

International Energy Agency (2010). *World Energy Outlook 2010*, ISBN 978-92-64-08624-1

14

Design and Fabrication of a Novel MEMS Silicon Microphone

Bahram Azizollah Ganji
Department of Electrical Engineering, Babol University of Technology, Iran

1. Introduction

A microphone is a transducer that converts acoustic energy into electrical energy. Microphones are widely used in voice communications devices, hearing aids, surveillance and military aims, ultrasonic and acoustic distinction under water, and noise and vibration control (Ma et al. 2002). Micromachining technology has been used to design and fabricate various silicon microphones. Among them, the capacitive microphone is in the majority because of its high achievable sensitivity, miniature size, batch fabrication, integration feasibility, and long stability performance (Jing et al. 2003; Miao et al. 2002; Li et al. 2001). A capacitive microphone consists of a variable gap capacitor. To operate such microphones they must be biased with a dc voltage to form a surface charge (Pappalardo and Caronti 2002; Pappalardo et al. 2002).

Typically, a cavity is etched into a silicon substrate by slope (54.74 deg) etching profiles using KOH etching to form a thin diaphragm or perforated back plate (Kronast et al. 2001; Pedersen et al. 1997; Bergqvist and Gobet 1994; Torkkeli et al. 2000; Kabir et al. 1999). The forming of a cavity or back chamber from the backside of a wafer by KOH etching is slow and boring in that several hundred micrometers of substrate must be etched to make the chamber. Moreover, the KOH etching process is not compatible with the CMOS process. Additionally, since the back plate requires acoustic holes that must be etched from the backside in the deep back volume cavity, a nonstandard photolithographic process must be used that requires the electrochemical deposition of the photoresist and an aluminum seed layer. Most surface and bulk micromachined capacitive microphones use a fully clamped diaphragm with a perforated back plate (Ning et al. 2004; Ning et al. 1996). The fabrication process is typically long, cumbersome, expensive, and not compatible with high volume processes. Furthermore, they are not small in size (Hsu et al. 1988; Chowdhury et al. 2000).

An important performance parameter is the mechanical sensitivity of the diaphragm. The mechanical sensitivity of the diaphragm is determined by the material properties (such as Young's modulus and the Poison ratio), thickness, and the intrinsic stress in the diaphragm. Very thin diaphragms are very fragile. In microfabrication, it is difficult to control the intrinsic stress levels in materials. In a prior paper (Rombach et al 2002), the stress problem was addressed by using a sandwich structure for diaphragms, in which layers with compressive and tensile stress were combined. If the diaphragm is composed of more than one material, this may induce a stress gradient because of the mismatch of the thermal expansions in the different materials. Any intrinsic stress gradient in the diaphragm

material will cause diaphragm to bend, leading to a change of the air gap in the device, and therefore the sensitivity and cut-off frequency.
The objective in this research is to overcome the disadvantages of the prior works by designing a novel MEMS capacitive microphone that utilizes a perforated diaphragm; thus achieving small size and improved microphone sensitivity by decreasing the mechanical stiffness of the diaphragm.

2. Microphone design

Capacitive microphones generally consist of a diaphragm that is caused to vibrate by impinging waves of acoustic pressure, a back plate and air gap. In its simplest form, a diaphragm is stretched over a conductive back plate and supported by post so that there is a gap between the membrane and the back plate. Figure 1 shows the basic structure of the condenser microphone. A diaphragm is stretched by a tensile force, *T*, is put in front of a fixed conducting back plate by means of a surrounding border which assures a separation distance, *d*, to create a capacitance with respect to the back plate and biased with a DC voltage. An acoustic wave striking the diaphragm causes its flexural vibration and changes the average distance from the back plate. The change of distance will produce a change in capacitance and charge, giving rise to a time varying voltage, *V*, on the electrodes.
This structure works as a condenser whose static capacitance is (Pappalardo et al. 2002):

$$C = \varepsilon_0 \frac{A}{d} \tag{1}$$

where ε_0 is the dielectric constant of the air and A is the surface area of the metallized membrane.

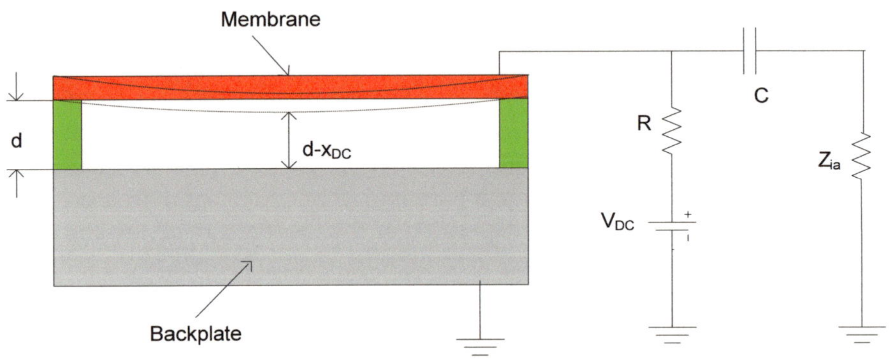

Fig. 1. Basic structure of the condenser microphone

When a DC voltage V_{DC} is applied between the two electrodes, an electric charge $Q_{DV} = C_0 V_{DC}$ appears on the surface of the membrane, where

$$C_0 = \frac{\varepsilon_0 A}{(d - x_{DC})} \tag{2}$$

accounting for the gap height variation due to the bias voltage, and x_{DC} is the static average displacement due to the DC electrostatic force. In reception, an acoustic wave striking the membrane causes its flexural vibration and changes the average distance from the back plate, which becomes

$$x = d - x_{DC} + x_{ac} = d_0 + x_{ac} \tag{3}$$

where x_{ac} is the dynamic average displacement of the vibrating membrane. As a consequence, the change of distance will produce a change in capacitance and charge, giving rise to a time varying voltage V on the electrodes.

$$V = \frac{Q}{C} = \frac{Q.x}{\varepsilon_0.A} \tag{4}$$

In the small signal approximation, using first order Taylor expansion around the bias point (V_{DC}; d_0) we have,

$$V_{ac} = \left|\frac{\partial V}{\partial Q}\right|_{bias} Q_{ac} + \left|\frac{\partial V}{\partial x}\right|_{bias} x_{ac} = \frac{d_0}{\varepsilon_0 A} Q_{ac} + \frac{Q_{DC}}{\varepsilon_0 A} x_{ac} \tag{5}$$

Where V_{ac} and Q_{ac} are voltage and charge signal components and Q_{DC} the polarization charge. For this reason the surface electrical charges are forced to move giving rise to a small alternating current which flows in the pre-amplifier input resistance Z_{ia}, through the condenser C.

In this research, 2 types of MEMS capacitive microphone have designed and fabricated on 4 inches silicon wafer. First design is microphone with clamped perforated diaphragm (see Fig. 2). The novelty of this method relies on diaphragm includes some acoustic holes to reduce air damping in the gap. Compared with previous works, the chip size of this microphone is reduced; the complex and expensive fabrication process can be avoided by making acoustic holes in diaphragm. Second design is microphone with slotted perforated diaphragm (see Fig. 3). The novelties of this method relies on the diaphragm includes some slots to reduce the effect of residual stress and stiffness of diaphragm and also includes some acoustic holes to reduce air damping in the gap. By this way, the microphone size was reduced, and the sensitivity was increased.

In next section, the behaviors of the microphones with clamped and slotted perforated diaphragms are analyzed using the finite element method (FEM).

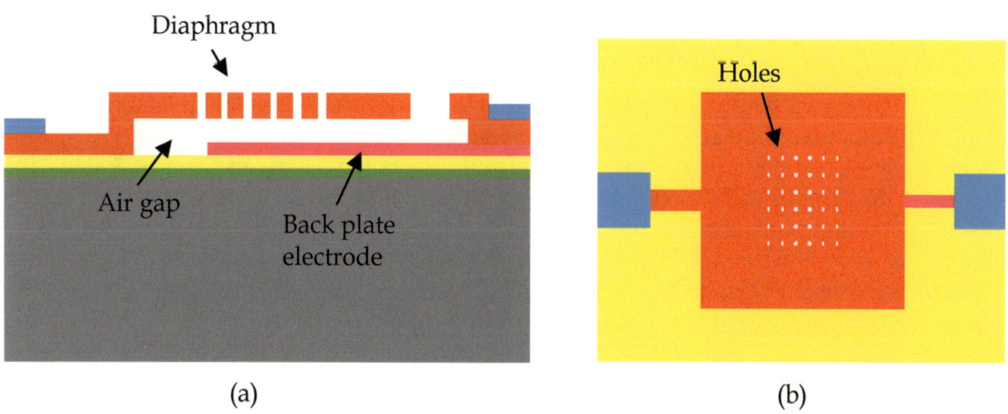

Fig. 2. (a) Cross-section, and (b) top view of clamped perforated microphone (Ganji and Majlis 2009)

Fig. 3. (a) Cross-section, and (b) top view of slotted perforated microphone (Ganji and Majlis 2009)

3. Finite element analysis (FEA) of the microphone

The analysis objectives are:

1. To verify the deformation of the diaphragm due to the electrostatic attraction force between the diaphragm and backplate, and the mechanically applied force
2. To verify the capacitance between the diaphragm and the back plate

The analysis options are nonlinear analysis, accuracy of convergence that is 0.001 µm, and a maximum mesh size that is 2.4% of *X-Y* dimension. Figure 4a shows the simulation setup of the microphone with clamped diaphragm. Silicon wafer faces and 4 lateral faces of the poly silicon diaphragm are fixed. Figure 4b shows the simulation setup for the microphone with slotted diaphragm. Silicon wafer faces and 8 lateral faces of arms are fixed. A DC bias voltage is provided between the diaphragm and the back plate.

Figure 5 show the stress distribution over of the clamped diaphragm (Fig. 5a) and the slotted diaphragm (Fig. 5b) using the FEM. We can see that the stress concentration is found at the edges of the clamped diaphragm. For the slotted diaphragm, however, the value of stress at the center and edges of the diaphragm is very low and it increases as it goes to the suspending area.

Figure 6 shows deformation in the Z axis of the diaphragm with a thickness of 3 µm and an initial stress of 20 Mpa at an applied pressure of 1.5 kPa. Figure 6a shows the maximum central deflection of clamped diaphragm is 0.245 µm and Figure 6b shows the maximum deflection of slotted diaphragm is 0.6643 µm. We can see that the slotted diaphragm has more deflection than the clamped one under same load.

Figure 7 shows the simulated diaphragm deflection versus voltage and Figure 8 show the simulated diaphragm deflection versus pressure for the clamped diaphragm (2.43 x 2.43 mm^2) and the slotted diaphragm (1.5 x 1.5 mm^2). According to the results, both microphones have the same pull-in voltage (7 V) and the same high mechanical sensitivity (53.3 nm/Pa), however the slotted microphone is at least 1.62 times smaller than the clamped structure.

Figure 9 shows the central deflection versus bias voltage of the clamped and slotted microphones using a 0.5-mm square diaphragm with a thickness of 3 µm, an air gap of 1 µm, and a diaphragm stress of 1500 MPa (Ganji and Majlis 2009). We can see that the pull-in voltage for the clamped diaphragm is 105 V, and that for the slotted diaphragm is 49 V. We can see that, by introducing slots in microphone, the diaphragm stiffness decreased, therefore the pull-in voltage decreased about 53%.

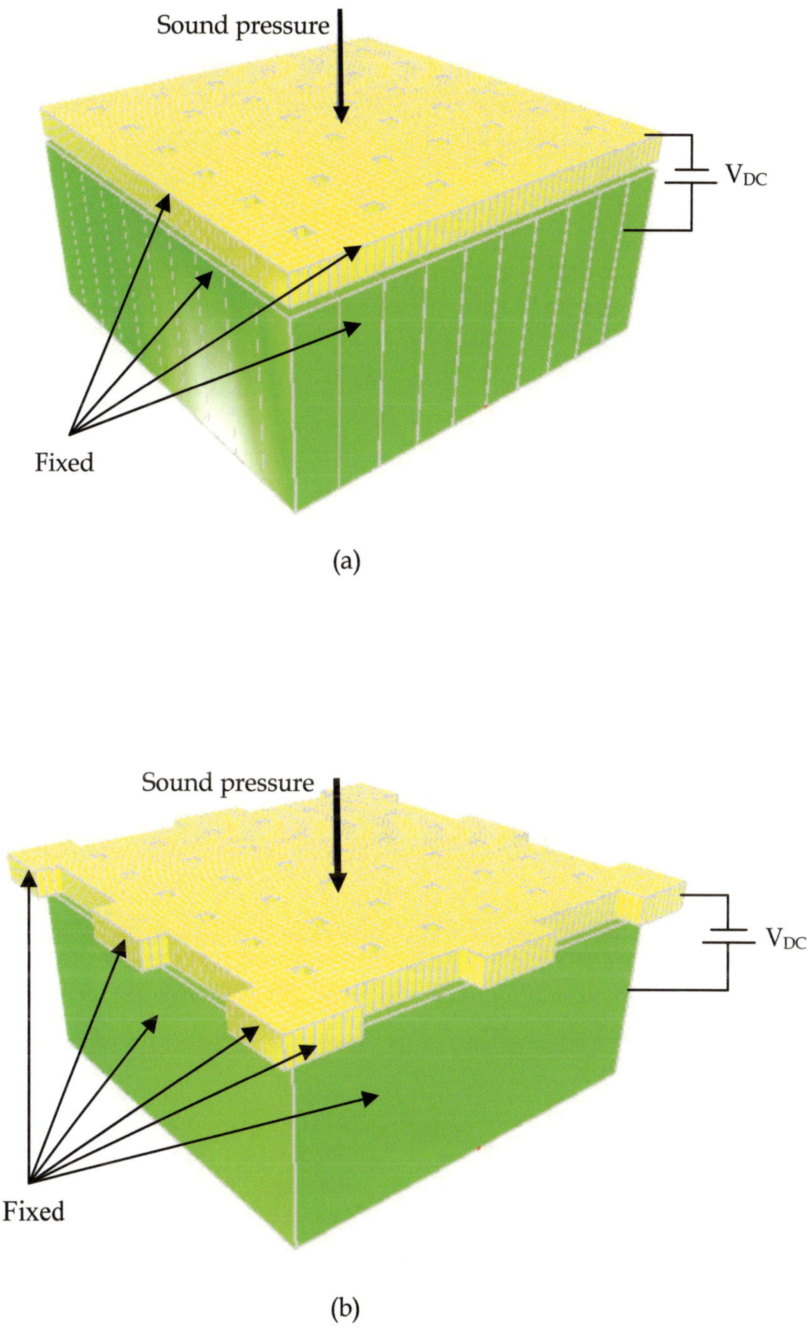

Fig. 4. Simulation setup for (a) clamped microphone, (b) slotted microphone

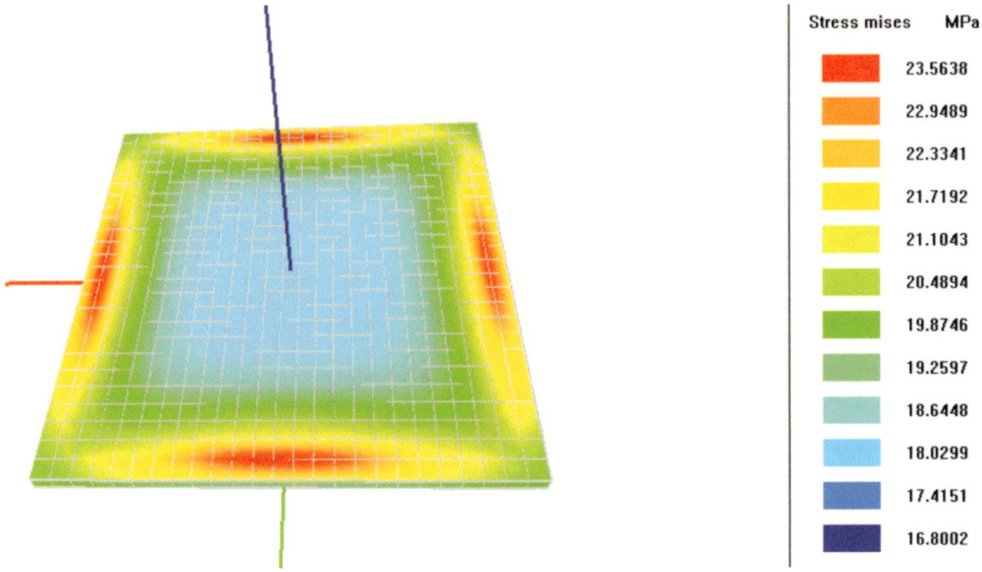

(a)

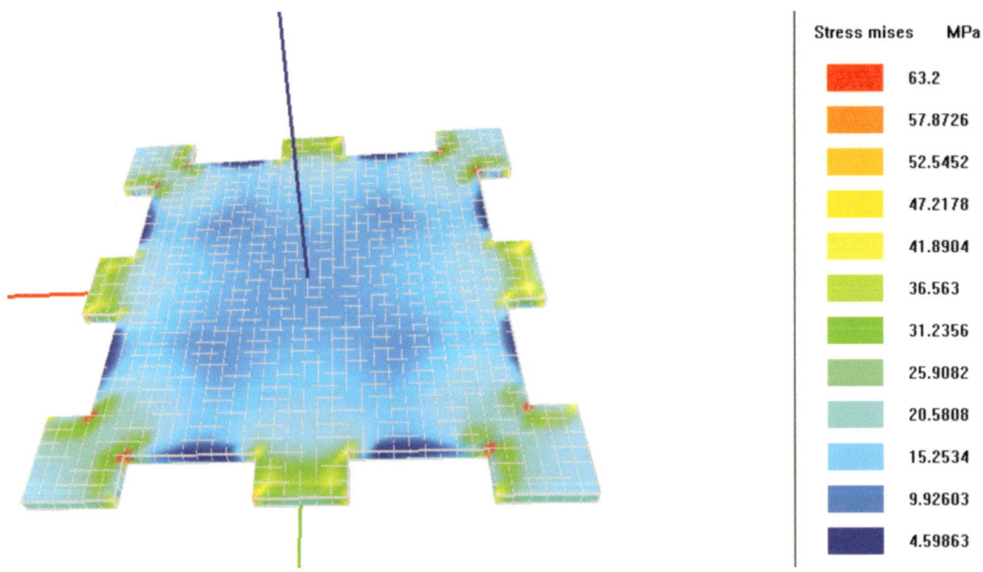

(b)

Fig. 5. Stress distribution on the (a) clamped diaphragm and (b) slotted diaphragm

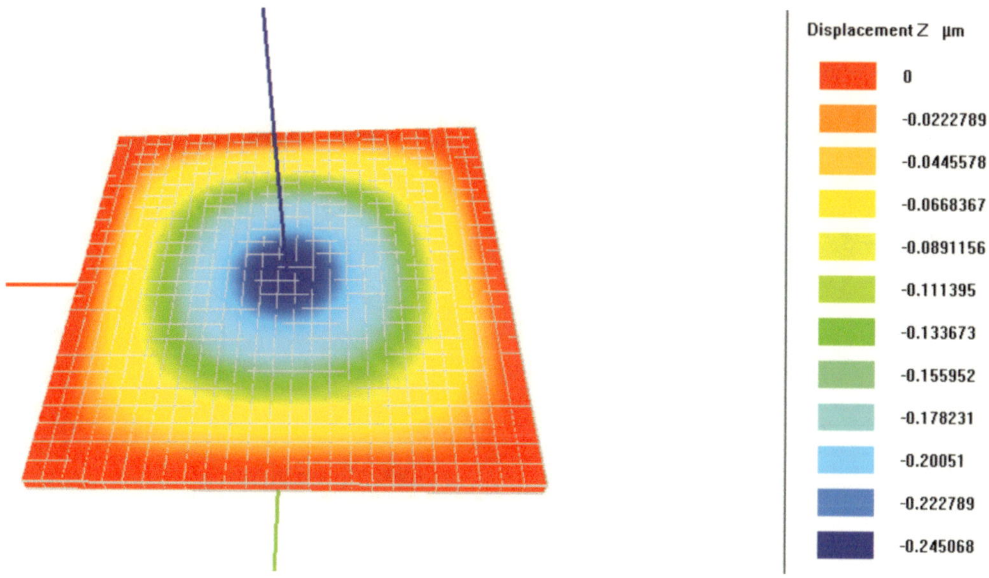

(a)

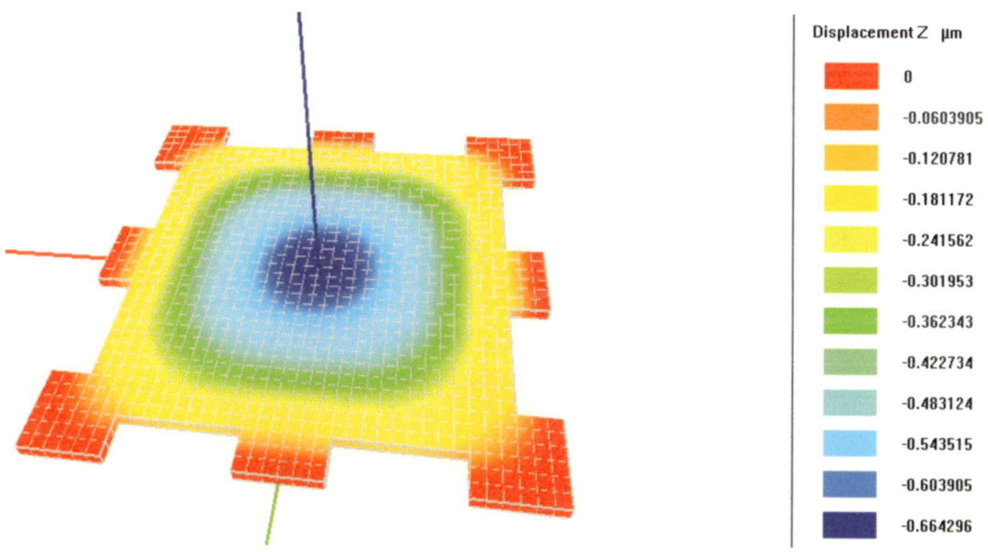

(b)

Fig. 6. Diaphragm deformation on the *Z* axis of the (a) clamped diaphragm and (b) slotted diaphragm

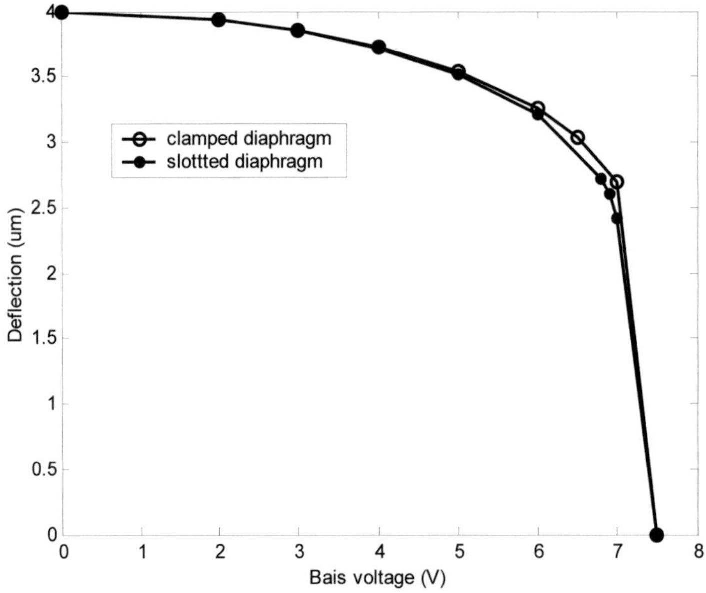

Fig. 7. Diaphragm deflection versus voltage

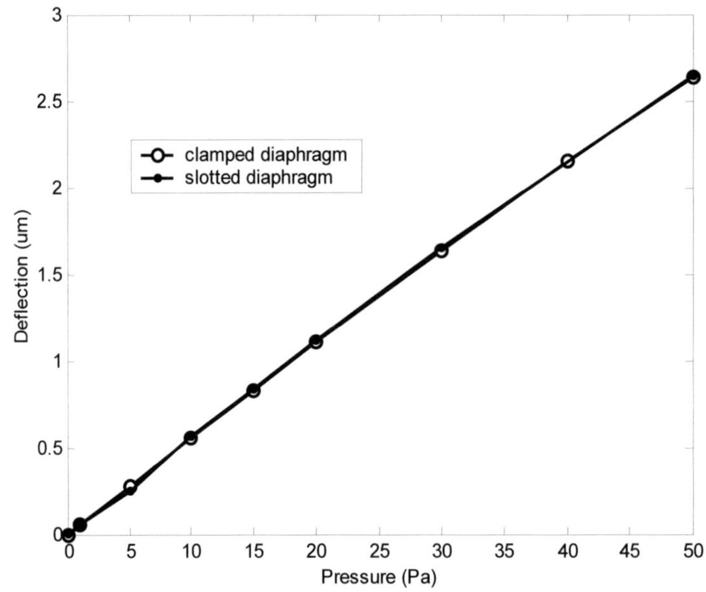

Fig. 8. Diaphragm deflection versus pressure

Figure 10 shows the relation between capacitance and pressure for clamped and slotted microphones under 60% of pull-in voltage. The results yield a sensitivity (S=dC/dP) of 5.33x10^{-6} pF/Pa for the clamped and 3.87x10^{-5} pF/Pa for the slotted microphones. By introducing the slots in the diaphragm, the sensitivity's increased 7.27 times. The first resonance frequency of the diaphragm is 1.11 MHz for the clamped and 528.57 kHz for the slotted microphones. From the preceding analysis, we can conclude that there is a dilemma between the high sensitivity and high resonance frequency. For all the diaphragms, to satisfy most of the microphones, the first resonance frequency of the diaphragm should be well above 20 kHz (hearing range).

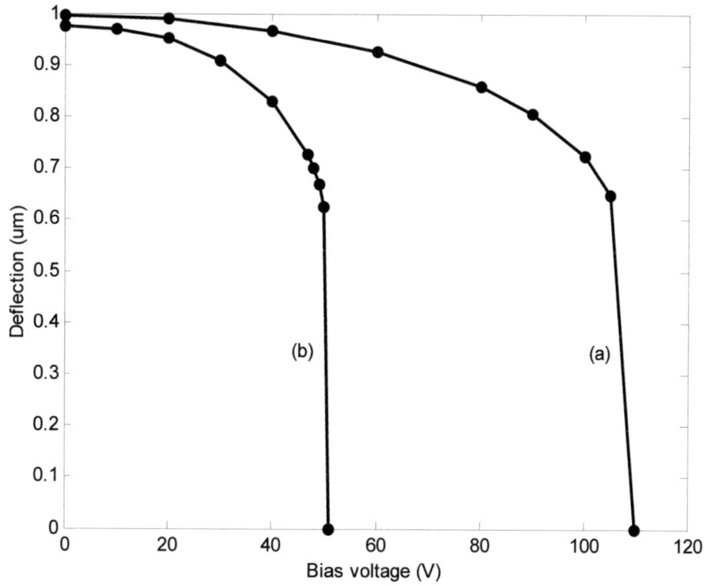

Fig. 9. Central deflection of a (curve a) clamped and (curve b) slotted diaphragm versus bias voltage

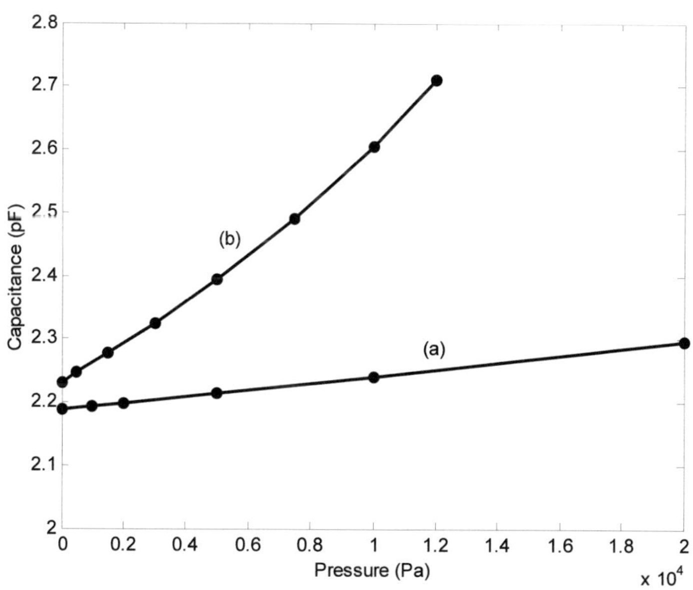

Fig. 10. Capacitance versus pressure for (a) the clamped and (b) the slotted microphones

4. Fabrication of microphone

This section will describe how the microphone was fabricated on silicon wafer. In this process, sputtered aluminum is used as a diaphragm and back plate electrode, resist (AZ1500) as a sacrificial layer, and sputtered silicon oxide as an insulation layer. The whole process sequence uses three masks and several deposition, and etching processes. The process starts with a single side polished silicon wafer as a substrate. The major fabrication steps are shown in Figure 11, and described as follows:

First a 4-inch silicon wafer should be cleaned using standard cleaning procedure to remove organic contaminants such as dust particles, grease or silica gel and then remove any oxide layer from the wafer surface prior to processing. The first step in the cleaning process is to clean the wafer using ultrasonic in the acetone solution for 5 minutes. The second step is to put the wafer into the methanol solution using ultrasonic for 5 minutes. Final step is to dip the sample in a 10:1 DI water-HF solution (10% HF) until hydrophobic (i.e. no water can stick to wafer). This will remove native oxide film (see Fig. 11a).

Then a 2 μm thick silicon oxide is sputtered on clean silicon wafer as an insulation layer (see Fig. 11b). Next, a 0.5 μm Al has been sputtered on silicon oxide as a back plate electrode. It was then patterned using photoresist mask and etched by Al etchant for 5 minutes (see Fig. 11c). The etch rate of sputtered Al in Al etchant is 60 nm/minute. Etchant for aluminum is 16:4:1 of phosphoric acid (H_3PO_4), DI water, and nitric acid (HNO_3). After that, a 1.3 μm thick resist (AZ1500) was deposited and patterned in order to form a sacrificial layer (see Fig. 11d). Resist can be easily deposited and removed using acetone. Moreover, acetone has a high selectivity to resist compared to silicon oxide and Al, thus it completely removes sacrificial resist without incurring significant damage silicon oxide and Al. Sacrificial resist is usually deposited by spin coater. Baking is the most important. The main purpose of baking is to remove solvent from resist. A few minutes of hot plate baking temperature of at least 100°C is required to evaporate the solvent. The samples are then heated at145°C for 3 minutes.

Then, a 3 μm thick layer of aluminum is sputtered on resist sacrificial layer as a material of diaphragm (see Fig. 11e). The Al layer is then patterned using positive resist mask to define the geometry of the diaphragm, contact pad, and anchors. After that the structure was immersed in Al etchant for 35 minutes to etch the Al for making diaphragm structure. The approximate etch rate of Al in acetone in room temperature is zero. Therefore acetone shows a high selectivity against Al.

Finally, the sacrificial resist layer is etched using acetone to release the diaphragm (see Fig. 11f). The fabrication process is completed by immersing it in deionized water (DI) and then acetone. Next, the whole structure is dried on hot plate at 60°C for 90 seconds to protect the diaphragm from sticking to the back plate.

After all processing on the wafers were completed, the last step was to determine if the fabrication process had been successful. It is important to observe the silicon membrane and check to ensure that the resist layer was removed. All testing was performed by using a Scanning Electron Microscope (SEM) and optical microscope to capture images of the membrane surface and images of the cross-section. Figure 12 shows the optical microscopy top view of Al back plate electrode and photoresist (AZ1500) sacrificial layer on silicon oxide.

Figure 13(a) shows the surface of the fabricated clamped microphone and Figure 13(b) shows the close up view of the Al diaphragm surface (0.5x0.5 mm^2) with acoustic holes using SEM. Figure 14 shows the SEM image of slotted microphone with 8 slots and 8 arms. Figure 15 show the sacrificial layer etching with diaphragm thickness of 3µm, and air gap of 1.3µm. It can be seen that, sacrificial layer has been removed under Al membrane completely, and Al membrane has been released.

The measured pull-in voltage for clamped microphone is 51 V, however the measured pull-in voltage of slotted microphone with sputtered aluminum diaphragm is 25 V. It can be seen that, by introducing slots in microphone, the diaphragm stiffness decreased, therefore the pull-in voltage about 50% decreased. Consequently, it causes the microphone sensitivity is increased.

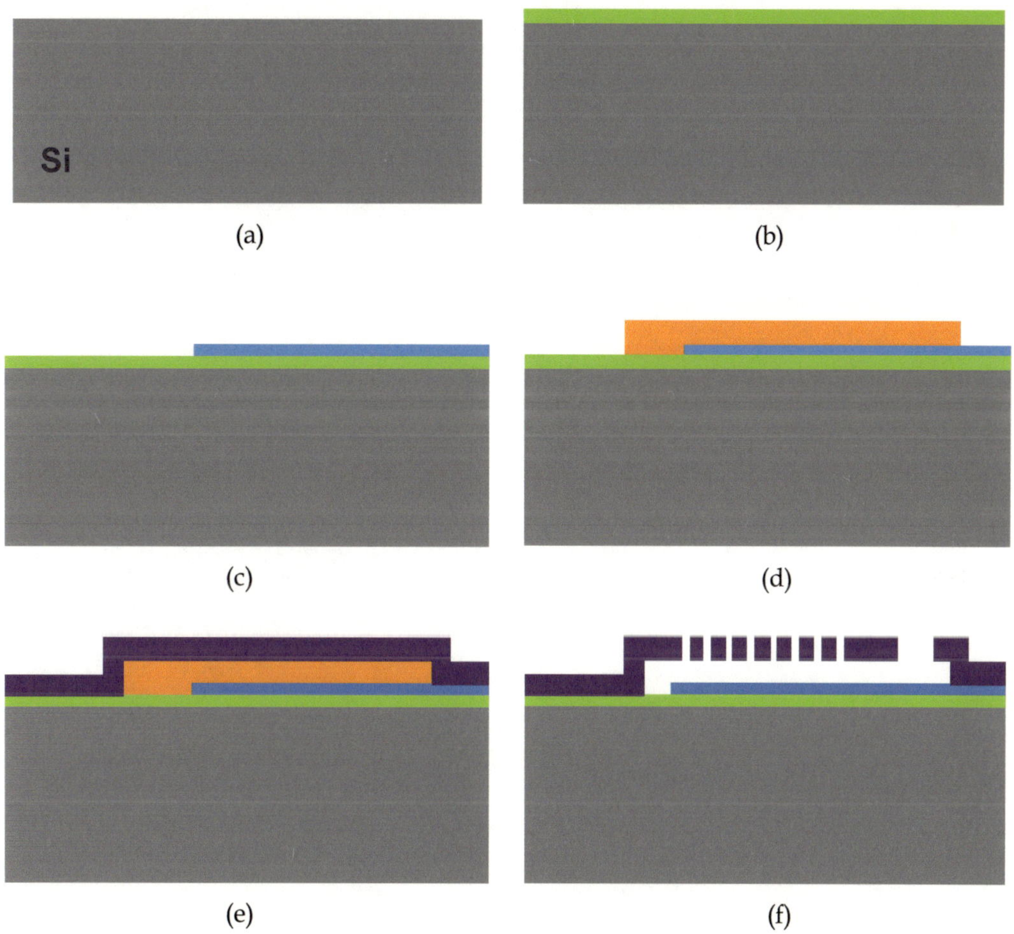

Fig. 11. Process flow of the microphone (Ganji and Majlis 2010)

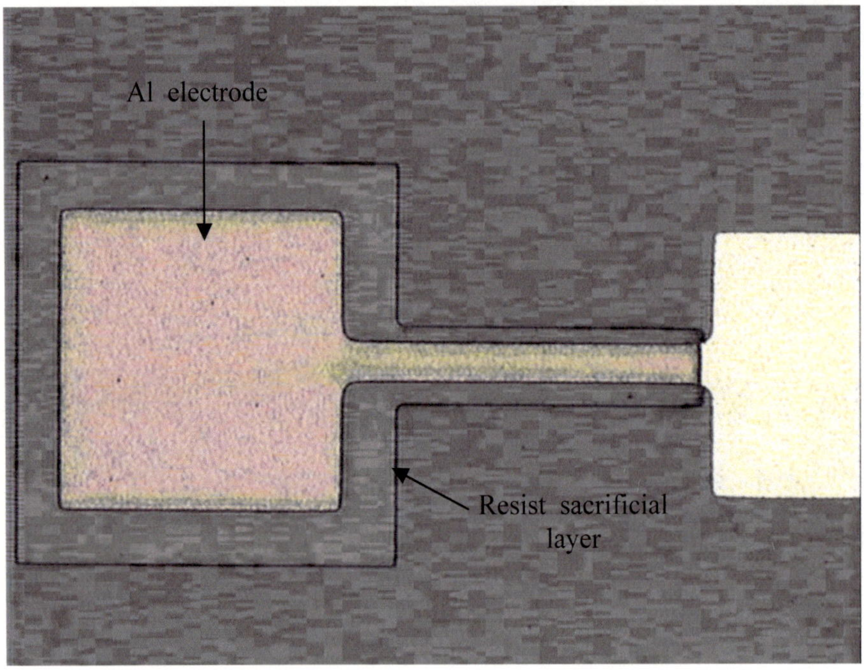

Fig. 12. Top view of Al back plate electrode and photoresist sacrificial layer on silicon oxide

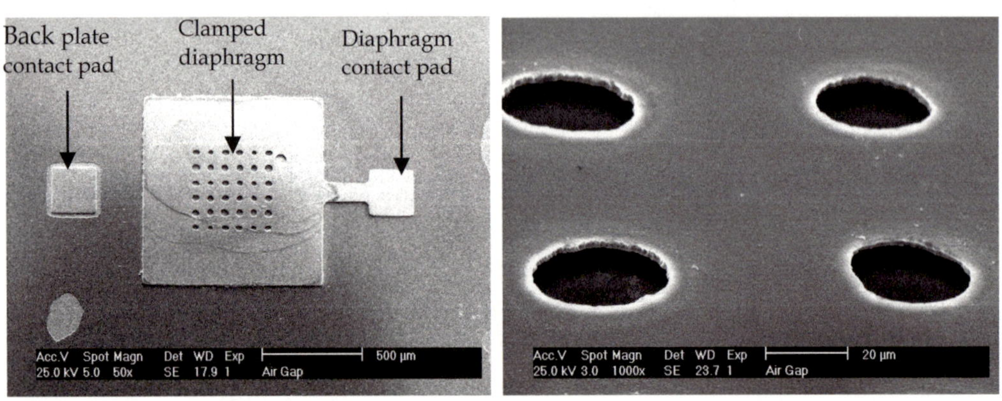

(a) (b)

Fig. 13. (a) Surface of the clamped microphone, (b) close up view of the diaphragm

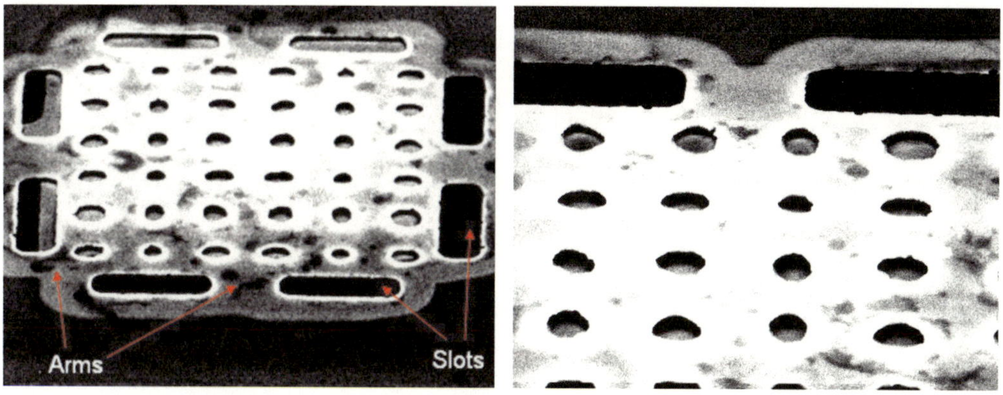

(a) (b)

Fig. 14. SEM picture of (a) slotted microphone, (b) close up view of diaphragm

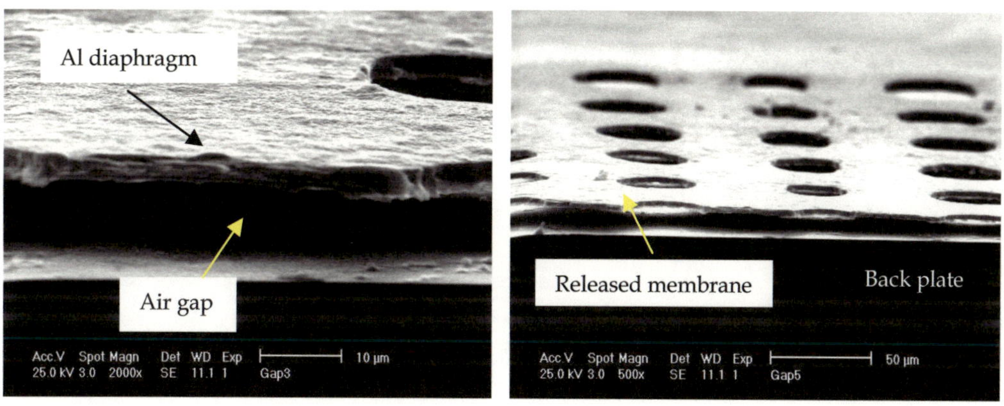

(a) Air gap of microphone (b) Released membrane structure

Fig. 15. Cross-section view of the microphone structure using SEM machine (Ganji and Majlis 2009)

5. Test of microphone

Figure 16 shows the MEMS capacitive microphone has been connected to amplifier, power amplifier and speaker. The bias voltage of microphone, V_b, is 3 V, and bias resistance, R_b, is 100MΩ. The amplifier consists of an operational amplifier LF347 with high input impedance of 10^{12} Ω, R_f of 1 MΩ, R_s of 1.25 KΩ, and Vcc of 9 V battery. The voltage gain of amplifier, ($A_{v1} = R_f/R_s$) is 800. The power amplifier is a mini amplifier-speaker CAT. No. 277-1008C. The voltage gain of power amplifier, A_{v2}, is 50. The total voltage gain of external amplifier, ($A_{vtot} = A_{v1}.A_{v2}$) is 40000. Figure 17 shows the 2 seconds of a speech signals are applied to the microphone. It can be seen that the external amplifier was able to detect the sound waves from microphone on oscilloscope. From the figure, the maximum amplitude of output speech signal of amplifier is 45 mV, thus the maximum output of microphone is 1.125 µV.

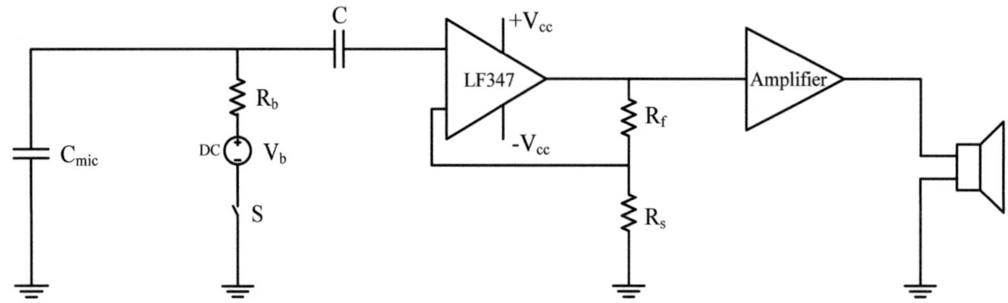

Fig. 16. Circuit diagram of external amplifier which connected to microphone (Ganji and Majlis 2010)

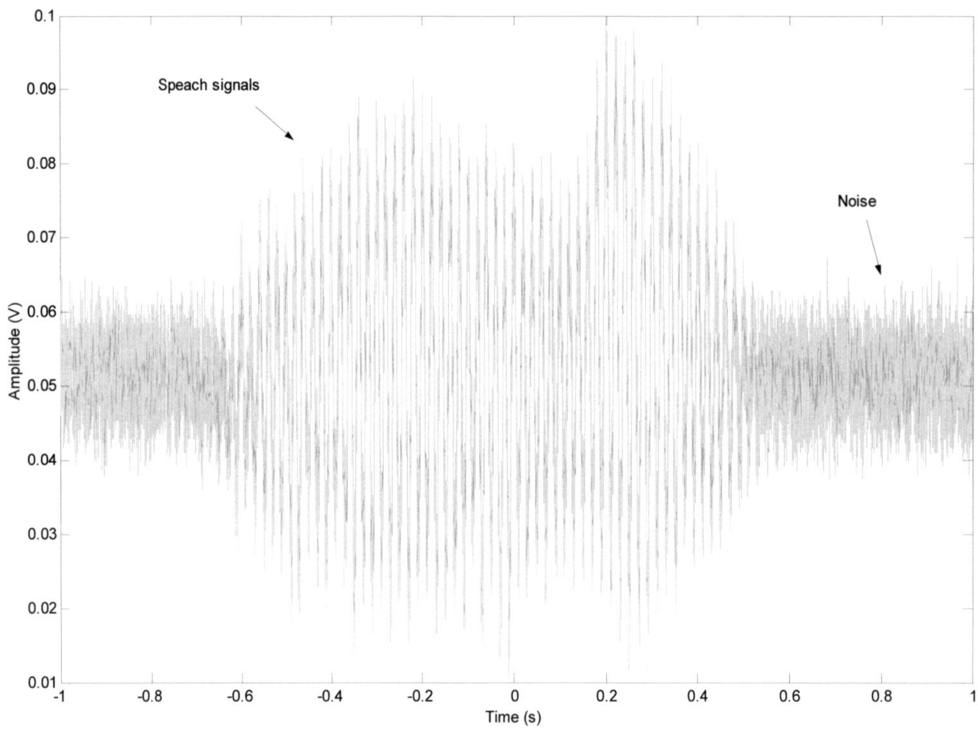

Fig. 17. 2 seconds of speech signals are applied to the microphone

6. Conclusion

A novel MEMS capacitive microphone was designed, and fabricated with a small size and a high sensitivity. The device used a perforated diaphragm, mono crystalline silicon back plate, and resist as a sacrificial layer. The results show the obvious improvement in size and sensitivity of the slotted microphone compared with the clamped one. According to the results, the slotted microphone with a 1.5-mm diaphragm width, at least 1.62 times is smaller than the clamped structure with a 2.43-mm diaphragm width. The results also yield a sensitivity of 5.33 $x10^{-6}$ pF/Pa for the clamped and 3.87 $x10^{-5}$ pF/Pa for the slotted microphones using a 0.5-mm square aluminum diaphragm with a thickness of 3 μm and an air gap of 1 μm. We can see that, by introducing the slots in the diaphragm, the microphone sensitivity was increased 7.27 times. The measured pull-in voltage for the clamped microphone with sputtered aluminum diaphragm is 51 V, however, the pull-in voltage of the slotted microphone is 25 V. This means that the slotted diaphragm stiffness has been decreased; consequently, the pull-in voltage decreased about 50%. The microphone has been tested with external amplifier and speaker, it can be seen that the external amplifier was able to detect the sound waves from microphone on speaker. The maximum amplitude of output speech signal of amplifier is 45 mV, and the maximum output of microphone is 1.125 μV.

7. References

Bergqvist, J., Gobet, J. (1994). Capacitive microphone with a surface micromachined backplate using electroplating technology. *J. Microelectromech. Syst.* 3(2): 69–75.

Chowdhury, S., Jullien, G. A., Ahmadi, M. A., Miller, W. C. (2000). MEMS acousto-magnetic components for use in a hearing instrument. *Presented at SPIE's Symposium on Design, Test Integration, and Packaging of MEMS/MOEMS, Paris.*

Ganji, B. A. and Majlis, B. Y. (2009). Design and fabrication of a new MEMS capacitive microphone using a perforated aluminum diaphragm. *Sensors and Actuators A*: Physical, 149: 29–37.

Ganji, B. A. and Majlis, B. Y. (2009). Design and fabrication of a novel single-chip MEMS capacitive microphone using slotted diaphragm. *J. Micro/Nanolith. MEMS MOEMS* 8(2), DOI: 10.1117/1.3091941, pp.021112 (1-7).

Ganji, B. A. and Majlis, B. Y. (2009). Fabrication and Characterization of a New MEMS Capacitive Microphone using Perforated Diaphragm. *International journal of Engineering,* Vol. 22, No. 2, pp. 153-160.

Ganji, B. A. and Majlis, B. Y. (2009). High Sensitivity and Small Size MEMS Capacitive Microphone using a Novel Slotted Diaphragm, *Microsystem Technology,* Vol. 15, Issue 9, pp: 1401-1406.

Ganji, B. A. and Majlis, B. Y. (2010). Slotted capacitive microphone with sputtered aluminum diaphragm and photoresist sacrificial layer, *Microsystem Technology,* Vol. 16, pp: 1803–1809.

Hsu, P. C., Mastrangelo, C. H., Wise, K. D. (1988). A high density polysilicon diaphragmcondenser microphone. *In Conf. Record IEEE 11th Int. Workshop on MicroElectro Mechanical Systems (MEMS),* pp. 580–585.

Jing, C., Liu, L., Li, Z., Tan, Z., Xu, Y., Ma, J. (2003). On the single-chip condenser miniature microphone using DRIE and back side etching techniques. *Sens. Actuators, A* 103: 42–47.

Kabir, A. E. , Bashir, R., Bernstein, J., De Santis, J., Mathews, R., O'Boyle, J. O., Bracken, C. (1999) Very High Sensitivity Acoustic Transducers with Thin P^+ Membrane and Gold Back Plate Sensors and Actuators-A, 78: 138-142.

Kronast, W., Muller, B., Siedel, W., Stoffel, A., (2001). Single-chip condenser microphone using porous silicon as sacrificial layer for the air gap. *Sens. Actuators, A* 87: 188–193.

Li, X., Lin, R., Kek, H., Miao, J., Zou, Q. (2001). Sensitivity- improved silicon condenser microphone with a novel single deeply corrugated diaphragm. *Sensors and Actuators A* 92: 257-262.

Ma, T., Man, T.Y., Chan, Y. C., Zohar, Y., Wong, M. (2002). Design and fabrication of an integrated programmable floating-gate microphone. *Proceedings of the Fifteenth IEEE International Conference on Micro Electro Mechanical Systems*, pp. 288–291.

Miao, J., Lin, R., Chen, L., Zou, Q., Lim, S. Y., Seah, S. H. (2002). Design considerations in micromachined silicon microphones. *Microelectronics Journal* 33: 21-28.

Ning, J., Liu, Z., Liu, H., Ge, Y. (2004). A silicon capacitive microphone based on oxidized porous silicon sacrificial technology. *Proc. 7th Int. Conf. on Solid-State and Integrated Circuits Technology, IEEE*, 3: 1872–1875.

Ning, Y. B., Mitchell, A. W., Tait, R. N. (1996). Fabrication of a silicon micromachined capacitive microphone using a dry-etch process. *Sens. Actuators, A* 53: 237–242.

Pappalardo, M, Caliano, G, Foglietti, V, Caronti, A, Cianci, E (2002) A new approach to ultrasound generation: the capacitive micromachined transducers. *University Roma, Rome, Italy*.

Pappalardo, M., A. Caronti (2002). A new alternative to piezoelectric transducer for NDE and medical applications: the capacitive ultrasonic micromachined transducer (cMUT). *University Roma, Rome, Italy*.

Pedersen, M., Olthuis, W., Bergveld, P. (1997). A silicon condenser microphone with polyimide diaphragm and back plate. *Sens. Actuators, A* 63: 97–104.

Rombach, P., Mullenborn, M., Klein, U., Rasmussen, K. (2002). The first low voltage, low noise differential silicon microphone, technology development and measurement results. *Sens. Actuators, A* 95: 196– 201.

Torkkeli, A., Rusanen, O., Saarilahti, J., Seppa, H., Sipola, H., Hietanen, J. (2000). Capacitive microphone with low- stress polysilicon membrane and high-stress polysilicon backplate. *Sens. Actuators* 85: 116–123.

15

Global Flow Analysis of Crystalline Silicon

Hiroaki Takiguchi[1] and Kazuki Morita[2]
[1]Ministry of the Environment
[2]The University of Tokyo
Japan

1. Introduction

Silicon is a unique material. Next to oxygen, it is the second most abundant element in the Earth's crust. Its abundance is one of the reasons it is used for a wide range of purposes. One of the most important uses of silicon is as a core element of microchips. To manufacture microchips, the microelectronics industry requires silicon with an impurity level of 10^{-11}. Since silicon forms a stable compound with oxygen (silicon oxide, SiO_2), the deoxidization of silicon oxide needed to reach this high level of purity consumes a substantial amount of energy, which, in turn, affects the environment through emissions of carbon dioxide (CO_2).

In the past decade, there has been a dramatic increase in the global supply and demand of crystalline silicon. This is because of a drastic increase in the demand for crystalline silicon for photovoltaic (PV) cells. While a core element for renewable energy, the purification process of crystalline silicon is energy-intensive. Therefore, the sustainability of crystalline silicon feedstock is an interesting area for research. The effective use of crystalline silicon could contribute to the sustainability of global environmental systems.

Reflecting its importance to sustainability, there has been a growing literature on material flows of crystalline silicon. Some studies highlight energy use and environmental impacts in the process of crystalline silicon production. For instance, Williams conducted a quantitative systems analysis of global production chains for silicon (Williams, 2003). He estimated global material flows of silicon in 1998 and then forecast how these flows would project out to 2020. In analyzing material flows, he chose different physical units such as weight and area, according to the characteristics of different outputs. Using this approach, Williams et al. were able to determine the energy required for a microchip based on the calculation on energy use at each production stage (Williams et al., 2002). Their analysis further was able to show in a quantitative manner that the purification to electronic grade silicon (EG-Si) is an energy intensive process.

Other studies have focused on the supply of crystalline silicon for solar cells. Woditsch and Koch expressed concern about the shortage of crystalline silicon for solar cells which were dependent on off-grade silicon from the EG-Si production process (Woditsch & Koch, 2002). They concluded that new routes for solar-grade silicon production are urgently needed. Sarti and Einhaus proposed short- and long-term solutions to respond to the demand for polycrystalline silicon (pc-Si, also called multicrystalline silicon) for solar cells. Their recommendations included a reduction in the amount of crystalline silicon in the short-term and the establishment of solar-grade silicon production from metallurgical-grade silicon (MG-Si) (Sarti & Einhaus, 2002).

As for major metals such as iron, copper and nickel, the Yale Stocks and Flows project conducted a material flow analysis of substances on national, regional and global scales (for example, Graedel et al., 2004). Their approach focused on illustrating anthropogenic metal cycles through four processes: production, fabrication and manufacturing, use and waste management. They then offered useful suggestions on the benefits and challenges of a material flow analysis.

In a previous paper, Takiguchi and Morita presented a material flow analysis of silicon in Japan from 1996 to 2006 (Takiguchi & Morita, 2009). The analysis tracked the input and output of silicon in a series of purification process in units of weight and found that rapid growth in demand for pc-Si and single crystalline silicon (sc-Si, also called monocrystalline silicon) changed the structure of the crystalline silicon supply. Takiguchi and Morita also developed the Resource Effective-use Index (REI) to demonstrate how effectively crystalline silicon is used. The analysis of the REI indicated that the effective use of pc-Si had reached its outer limits. At the same time, the paper found a domestic analysis was limited in what it could demonstrate because crystalline silicon is traded globally.

This chapter analyzes and discusses the *global flow* of crystalline silicon to assess the sustainability of silicon feedstock. The chapter begins by reviewing how crystalline silicon is produced as well as recent trends in crystalline silicon supply. The next section provides a material flow analysis of silicon on a global scale, focusing mainly on crystalline silicon for solar cells. The "results and discussion" section describes the results of the material flow analysis, followed by a discussion of the sustainability of silicon feedstock.

2. Crystalline silicon and solar cells

Before conducting a material flow analysis, this section observes the recent trend of solar cell production and explains how crystalline silicon is produced, followed by sources of crystalline silicon for solar cells.

2.1 Expansion of solar cell production

Nowadays, many countries have promoted the use of renewable energy to reduce carbon dioxide (CO_2) for climate change mitigation and diversify energy sources. In particular, the installation of PV systems is expanding in many parts of the world. Under these circumstances, the production of solar cells, core elements of PV systems, is increasing rapidly. Fig. 1 illustrates trends in solar cells production by types. The total production of solar cells in the world sharply increased from 126 MW in 1997 to 10,660 MW in 2009. The share of sc- and pc- Si solar cells grew the most at 32% and 45%, respectively. This can be attributed to the ease of mass production using highly developed silicon technology, a good balance between energy conversion efficiency and cost, and the fact that their products are non-toxic. These advantages suggest that crystalline silicon (pc- and sc- Si) solar cells will be likely to continue to expand in the future.

2.2 Process for crystalline silicon production

A typical production process for crystalline silicon for semiconductors is illustrated in Fig. 2. In the first step, the reduction of silica (quartz, SiO_2) produces metallurgical-grade silicon (MG-Si) with a purity of approximately 98% in electronic furnaces.

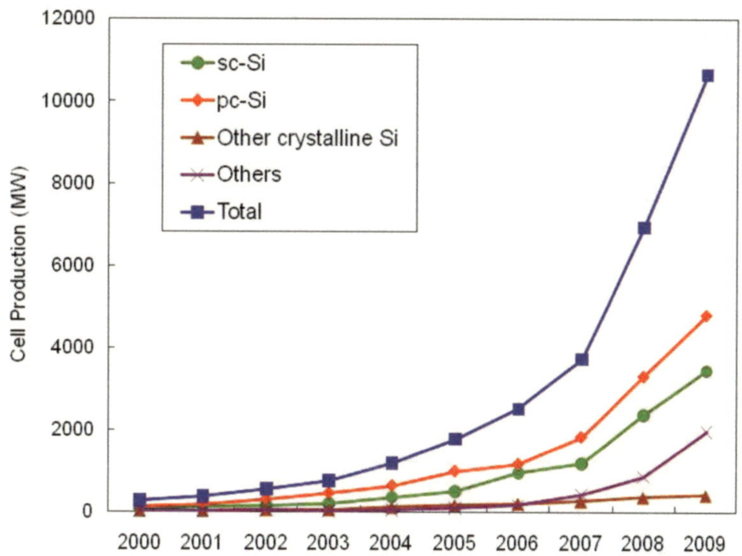

Fig. 1. Trends in solar cell production by types.

The second step is purification from MG-Si to pc-Si. Usually, a chemical gaseous purification technique known as the Siemens method is used for purification. This method involves reacting MG-Si with hydrochloric acid (HCl) to convert purified trichlorosilane ($SiHCl_3$) by distillation ($Si + 3HCl \rightarrow SiHCl_3 + H_2$). The trichlorosilane is then decomposed with hydrogen on pure silicon surfaces and converted back into pc-Si ($SiHCl_3 + H_2 \rightarrow Si + 3HCl$). After the reaction, pc-Si with an impurity level of 10^{-11} is obtained. In this process, the formation of silicon tetrachloride ($SiCl_4$) as a by-product of the production of trichlorosilane lowers the yield of pc-Si.

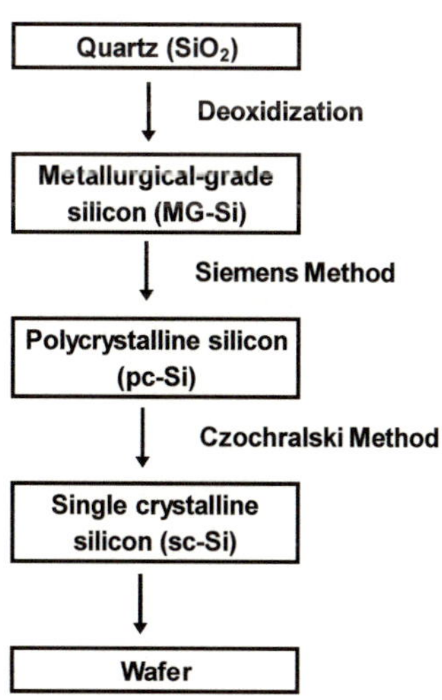

Fig. 2. Typical process flow for purified Si production.

The third step is to convert pc-Si to sc-Si. For the conversion process, the semiconductor industry usually employs the Czochralski (CZ) method. In the CZ method, sc-Si grows on a seed crystal drawn up from molten pc-Si in a crucible, producing a sc-Si ingot. The cylindrical ingot of sc-Si is sliced into wafers followed by a series of processes such as wrapping and etching. Finally, the wafers are processed into integrated circuit (IC) chips.
Table 1 provides the energy consumption at each stage from quartz to wafers (Frankl etc., 2004). It is obvious that the process of producing pc-Si from MG-Si consumes the most energy of all of the processes. While it is the established and most frequently used method of producing pc-Si, the production method presents several issues that need to be resolved, such as the limited velocity of the chemical reaction, the considerable initial investment required to enlarge the process, and the high cost in proportion to high consumption of energy.

Product	Electrical energy input/ kg silicon out (kWh/kg)	Silicon yield
Quartz → MG-Si	11	0.79
MG-Si → pc-Si	149.9	0.76
pc-Si → sc-Si ingot (Czochralski)	31.1	1
sc-Si ingot → wafer	42.5	0.37

Table 1. Energy required for Si purification.

2.3 Crystalline silicon for solar cells

Crystalline silicon for solar cells does not require a purity level as high as the level required for semiconductors. While the impurity level required for EG-Si is 10^{-11}, the level for pc-Si solar cells is 10^{-7} to 10^{-6}. Therefore, off-grade silicon produced as a by-product of the EG-Si production process has been used for solar cells. The following describes the process typically used to generate off-grade silicon. In the process of crushing pc-Si, approximately 5-10% of the produced pc-Si is lost; this is used as off-grade silicon. In the CZ process of producing sc-Si, nearly 40% of the pc-Si is lost as pot scrap, tops and tails, kerf loss and test wafers (Fig. 3) and used as off-grade silicon (New Energy and Industrial Technology Development Organization [NEDO], 2001). Scrap wafers are also used as off-grade silicon. The off-grade silicon is melted and turned into a pc-Si ingot via castings in a crucible mold or a sc-Si ingot via another CZ process. Polycrystalline silicon or sc-Si produced from the off-grade silicon ingot is sliced into wafers with a wire saw and used for solar cells. Depending on its purity level, off-grade silicon is used for other low-grade purposes such as making aluminium alloy.
In the 1990s, off-grade silicon obtained from the EG-Si production process satisfied the demand for silicon for solar cells. The recent increase in demand for solar cells, however, resulted in production of crystalline silicon independently, not as scraps of EG-Si. Since the purification level of pc-Si for solar cells is lower than that for semiconductors, pc-Si for solar cells is produced by a simplified Siemens method which increases the speed of trichlorosilane decomposition. It should be noted that the simplified method still consumes considerable energy.
Thus, crystalline silicon for solar cells is currently obtained from two sources: 1) off-grade silicon produced as a by-product of the EG-Si production process and 2) silicon produced independently for solar cells.

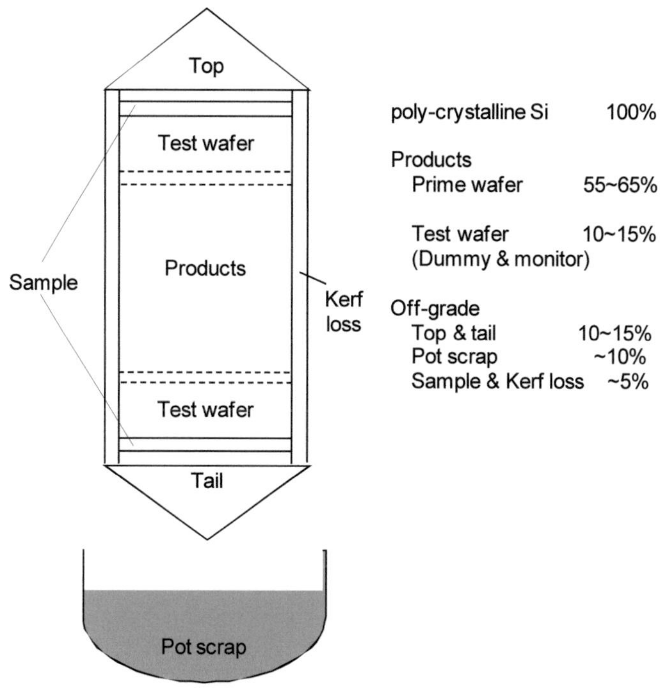

Fig. 3. Products and off-grade crystalline Si of Czochralski (CZ) ingot.

3. Global flow analysis of crystalline silicon

This section presents a material flow analysis of crystalline silicon. After explaining the scope and methodology of the analysis, the material flows are shown and discussed.

3.1 Scope of material flow analysis

A material flow analysis tracks flows of materials at a particular scale in a quantitative manner. Possible scales include the global, regional, national, community, or factory scale. A material flow analysis for a specific material is called a "substance flow analysis." This chapter analyzes the material flow of crystalline silicon, clarifying the input and output of the material at each phase of the production process.

In undertaking a material flow analysis, the scope of the material flow should be clarified. This chapter analyzes the material flow of silicon on a global scale from 1997 to 2009. The material flow also focuses on the stages of production, fabrication and manufacture (Fig.4), because it aims to demonstrate the sustainable supply of crystalline silicon. The global scale was chosen as the geographical boundary for the material flow analysis since the objective of this analysis is to understand the sustainability of flows. Nowadays, material flows at the national level would not be closed due to the export and import of the materials. A global material flow can offset export and import and hence capture the entire flows. On the other hand, a drawback of the global material flow is the quality of data. A global material flow requires global data. While some countries have robust data, others do not, which can influence the accuracy of the analysis. Therefore, these advantages and disadvantages should be considered carefully in analyzing global material flows.

The timeframe for the material flow analysis is 1997 to 2009. This twelve year period starts when relevant solar cell first became available and goes to the latest available data. Whereas a material flow in a single year is like a "snapshot," a time series analysis of material flows over the period enable illustrates the changes of flows over time.

In the material flow analysis in this chapter, silicon used for making thin-film and amorphous types are not considered, because silicon used for these types seems to be less than 1% of that used for making the crystalline types. This point is discussed in greater detail in the "results and discussion" section.

3.2 Methodology

To determine the material flow of silicon on a global scale, data used for the analysis and the respective sources are listed in the Table 2. The assumptions made in developing the material flow and methodologies to estimate values are based mainly on Takiguchi and Morita (Takiguchi & Morita, 2009). Most of the data for the material flows are gathered from the Japanese journal Industrial Rare Metal, which reviews annual trends in industrial materials (Industrial Rare Metal, 1998-2010).

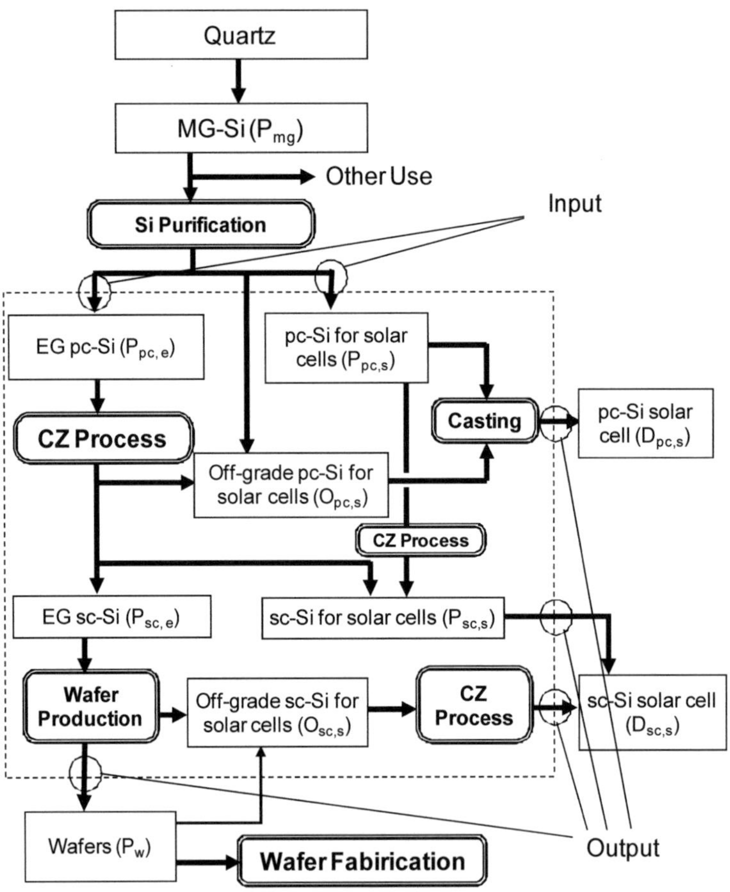

Fig. 4. Analyzed process flow of Si production.

Product	Category	Symbol	Method	References
MG-Si	Production	P_{mg}	Primary	Industrial Rare Metal
pc-silicon	EG Production	$P_{pc,\ e}$	Primary	Industrial Rare Metal
	Production for solar cells	$P_{pc,\ s}$	Primary	Industrial Rare Metal
	Off-grade for solar cells	$O_{pc,s}$	Not available	-
	Demand for solar cells	$D_{pc,s}$	Estimated	Prometheus Institute
sc-silicon	EG production	$P_{sc,\ e}$	Estimated	Industrial Rare Metal
	Production for solar cells	$P_{sc,\ s}$	Not available	-
	Off-grade for solar cells	$O_{sc,s}$	Not available	-
	Demand for solar cells	$D_{sc,s}$	Estimated	Prometheus Institute
Wafer	Production	P_{w}	Primary and Estimated	SEMI

Table 2. Data sources and calculation methods.

3.2.1 Metallurgical-grade silicon

MG-Si is supplied mainly by China, Norway, and Brazil with China increasingly recognized as the dominant supplier. MG-Si is used for various purposes: production of crystalline silicon; deoxidization of steel; and production of aluminium alloy and silicon resin etc. Data on the supply of MG-Si on a global scale are available (Industrial Rare Metal, 1997-2010). It is assumed that the supply of MG-Si is equal to the figures in that data. While data regarding the amount of MG-Si used for crystalline silicon were not available, it is estimated that nearly 1.3 kg MG-Si is used to produce 1 kg pc-Si.

3.2.2 Polycrystalline silicon

Polycrystalline silicon as the primary products is divided into electronic-grade pc-Si (EG pc-Si, $P_{pc,e}$) and pc-Si produced independently for solar cells ($P_{pc,s}$). Data on the production of $P_{pc,e}$ and $P_{pc,s}$ are available (Industrial Rare Metal, 1998-2010). In 2009, 92,100 tons of pc-Si was produced, the ratio of $P_{pc,s}$ to $P_{pc,e}$ being approximately 3:1. The data are the sum of production by manufacturers located mainly in the United States, Japan and Germany. Off-grade pc-Si (O_{pc}) is generated in the pc-Si production process and the CZ process.

The total demand for pc-Si for solar cells ($D_{pc,s}$) is satisfied by two sources: off-grade pc-Si for solar cells ($O_{pc,s}$) and pc-Si produced independently for solar cells ($P_{pc,s}$). Based on the assumption that there is no loss in casting, this supply and demand relationship is expressed as:

$$P_{pc,s} + O_{pc,s} = D_{pc,s} \tag{1}$$

Because pc-Si ($P_{pc,s}$) has been produced independently for solar cells since around 2000, the chapter assumes that $P_{pc,s}$ was zero before the year 2000. Demand for pc-Si for solar cells ($D_{pc,s}$) has been calculated by multiplying the global production of pc-Si solar cells (in Watts) by the amount of pc-Si used for the production of 1 W. Since the amount of pc-Si consumed per Watt steadily decreased from 20 g in 1995 to 15 g in 2000 to 10 g in 2005 (Industrial Rare

Metal), it is estimated from this linear relationship from 1997-2005. After 2006, it is estimated at 9 g per Watt.

It should be noted that there is a loss of crystalline silicon in the wafer saw process. Given 15% for the cell efficiency and 180 micrometers for the wafer thickness, 1 W of the solar cell includes 2.8 g of pc-Si inside. The difference between the demand of pc-Si for solar cells (9 g/W in 2009) and 2.8 g is the loss.

Although the amount of off-grade pc-Si for solar cells is not given, it can be estimated by equation (1). Data for the average price of EG pc-Si per ton each year was available (Industrial Rare Metal, 1998-2010).

3.2.3 Single crystalline silicon

Single crystalline silicon as the primary product is divided into EG sc-Si ($P_{sc,e}$) and sc-Si produced independently for solar cells ($P_{sc,s}$). While data on these variants of silicon are not available, EG sc-Si can be estimated by multiplying the amount of EG pc-Si ($P_{pc,e}$) by the yield rate of EG sc-Si. In this chapter, the rate was set at 0.7. Single crystalline silicon is produced independently for solar cells, partly using pc-Si for solar cells. Off-grade sc-Si (O_{sc}) is generated in the wafer production process or as wafer waste and presumably used for solar cells. Data on the off-grade sc-Si are also not available. As well as pc-Si for solar cells, off-grade sc-Si was used as a proxy for the demand for sc-Si for solar cells until around the year 2000.

The same approach to estimating the demand of pc-Si for solar cells ($D_{pc,s}$) is applied to estimating demand of sc-Si for solar cells ($D_{sc,s}$), assuming that consumption of sc-Si per Watt is identical to that of pc-Si for solar cells. The amount of sc-Si per Watt for the combined type of sc-Si and amorphous silicon in heterostructures was assumed to be 60% of other sc-Si cells.

3.2.4 Wafers

In Fig. 4, "wafers" refers to those used for semiconductors not solar applications. Data regarding the global shipments of wafers for semiconductors are available in units of area (Semiconductor Equipment and Materials International [SEMI], 2011) and are assumed to be equal to production of the wafers (P_W). The calculation of weight of the wafers requires currently unavailable thickness data. Therefore, composition of wafers in diameter has been assumed to be identical to the case in Japan (Ministry of Economy, Trade and Industry, Japan, 1997–2010). Using this assumption, the composition of wafers in 2009 is 4.2% for a wafer under 5 inches (in.), 10.4% for 6 in., 22.6% for 8 in., and 62.7% for 12 in. The weight of products, in turn, has been calculated based on the assumptions that the density of a wafer is equal to that of silicon (2,330 kg/m^3) and that the thickness is 0.625 mm for a wafer under 5 in., 0.675 mm for 6 in., 0.725 mm for 8 in., and 0.775 mm for 12 in.

3.2.5 Resource effective-use index

As in Takiguchi and Morita, the chapter uses the resource effective-use index (REI), which is the ratio of resource input to output required for a given product (Takiguchi & Morita, 2009). The REI enables one to measure quantitatively the extent to which resources are used effectively. The trend in the REI values, therefore, explains how effectively the materials in question have been used over time. The increase of the REI value means that the resource in question is being used more effectively.

In this chapter, the input is the sum of production of electronic-grade pc-Si ($P_{pc,e}$) and pc-Si produced independently for solar cells ($P_{pc,s}$), while the output is the sum of demand of pc-Si and sc-Si for solar cells ($D_{pc,s}$ and $D_{sc,s}$) and production of wafers (P_w). The REI can be defined as follows:

$$REI = \frac{P_w + D_{pc,s} + D_{sc,s}}{P_{pc,e} + P_{pc,s}} \quad (2)$$

To be exact, reuse of wafers as off-grade silicon should be added into the inputs in the calculation of the REI. However, data on the amount of reuse are not available, and therefore they are excluded.

3.3 Results and discussion

In the period of interest, the amount of crystalline silicon supply has expanded. Fig. 5 illustrates that the growing demand of crystalline silicon for solar cells brought about the increase in pc-Si supplies. The supply increased from 16,050 tons in 1997 to 92,100 tons in 2009. The production level of wafers decreased in 2009 from the previous year, probably due to the financial crisis and global economic slump. Nevertheless, it merits attention that the supply of pc-Si did not decline despite of downward trends in the global economy.

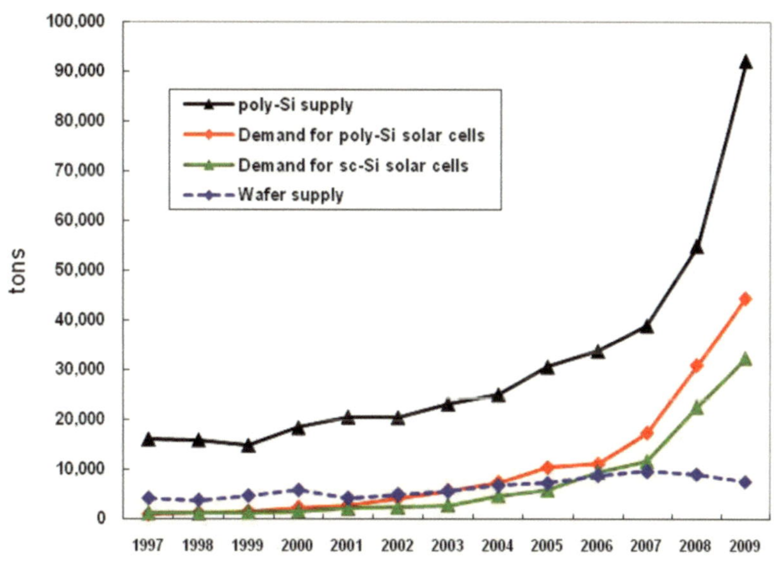

Fig. 5. Production of pc-silicon and wafers.

As Fig. 6 clearly shows, the growing demand for crystalline silicon boosted its price. The price of pc-Si increased after 2004 and approached nearly 10,000 Japanese Yen (JPY) per kg (approximately, 100 US dollars, using the exchange rate: 1 US$ = 100 JPY). In 2008 and 2009, the trend remained stable mainly because of sufficient pc-Si supply.

The main objective of this chapter is to track the material flow of silicon on a global scale. Figs. 7 and 8 show the global material flows of silicon in 1997 and 2009 respectively.

Comparing the two figures, there is a remarkable increase in the amount of silicon at each stage. The 1997 material flow is relatively simple, because demand for crystalline silicon for solar cells was covered by the off-grade silicon from the EG-Si production process. In 2009, however, pc-Si produced independently for solar cells was much larger than the off-grade silicon.

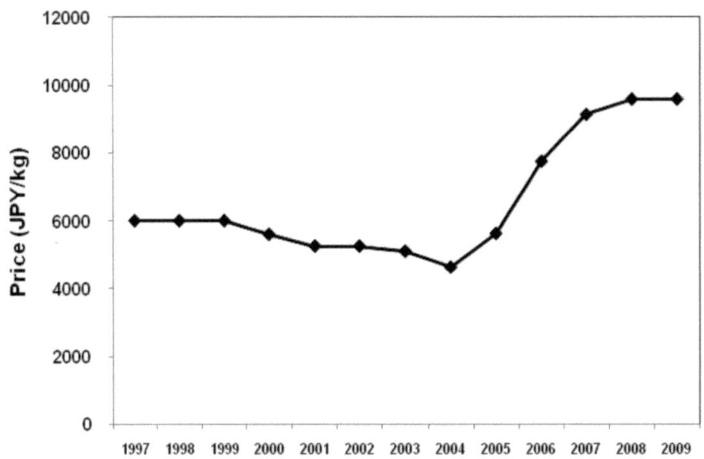

Fig. 6. Price of pc-silicon.

While the global material flow expanded over the period of interest, the question is how effectively crystalline silicon had been used. An analysis of trends in the REI can help answer this question. Fig. 9 describes the trends in the REI. From 2001 to 2008, the values of REI consistently increased. This trend implies progress in the effective use of crystalline silicon. In 2009, however, the REI fell to 0.92 from 1.14 in 2008, partly because the global economic downturn created some slack between supply and demand.

Effective use of crystalline silicon was probably achieved by improvements in the yield rate at each stage, reductions in wafer thickness and kerf loss, and enhanced use of off-grade silicon. Off-grade silicon, which was used as a cheap additive to aluminium alloy in the past, is now used for more valuable products—i.e. solar cells.

It is interesting to note that the REI value exceeded 1.0 in 2008. This implies the mass balance was not achieved in that year, partly because of the reuse of test wafers, changes in stocks, and inaccuracy of data. Nevertheless, the upward trend in the REI values is apparent.

Fig.10 plots the REI values as a function of the price over the period of interest. The REI values seem to rise in response to the pc-Si's increasing price, because the increasing value leads to the more efficient use of materials. This applies to the second half of the analyzed period. In the first half of the period, the REI values increased despite price fluctuations. This is not surprising, given the fact that advanced technologies for effective use of pc-Si would be used regardless of price fluctuations once they were built. According to Tilton, the supply of scraps generated in the course of producing new goods is unresponsive to changes in the market price because of the ease of its collection, high quality, and low recycling cost (Tilton, 1999).

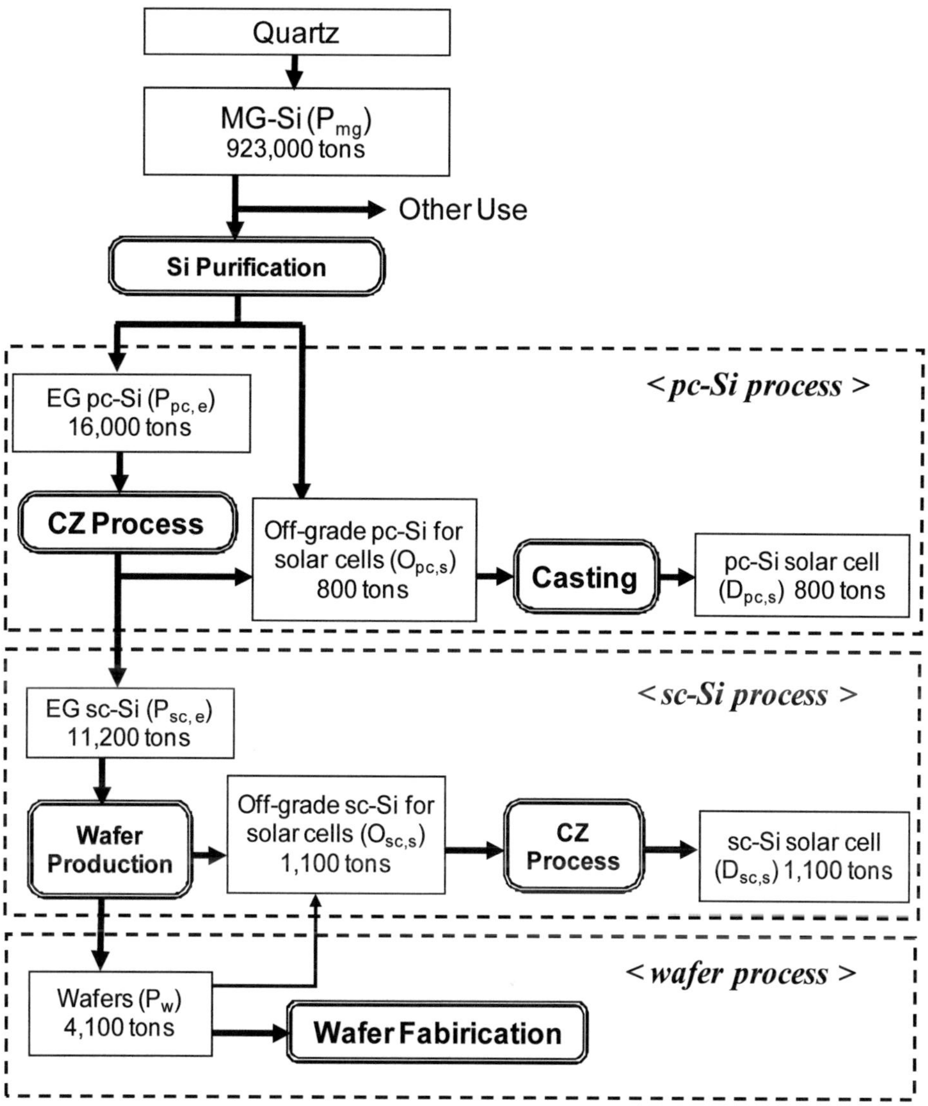

Fig. 7. Global silicon material flow (1997).

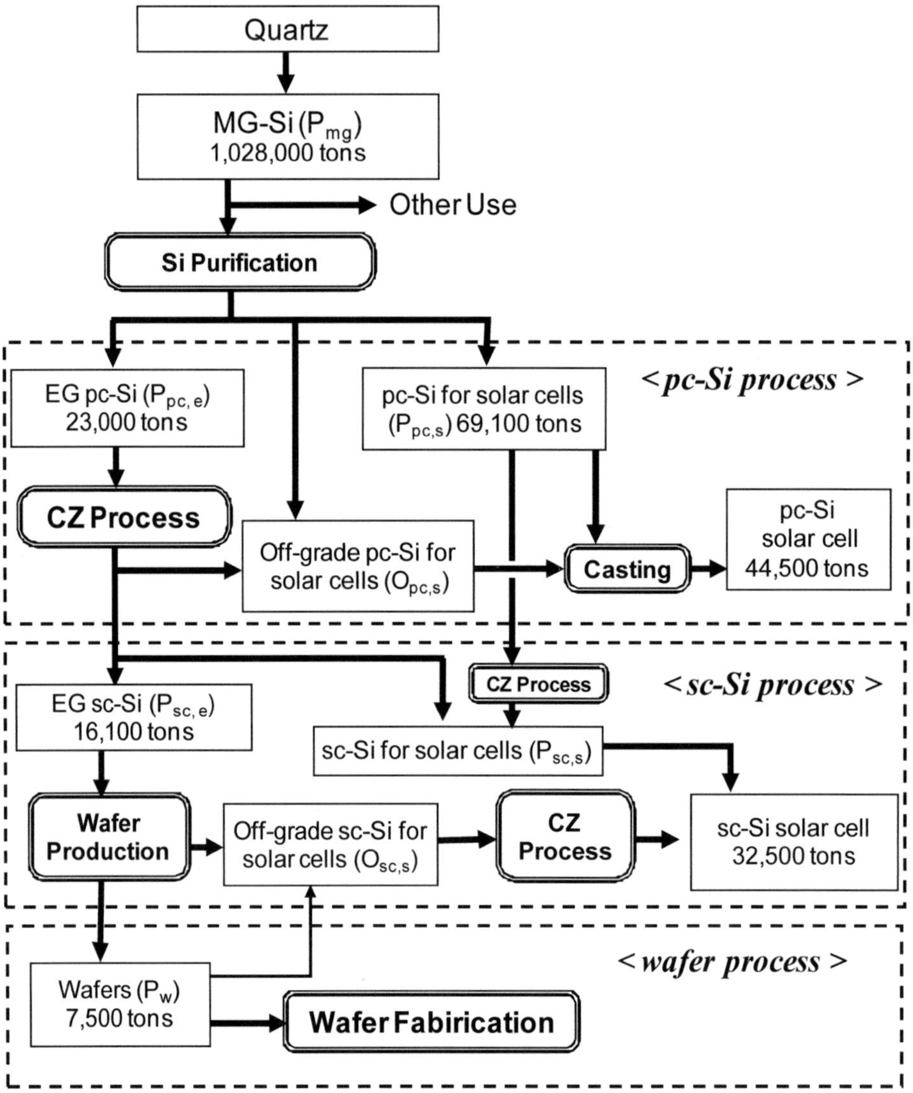

Fig. 8. Global silicon material flow (2009).

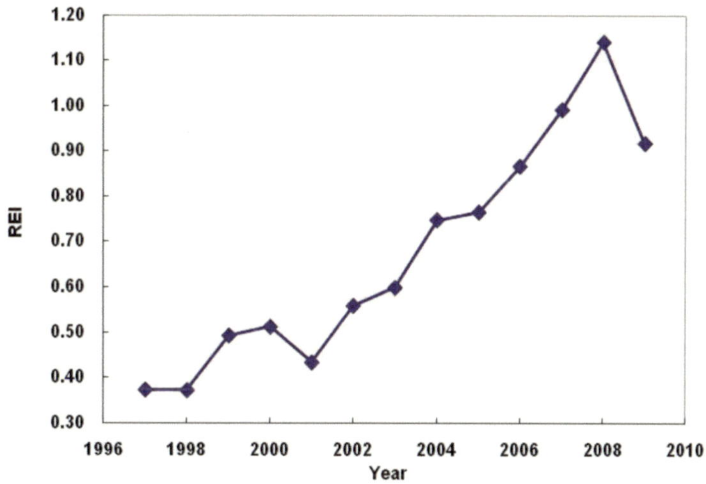

Fig. 9. Trend in Resource Effective-use Index.

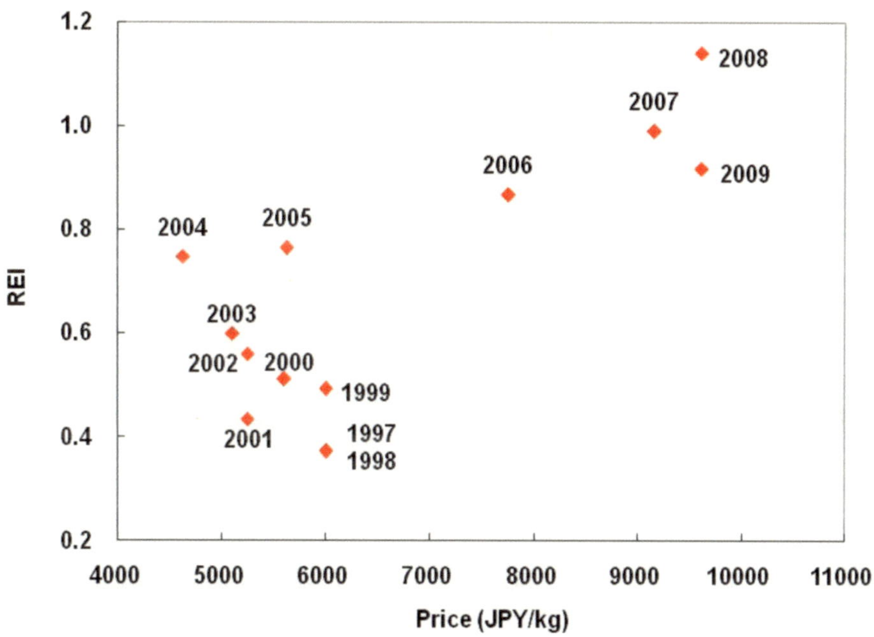

Fig. 10. Relationship between REI and pc-silicon price.

Although it is important to pay attention to trends in 2010 and after, concerns over the shortage of crystalline silicon for solar cells are rarely raised recently, due to an expansion in the supply. Progress in the effective use of crystalline silicon has been demonstrated by a material flow analysis of silicon on a global scale. However, pc-Si for solar cells is produced independently by conventional energy-intensive methods. Taking into consideration the continuous expansion of solar cells, a sustainable supply of crystalline silicon should be achieved by low-energy and low-cost methods.

In Takiguchi and Morita, four solutions are proposed to ensure a sustainable supply of silicon feedstock (Takiguchi & Morita, 2009): (1) production of solar-grade pc-Si by a less costly and less energy-intensive method, (2) reduction of the amount of pc-Si per Watt in solar cells, (3) acceleration of the development and deployment of other PV types, and (4) reuse and recycling of solar cells in the future. With the exception of the third recommendation which is predicated on diversifying the materials used for solar cells into non-silicon, the other three suggestions are applicable to global supply of crystalline silicon. Less costly and less energy-consuming silicon refining processes for solar cells are currently being developed, including a process that develops the refining solidification of silicon using the Si-Al solvent under low temperatures (Morita & Yoshikawa, 2007). Furthermore, in Japan, the JFE steel company produces solar-grade silicon directly from MG-Si using a pyrometallurgical process at a production scale of 400 tons per year (Yuge et al. 2001).
There have been achievements thus far in reducing the amount of crystalline silicon per Watt in solar cells. More significant reductions of silicon could be realized by new types of silicon solar cells. For example, thin-film silicon has been introduced for solar cells, typified by tandem-type silicon composed of stacked amorphous silicon and microcrystalline silicon, also known as nanocrystalline silicon. Tandem-type silicon with a thickness layer less than one hundredth that of bulk types can contribute to meaningful reductions of silicon used for solar cells. In this regard, the material flow in unit of weight may not be the best indicator of resource efficiency since small but important flows, such as development of thin-film silicon, are likely to be neglected. In analyzing the material flow, therefore, attention should be paid to important trends behind the flow.
The reuse and recycling of solar cells will gain in significance in the near future. "Reuse" implies the second use of end-of-use PV modules, while "recycling" refers to use of the material recovered from decomposed PV modules. Needless to say, the reuse of PV modules would reduce energy consumption and CO_2 emissions, compared to newly produced modules. With regard to recycling end-of-use PV modules, a quantitative analysis showed that the recycling can reduce energy and CO_2 emissions when inputting recovered silicon into the process after purification (Takiguchi & Morita, 2010). According to the NEDO report, modules of crystalline silicon did not show any deterioration in performance even after being in use for more than 15 years (NEDO, 2006). In the reuse and recycling of PV modules, a robust system to collect end-of-use modules will be a key to success, because unintentional incorporation of impurities into the reuse and recycling process will make reuse more difficult. Recycling is not limited to PV modules. As described in 3.2.2, there is loss of crystalline silicon in the wafer saw process. Dong et al. conducted a beneficial and technological analysis for solar grade silicon wastes demonstrating it is feasible to recycle silicon ingot top-cut scraps and sawing slurry wastes (Dong et al., 2011).
Overall, the material flow analysis on a global scale was found to be a useful approach to examine the sustainability of crystalline silicon supply. As described in the sub-section of methodology, uncertainty of the data on a global scale is a drawback to the analysis. Nevertheless, global flow analyses are meaningful to overview a worldwide picture.

4. Conclusions

This chapter discussed the sustainability of crystalline silicon supply. The discussion focused on the material flow analysis of silicon on a global scale. The results showed significant changes in crystalline silicon supply due to growing demand for solar cells. The global supply chains not only expanded but became more complicated. While the analysis of the REI values showed progress in the effective use of crystalline silicon, pc-Si for solar cells is being produced through an energy-intensive method. To ensure a sustainable supply of silicon feedstock, three recommendations were made: 1) solar-grade pc-Si should be produced through a less costly and less energy-intensive method; 2) the amount of pc-Si per Watt in solar cells should be reduced; and 3) solar cells should be reused and recycled. The demand for solar cells is still strong. Crystalline silicon supply in the future will be integral to the sustainability of global environmental systems.

5. Acknowledgement

Thanks are due to staff members of the Material Production and Recycling Engineering Laboratory, the Institute of Industrial Science, the University of Tokyo. Furthermore, we thank the reviewers of this chapter for their valuable comments on the manuscript. The contents of the chapter do not necessarily reflect the views of the Ministry of the Environment, Japan.

6. References

Dong, A.; Zhang, L. & Damoah, L. (2011). Beneficial and technological Analysis for the Recycling of Solar Grade Silicon Waste. JOM, Vol.63, No.1, (January 2011), pp.23-27

Frankl, P.; Corrado, A. & Lombardelli, S. (2004). Photovoltaic (PV) systems. Final Report. Environmental and ecological life cycle inventories for present and future power systems in Europe Project. (January 2004)

Gradel, T.; Van Beers, D.; Bertram, M.; Fuse, K.; Gordon, R.; Gritsinin, A.; Kapur, A.; Klee, R.; Lifset, R.; Memon, L.; Rechberger, H.; Spatari, S. & Vexler, D. (2004). Multilevel Cycle of Anthropogenic Copper. Environmental Science and Technology, Vol.38, No.4, pp.1242-1252

Industrial Rare Metal. (1998-2010) Annual Review, No.113-126, Alm Shuppansha, ISSN 0368-654X (in Japanese)

Ministry of Economy, Trade and Industry Japan (METI). (1997-2010). Preliminary report on iron and steel, non-ferrous metal, and fabricated metals and products industry (in Japanese)

Morita, K. & Yoshikawa, K. (2007). Problems and new solutions in refining of solar grade silicon. Materia Japan, Vol.46, No.3, pp.133-136 (in Japanese)

New Energy and Industrial Technology Development Organization (NEDO). (2001). Report on Research and study on commercialization of production process of silicon materials for solar cells (in Japanese)

New Energy and Industrial Technology Development Organization (NEDO). (2006). Report on the results of research and development of photovoltaic power generation systems. (in Japanese)

Prometheus Institute. (2007). PV News, Vol.26, No.4

Sarti, D. & Einhaus, R. (2002). Silicon feedstock for the multi-crystalline photovoltaic industry. Solar Energy Materials & Solar Cells, Vol.72, pp.27-40

Semiconductor Equipment and Materials International (SEMI). (February 2011). Worldwide Silicon Shipment Statistics, 26.02.2011, Available from

http://www.semi.org/jp/MarketInfo/SiliconShipmentStatistics
Takiguchi, H. & Morita, K. (2009). Sustainability of Silicon Feedstock for a Low-Carbon Society. Sustainability Science, Vol.4, No.1, (April 2009) pp.117-131, ISSN 1862-4065
Takiguchi, H. & Morita, K. (2010). Model Development of Assessing 3Rs for Photovoltaic Cells, Environmental Science, Vol.23, No.2, pp.81-95, ISSN 0915-0048 (in Japanese)
Tilton, J. (1999) The future of recycling. Resource Policy, Vol.25, pp.197–204
Williams, E.; Ayres, R. & Heller, M. (2002). The 1.7 kilogram microchip: energy and material use in the production of semiconductor devices. Environmental Science and Technology, Vol.36, No.24, pp.5504–5510
Williams, E. (2003). Forecasting material and economic flows in the global production chain for silicon. Technological Forecasting and Social Change, Vol.70, pp.341–357
Woditsch, P. & Koch, W. (2002) Solar grade silicon feedstock supply for PV industry. Solar Energy Materials & Solar Cells, Vol.72, pp.11–26
Yuge, N.; Abe, M.; Hanazawa, K.; Baba, H.; Nakamura, N.; Kato, Y.; Sakaguchi, Y.; Hiwasa, S. & Aratani, F. (2001). Purification of metallurgical-grade silicon up to solar grade. Progress in Photovoltaics: Research and Applications, Vol.9, pp.203–209